MACROPHAGES AND NATURAL KILLER CELLS

Regulation and Function

ADVANCES IN EXPERIMENTAL MEDICINE AND BIOLOGY

Recent Volumes in this Series

MACROPHAGES AND NATURAL KILLER CELLS
Regulation and Function

Edited by

Sigurd J. Normann

College of Medicine
University of Florida
Gainesville, Florida

and

Ernst Sorkin

Department of Medicine
Swiss Research Institute
Davos, Switzerland

PLENUM PRESS • NEW YORK AND LONDON

Library of Congress Cataloging in Publication Data

International RES Congress on Macrophages and Natural Killer Cells (9th: 1982:
  Davos, Switzerland)
  Macrophages and natural killer cells.

  (Advances in experimental medicine and biology; v. 155)
  Includes bibliographical references and indexes.
  1. Macrophages—Congresses. 2. Phagocytes—Congresses. I. Normann, S. J. II.
Sorkin, Ernst, 1920-   . III. Series. [DNLM: 1. Killer cells—Physiology—Con-
gresses. 2. Macrophages—Physiology—Congresses. W1 AD559 v. 155/WH 650
M1744 1982]
QR185.8.M3I57 1982   616.07′9   82-16158
ISBN 0-306-41180-6

Proceedings of the Ninth International RES (Reticuloendothelial Society) Congress,
held February 7-12,1982, in Davos, Switzerland

©1982 Plenum Press, New York
A Division of Plenum Publishing Corporation
233 Spring Street, New York, N.Y. 10013

Printed in the United States of America

TO
ELIE METCHNIKOFF
ON THE 100TH ANNIVERSARY OF HIS DISCOVERY OF
MACROPHAGES

PREFACE

This book is the outcome of a meeting held in Davos, Switzerland, February 7-12, 1982 focused primarily on mononuclear phagocytes and on natural killer (NK) cells. This IX International RES Congress was attended by 489 scientists from 31 countries and there were 340 scientific presentations in oral or poster session.

The essential purpose of the Congress was to bring together scientists representing various aspects of mononuclear phagocyte biology to review and examine critically the effects and mechanisms of macrophage growth control as well as the participation of these cells in the afferent and efferent limbs of the immune response. Additional topics included the production and distribution of mononuclear phagocytes; the intrinsic and extrinsic regulation of these cells; and the origin, nature, function and regulation of NK cells. The ultimate goal of the Congress was to enhance communication between scientists in various countries and disciplines so that new research directives could be defined with which to explore basic aspects of macrophage and NK cell participation in the control of cancer and infection.

The macrophage is receiving increased scientific attention which is a proper reflection of the central role this cell plays in homeostasis. The Congress was designed to emphasize the following unique and timely aspects of macrophage biology: (a) The various growth regulating aspects of macrophages were examined with respect to both normal and abnormal cellular proliferation. Discussion focused upon the regulatory role of macrophages in hematopoiesis and lymphocytic and fibroblastic proliferation as examples of normal cell processes, and these effects were contrasted with those on neoplastic and transformed cells. (b) Regulation of macrophage function by prostaglandins, complement, and interferon was analyzed from the point of view of an auto-regulatory network. (c) The parallel between chemotaxis and phagocytosis was explored with reference to recognition events and cytoskeletal function. (d) The role of macrophages in the af-

ferent and efferent limbs of the immune response focused on the nature
of lymphocyte-macrophage communication and the role of I region associated determinants.

This focus on macrophages and homeostasis generated certain
questions which formed the basis of the Congress and which are addressed in the papers collected in this volume. Among the questions
are the following:

By what means does the macrophage plasma membrane recognize for-
    eigness? Is chemotaxis a form of directional phagocytosis? How
    are trans-membrane signals utilized to effect microfilament and
    microtubular functions essential to chemotaxis and phagocytosis?

Can monoclonal antibodies or physical properties be used to define or
    isolate subsets of macrophages with restricted activities?

Do all macrophages have the capacity to express Ia antigens despite
    the fact that only a minority of the cells express the product
    of the Ir gene at any given time?

How do lymphokines activate macrophages?

Is a cell which produces colony stimulating factor capable of respond-
    ing to lymphokine and becoming cytotoxic to tumor cells? Do
    monokines and NK cells act as suppressor cells to certain lympho-
    cyte functions? Are the prostaglandins and colony stimulating
    factors that participate in hematopoiesis also stimulating or
    retarding growth of transformed cells?

It is becoming increasingly evident that NK cells and macrophages
share the common property of non-specific cytotoxic and bactericidal
activities. Although the spectrum of susceptible target cells may
differ and the two cells may be derived from different cell lineages,
it appeared appropriate to include in the Congress and in this volume
a discussion of natural resistance mechanisms.

Spontaneously occurring cytotoxic cells may influence cancer
emergence and hematopoietic differentiation as well as graft rejection
and viral infection. The activity of these cells is modulated by
interferon and other agents which also activate macrophages. In the
discussion oriented symposium which concluded the Congress and which
is reproduced here, the critical issues were addressed of whether or
not the cells responsible for natural resistance to grafts of renal,
hematopoietic and leukemic cells, to carcinoma, sarcoma and melanoma
cells and to certain viruses are related and which kinds of molecules
are recognized on the surface of neoplastic and non-neoplastic tar-

gets. In addition, the nature of surface markers on effector cells
and the regulation of natural resistance mechanisms were presented,
including the interaction of effector and regulatory accessory cells.

     We hope that this volume reproduces the essence of the scientific
exchange that characterized the IX International RES Congress.

                                   S. Normann
                                   Gainesville, Florida, USA

                                   E. Sorkin
                                   Davos, Switzerland

ACKNOWLEDGEMENTS

IX INTERNATIONAL RES CONGRESS

SPONSORING SCIENTIFIC ORGANIZATIONS

    American Reticuloendothelial Society

    European Reticuloendothelial Society

    Japanese Reticuloendothelial Society

    Swiss Society of Allergology and Immunology

CONTRIBUTORS

Registration fees rarely cover the costs of a meeting and generally arrive late after certain expenses have been incurred. Thus, the financial health of the Congress would not have been possible without the timely and generous support of our friends in industry, government, and private foundations. We express sincere gratitude for the financial support provided by the following:

    Accurate Chemical & Scientific Corp., Westbury, NY

    Bayer AG, Leverkusen, Germany

    Centre de Recherche Merrell International, Strasbourg, France

    Ciba-Geigy AG, Basel, Switzerland

    Eli Lilly and Company, Surrey, England

    Gruppo Lepetit, Milano, Italy

    Hoffmann-La Roche AG, Basel, Switzerland

    Imperial Chemical Ind. Ltd., Macclesfield, Cheshire, U.K.

Merck Sharp & Dohme, Rahway, NJ, USA

Monsanto Co., St. Louis, MO, USA

Nestlé S.A., Vevey, Switzerland

Nyegaard Cie., Oslo, Norway

Sandoz AG, Basel, Switzerland

Solco AG, Basel, Switzerland

Schweizerische Kreditanstalt, Davos, Switzerland

Swiss Cancer League, Berne, Switzerland

Swiss National Science Foundation, Berne, Switzerland

Swissair, Zürich, Switzerland

The American Reticuloendothelial Society

The European Reticuloendothelial Society

The Japanese Reticuloendothelial Society

Upjohn, Kalamazoo, MI, USA

US Pharmaceutical Corp., Revlon Health Care Group
Tuckahoe, NY, USA

CONGRESS ORGANIZING COMMITTEE

Sigurd Normann,
Department of Pathology, University of Florida, Gainesville,
Florida 32610/USA

Ernst Sorkin,
Schweizerisches Forschungsinstitut, Medizinische Abteilung,
7270 Davos-Platz/Switzerland

David Wilkins,
Centre de Recherche Merrell, 16, rue d´Ankara,
67084 Strasbourg-Cédex/France

SPECIAL APPRECIATION

Special appreciation is given to Ciba-Geigy, Basel, for the printing of the circular, program, and the abstracts; to Miss Marianne Furlenmeier, secretary at the Swiss Research Institute, Davos, who helped prepare the program and who coordinated many of the administrative details essential to a smooth running meeting; and to Miss Furlenmeier and Miss Crystal Grimes, Gainesville, Florida for transcribing the tapes of the Congress.

The organizers thank the Swiss Institute for Medical Research, Davos and the Department of Pathology at the University of Florida, Gainesville for making available their respective institutional facilities.

And a very sincere expression of gratitude is extended to Ms. LeJene Martin who unselfishly devoted many hours to the preparation and typing of the manuscripts for this volume.

COLLAGE

The collage was designed by E. Sorkin, Davos. The electron micrograph of the human peripheral blood polymorphonuclear leucocyte was kindly provided by Virginia Hartwig of the Experimental Hematology Department, Armed Forces Radiology Research Institute, Bethesda, MD/ USA.

MONUMENT TO A PHAGOCYTE

CONTENTS

SECTION 3
PROMONOCYTES AND MONOCYTES

SECTION 4
GROWTH REGULATION BY MACROPHAGES

CONTENTS xix

SECTION 5
LEUKOCYTE INFLAMMATORY RESPONSES

SECTION 6
DEFINING MACROPHAGE HETEROGENEITY

SECTION 7
MACROPHAGE REGULATION

SECTION 8
ANTIGEN PRESENTING MACROPHAGES

SECTION 9
ROLE OF MACROPHAGES IN HUMAN DISEASE

SECTION 10
ACTIVATION AND EFFECTOR MECHANISMS AGAINST
MICROBES AND TUMORS

SECTION 11
NATURAL RESISTANCE MECHANISMS

MEETING REPORT

CONTENTS

OPENING ADDRESS

IX INTERNATIONAL RES CONGRESS, DAVOS, SWITZERLAND

THE IMPACT OF IMMUNOLOGY ON MEDICINE

J. H. Humphrey

Department of Medicine
Royal Postgraduate Medical School
Hammersmith Hospital   London   W12   U.K.

If Elie Metchnikoff, who must be regarded as the ancestral father of Reticuloendothelial Societies and Congresses, were to return to life and preside over this IX Congress he would be pleased and astonished at the size of the audience and the number and range of topics for discussion. He would probably also be surprised that the discussions were to be in English rather than in French and German as well as at the general atmosphere of amity which pervades - for he himself was accustomed to and, I suspect, enjoyed polemical arguments with those opponents who did not accept the essential rightness of his view that microphages and above all macrophages provided the one true mechanism of immunity.

He would have no difficulty in approving several of the themes which will be discussed, for they address problems which he himself recognised as important 80 or 90 years ago: surface receptors, cell movement, the mechanisms of intracellular digestion and killing of microbes, and the origin and turnover of macrophages, and he would appreciate the technologies with which these can nowadays be studied. He might wonder why human and laboratory rodent cells receive more attention than those of invertebrates or reptiles - but then he would be unaccustomed to writing grant applications, and unaware of the availability of inbred strains of laboratory animals.

I am also sure that he would be delighted to learn that macrophages besides making hydrolytic enzymes can synthesize many of the most important complement components and other factors, though it was microphages which he endowed with the property of making what he called "fixatives" - extracellular enzymes which showed some specificity, and which he later tentatively identified with Erlich's amboceptors. Of course he would have expected macrophages to be able to destroy tumors!

3

What would be wholly unfamiliar to him would be the idea that
macrophages should be intimately involved in the presentation of
antigens to lymphocytes so as to stimulate these to make antibodies
or to become themselves capable of killing target cells, secreting
lymphokines, or even stirring macrophages into greater activity.
The importance of lymphocytes would have been as novel to him as it
was to me and to most people until their role as the immunologically
competent cell was proposed by Burnet and established by Gowans
and his colleagues a mere 23 years ago - though it must be admitted
that if the intellectual climate had been more receptive, the experiments of Harris and Ehrlich in 1945 would have given a very clear
indication that lymphocytes were the likely candidates. As for K and
NK cells, whose place in the scheme of lymphocytes and mononuclear
phagocytes is still unsettled - he would perhaps have expressed polite
interest only, since they are not obviously phagocytic. Although
he would have approved the subdivision of macrophages into functionally different categories of mononuclear phagocytes, I doubt whether
he would have known any more than we do where to place Langerhans
cells and their congeners.

The preamble about Metchnikoff leads me into the theme of my
address. The reticuloendothelial system now embraces much of modern
immunology, including many aspects of the perennial problems of
infectious diseases and inflammation, as well as autoimmune diseases,
immunodeficiency and even oncology. What impact has all this had on
medicine?

The most obvious and most important has certainly been due to
active and passive prophylaxis against infectious diseases. In 1980
the World Health Organization was able to announce a great triumph,
the final conquest of smallpox and its abolition from the face of
the earth, except for samples of the virus locked up in a few spe-
cially safe laboratories. This was largely achieved by mass vaccination with vaccinia virus of the population in endemic areas. It
relied upon the fact that vaccinia virus elicited effective and last-
ing immunity against variola virus, that the vaccine strain was stable
and easily propagated, and that there were no feral hosts to act as
reservoirs other than man himself. The ten years long elimination
campaign cost less than the amount of money currently spent by the
nations of the world in 6 hours on their armed forces! We shall
shortly be celebrating the bicentenary of the proof by Dr. Edward
Jenner that cowpox protected against smallpox. Jenner is often referred to as the founder of immunology, and in some ways this is
right, since his experiments (which would never have got past an
ethical committee nowadays!) were well designed, and his observations were acute and clearly presented. But it is only fair to point
out that rather effective control of smallpox had been already
achieved by the mid-eighteenth century in some countries of Europe
through the practice of variolation. This practice was dangerous in

the hands of most professional physicians, who could not resist
accompanying it by blood letting and purging, but had a mortality of
less than 1:1000 in the hands of skilled practitioners. The reduction in childhood mortality consequent upon widespread variolation
was almost certainly the immediate cause of the 60% rise in population
which occurred in Britain during the 18th century. However the vario-
lators and Jenner himself had no idea how vaccination worked, other
than the general concept that a mild attack of a disease could prevent
subsequent serious infection by a similar disease. Any extension of
Jenner's achievement required the development of a germ theory of
disease.

The true founders of immunology, to my mind, were Pasteur, Koch,
Ehrlich, Roux, von Behring, Kitasato, Buchner, Bordet and others,
whose work around the turn of the century led them to recognize that
foreign materials - not only bacteria and toxins but many harmless
substances - introduced into the body caused the appearance of some
new property in the blood whereby it could detectably and specifically
interact with the same foreign materials, and protect against some
toxins and some infections. This new property became attributed to
'antibodies', and it was soon recognized that they could also increase
the efficiency of phagocytes, especially in the presence of a heat-
labile entity in fresh blood termed 'complement'. Shortly afterwards
Portier and Richet, Arthus, Browning, von Pirquet and others also
recognized that administration of foreign substances led to a state
of altered reactivity of the body to them - appropriately named
"allergy" by von Pirquet - which could produce harmful reactions when
they were reintroduced.

The great excitement produced by these findings and by the early
successes in passive protection against death from diptheria or in
active immunization of sheep against anthrax gradually died away.
Although antitoxins made in horses saved many lives, most crude toxins
proved too toxic for active immunization of man and even of domestic
animals, and active immunization against many microbes was not, or
not reliably successful. Too little was known about antibodies,
antigens and complement to make sense of the phenomena described.

Between 1910 and about 1935 progress was relatively slow. Immunology was mainly studied in bacteriological laboratories. The remarkable specificity of antibodies was exploited to help to classify
strains of microbes and even of animal species according to distinct
antigens associated with them. Prophylactic vaccines were greatly
improved - especially after it was shown by Ramon and Glenny in 1923
that several toxin preparations could be rendered non-toxic by formaldehyde without losing their capacity to elicit antibodies against the
toxins themselves. Some were suitable for use in man, but their
usefulness for civilian populations was only slowly accepted - despite
the fact that diphtheria was actually abolished in Hamilton, Ontario,

after 1933 as the result of a vigorous immunization campaign. In
Britain in 1940 there were 50,000 cases of diphtheria with 2500
deaths: by 1955 there were almost none. The use of antitoxin to
treat dangerous bacterial infections, such as gas gangrene, tetanus,
or pneumonia also became widespread. There were some major theoretical advances. Karl Landsteiner showed that very specific antibodies
could be elicited against many defined chemical groups attached to
proteins, including compounds most unlikely to occur in Nature.
Accurate quantitative methods were developed for measuring the amounts
of antibody in serum which could combine with a given antigen, and confirmed that when antigens were reintroduced into an animal months
or even years later the specific antibody elicited appeared sooner
and in much larger amounts than on the first occasion - i.e. the
animal had a kind of 'memory' for the antigen. J. R. Marrack using
these quantitative methods, proposed that antibodies interact with
antigens in solution to build up polymeric lattices, and even that
antibodies probably have two identical combining sites which behaved
like non-covalent valencies.

    All this had been learned about antibodies before anyone knew
what they were, let alone what cells made them. The reason why the
prophylactic immunization methods effective up to that time - and
indeed most of those introduced since - could have been developed
by skilled empiricism, and without the knowledge which we now possess, was that they essentially involved imitating a natural infection
in a controlled way, and were likely to set in train the complicated
system of checks and balances which are the normal outcome of the
body's response to microbial invasion. During the course of evolution, over the last 300 million or more years, the immune system must
have optimized its capacity to deal with all sorts of bacteria,
viruses, protozoa and other parasites. It is not until we try to
interfere with or modify this system that we really have to understand
it.

    I first became interested in immunology in 1938, while still a
medical student, because antibodies appeared to be easier to obtain
and purify and just as specific as enzymes, at a time when the specificity of proteins seemed to me to be the most exciting problem in
biochemistry. Yet it was only the following year that antibodies
were shown to be proteins belonging to the gamma globulins first
identified in serum by Tiselius in 1937 and 8 years before Astrid
Fagräus provided convincing evidence that plasma cells made them.

    Immunology at that time was obviously important and even useful,
but did not excite much general interest among the medical profession,
apart from military and public health physicians, and a small number
of enthusiastic bacteriologists, biochemists, physiologists and pa-
thologists. In 1940 there was only one immunological society (the
American Association of Immunologists) and four international journals
devoted to immunology. By 1960 there were two immunological soci-

eties, the British Society having been founded in 1956, and now six
journals and three regular review volumes. By 1970 there were 12
immunological societies, 13 international journals and nine review
volumes. Ten years later (1980) there were 35 societies, 26 journals,
and nine review volumes. The 4th International Congress of Immunology
in Paris that year was attended by 5,500 persons. Since then nine
more journals and at least one more review series have been started -
and none of them has yet become bankrupt. A market research carried
out for "Immunology Today" identified some 600 journals covering
immunology to a significant extent.

I shall now try to account for this extraordinary burst of
interest in a subject which concerns itself largely with the function,
products and interactions of one kind of rather dull looking cell -
the lymphocyte - whose importance was not recognized until 1960.

During the 1939-45 World War, prophylactic immunization was
widely used and extended to cover a new range of diseases such as
dysentery, tetanus, pneumonia, and even mumps, but the newly developed sulphonamide drugs and penicillin, new antimalarials, and prevention of insect-born infections with DDT were regarded as more
important. The war years also saw an enormous increase in the use
of blood transfusion, bringing in turn renewed interest in blood
group antigens and the recognition that there were many more blood
groups than the three major ones discovered by Landsteiner in 1902.
Another byproduct was great improvement in methods for purification
of blood plasma proteins, which made relatively pure proteins available for the first time on a large scale. Problems of skin grafting
burned patients emphasized the known but unexplained fact that as a
rule only autografts would take, and led to experiments which clearly
indicated that rejection was due to immunological mechanisms. At
the same time the techniques for really accurate quantitative analysis
of amino acids began to be developed, and - once the war was over -
radioactive isotopes became available. These were part of the foundations for the new developments in immunology which were yet to come.

The next ten years saw a further interest in immunology. Antibiotics, immensely important though they were, did not cure viral nor
all bacterial diseases, and this emphasized the importance of studying the natural protective mechanisms and of prophylactic immunization. Additional effective vaccines, such as against whooping cough
and the Salk vaccine against poliomyelitis, were introduced on a wide
scale. Certain diseases of hitherto unknown etiology were shown to
be associated with and perhaps due to antibodies directed against the
body's own components - thereby emphasizing the largely neglected
problem of why animals of any species did not make antibodies against
their own constituents whereas the same constituents readily evoked
antibodies in another species. Experimental models were developed
in which diseases such as nephritis could be reproduced by immunological means, and the mechanisms of damage in acute allergies -

namely to extrusion of mast cell granules - were partly clarified.
Tumors transplanted between inbred strains of mice were shown to be
rejected by immunological mechanisms, raising the possibility that
spontaneous tumors could also be rejected in a similar way. This
period also saw important technical advances. One was to allow antibodies and antigens to diffuse toward one another in a stationary
gel so as to form linear precipitates where they met in the right
proportions. In this way the resolving power of antibodies to distinguish different antigens in a mixture was enormously increased and
this simple technique was able not only to show the presence of many
unsuspected proteins in blood plasma, but also to indicate whether or
not two substances were antigenically identical. Another was the use
by A. L. Coombs of specific antibodies to which a fluorescent label
had been attached to detect antigens or antibodies in single cells.
Besides proving that antibodies were made by cells with a characteristic morphology called plasma cells, which had long been recognized
but whose lineage was unknown, the method provided a sensitive and
entirely new way of examining the presence of any material against
which specific antibodies were available, such as viruses or spe-
cialized proteins, within cells.

        Such advances were of course based on hypotheses about the nature
of immune responses but they were largely ad hoc and a coherent theory
was lacking. Between 1953 and 1960 this changed. Firstly Medawar
and his colleagues tested and showed to be correct a hypothesis,
proposed by Burnet and Fenner in 1949, that animals would not respond
to - i.e. they would be tolerant of - antigens with which they had
come in contact during fetal or early neonatal life. This provided
an explanation - satisfactory, though later shown to be only partly
correct - for self tolerance, and Burnet and Medawar later received
Nobel prizes for it. Then in 1959 Talmage, and Burnet and Lederberg
put forward a Clonal Selection hypothesis to explain the biology of
immune responses, with the guess that lymphocytes were the cells
responsible. In the same year, Gowans published the first of a
series of papers which proved that lymphocytes actually were the
cells responsible. Lastly 1960 saw the elucidation of a general
structure for antibodies as molecules composed of two pairs of polypeptide chains, one heavier (H) and one lighter (L), the four chains
being joined symmetrically by disulphide bonds so as to form a flex-
ible Y shape. However even preparations of antibodies specific for
a given antigen were mixtures, which precluded detailed chemical
analysis.

        To resolve this, it was necessary to recognize that myeloma
proteins were not abnormal or 'paraproteins' and to use them on the
one hand to define Ig classes and subclasses and on the other to
permit sequence analysis of the variable regions. The detailed
structure of these molecules now became more clear - namely variable
regions at the amino terminus of both light and heavy chains which
embraced the specific combining sites, and constant regions which

endowed the molecules with their biological properties - activating complement, crossing membranes, occurrence in secretions, binding selectively to mast cells and so on. Evolution had devised an extra-ordinarily ingenious way in which different molecules with combining sites posessing an extraordinary number of different specificities could exert their biological effects through a limited number of final biological pathways. This was clearly intelligible to medical people. The subsequent developments concerned with the generation of diver-sity, the discovery of separate genes controlling V, D, J and con-stant regions are fascinating for immunologists but at present have little impact on medicine.

Concomitant discoveries about the nature and function of comple-ment and the possibility of explaining many of the manifestations of inflammation in terms of its activation, and of some rare cases of susceptibility to infection in terms of complement deficiencies, like-wise brought about a great apparent simplification and unification of the pathology underlying various inflammatory diseases and gave a great boost to immunopathology.

The clonal selection theory and knowledge of the structure and properties of immunoglobulins, and of their interaction with comple-ment were sufficient to attract the attention of the medical profes-sion to what immunology could offer in explaining the pathology of some diseases - especially those associated with circulating antigen antibody complexes. But of course it left much unexplained - autoimmunity, allergies due to delayed type hypersensitivity, aber-rations of lymphocyte proliferation other than myelomas - i.e. leukae-mias and lymphomas - and so on. A completely new insight was provided by the discovery of the fundamental distinction between thymus-dependent T-lymphocytes and thymus independent B-lymphocytes, and the recognition that only the latter have immunoglobulin receptors and secrete Ig, whereas the former have quite separate functions. It is astonishing to me how recent is our knowledge of T-cells - so recent that, although knowledge of natural and experimental thymus deficiency syndromes had existed for several years, when Prof. R. G. White and I wrote a third edition of a text book in 1969 we regarded the evidence of distinct functions as sufficiently uncertain not to include it! Analysis of the role of T-cells, which had no easily analyzable secretory product and no known markers, and depended there-fore solely on recognizable biological activities, depended initially almost entirely on experiments using inbred mice. (This is an ap-propriate occasion to pay a tribute to the invention of the inbred and congenic mouse as one of the most important contributions ever made to medical science, to immunology and to genetics, and I am glad that the Committee which awarded last year's Nobel prizes for Medicine recognized this invention.)

As a result of imaginative and ingenious experiments, which have been made possible by improved techniques of culture in vitro, we now

know rather a lot about T-cells - though admittedly not enough - and
much of this putative knowledge has rapidly penetrated the medical
literature. This has been greatly assisted by the development of
specific allo-antibodies which can distinguish T-cells and most recently by monoclonal antibodies which distinguish subsets of them
even in Man. Many clinicians know, or think that they ought to
because their younger colleagues do, that T-helper cells are needed
to enable B-memory cells to respond to thymus dependent antigens;
that T-helper cells or something very like them are responsible for
the manifestations of delayed hypersensitivity reactions, and that
when activated they secrete a variety of lymphokines which include
macrophage activating factors; that T-helper cells can induce other
T-cells which are specifically cytotoxic for allogeneic target cells
or for autologous cells with surface antigens which are altered in
some way by microbial or even tumor antigens; and that similar cells
can act as suppressors of antibody formation or of cell mediated
immunity, putatively by inhibiting the T-helper cells themselves.
All these T-cells have entered into common parlance among physicians.
How they act is another question - immunologists have not yet taught
them, for they do not know themselves! But conceptually this does
not matter. Even Jerne's Network Hypothesis is listened to with
interest, and clinicians are learning to grapple with the notion
of idiotypes, and anti-idiotypes and anti-anti-idiotypes etc. as the
mechanism whereby the immune system may ultimately be regulated.

The relevant question is does all this help rather than complicate understanding, and has it a real impact on Medicine. Naturally
I think it does. Let me take some examples:- Prophylactic immunization: I pointed out earlier that no basic understanding of immunology is needed to imitate natural infections. However it certainly
helps to understand that preformed antibodies can neutralize microbial toxins, prevent microbes from attaching to sites of entry into
the body or into target cells, and can destroy many microbes with the
help of complement or by opsonization and phagocytosis. At one time
it was reckoned that antibodies were all that were needed for protection. But it is now recognized that once microbes capable of
intracellular multiplication have become established, cellular immunity (cytotoxic lymphocytes or activated macrophages) are required
to destroy them along with the infected cells. Recognition that both
are needed has greatly altered the strategy of prophylactic immunization, especially against viruses and intracellular bacteria such as
salmonella. Much effort is currently being put into the use of
highly purified or even synthetic microbial antigens, but the lesson
has been learned that acceptable adjuvants will also have to be
devised which will ensure that the correct responses will be made
and that they will endure. Furthermore the discovery that during
infections with some important parasites effective immunity is often
circumvented not only by changes in the surface antigens of the parasites but by the development of profound immunosuppression, has at
least clarified what aspects of the interplay between host and para-

site must be understood better before immunization is likely to be
effective. I suppose that if the effort were considered worthwhile
mumps, measles and rubella could be eliminated in the same way as
smallpox. But we could never hope to eliminate diphtheria, since
lysogenic phages will continue to convert commensal diphtheroids into
potential toxin producers, and I doubt whether we could eliminate
influenza. Prevention of rhesus isoimmunization by anti-D will also
remain a continuing need, since rhesus negative genes will also persist, but we may see monoclonal anti-D produced in vitro replacing
human donors.

     Antibodies as reagents for serological classification, diagnostic
and epidemiological purposes have been important for a long time.
However the ability to distinguish antibody classes, and the introduction of technical improvements such as ELISA tests has enormously
increased their scope. The brilliant discovery that hybridomas can
secrete monoclonal antibodies has extended the usefulness of such
reagents even further. Endocrinology and pharmacology, for example,
have been revolutionized by the ability to use   such   techniques
for the measurement of physiological concentrations of hormones and
of drugs at trace levels.

     Immunodeficiency diseases due to defects in the lymphocyte can
now be understood, in the sense that we can ascertain what is defective (though only sometimes why) and know in principle how to correct
it. Rational treatment often proves gratifyingly effective.

     Proliferative disorders of the lymphoid system - i.e. leukaemias
and lymphomas - are more intelligible. At any stage in their development from stem cells it appears that malignant transformation of a
clone of lymphocytes and uncontrolled proliferation of the transformed clone can occur. Now that markers are available to permit
identification of the state at which transformation took place,
classification and prognosis are much improved. Treatment with cy-
toxic drugs and radiation remains empirical, but it is not far fetched
to hope that increased understanding of what controls normal differentiation may eventually produce means of bypassing the maturation
arrest or of devising chemotherapy specific for the arrested stage.
When there is an associated chromosome abnormality, whose function
can be ascertained, the prospects of doing this seem to me particularly good.

     Cancers of various kinds express surface antigens peculiar to
the tumor cells - though more often than not there is loss rather
than gain of antigens. In principle tumor-specific antigens should
evoke an immune response against them - and in experimental animals
(when the antigens are often coded for by viruses) they do so. Even
in man circulating immune complexes, which indicate that an antibody
response must be taking place, are often present. But effective cellular immunity sufficient to prevent tumor growth is rarely detect-

able even though many tumors can be shown to contain lymphocytes
specifically cytotoxic for the tumor cells. However cellular immunity
in general may be depressed, and this is perhaps because tumor cell
products actually inhibit cellular immune mechanisms or because they
fail to evoke them. Much effort is being expended on devising means
to stimulate effective anti-tumor immunity. In laboratory animals
it can be done, so the effort is surely justified.

Auto immunity was proposed as a cause of disease, and indeed
demonstrated, long before it became explicable theoretically. It
proved remarkably easy to provoke tissue specific auto immune diseases
in animals, more or less resembling those in man, by immunization
with the appropriate tissues combined with an adjuvant containing
dead tubercle bacilli, which is a potent evoker of T-cell help. The
experimental diseases are nearly always transient, the most likely
reason for this being that in the end suppressor cells become predomi-
nate. A generally accepted explanation for the development of tissue
specific auto immunity (such as to thyroid antigens or pancreatic
islet cells) is that although potentially reactive B cells are present, appropriate helper T-cells able to cooperate with them are
normally absent or suppressed. These can be substituted for by helper
T-cells evoked by the tissue antigen altered in some way, such as
by association with viral antigens. Since in some animal models
(such as obese strains of fowls or rats which develop spontaneous
thyroiditis) there is notable deficiency of suppressor T-cells, such
deficiency may have a similar result.

Another mechanism is the development of antibodies against self-
components which are normally not expressed or not available to
interact with lymphocytes. The most important example is Rheumatoid
Factors, which are antibodies against normal immunoglobulin molecules
distorted by interaction with antigens. Rheumatoid Factors are regu-
larly produced when antigenic stimulation is prolonged and immune
complexes are formed, but their appearance is transient. In rheumatoid arthritis there must be some additional factor which maintains
both the production of antibodies and the distortion of immunoglobulins to create a vicious circle - one possibility is that this is
the inflammatory reaction set up by rheumatoid factors interacting
with normal immunoglobulins and with themselves. Less common examples
are auto-antibodies to sperm antigens following spermatic cord ligation or, sometimes, mumps, and to uveal tract pigment following damage
to one eye leading to sympathetic opthalmia in the other.

Systemic lupus erythematosus (SLE) and its congeners, characterized by auto antibodies against a variety of nuclear constituents,
remain a puzzle. There are now several animal models which develop
spontaneous auto antibodies of a similar nature, some presenting a
disease picture with a sex incidence very like that in Man. In all
these models the disease is genetically determined, but none of the
proposed explanations for S.L.E. - such as early loss of suppressor

cells - is common to all the models. The only common feature is
heavy infection with a C-type virus, and this has not been found in
the human disease. We have to acknowledge that clones of auto reactive cells, which we know to be normally present, can be stimulated
for reasons which we do not know, though there is clearly a strong
genetic predisposition for this to happen. Sometimes a provocative
factor is prolonged administration of certain drugs - e.g. hydral-
azine leading to antibodies similar to those in S.L.E. and α-methyl
dopa to auto antibodies against erythrocytes (usually involving Rh
antigens). But these are reversible when the drug is stopped, and
we do not know why or how they have this effect. Methyl dopa has
been claimed specifically to inhibit suppressor T-cells, but this
hardly explains why the auto-antibodies should be confined to erythrocytes. One of the greatest puzzles is why the auto-immune response
in different diseases is confined to particular antigens. The answer
to this would shed light on the whole problem.

There are several auto immune diseases in which, if the patients
are kept alive and their auto antibody levels reduced, e.g. by plasmapheresis, and kept down by immuno-suppressive drugs for months or
even years, the patients are apparently cured. Goodpasture's disease
caused by antibody against glomerular basement membranes, and some
forms of myasthenia gravis, caused by antibodies against acetyl
choline receptors, are examples. One possibility is that the clones
of antibody producing cells have become exhausted, but a more interesting possibility - supported by evidence in experimental animals -
is that the balance of the idiotype anti-idiotype network has been
altered and that suppressor cells have become dominant. How to stimulate suppressor cells selectively is a problem which deserves, and is
getting, much further attention.

Immune complex diseases and Glomerulonephritis. It may seem
surprising for someone coming from Hammersmith Hospital not to have
put glomerulonephritis and other nephropathies and vasculities at
the top of the list of diseases in which immunological concepts have
had a profound influence. A good reason is that the more one learns
about clinical conditions of this sort the harder it becomes to make
satisfactory broad generalizations! The concepts however, are
simple - namely that antigen-antibody complexes which appear in the
circulation (having failed to be removed and catabolized by the
reticuloendothelial system) are liable to become deposited on the
basement membrane of renal glomeruli and/or in the walls of arteries,
particularly small arteries elsewhere. When complement is activated
by the complexes, fragments are released which increase vascular
permeability, platelets are activated, granulocytes are attracted
and release their lysosomal enzymes and the result is an acute local
inflammatory response in and around the vessel wall. In the kidney,
where there is a continuous outflow of small molecules into the urine,
any acute inflammatory response is shortlived but complexes and complement components continue to accumulate, causing thickening of the

glomerular basement membrane, disruption of the epithelial foot processes and leaky glomeruli, and eventually their destruction. This
is a very simplified account of a much more complicated series of
events. But the concept that antigen-antibody complexes, which are
present in the circulation more than transiently and above some
threshold concentration, for whatever reason, are liable to cause
vasculitis and especially to damage the kidney has been enormously
important for understanding the pathology and guiding treatment.

Without attempting to compile an exhaustive list, there are
two other areas of medicine which must be mentioned in which immunological concepts have proved essential, namely allergy and organ
transplantation.

Allergies. The immunological basis has long been recognized,
but distinction between immediate type acute and delayed type allergic
reactions was essential for their understanding, even though the
two may coexist. So far as immediate type reactions are concerned,
the basic concept is that IgE binds to specific receptors on mast
cells, and that cross linking of these activates a series of reactions which lead to extrusion of the mast cell granules. What happens
thereafter is fascinating, but essentially in the realm of pharmacology. Why some persons but not others develop such allergies remains
a mystery, despite the evidence that those with an "atopic" genetic
background make IgE more readily. Progress will depend upon defining
the allergens responsible. Delayed type hypersensitivity reactions
are conceptually more complex. They involve interaction between
antigens and part of the circulating helper T-cell population; the
visible effects are mediated by activated macrophages, and the other
lymphokines which I mentioned, but may be modulated to a variable
extent by concomitant stimulation of cytotoxic and suppressor T-cells
which also arrive on the spot. This means that their duration and
intensity depends greatly on the balance achieved, and I fear that
rational specific treatment - other than removing the offending allergen - may be a long way off.

Organ or Tissue Transplantation, except between identical twins,
involves not only skilled surgery but permanent acceptance by the
body of an organized tissue which carries at least some foreign
antigens of the sort which are particularly effective at stimulating
an immune reaction and causing rejection of the graft by the host.
In the case of bone marrow transplantation, in which the aim is to
replace by donor stem cells the whole of the haemopoietic system of
the recipient - either because the recipient's own system is faulty
or has been purposely destroyed by radiation or cytotoxic drugs -
there is the added complication that the donor lymphocytes may react
against and destroy the host. Success in both these procedures
depends upon having sufficient understanding of immunological concepts to minimize the stimuli which will cause rejection, and to damp
down by immunosuppression the severity of the immune response until

a state of tolerance has been achieved. This is now reckoned to be
due to induction of suppressor mechanisms rather than to elimination
of lymphocytes potentially reactive against the graft of the host
as the case may be.

What about Therapeutic Measures? Several cytotoxic drugs are
useful in suppressing unwanted immune response by lymphocytes. They
succeed because they act on dividing lymphocytes, which are especially
sensitive to them, if the right drug or combination of drugs is used.
The lymphocytes include those responsible for the unwanted immune
responses. Although other useful responses are also suppressed, it
is hoped that those responses already established (and not requiring
cell division at the time when the drugs are administered), will
suffice to protect the patient from microbial invasion, together
with the various important non-specific mechanisms which I have not
discussed. Corticosteroids, besides selectively inhibiting lymphocytes, have the added advantage that they inhibit other activities,
such as those of macrophages, involved in inflammatory reactions.
I have to admit that none of these drugs were developed with immunological concepts in mind, and we do not really know how corticosteroids exert their effects. However certain special properties
have been discovered because they were actually looked for. One is
that a metabolite of cyclophosphamide makes mature B-cells behave
like 'early' B-cells and become particularly susceptible to switching
off by low concentrations of free antigens; and another that cyclo-
sporin-A has a selective action on multiplication of T-helper cells.

Plasmapheresis to remove unwanted circulating antibodies or
immune complexes is one therapeutic measure which does spring directly from immunological concepts and so does the use of anti-lymphocyte
antibodies. Between them and used judiciously, the existing measures
are effective in tiding patients over many immunologically based
diseases until the cause has been eliminated or the balance within
the immune system has become stabilized. There are also drugs which
are claimed to act positively to increase one or another arm of the
immune response:- thymopoietin, interferon, the mysterious leucocyte
extract 'Transfer Factor', levamisole and several other compounds.
Convincing controlled therapeutic trials are to my mind needed to
prove their real worth in clinical practice.

The main point which I want to make in this connection is that
we are now in a position to know (more or less) what to aim for
and perhaps equally importantly what concepts may be useless or
"therapies" even harmful. I will give one example of an aimed approach which could provide a severe test of the adequacy of current
ideas. In experimental bone marrow transplantation, it has been
shown that if all mature T-cells are removed (by specific antibodies
and complement or other devices) and only haemopoietic stem cells
are left viable, such cells can completely reconstitute lethally
irradiated animals even across a major histocompatibility barrier

without causing any graft versus host disease. If monoclonal antibodies can be developed which can selectively remove mature T-cells
from human marrow while leaving stem cells intact - and there are
several groups working to this end - and if man hopefully resembles
the rat, the ease and available range of donors in bone marrow transplantation should be greatly increased. But there are two questions
to ask:- will it really prove practicable, and if it does, what will
be the cost?

A major consequence of the widespread acceptance of immunological
concepts in medicine has been to alter the way in which disease processes are thought about. There is even a bias among well taught
clinicians towards seeking an immunological explanation for puzzling
clinical conditions. To illustrate the effect of preconceived ideas
it may be worth showing two examples, often used by psychologists.
Figure 1 at first sight looks like a Rorschach blot test - or perhaps
an abstract painting by Prof. Sorkin, or the clues to the etiology
of multiple sclerosis - and probably signifies nothing special.

Figure 1.

If it is now explained that the pattern of blots is based upon
a picture of the head and shoulders of the bearded Christ, the pattern
becomes clearly recognizable. The point of this illustration is
that when this figure was shown and explained to a group of students,
and then shown again to the same persons 15 years later, they all
recognized the hidden figure without hesitation. Even this trivial
conditioning had lasted for all those years.

A different example of prejudice is illustrated by Figure 2. Those of you who think about rats will have no difficulty in seeing a rat, whereas those who think about people are more likely to recognize a comic face.

Figure 2.

My point is that a bias towards an immunological approach may well be an asset, provided that other approaches are not excluded. For example, if phenomena apparently explicable in immunological terms were secondary, and the psyche really more important (they can of course interact), or indications suggesting a primarily immuno- logical explanation were misleading, such bias could impede reaching the right conclusions. Although I have dabbled in immunology for more than 40 years, I still prefer to regard myself as an experimental pathologist, since this helps to avoid the prejudices that go with a label. The abstracts of the presentations at this Congress indicate to me that most of the participants are also experimental patholo- gists!

SECTION 1

MAJOR ADDRESSES

MEMBRANE RECEPTORS AND THE REGULATION OF

MONONUCLEAR PHAGOCYTE EFFECTOR FUNCTIONS

Samuel C. Silverstein

Laboratory of Cellular Physiology and Immunology
Rockefeller University
New York, New York   10021 USA

Mononuclear phagocytes form a body-wide system of cells (1)
that originate from precursors in the bone marrow, circulate in the
blood, and emigrate from the vascular compartment into the tissues
where they spend the major portion of their life span. At the time
of their emigration from the blood into the tissues mononuclear
phagocytes are extremely immature cells. In the tissues they differentiate into their adult form, the macrophage. Macrophages in
different tissues develop characteristic metabolic and structural
properties. Thus alveolar macrophages develop a high capacity for
oxidative metabolites (2), while splenic and peritoneal macrophages
derive the major proportion of their metabolic energy from anerobic
glycolysis (3). Inflammatory macrophages may develop into sheets of
interlocking epithelioid cells, or fuse to form multinucleate giant
cells (4). In the bones they fuse to produce osteoclasts (5).

Most investigators believe that monocytes emigrate at random
from the vascular compartment and differentiate into specific types
of tissue macrophages (e.g. alveolar, splenic, etc.) in response to
trophic influences in the tissues. However, formal proof for this
belief is lacking. It is possible that specific bone marrow clones
are the precursors of subclasses of monocytes, each with a unique
tissue destination and a different capacity for differentiation.

Mononuclear phagocytes are not merely isolated sentinels on
frontier duty. They have the capacity to influence the physiology of
the entire organism. For instance, when stimulated to phagocytose
infectious microorganisms macrophages synthesize and secrete endogenous pyrogen (6). This hormone is absorbed into the blood and carried to the brain where it stimulates the thermoregulatory center in
the anterior hypothalmus. The result is fever (6).

     In this lecture I will touch upon some of the ways the mono-
nuclear phagocyte system regulates essential physiological functions,
and suggest areas which I believe will expand rapidly in the future.
The advances of the past decade have been so numerous that it is hard
to know exactly where to begin. Alice had the same difficulty in
describing her Adventures in Wonderland and was sagely advised to
"begin at the beginning". That seems an appropriate point of depar-
ture on this occasion as well.

PHAGOCYTOSIS

     It is now one hundred years since the Russian Zoologist Elie
Metchnikoff recognized the central role of phagocytic leukocytes in
host defense, thereby establishing himself as the first cellular
immunologist. In the twenty years that followed this seminal insight
Metchnikoff labored continuously to convince the protagonists of
humoral immunity, von Behring, Ehrlich and others, of the importance
of cellular defense mechanisms. In retrospect, the controversy was
stilled in 1904 by Wright and Douglas who demonstrated the inter-
dependence of humoral and cellular immune systems in the phagocytosis
of pathogenic bacteria. By separating leukocytes from immune serum
and performing reconstruction experiments they established two fundamental principles. First, the phagocytosis promoting factors
in serum act upon the bacteria and not upon the white cells; and
second, the factors responsible for acquired antibacterial immunity reside in serum and not in the phagocytes. Wright and Douglas
termed these serum factors opsonins to denote their action in preparing bacteria for ingestion (7).

     We now know that these opsonins are antibodies of the IgG class
and the cleaved third component of complement, that they promote
ingestion of opsonized particles by binding to receptors for the Fc
fragment of IgG and for the third component of complement on the
phagocyte's plasma membrane, and that these receptors regulate the
movement of the plasma membrane around the particle (8). Several
lines of evidence indicate that prior to receptor ligation phagocytosis promoting Fc and C3b receptors are mobile in the plane of the
macrophage membrane (9). Whether ligation promotes receptor linkage
to the cytoskeleton is an important unresolved issue. The observed
cooperation between different types of membrane receptors in particle
interiorization (10), the capacity of different phagocytosis promoting receptor systems on the same cell to operate independently of
one another (11), and the ability of the macrophage membrane to ac-
comodate to particles of all shapes and sizes suggest to me that
receptor ligation signals assembly of the cytoskeleton without requiring physical coupling between receptor and cytoskeletal proteins.
It is clear that cytoplasmic contractile proteins (actin etc.) provide
the locomotive force to move the cell's plasma membrane around the
particle (12), but the ways they accomplish this task remain to be

clarified. For instance, is actomyosin contraction, as occurs in
skeletal muscle, required, or does actin filament assembly suffice
to move the advancing pseudopods as is the case in thyone sperm (13)?
Also unresolved is the nature of the signal(s) transmitted by phagocytosis promoting receptors to the underlying cytoplasm. Several
investigations have stressed the importance of changes in transmembrane potential (14,15) and of the release of bound calcium (16),
but there is no coherent explanation for the mechanism by which
receptor ligation coordinates the assembly and disassembly of the
cytoskeleton. Yin et al. have suggested that release of bound calcium
produces solation of the actin underlying the forming phagosome (16,
17). If so, then receptor ligation must generate an earlier signal
that initiates the formation of the actin gel and sets in motion a
cascade of reactions of which calcium release may be one component.
A major immediate challenge in the field of phagocytosis is to identify the ways receptor-ligand interactions of the cell surface coordinate the movement of membranes and cytoplasm.

SECRETION

        A second major theme of mononuclear phagocyte physiology emerged
with the recognition of the importance of their secretory activity
by Gordon and his colleagues (18). They showed that these cells
constitutively secrete lysozyme (19); and can be induced to secrete
several neutral proteases including plasminogen activator, elastase
and collagenase (20). Subsequent studies by many investigators have
shown that mononuclear phagocytes secrete an incredible variety of
biologically active molecules, from nucleosides (21), arachidonic
acid metabolites (22), superoxide anion (23), and H_2O_2 (24), to
complement components (25), procoagulants (26), fibronectin (27,28)
and hormones (interleukin 1 (29), interferon (30)).

        The kinds and amounts of products secreted are dependent upon
the state of differentiation or activation of the macrophages. For
example, resident murine macrophages secrete large amounts of prostaglandin E when stimulated with zymosan or IgG coated particles,
yet they secrete little or no H_2O_2 (31), plasminogen activator or
collagenase (18). Thioglycollate broth elicited macrophages secrete
large amounts of plasminogen activator and collagenase without further
stimulation (18), but do not secrete H_2O_2 (31) or arachidonic acid
metabolites (31) even when stimulated with zymosan or immune complexes. Macrophages elicited by in vivo immunization with BCG can be
induced to secrete large amounts of H_2O_2 by immune complex coated
particles or phorbol esters (31), and of plasminogen activator by
latex phagocytosis (18); but compared with resident macrophages they
secrete markedly reduced amounts of arachidonic acid metabolites (32).

        Mononuclear phagocytes express surface receptors for many dif-
ferent types of ligands (Table 1). Mouse macrophages express differ-

Table 1.   Monocyte and Macrophage Plasma Membrane
           Receptors

Ligand	Species
Fc Ig2a	Mouse
Fc Ig1 & 2b	Mouse
Fc Ig3	Mouse
Fc IgE	Mouse and Rat
C3b	Mouse and Human
C3bi	Human
Mannose Oligosaccharides	Mouse and Human
Insulin	Human
CSF	Mouse
Acetyl LDL	Mouse and Human
LDL	Human
Fibronectin	Mouse and Human
α2Macroglobulin Trypsin Complex	Rabbit
Fibrin	Guinea Pig

ent Fc receptors for mouse immunoglobulins of the Ig2a (33), Ig1 and
Ig2b (34), and Ig3 (35) subclasses, and human monocytes have separate
receptors for C3b and C3bi (36). It seems unlikely to me that all
of these receptors evolved to serve the same functions under all
physiological conditions. Rather, I suggest that some of these
receptors may mediate special functions in macrophages at unique
anatomical sites or at specific stages of macrophage differentiation.
For instance, the C3b receptors on resident mouse macrophages (37),

or the C3b (38) and C3bi receptors on human monocytes (39) mediate binding but not ingestion of particles coated with the corresponding ligands. C3b receptors on inflammatory (40) or lymphokine treated mouse macrophages (41), and C3b and C3bi receptors on phorbol ester treated human macrophages (39) promote phagocytosis of C3b and C3bi coated particles. In contrast, Fc receptors on all types of monocytes and macrophages mediate phagocytosis. Thus these two potentially phagocytosis producing receptor systems (the Fc and C3 receptors) express different effector functions in a homogeneous macrophage population. It will be of great interest to know whether similar differences exist in the capacities of these or other receptor systems to signal the release of unique secretory products.

HOMEOSTATIC FUNCTIONS

     A third major theme of mononuclear phagocyte function, and one that I believe will claim a much larger share of our attention in the future, is the role these cells play in homeostatic processes other than immune defense. In terms of the mass of material processed, one of the major daily functions of the mononuclear phagocyte system is the removal of senescent red blood cells. A 60-70 kg adult has about 6 liters of blood, of which about 2.4 liters are red blood cells. Each day about 20 mls of red blood cells reach the end of their 120 day lifespan and are removed from the circulation by mononuclear phagocytes in the liver and spleen. Within these mononuclear phagocytes the red cells are digested to their constituent amino acids, fats and sugars; the heme group is converted to biliverdin and the iron is returned to the body stores. In the course of a year a normal adult's mononuclear phagocytes process about 2.7 kilograms of hemoglobin.

     In accomplishing this task the mononuclear phagocytes must perform an enormous screening function in order to separate the 3×10^{11} red cells destined for destruction from the 5×10^{13} red cells in the circulation. We assume that changes in the red cells' surface cause their removal but we have no direct information concerning the specific change that is responsible for their denouement. Loss of surface area, decreased sialic acid content, and auto-antibodies have all been suggested.

     Another major macrophage function is the turnover of surfactant in the lung. Surfactant is produced by type II pneumocytes and is released into the alveolar space where, in the rat, it has a half life of about 9 hours. Surfactant turnover is regulated by alveolar macrophages which phagocytose it (42). Failure to process surfactant might lead to obstruction of the alveolar spaces. We have no information about the changes in surfactant structure that identifies it for ingestion by the alveolar macrophages. What macrophage membrane receptors promote surfactant uptake?

Earlier I alluded to the production of endogenous pryogen by macrophages. Several lines of evidence indicate that endogenous pyrogen and interleukin 1 (IL 1) are the same molecule (43) and that IL 1 effects not only the thermoregulatory center in the brain but also stimulates hepatocyte biosynthesis of fibrinogen and C reactive protein. Studies by Fuller and his colleagues (44,45) show that both monocytes and Kupffer cells produce secretory products that stimulate hepatocyte fibrinogen synthesis. Moreover, these investigators have documented that fibrin degradation products induce monocytes to secrete these hepatocyte stimulating factors (44). Whether Kupffer cells also respond to fibrin degradation products by secreting these factors, and whether IL 1 is the monocyte factor responsible for stimulating hepatocyte fibrinogen biosynthesis are unresolved issues. Nevertheless, it is already evident that the biosynthetic potential of hepatocytes is coupled, via mononuclear phagocyte secretion products, to the degradation of fibrin. I think it likely that this is but one example of a general control mechanism. It is possible that Kupffer cells and other mononuclear phagocytes sense the plasma levels of a variety of plasma proteins and secrete hormones that, like IL 1, stimulate synthesis of these proteins by hepatocytes. A similar feedback loop may control the production and release of monocytes by the bone marrow (46).

One has only to reflect on the roles of mononuclear phagocytes in bone remodeling and resorption (47,48), in lipid accumulation in the atherosclerotic plaque (49), and in wound healing to recognize that cells of this lineage play essential roles in many other homeostatic processes unrelated to immune defense.

MICROBICIDAL AND TUMORICIDAL EFFECTS

Although Metchnikoff recognized that facultative intracellular pathogens such as tubercle and leprosy bacilli elicit a mononuclear, and not a polymorphonuclear, leukocyte response; and Koch showed that tubercle bacilli produce a delayed type skin reaction, it took until the 1960's before Mackaness and his colleagues identified thymus dependent lymphocytes as essential for stimulation of macrophage bacteriocidal activity (50,51). At about the same time David (52) and Bloom and Bennett (53) identified migration inhibitory factor as a lymphocyte secretory product. This was followed by the recognition that lymphocytes secrete a variety of protein hormones (lymphokines) and that lymphokine treated mononuclear phagocytes have potent bacteriostatic (54), parasitocidal (55) and tumoricidal (56,57) activities. There is an excellent correlation between the capacity of lymphokine stimulated mononuclear phagocytes to secrete H_2O_2 and their parasitocidal and tumoricidal activities (58,31); and in several instances there is direct evidence that oxygen metabolites mediate the parasitocidal (59) and tumoricidal (60) activities of these mononuclear phagocytes. It is ironic that we have comparatively less

information concerning the mechanisms of macrophage mediated resistance to intracellular bacterial pathogens.

My colleague, Marcus Horwitz, and I have identified an intracellular bacterium that seems ideal for analyzing these questions; it
is Legionella pneumophila, the causative agent of Legionnaires'
disease. Legionella pneumophila grows intracellularly in human monocytes but not in serum containing cell culture media or in lymphocytes. Thus the extracellular organisms do not overgrow the culture
and obscure intracellular events. The bacteria grow in unique ribo-
some studded vacuoles within the monocytes' cytoplasm (61).

Intracellular growth of Legionella pneumophila is not inhibited
even when the bacteria are incubated with fresh human immune serum
prior to infection of the monocytes (62). However, monocytes treated
for 24-48 h with cell-free supernatants of Concanavalin A stimulated
lymphocytes suppress the growth of Legionella pneumophila. It is
of interest that the age of the monocytes is a critical determinant
of the effectiveness of the lectin induced supernatant factors in
suppressing growth of the bacteria. Freshly explanted supernatant
treated monocytes were highly effective suppressants of Legionella
pneumophila growth, while monocytes that had been maintained in cell
culture for 7-10 days prior to treatment with supernatant factors
grew these bacteria at nearly control rates (63). Similar results
were obtained by Poste et al. (64) in their studies of lymphokine
stimulation of macrophage tumoricidal activities.

The observation that freshly explanted human monocytes (58) or
relatively immature mouse macrophages (59) are required for efficient
in vitro induction by lymphokines of their bacteriostatic and tumoricidal properties has important implications for the way the mononuclear phagocyte system functions in vivo. These findings indicate
to me that to achieve eradication of microbes and of tumors there must
be a continued influx of blood monocytes into the lesion. Drugs,
treatments, or immunologically regulated suppressive factors that
inhibit the influx of monocytes into the lesion will likewise inhibit
eradication of the infection or destruction of the tumor.

In his Lectures on the Comparative Pathology of Inflammation
Metchnikoff reminded us to look upon more primitive life forms as
illustrative of more complex processes of higher animals (65). This
is especially relevant to the Legionnaires' disease bacillus, a fresh-
water organism (found in the wild in streams and ponds) that has adapted to the intracellular milieu of human monocytes and macrophages.
Clearly man is an occasional host for this organism and not a major
reservoir. Rowbotham (66) has suggested that Legionella pneumophila
grow intracellularly in fresh water soil amoebae, a predator they
presumably encounter daily in their natural environment. Despite the
phylogenetic distance between amoebae and man the vacuolar systems
of human monocytes and soil amoebae may be remarkably similar, thus

explaining how this bacterium acquired the capacity to grow in human mononuclear phagocytes.

I believe that the Legionella pneumophila system offers unique opportunities to identify novel receptors, both within the macrophage cytoplasm and on its surface, that govern the uptake, intracellular sorting and microbistatic effector functions of mononuclear phagocytes. By searching for receptors that evolved in response to microbial pathogens we may identify those that govern the tumoricidal effector functions of "activated" macrophages as well.

The remarkable adaptive and differentiative capabilities of mononuclear phagocytes reflects our evolutionary past. We evolved from free-living single-celled organisms that must have resembled amoebae in many respects. As these ancestral cells evolved they acquired the capacity to aggregate like epitheliod cells, to adhere to, to invade, and to degrade many types of surfaces, to communicate with one another by means of secretory products; and to fuse with one another. Perhaps our mononuclear phagocytes are the only still recognizable cellular relics of this distant past. Perhaps they alone retain the phenotypic flexibility of our single celled ancestors. Viewed from this perspective mononuclear phagocytes achieve a hitherto unrecognized importance as the prototype from which all more differentiated cells evolved.

REFERENCES

1. van Furth, R., in "The Mononuclear Phagocytes" (R. van Furth, ed.), pp. 1-30, Martinus Nijhoff Publishers, Inc., The Hague, 1980.
2. Axline, S. G., Simon, L. M., Robin, E. D. and Pesanti, E. L., in "The Mononuclear Phagocytes" (R. van Furth, ed.), pp. 1247-1260., Martinus Nijhoff Publishers, Inc., the Hague, 1980.
3. Loike, J. D., Silverstein, S. C. and Sturtevant, J. M., Proc. Natl. Acad. Sci. U.S.A. 78:5958, 1981.
4. Adams, D. O., in "Basic and Clinical Aspects of Granulomatous Diseases" (D. Boros and T. Yoshida, eds.), pp. 153-169, Elsevier/North Holland, Inc., New York, New York, 1980.
5. Erickson, J. L. E., in "The Mononuclear Phagocytes" (R. van Furth, ed.), pp. 203-230, Martinus Nijhoff Publishers, Inc., The Hague, 1980.
6. Dinarello, C. A., Fed. Proc. 38:52, 1979.
7. Wright, A. E. and Douglas, S. R., Proc. Roy. Soc. Lond. B. Biol. Sci. 72:357, 1904.
8. Silverstein, S. C., Steinman, R. M. and Cohn, Z. A., Ann. Rev. of Biochem. 46:669, 1977.
9. Silverstein, S. C., Michl, J. and Loike, J., in "International Cell Biology" (Schweiger, H. G., ed.), pp. 604-612, Springer-Verlag, Berlin, 1981.

10. Shaw, D. R and Griffin, F. M.Jr., Nature 289:409, 1981.
11. Michl, J., Pieczonka, M. M., Unkeless, J. C. and Silverstein, S. C., J. Exp. Med., 150:607, 1979.
12. Hartwig, J. H., Yin, H. L. and Stossel, T. P., in: "The Mononuclear Phagocytes" (R. van Furth, ed.), pp. 971-996, Martinus Nijhoff Publishers, Inc., The Hague, 1980.
13. Tilney, L. G., Kiehart, D. P., Sardet, C. and Tilney, M., J. Cell Biol. 77:536, 1978.
14. Hoffstein, S., J. Immunol. 123:1395, 1979.
15. Korchak, H. M. and Weissmann, G., Proc. Natl. Acad. Sci., U.S.A. 75:3818, 1978.
16. Gallin, E. K. and Livengood, D. R., Am. J. Physiol. 241:C9, 1981.
17. Yin, H. L., Hartwig, T. H., Maruyama, K. and Stossel, T. P., J. Biol. Chem. 256:9693, 1981.
18. Gordon, S., in: "The Mononuclear Phagocytes" (R. van Furth, ed.), pp. 1273-1294, Martinus Nijhoff Publishers, Inc., The Hague, 1980.
19. Gordon, S., Todd, J., and Cohn, Z. A., J. Exp. Med. 137:1228, 1974.
20. Werb, Z., and Gordon, S., J. Exp. Med. 142:346, 1975
21. Stadecker, M. J., Calderon, J., Karnovsky, M. L., and Unanue, E. R., J. Immunol. 119:1738, 1977.
22. Danes, P., Bonney, R. J., Humes, J. L., and Kuehl, F. F. Jr., in: "The Mononuclear Phagocytes" (R. van Furth, ed.), pp. 1317-1345, Martinus Nijhoff Publishers, Inc., The Hague, 1980.
23. Johnston, R. B. Jr., Lehmeyer, J. E., and Guthrie, L. A., J. Exp. Med. 143:1551, 1976.
24. Nathan, C. F., and Root, R. K., J. Exp. Med. 148:1648, 1977.
25. Whaley, K., J. Exp. Med. 151:501, 1980.
26. van Ginkel, R. J. W., and van Aken, W. G., in: "The Mononuclear Phagocyte" (R. van Furth, ed.), pp. 1351-1365, Martinus Nijhoff Publishers, Inc., The Hague, 1980.
27. Alitalo, K., Honi, T., and Vaheri, A., J. Exp. Med. 151:602, 1980.
28. Johansson, S., Rubin, K., Hook, M., Allgran, T., and Seljelid, R., FEBS Lett. 105:313, 1979.
29. Mizel, S. B., and Farrar, J. F., Cell. Immunol. 48:433, 1979.
30. Smith, T. J., and Wagner, R. M., J. Exp. Med. 125:559, 1967.
31. Nathan, C. F., Brukner, L. H., Silverstein, S. C., and Cohn, Z. A., J. Exp. Med. 149:84, 1979.
32. Scott, W. A., Pawlowski, N. A., Murray, H. W., Andreach, M., Zrike, J., and Cohn, Z. A., J. Exp. Med. 155:535, 1982.
33. Unkeless, J. C., and Eisen, H. N., J. Exp. Med. 142:1520, 1975.
34. Unkeless, J. C., Fleit, H., and Mellman, I. S., Advances Immunol. 31:247, 1981.
35. Diamond, B., and Yelton, D. E., J. Exp. Med. 153:514, 1981.
36. Ross, G. D. and Lambris, J. D., J. Exp. Med. 155:96, 1982.
37. Griffin, F. M. Jr., Bianco, C., and Silverstein, S. C., J. Exp. Med. 141:1269, 1975.

38. Ehlenberger, A. G., and Nussenzweig, V., J. Exp. Med. 145:357,
    1977.
39. Wright, S. D., and Silverstein, S. C., J. Cell Biol. 91:255a,
    1981.
40. Bianco, C., Griffin, F. M. Jr., and Silverstein, S. C., J. Exp.
    Med. 141:1278, 1975.
41. Griffin, F. M. Jr., and Mullinax, P. J., J. Exp. Med. 154:291,
    1981.
42. Nichols, B. A., J. Exp. Med. 144:906, 1976.
43. Murphy, P. A., Simon, P. L., and Willoughby, W. F., J. Immunol.
    124:2498, 1980.
44. Fuller, G. M., and Ritchie, D. G., Annals of the New York Academy
    of Sciences, 1981, in press.
45. Sanders, K. D., and Fuller, G. M., J. Cell Biol. 91:401a, 1981.
46. Sluitter, W., van Waarde, D., Hulsing-Hesselink, E., Elzenga-
    Claasen, I. L., and van Furth, R., in: "The Mononuclear
    Phagocytes" (R. van Furth, ed.), pp. 325-339, Martinus
    Nijhoff Publishers, Inc., The Hague, 1980.
47. Ash, P., Loutit, J. F., and Townsend, K. M. S., Nature 283:669,
    1980.
48. Coccia, P. F., Krivit, W., Cervenka, J., Clauson, C. Kersey,
    J. H., Kim, J. H., Nesbit, M. E., Ramsay, N. K. C., Warkentin,
    P. I., Teitelbaum, S. L., Kah, A. J., and Brown, D. M., New
    Eng. J. Med. 302:701, 1980.
49. Fowler, S., Shio, H., and Haley, N. J., Lab. Invest. 41:372,
    1979.
50. Mackaness, G. B., in: "Mononuclear Phagocytes" (R. van Furth,
    ed.), Blackwell Scientific Publications, Oxford, 1970.
51. North, R. J., in: "Mechanisms of Cell-mediated Immunity" (Mc-
    Clusky, R. T. and Cohen, S., eds.), p. 185, John Wiley and
    Sons, Inc., New York, 1974.
52. David, J. R., Proc. Natl. Acad. Sci. U.S.A. 56:72, 1966.
53. Bloom, B. R., and Bennett, B., Science 153:80, 1966.
54. Sheagren, J. N., Simon, H. B., Tuazon, C. U., and Mehrotra, P.
    O., in: "Mononuclear Phagocytes" (R. van Furth, ed.), Black-
    well Scientific Publications, Oxford, 1970.
55. Nogueira, N., Kaplan, G., and Cohn, Z. A., in: "The Mononuclear
    Phagocytes" (R. van Furth, ed.), pp. 1587-1606, Martinus
    Nijhoff Publishers, Inc., The Hague, 1980.
56. Fidler, I. J., J. Natl. Canc. Inst. 55:1159, 1975.
57. Piessens, W. F., Churchill, W. A., and David, J. R., J. Immunol.
    114:293, 1975.
58. Nathan, C. F., Nogueira, N., Juangbhanich, C. W., Ellis, J., and
    Cohn, Z. A., J. Exp. Med. 149:1056, 1979.
59. Murray, H. W., Juangbhanich, C. W., Nathan, C. F., and Cohn,
    Z. A., J. Exp. Med. 150:950, 1979.
60. Nathan, C. F., Silverstein, S. C., Brukner, L. H., and Cohn,
    Z. A., J. Exp. Med. 149:100, 1979.
61. Horwitz, M. A., and Silverstein, S. C., J. Clin. Invest. 66:441,
    1980.

62. Horwitz, M. A., and Silverstein, S. C., _J. Exp. Med._ 153:398, 1981.

63. Horwitz, M. A., and Silverstein, S. C., _J. Exp. Med._ 154:1618, 1982.

64. Poste, G., and Kirsch, R., _Cancer Res_. 39:2582, 1979.

65. Metchnikoff, E., Lectures on the Comparative Pathology of Inflammation, Dover Publications, Inc., New York, 1968.

66. Rowbotham, T. J., _J. Clin. Pathol._ 33:1179, 1980.

REGULATION OF MACROPHAGE PRODUCTION

Donald Metcalf

Cancer Research Unit
The Walter and Eliza Hall Institute of Medical Research
Post Office Royal Melbourne Hospital
Victoria 3050, Australia

In parallel with the production of other hemopoietic cells, macrophage production is continuous throughout adult life. As shown in Figure 1, the production of tissue macrophages is the end result of a sequential series of events in which, (a) multipotential stem cells (colony forming units, spleen, CFU-S) generate progenitor cells committed to granulocyte and monocyte production (GM-colony forming cells, GM-CFC), (b) individual progenitor cells generate clones of progeny cells that progressively lose the capacity for further division and generate maturing monocytes, (c) monocytes are released to the circulation from the marrow and spleen, (d) circulating monocytes seed in the tissues either to die or to form relatively long-lived tissue macrophages, some of which have a limited proliferative capacity and (e) limited recirculation of tissue macrophages occurs between different organs.

For the mouse, these various events can be monitored with reasonable precision by the variety of techniques listed in Table 1. With the exception of the CFU-S assay, comparable techniques are available for the analysis of macrophage production in man. It should be emphasized that no studies have so far been made in which all techniques have been combined to accurately measure and quantify macrophage production either in normal health or in any disease state. Because of the complex life history of monocytes, circulating monocyte levels are an unreliable estimate of monocyte production (for the same reasons making circulating neutrophil levels an invalid index of granulocyte production) (11).

Calculation of the total number of GM (granulocyte-macrophage) progenitor cells in various locations in the mouse indicates that

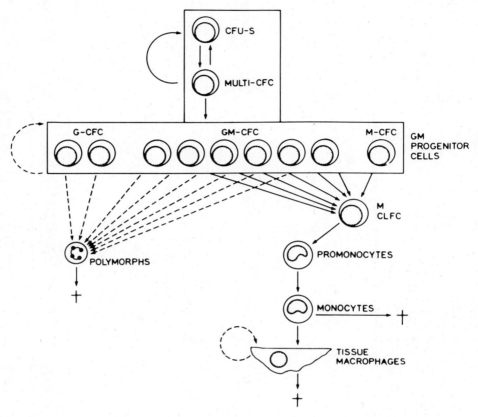

Fig. 1.   Schematic representation of the mechanisms by which tissue
           macrophages are produced.

in the normal animal 99% are located in the bone marrow and only 1%
in the spleen (Table 2).   From this it can be concluded that in the
mouse and presumably in man most monocyte formation occurs normally
in the bone marrow.   In most situations perturbing monocyte formation,
e.g. endotoxin injection, antigenic stimulation, active infections,
etc. this distribution can alter radically since the spleen is a
highly reactive organ able not only to increase in cellularity but
also to become the site of active hemopoiesis.   Thus, at the height
of a response to an injection of endotoxin or lipid A purified from
bacterial cell walls, spleen cellularity may double and the total
number of GM progenitor cells in the spleen can equal those in the
marrow (12,13) (Table 2).   Although GM progenitor cells do enter the

Table 1. Methods for Monitoring the Origin and Maturation
          of Mouse Macrophages

Cell Type or Process	Assay	Reference
Multipotential stem cells (CFU-S)	Colony formation in spleen of irradiated recipients	1
Multipotential colony forming cells (MULTI-CFC)	Clonal assay in agar culture	2,3
Granulocyte-macrophage progenitor cells (GM-CFC)	Clonal assay in agar culture	4,5,6
Monocyte formation and maturation	Autoradiography and cell marker studies	7
Monocyte migration to tissue	Autoradiography, parabiosis, skin windows	8,9
Macrophage recirculation	Radiogold labeling and parabiosis	10

circulation and blood GM progenitor levels are elevated in the above
situations, it is likely that the elevated levels in the spleen are
not due to migration of cells from the marrow but to the increased
formation of progenitors by stem cells resident in the spleen.

   In more extreme situations, e.g. following whole body irradiation
or in some chronic disease states, significant levels of GM progenitor
cells and presumably of monocyte formation can also develop in the
liver (14,15), a situation recapitulating the active hepatic hemopoiesis occurring normally in fetal development (16,17,18).

CONTROL OF PROGENITOR CELL FORMATION FROM STEM CELLS

   As indicated above, the ultimate source of macrophages in the
mouse is the population of multipotential hemopoietic stem cells.
(CFU-S). Normally, most of these cells are not in active cell cycle
and the initial event in macrophage formation therefore requires
CFU-S to be activated in cell cycle. Several different factors have
been described that are able to achieve this:

(a) A high molecular weight factor in medium conditioned by actively
     cycling marrow cells able to induce cycling in G_o CFU-S (19).

(b) A factor in mitogen-stimulated lymphocyte-conditioned medium or
     fibroblast-conditioned medium able to induce cycling in G_o CFU-S
     and to stimulate modest increases in the net total number of
     CFU-S in suspension cultures of marrow or fetal liver cells (20-
     22).

(c) A factor from cyclophosphamide-treated marrow able to activate
     CFU-S into cycle and to modify the type of spleen colonies gen-
     erated by them (23).

(d) A macromolecule appearing in sera of mice following the injection
     of endotoxin or lipid A that increases total CFU-S and progenitor
     cell numbers in the spleen on injection into recipient mice (13,
     24,25).

It has yet to be determined which, if any, of the above represent the
normal regulator of CFU-S cycling and self generation, but the mediator involved and its cellular origin might well vary according to the
nature of the situation perturbing the steady state.

     Little is known of the cellular processes involved in the forma-
tion of committed progenitor cells from multipotential CFU-S. It has
been suggested that the formation of progenitor cells is a stochastic
process in which different progenitor cells are generated randomly

Table 2.   Tissue Distribution of Granulocyte-Macrophage
           Progenitor Cells in Young Adult Mice

	Total Granulocyte-Macrophage Progenitor Cells	
Organ	Normal Mice	Lipid A-injected Mice
Bone marrow	480,000	580,000
Spleen	3,000	120,000 - 500,000
Blood	50	500

Calculations for 3 month old C57BL mice from data published in Ref.
12,13.

(26), or that this might involve processes in which restricted cells
are formed asymmetrically in a definable sequence by dividing multipotential cells (27). In many situations, e.g. following the injection of endotoxin, all types of progenitor cells are produced in increased numbers (12), suggesting that there may be no selective
mechanism able to direct the formation only of one particular type
of progenitor cell. There is also disagreement as to the control
mechanisms involved in this process. Studies on spleen colony formation by CFU-S suggested that different microenvironmental areas in
the spleen were able to commit CFU-S to a restricted e.g. erythro-
poietic lineage (for reviews see Ref. 28,29). However subsequent
studies showed that colonies containing a restricted population of
differentiating cells e.g. erythroid, in fact also contain many types
of progenitor cells (30,31), and there is little evidence now to
support the concept (a) that microenvironmental cell contact is
necessarily the only committing control mechanism or (b) that in such
colonies restricted committment to form only one type of progenitor
cell occurs. It seems more likely that progenitor cells may be generated at random in an early developing spleen colony and that the
development of a single population of differentiating cells is the
consequence of local control systems stimulating selected progenitors
in this population to generate a pure population of progeny cells
of one particular type. Certainly multipotential CFC grown in isolation in agar are able to generate a variety of progenitor cells (32,
33) and the serum macromolecule in mice injected with endotoxin is
also able to stimulate marked elevations in spleen progenitor cells
of all types following injection into normal mice (13,24).

CONTROL OF GM PROGENITOR CELLS

The formation of maturing monocytes and macrophages by GM progenitor cells is reasonably well understood because of the ability
to analyze this process in semi-solid cultures in which individual
GM progenitor cells (GM-colony forming cells, GM-CFC) generate clones
of progeny cells (4,5,6,34). Committed GM progenitor cells have been
purified and identified as immature blast cells with no morphological
features either of granulocytes or monocytes (35). This cell population exhibits heterogeneity in size, density and cell membrane markers (35-38) and in responsiveness to regulatory factors (36) but in
C57BL bone marrow three functional subsets of GM-CFC have been identified (a) cells committed solely to G formation (approx. 20%), bipotential cells able to form both G and M progeny (70%) and cells
committed solely to M formation (10%) (39). GM-CFC have also been
subdivided according to their varying dependency on additives to the
culture medium or on various stimulating factors (40) and it is possible but unproved that a hierarchy of GM-CFC exists with some CFC's
being progeny of others and having some limitation in proliferative
potential.

GM-CFC generate cells with a restricted capacity for futher proliferation (cluster forming cells) and from these are generated the
final populations of maturing granulocytes and monocyte-macrophage
cells. GM-CFC vary in their proliferative capacity, but can generate
from 50 to 10,000 progeny. The number of progeny generated by individual CFC's is in part determined by the intrinsic limitations in
proliferative capacity, but can be made to vary enormously according
to the regulatory factors to which the CFC and their progeny are
exposed. GM-CFC should therefore be visualized as a heterogeneous
population possessing an extreme degree of flexibility in their capacity to generate G and/or M progeny by virtue of variations in the
relative proportions of different CFC's and variations in the number
and type of progeny produced by each CFC.

In vitro, all cell proliferation in the GM pathway is dependent
absolutely on the continuous presence of adequate concentrations of
granulocyte-macrophage colony stimulating factor (GM-CSF). While
many factors can restrict this GM-CSF mediated proliferation (e.g.
nutritional limitations, inhibitors, chalones, etc.) GM-CSF is the
only factor able to stimulate proliferation and this factor has no
proliferative effects for other hemopoietic cells or non-hemopoietic
cells. Since purified GM-CSF can be shown to be active in vitro
on single cultured CFC (41) and is effective in serum-free cultures
(42) it is evident that CSF acts directly on target CFC's and does
not act indirectly through some other cell type or by activating some
other mediating factor.

GM-CSF is necessary for survival in vitro of GM-CFC (43,44), is
able to force non cycling CFC to the S phase (45) and has a concentration-dependent shortening effect on total cell cycle times (46). As
a consequence of these actions, the total number of progeny generated
by individual CFC's is directly determined by GM-CSF concentration.

Analysis of GM colony formation in vitro after stimulation by
purified GM-CSF has shown that in the presence of low CSF concentrations (10^{-11}M) most progeny produced are macrophages whereas in cultures containing high CSF concentrations (10^{-10}M) a high proportion
of the progeny are polymorphs (47). This relationship between CSF
concentration and the type of progeny produced is seen also in the
disease cyclic neutropenia in which it has been shown that cyclic
fluctuations in neutrophil levels are accompanied by inverse cycles
of elevated monocyte levels and cyclic fluctuations in both serum
CSF levels and the capacity of peripheral blood monocytes to produce
CSF (48).

This interesting phenomenon appears to be based on two quite
distinct processes: (a) CFC preprogrammed to form only G progeny
require high concentrations of CSF before being stimulated to proliferate, while CFC preprogrammed to form only M progeny can be stimulated by relatively low CSF concentrations (36), and (b) where CFC

are bipotential and are able to form both G and M progeny, high GM-CSF concentrations force the formation of G progeny and low GM-CSF concentrations the formation of M progeny (46).

Biochemical studies have shown that in the mouse, three major subtypes of GM-CSF exist (Table 3). These are:

(a)   M-CSF (Syn. CSFI, MGI1, macrophage growth factor). Two forms have been described - a neuraminic acid containing glycoprotein that has been purified from L cell conditioned medium of MW 70,000 (49,50) and a functionally and antigenically similar molecule of MW 49,000 purified from pregnant mouse uterus extract (Wagemaker, G., unpublished data).

(b)   GM-CSF (Syn. CSFII, MGI2, CSA) a neuraminic acid-containing glycoprotein of MW 23,000 that has been purified from mouse lung conditioned medium (51). This GM-CSF is antigenically non cross-reactive with M-CSF (52,53). GM-CSF's of similar biological activity and molecular weight have been characterized in a variety of other mouse organ conditioned media e.g. heart, striated muscle, thymus, kidney, bone marrow (54) but these are antigenically crossreactive with M-CSF).

(c)   G-CSF, a polypeptide of MW 25,000 to 28,000, antigenically distinct from M-CSF (55-57).

Functionally, M-CSF stimulates selectively the formation of M cells, GM-CSF the formation of both G and M cells while G-CSF stimulates mainly the formation of G cells.

Studies using reciprocal colony transfers and sequential exposure of CFC and their immediate progeny to M-CSF or GM-CSF have shown that in C57BL bone marrow two thirds of GM-CFC are both bi-responsive and bipotential. These cells are able to be stimulated either by M-CSF (to form M cells) or GM-CSF (to form both G and M cells) (39, 58-60). This stimulation of bipotential CFC's ultimately involves an irreversible qualitative change (commitment) in the responding cells. Thus, for most such CFC's, after more than two divisions in the presence of M-CSF, the progeny cells are committed irreversibly to macrophage formation. After this time, proliferation can be supported by GM-CSF but only M progeny will be produced. Conversely, GM-CSF irreversibly commits CFC to both G and M formation and thereafter G formation can be supported by continuing stimulation using M-CSF (39).

In the context of macrophage production, two classes of GM-CSF molecules (M-CSF and GM-CSF) can be involved in stimulating macrophage formation and fluctuations in their relative concentrations can result in major changes in the number of macrophages being produced from a given population of GM-CFC.

Table 3.   Colony Stimulating Factors Active on Granulocyte-
           Macrophage Precursor Cells

Species/ Factor	Molecular Weight	Purified	Target Progenitor Cells	End Cells Produced
Mouse				
M-CSF	70,000	Yes	M-CFC	M
	51,000	Yes	M-CFC	M
GM-CSF	23,000	Yes	MULTI-CFC	GM
			E-CFC	
			EO-CFC	
			MEG-CFC	
			G-CFC	
			GM-CFC	
			M-CFC	
G-CSF	25,000– 28,000	±	G-CFC	G
Human				
GM-CSFα	30,000	±	GM-CFCα	GM
GM-CSFβ	30,000	±	GM-CFCβ	G

G = neutrophilic granulocytes.   M = macrophages.

A number of factors have been shown to modulate GM-CSF – stimulated production of G and M cells:

(a)   Strain variations.   GM-CFC from different mouse strains vary
      widely in their responsiveness to stimulation by GM-CSF, e.g.
      BALB/c and CBA cells are relatively unresponsive while C57BL
      cells are highly responsive (61).   This overall responsiveness
      can be depressed by prior injection of corticosteroids (61).

(b)   Serum lipoproteins.   Addition of high molecular weight, light
      density lipoproteins to cultures of mouse bone marrow cells
      selectively suppresses G colony formation and preincubation of
      GM-CFC with such lipoproteins appears to commit CFC to M forma-

tion (62-64). Lipoprotein levels are high in certain strains, e.g. BALB/c and low in others e.g. C57BL (64).

(c) Chalones. It has been reported that media from washed polymorphs or short-term incubation of polymorphs inhibits GM-CSF-stimulated colony formation (65,66).

(d) Prostaglandin E. This agent has a selective suppressive effect on M colony formation at low molar concentrations, leaving G colony formation unaltered (67,68).

(e) Age. Serum GM-CSF levels are high in fetal mice and relatively low in normal adult mice.

In general, serum GM-CSF levels reflect recent or current exposure to microorganisms being low or undetectable in germ-free mice (69) and high in mice with active bacterial, viral or protozoal infections (70,71,72). Serum GM-CSF levels rise sharply following the injection of endotoxin (73,74), purified bacterial cell wall components (75) or a variety of microbial antigens and are also elevated in GVH responses (76,77). Although all tissues have been shown to synthesize GM-CSF, there is some evidence to suggest that GM-CSF produced locally within the marrow cavity can be of special importance in regulating GM proliferation (78).

Not many pure cell populations have been tested for their capacity to synthesize GM-CSF. However cells known to synthesize one form or other of GM-CSF are L cells (79), mitogen stimulated T-lymphocytes (80,81), endothelial cells (82) and monocyte-macrophages (83,84,85).

It is of some interest that several interactions have been described between polymorphonuclear leukocytes (PMNs) and macrophages on the one hand and the process of GM production:

(a) Monocytes and macrophages actively synthesize and secrete GM-CSF. This activity is stimulated by exposure of the cells to endotoxin (86-89). No biochemical studies have been performed on the types of CSF produced but from the biological activity of media conditioned by macrophages it can be concluded that at least GM-CSF and G-CSF must be produced. Similar studies on media conditioned by human monocytes indicates that both GM-CSF α and β must be produced as well as the corresponding factor for eosinophils, EO-CSF (89-91).

(b) Production of GM-CSF by monocyte-macrophages is suppressed by a macromolecule released from polymorphs and identified as lactoferrin (92).

(c) Macromolecules released from PMNs have an inhibitory effect on GM-CSF-stimulated GM production (93).

(d)   Macrophages produce PGE1 an agent having a selective suppressing
      effect on macrophage production stimulated by GM-CSF (67,68,87).

These activities represent positive and negative feedbacks from end
cells that could, in principle, serve to modulate the production rate
of PMNs and macrophages but it must be emphasized that it is improba-
ble that these feedbacks provide the major mechanism controlling GM
production since other cell types are known to actively produce CSF
and to respond to bacterial products.

     From the cell biological viewpoint, it is of great interest that
monocyte-macrophages synthesize GM-CSF, a factor necessary for in
vitro survival at least of macrophage precursors, for the proliferation of macrophage precursors and for stimulating macrophage functional activity (see below). From the biological action of GM-CSF
on macrophages it can be concluded that these cells possess receptors
for GM-CSF and indeed binding of radiolabeled CSF to macrophages has
been observed experimentally (53). Why then are not all macrophages
in a permanently activated state? Furthermore, from the work of Lin
and Stewart (94) it seems likely that at least some macrophages have
a latent proliferative capacity. Why do these cells normally not
spontaneously proliferate or behave as neoplasms if they synthesize
their own growth factor? One possibility may be that the GM-CSF
synthesized within the cytoplasm is unable to find access to intracellular or intranuclear receptors mediating GM-CSF action. If the
action of GM-CSF can only be triggered by an initial interaction
between GM-CSF and a membrane receptor, it may be that synthesized
and secreted GM-CSF diffuses away too rapidly to bind efficiently
to such receptors. From this it should follow that if sufficiently
high concentrations of macrophages were cultured, such macrophages
should eventually secrete sufficiently high concentrations of GM-CSF
to stimulate macrophage division. This may indeed be the mechanism
permitting the establishment of macrophage monolayers and the low
level of proliferative activity they exhibit.

TISSUE MACROPHAGE COLONY FORMING CELLS

     The above description of macrophage production by marrow and
spleen GM-CFC presents a process with obvious parallels to the production of erythroid, eosinophil or megakaryocytic cells by precursors
in these organs.

     However the generation of macrophages may potentially be more
complex. Lin, Stewart and colleagues have demonstrated that there
is a quite high proportion (5-10%) of cells in peritoneal exudates
that are capable of generating macrophage colonies when cultured in
semi-solid medium in the presence of GM-CSF (94). Subsequent studies
have shown that similar cells exist in low frequency in a number of
other tissues e.g. blood, lung, liver, thymus, lymphnodes, etc. (95-

97). As reported, the biology of these colony forming cells appears to exhibit a number of distinctive characteristics: (a) the colony forming cells can survive in the initial absence of GM-CSF and only proliferate in vitro after a delay of one to two weeks, (b) all proliferation is GM-CSF dependent, (c) all colonies are of small to medium size and (d) all colonies are composed solely of macrophages. From these characteristics, the tissue macrophage colony forming cells (M-CFC) appear to be more mature cells than the marrow GM-CFC population, to be already committed solely to macrophage production, and to have a relatively restricted proliferative capacity. The observations raise several important questions: (a) do tissue M-CFC originate from marrow GM-CFC? (b) do M-CFC have the morphology of macrophages and if so, are all macrophages potentially able to proliferate when suitably stimulated? (c) what role do M-CFC play in generating tissue macrophages?

Cell separation studies are needed urgently to determine the nature of M-CFC. At present it seems premature to conclude that all macrophages have a previously unsuspected capacity for extensive division since M-CFC could well be a subset of cells distinct from, but co-resident with, tissue macrophages.

Prior to these studies there was evidence indicating that cells in the peritoneal cavity (?macrophages) could be induced to proliferate following suitable regimes of antigenic stimulation (98). The important question now at issue is whether local production of macrophages occurs in tissues at a significant level and, if so, under what abnormal conditions? Because mitotic figures are essentially never seen in sections of normal adult mouse lung or liver it seems valid to conclude that in normal adult animals local production of macrophages in the lung or liver is probably not an important source of such cells. The situation in chronic inflammatory states or under conditions leading to granuloma formation or extramedullary hemopoiesis is quite unclear since mitotic activity can exist in such tissues. It will require analysis of such tissues to establish the likely level of local macrophage production in situations where there probably exists a heightened demand for macrophages.

ACTIONS OF GM-CSF ON MATURE POLYMORPHS AND MONOCYTE-MACROPHAGES

Tests using injected GM-CSF make it unlikely that this factor influences the release from the marrow of either GM progenitor cells or of mature PMNs or monocytes.

There is, however, clear evidence that GM-CSF can stimulate the functional activity of mature PMNs and macrophages:

(a) Purified GM-CSF increases RNA and protein synthesis in polymorphonuclear leukocytes (47,99).

(b)   Purified GM–CSF increases the ability of mouse peritoneal macro-
      phages to phagocytose and kill Leishmania organisms (100).

(c)   M–CSF has been shown to stimulate the production by macrophages
      of PGE1 (87) and plasminogen activating factor (101).

(d)   Human placental GM–CSF α and β stimulate antibody-dependent
      killing of murine tumor cells by human polymorphonuclear leuko-
      cytes (Vadas, M., personal communication).

     It is of obvious importance to undertake further studies on the
effects of GM–CSF on macrophage function. From the data already
available, however, the effects of GM–CSF on end cell functional
activity appear to be falling into a pattern of biological effects
seen with at least two other hemopoietic regulators. Thus erythro-
poietin not only stimulates erythropoiesis but also hemoglobin synthesis (102) while EO–CSF is not only essential to stimulate the
proliferation of eosinophil progenitor cells but also stimulates the
killing of Schistosomes by eosinophils and a wide variety of functional activities of eosinophils (103, Vadas, M., Nicola, N. A. and
Metcalf, D., unpublished data).

GENERAL COMMENTS

     GM–CSF in its various molecular forms appears to play a central
role in regulating the production and functional activity of macrophages. The role comprises three quite separable activities: (a)
stimulation of the proliferation of macrophage progenitors (GM–CFC
and tissue M–CFC) to generate maturing populations of monocyte-macrophages, (b) commitment of bipotential GM–CFC to macrophage production
and (c) stimulation of end cell functional activity of tissue macrophages.

     The relative importance of these actions will clearly vary ac-
cording to the circumstances. In ensuring immediate responses to in-
vading microorganisms, activation of pre-formed macrophages is clearly
of high importance whereas in sustaining responses to more chronic
infections or neoplasms, the ability of GM–CSF to control the formation of new monocytes and macrophages would become progressively more
important.

Acknowledgement

     The work from the author's laboratory was supported by grants
from the Anti-Cancer Council of Victoria, the National Health and
Medical Research Council, Canberra and the National Institutes of
Health, Grant Nos. CA22556 and CA22972.

REFERENCES

1. Till, J. E., and McCulloch, E. A., Rad. Res. 19:213, 1961.
2. Johnson, G. R., and Metcalf, D., Proc. Natl. Acad. Sci. USA 74: 3879, 1977.
3. Johnson, G. R., J. Cell. Physiol. 103:371, 1980.
4. Bradley, T. R., and Metcalf, D., Aust. J. Exp. Biol. Med. Sci. 44:287, 1966.
5. Ichikawa, Y., Pluznik, D. H., and Sachs, L., Proc. Natl. Acad. Sci.USA 56:488, 1966.
6. Metcalf, D., "Hemopoietic Colonies. In vitro Cloning of Normal and Leukemic Cells," Springer-Verlag, Heidelberg, New York, 1977.
7. Van Furth, R., Goud, T. J. C. M., and Van Waarde, D., in "Experimental Hematology Today 1978" (S.J. Baum and G. D. Ledney, eds.), pp. 65-71, Springer-Verlag, New York, 1978.
8. Volkman, A., and Gowans, J. C., Brit. J. Exp. Path. 46:50, 1965.
9. Ryan, G. B., and Spector, W. G., Proc. Roy. Soc. B 175:269, 1970.
10. Roser, B., Aust. J. Exp. Biol. Med. Sci. 43:553, 1965,
11. Athens, J. W., Natl. Cancer Instit. Monograph 30:135, 1969.
12. Staber, F. G., and Johnson, G. R., J. Cell. Physiol. 105:143, 1980.
13. Staber, F. G., and Metcalf, D., Proc. Natl. Acad. Sci USA 77: 4322, 1980.
14. Testa, N. G., and Hendry, J. H., Exp. Hematol. 5:136, 1977.
15. Hays, E. F., Firkin, F. C.,Koga, Y., and Hays, M., J. Cell Physiol. 86:213, 1975.
16. Moore, M. A. S., and Williams, N., Cell Tissue Kinet. 6:431, 1973.
17. Johnson, G. R., and Metcalf, D., in "Differentiation of Normal and Neoplastic Hematopoietic Cells" (B. Clarkson and R. Baserga, eds.), pp. 49-62, Cold Spring Harbor: Cold Spring Harbor Laboratory, 1978.
18. Johnson, G. R., and Metcalf, D., Exp. Hematol. 6:246, 1978.
19. Lord, B. I., Mori, K. J., Wright, E. G., and Lajtha, L. J., Brit. J. Haematol. 34:441, 1976,
20. Ghio, R., Bianchi, G., Lowenberg, B., Dicke, K. A., and Ajmar, J., Exp. Hematol. 5:341, 1979.
21. Wagemaker, G., in "Hemopoietic Cell Differentiation" (D. W. Golde, M. J. Cline, D. Metcalf and C. F. Fox, eds.), pp. 109-118, Academic Press, New York, 1978.
22. Wagemaker, G., and Peters, M. F., Cell Tissue Kinet. 11:45, 1978.
23. Frindel, E., in "Stem Cells, Cell Lineages and Cell Determination" (N. Le Douarin, ed.), pp. 227-239, Elsevier/North Holland, Amsterdam, 1979.
24. Staber, F. G., and Metcalf, D., Exp. Hematol. 8:1094, 1980.
25. Staber, F. G., and Burgess, A. W., J. Cell. Physiol. 102:1, 1980.
26. Till, J. E., McCulloch, E. A., and Siminovitch, L., Proc. Natl. Acad. Sci. USA 51:29, 1964.

27. Johnson, G. R., in "Experimental Hematology Today" (S. J. Baum, G. D. Ledney and A. Kahn, eds.), pp. 13-20, Karger, Basel, 1981.
28. Trentin, J. J., in "Regulation of Hematopoiesis" (A. S. Gordon, ed.), pp. 161-186, Appleton-Century-Crofts, New York, 1970.
29. Metcalf, D., and Moore, M. A. S., "Haemopoietic Cells," North Holland, Amsterdam, 1971.
30. Johnson, G. R., and Metcalf, D., in "Stem Cells, Cell Lineages and Cell Determination" (N. Le Douarin, ed.), pp. 199-213, Elsevier/North Holland, Amsterdam, 1979.
31. Gregory, C. J., and Henkelmann, R. M., in "Experimental Hematology Today" (S. J. Baum and G. D. Ledney, eds.), pp. 93-101, Springer-Verlag, New York, 1977.
32. Metcalf, D., Johnson, G. R., and Mandel, T. E., J. Cell. Physiol. 98:401, 1979.
33. Humphries, R. K., Jacky, P. B., Dill, F. J., Eaves, A. C., and Eaves, C. J., Nature 279:718, 1979.
34. Metcalf, D., in "Clinics in Haematology", 8:263, Saunders, London, 1979.
35. Nicola, N. A., Burgess, A. W., Johnson, G. R., Metcalf, D., and Battye, F. L., J. Cell. Physiol. 103:217, 1980.
36. Metcalf, D., and MacDonald, H. R., J. Cell. Physiol. 85:643, 1975.
37. Byrne, P., Heit, W., and Kubanek, B., Cell Tissue Kinet. 10:341, 1977.
38. Bol, S., Visser, J., and van den Engh, G. J., Exp. Hematol. 7:541, 1979.
39. Metcalf, D., and Burgess, A. W., J. Cell. Physiol. in press, 1981.
40. Bol, S., and Williams, N., J. Cell. Physiol. 102:233, 1980.
41. Metcalf, D., Johnson, G. R., and Burgess, A. W., Blood 55:138, 1980.
42. Guilbert, L. J., and Iscove, N. N., Nature 263:594, 1976.
43. Metcalf, D., J. Cell. Physiol. 76:89, 1970.
44. Metcalf, D., and Merchav, S., J. Cell. Physiol., in press, 1982.
45. Moore, M. A. S., and Williams, N., in "Hemopoiesis in Culture" (W. A. Robinson, ed.), pp. 16-26, DHEW Publication No. (NIH) 74-205, Washington, 1973.
46. Metcalf, D., Proc. Natl. Acad. Sci. USA 77:5327, 1980.
47. Burgess, A. W., and Metcalf, D., in "Experimental Hematology Today" (S. J. Baum and G. D. Ledney, eds.), pp. 135-146, Springer-Verlag, Heidelberg, New York, 1977.
48. Moore, M. A. S., Spitzer, G., Metcalf, D., and Penington, D. G., Brit. J. Haematol. 27:47, 1974.
49. Stanley, E. R., Cifone, M., Heard, P. M., and Defendi, V., J. Exp. Med. 143:631, 1976.
50. Stanley, E. R., and Heard, P. M., J. Biol. Chem. 252:4305, 1977.
51. Burgess, A. W., Camakaris, J., and Metcalf, D., J. Biol. Chem. 252:1998, 1977.
52. Shadduck, R. K., and Metcalf, D., J. Cell. Physiol. 86:247, 1975.

53. Stanley, E. R., Proc. Natl. Acad. Sci. USA 76:2969, 1979.
54. Nicola, N. A., Burgess, A. W., and Metcalf, D., J. Biol Chem.
        254:5290, 1979.
55. Burgess, A. W., and Metcalf, D., Int. J. Cancer 26:647, 1980.
56. Lotem, J., Lipton, J. H., and Sachs, L., Int. J. Cancer 25:763,
        1980.
57. Nicola, N. A., and Metcalf, D., J. Cell. Physiol. 109:253, 1981.
58. Johnson, G. R., and Burgess, A. W., J. Cell. Biol. 77:35, 1978.
59. Horiuchi, M., and Ichikawa, Y., Exp. Cell Res. 110:79, 1977.
60. Metcalf, D., Merchav. S., and Wagemaker, G., in "Experimental
        Hematology Today 1982" (S. J. Baum, ed.), Karger, Basel, in
        press, 1982.
61. McNeill, T. A., and Fleming, W. A., J. Cell. Physiol. 82:49,
        1973.
62. Chan, S. H., Aust. J. Exp. Biol. Med. Sci. 49:553, 1971.
63. Chan, S. E., Metcalf, D., and Stanley, E. R., Brit. J. Haematol.
        20:329, 1971.
64. Metcalf, D., and Russell, S., Exp. Hematol. 4:339, 1976.
65. Paukovits, W. R., and Hinterberger, W., Blut 37:7, 1978.
66. Aardal, N. P., Laerum, O. D., Paukovits, W. R., and Mauer, H. R.,
        Virchows Archiv (Cell Pathol.) 24:27, 1977.
67. Williams, N., Blood 53:1089, 1979.
68. Pelus, L. M., Broxmeyer, H. E., Kurland, J. I., and Moore, M. A.
        S., J. Exp. Med. 150:277, 1979.
69. Chang, C. F., and Pollard, M., Proc. Soc. Exp. Biol. Med. 144:
        177, 1973.
70. Foster, R., Metcalf, D., and Kirchmyer, R., J. Exp. Med. 127:
        853, 1968.
71. Trudgett, A., McNeill, A., and Killen, M., Infect. Immunol. 8:
        450, 1973.
72. Bro-Jorgensen, K., and Knudtzon, S., Blood 49:57, 1977.
73. Metcalf, D., Immunology 21:427, 1971.
74. Quesenberry, P., Morley, A., Stohlman, F., Rickard, K., Howard,
        D., and Smith, M., New Engl. J. Med. 286:227, 1972.
75. Staber, F. G., Gisler, R. H., Schumann, G., Tarcsay, L.,
        Schlafli, E., and Dukor, P., Cell. Immunol. 37:174, 1978.
76. Singer, J. W., James, M. C., and Thomas, E. D., in "Experimental
        Hematology Today" (S. J. Baum and G. D. Ledney, eds.), pp.
        221-231, Springer-Verlag, Heidelberg, New York, 1977.
77. Hara, H., Kitamura, Y., Kawata, T., Kanamura, A., and Nagai, K.,
        Exp. Hematol. 2:43, 1974.
78. Chan, S. H., and Metcalf, D., Cell Tissue Kinet. 6:185, 1973.
79. Austin, P. E., McCulloch, E. A., and Till, J. E., J. Cell. Phy-
        siol. 77:121, 1971.
80. McNeill, T. A., Nature 244:175, 1973.
81. Parker, J. W., and Metcalf, D., J. Immunol. 112:502, 1974.
82. Quesenberry, P., and Gimbrone, M. A., Blood 56:1060, 1980.
83. Moore, M. A. S., and Williams, N., J. Cell. Physiol. 80:195,
        1972.
84. Chervenick, P. A., and Lo Buglio, A. F., Science 178:164, 1972.

85. Golde, D. W., and Cline, M. J., J. Clin. Investigation 52:2981, 1972.
86. Sheridan, J. W., and Metcalf, D., in "Hemopoiesis in Culture" (W. A. Robinson, ed.), p. 135, DHEW Publication No. (NIH) 74-205, Washington, 1973.
87. Kurland, H. I., Broxmeyer, H. E., Pelus, L. M., Bockman, R. S., and Moore, M. A. S., Blood 52:388, 1978.
88. Eaves, A. C., and Bruce, W. R., Cell Tissue Kinet. 7:19, 1974.
89. Chervenick, P., and Boggs, D. R., Blood 37:131, 1971.
90. Dresch, C., Johnson, G. R., and Metcalf, D., Blood 49:835, 1977.
91. Nicola, N. A., Metcalf, D., Johnson, G. R., and Burgess, A. W., Blood 54:614, 1979.
92. Broxmeyer, H. E., Smithyman, A., Eger, R. R., Meyers, P. A., and de Sousa, M., J. Exp. Med. 148:1052, 1978.
93. Broxmeyer, H. E., Bognacki, J., Dorner, M. H., de Sousa, M., and Lu, L., in "Modern Trends in Human Leukemia IV" (R. Neth, R. C. Gallo, T. Graf, K. Mannweiler and K. Winkler, eds.), pp. 243-245, Springer-Verlag, Berlin, 1981.
94. Lin, H-S., and Stewart, C. C., J. Cell. Physiol. 83:369, 1974.
95. Lin, H-S., Blood 49:593, 1977.
96. Chen, D-M., Lin, H-S., Stahl, P., and Stanley, E. R., Exp. Cell Res. 121:103, 1979.
97. Ledney, G. D., MacVittie, T. J., Stewart, D. A., and Parker, G. A., in "Experimental Hematology Today 1978" (S. J. Baum and G. D. Ledney, eds.), pp. 73-84, Springer-Verlag, New York, 1978.
98. Forbes, I. J., and Mackaness, G. B., Lancet 2:1203, 1963.
99. Burgess, A. W., and Metcalf, D., J. Cell. Physiol. 90:471, 1977.
100. Handman, E., and Burgess, A. W., J. Immunol. 122:1134, 1979.
101. Lin, H-S., and Gordon, S., J. Exp. Med. 150:231, 1979.
102. Krantz, S. B., Gallien-Lartigue, O., and Goldwasser, E., J. Biol. Chem. 238:4085, 1963.
103. Vadas, M., Dessein, A., Nicola, N. A., and David, J. R., Aust. J. Exp. Biol. Med. Sci. 59:739, 1981.

SYMBIOTIC RELATIONSHIPS

BETWEEN MACROPHAGES AND LYMPHOCYTES

Emil R. Unanue

Department of Pathology
Harvard Medical School
Boston, Massachusetts   02115 USA

This presentation summarizes our recent studies on the immunoregulatory function of the macrophage. Two issues will be analyzed:
the process of antigen presentation and the regulation of expression
of I-region-associated antigens (Ia) by the macrophage.

We recently discussed in a review that the relationship between
macrophages and the helper-inducer set of T lymphocytes is truly symbiotic (1). All expressions of T cell immunity, either in vivo or in
vitro- proliferation, mediator production, macrophage cytocidal activation, delayed sensitivity, etc.- require that antigen be presented
to the T cells by an accessory phagocytic cell (reviewed in 1). We
accept the premise that all responses to polypeptide antigens require
phagocytes to initiate them in an essential interaction modulated by
the I-region gene products expressed by the macrophage (for example,
2,3,4). The macrophage-lymphocyte interaction is also found in the
effector limb of the immune reactions- or delayed sensitivity stage-
in which the recuitment and activation of the macrophage for enhanced
cytocidal activity depends on the interaction with T cell lymphokines;
these are elaborated, however, only following an initial macrophage
presentation of antigen to the T cell (!). In vivo we envision the
following scenario: macrophages in tissue take up the antigen and
present it to T cells, initiating T cell activation. The T cell
then releases the mediators that call forth new phagocytes from the
bone marrow and which also activates them for heightened cytocidal
function. The cycle of stimulation stops with the disappearance of
antigen and by the suppressor T cell limb of the response.

ANTIGEN PRESENTATION

Insights into the mechanisms of antigen presentation have been

obtained primarily by culture experiments in which macrophages are
mixed with purified T cells in the presence of antigen. We have used
several protein antigens, although recently we have employed the
intracellular facultative bacteria Listeria monocytogenes.

Several sequential steps in the presentation can be dissected
as one mixes macrophages with antigen and specific immune T cells.
First, the T cells bind tenaciously to the macrophages bearing the
antigen molecules and form multicellular clusters. Second, the macrophage elaborates and releases a potent lymphostimulatory molecule,
the 15,000-dalton thymus mitogen also termed lymphocyte-activating
factor (LAF) or interleukin-1 (IL-1) (5,6). This protein was first
identified by Gery and Waksman (7,8) and later studied biologically
in detail in our laboratory (9-12). Third, the T cells now activated
elaborate lymphokines and start their proliferative activity. Formation of the T cell-macrophage conjugates is immediate; LAF appears
within six hours, lymphokine by about twenty-four hours. The process
of presentation of antigen given as a single pulse stops usually by
about twenty-four hours. At this time, the immunogenic antigen has
disappeared (6), Ia expression has decreased (13), and the production
of LAF is also diminishing (5,6). The whole process requires that
the specific antigen-committed T cells recognize antigen in the
context of macrophages bearing Ia of the correct histocompatibility
haplotype (5,6,14) in agreement with investigations by others.

Uptake

In trying to dissect the molecular events, one question to ask
is what the "receptor" for antigen in the macrophage is and its
relationship to the Ia antigens. Along these lines, a second issue
is the role of the classical opsonic receptors, the Fc and C3 receptors. All our evidence indicates that the Ia antigens are not involved in the ititial binding of antigen to the plasma membrane of
macrophages. First, most protein antigens bind directly to all macrophages regardless of the presence or absence of Ia antigens. As will
be discussed below, it is only a small proportion of the macrophages
that express Ia at a given time. Second, anti-Ia antibodies do not
inhibit the binding of antigen (14). Finally, Doctors Weinberg and
Elsas in my laboratory examined directly the protein on the macrophage
that binds Listeria monocytogenes. They have identified biochemically
a major molecular species of about 220,000 to 240,000 daltons unrelated to the classical Ia antigens. This binding protein is very
sensitive to proteases and, interestingly, is absent normally in the
murine alveolar macrophage (15).

Concerning the Fc receptor, we can state that antigen presentation to T cells can take place equally well- although it is not neces-
sary- by antigen molecules found as an antigen-antibody complex (15,
16). A very clear result to this effect came from studying alveolar
macrophages which, as mentioned above, lack the Listeria- binding

protein, although possessing normal Fc receptor activity. Such alveolar macrophages presented Listeria poorly to T cells, as would be
expected by their lack of uptake; yet Listeria bound to antibodies
was taken up and presented to T cells, which underwent proliferation
(Figure 1).

Handling of Listeria

Using Listeria monocytogenes, the events subsequent to binding
were relatively easy to trace. Thus, the surface-bound Listeria
were rapidly internalized in vesicles and underwent lysosomal catabolism within minutes (14). In trying to relate these antigen-handling
events to antigen presentation, a functional bioassay was devised for
monitoring the binding of the T cell to the macrophage (17). This
assay made use of the early observations from Alan S. Rosenthal's
laboratory that the immune T cells would bind directly to the macrophage (18). It consisted of lightly spinning the immune T cells onto
the macrophage monolayer, then incubating for a brief period of fifteen to thirty minutes, after which the cultures were shaken and the
non-adherent T cells harvested and assayed in a standard fashion.
With effective antigen presentation, the immune T cells remained
attached to the macrophage and were depleted from the non-adherent
population. Similar results were also obtained by Werdelin and
Shevach using a comparable assay (19). This simple binding assay
has the advantage that the time for effective Ia-dependent antigen
presentation is brief so that a number of experimental manipulations
can then be carried out. Thus, one experimental study consisted of
adding the T cells to the macrophages at various times after Listeria
uptake, i.e., five to sixty minutes. The results of several experiments indicated that optimal time for binding of T cells was from
thirty to sixty minutes after surface binding of Listeria, at which
time most bacteria were internalized and in the process of being
catabolized (Figures 2 and 3). Furthermore, no binding of T cells
took place if, following binding, the bacteria were not internalized,
either by using low temperatures or by placing inhibitors of energy
metabolism. Interestingly, macrophages that had taken up Listeria
for one hour could be fixed in paraformaldehyde and still bind the
T cells, clearly indicating that the macrophage membrane was serving
as a substrate for the recognition of the Ia-dependent immunogen.
Suffice to say that T cells did not bind to macrophages fixed shortly
after uptake of Listeria.

Ziegler and I next explored the relationship between catabolism
and the presentation of Listeria by making use of the lysosomotropic
compounds chloroquine and ammonium chloride. These drugs, by con-
centrating on the lysosomes, reduce catabolism of ingested proteins
(20). Other possible actions of the drugs are on lysosome-phagosome
fusion (21) and on receptor recycling (22). We found that these compounds did not affect the binding of Listeria to macrophages nor their
internalization. However, catabolism was inhibited and with it a

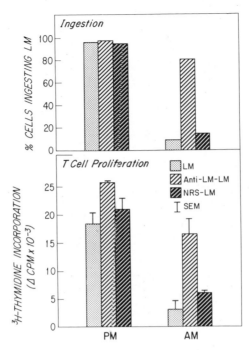

Fig. 1. Effect of opsonization on the uptake and antigen presentation
         of Listeria monocytogenes (LM) by alveolar (AM) or peritoneal
         (PM) macrophages. Both sets of cells were cultured for one
         hour with Listeria organisms, untreated, coated with antibody
         (anti-LM-LM), or with normal rabbit serum (NRS-LM) and the
         uptake measured (upper panel). The T cells were then added
         to the culture and their degree of proliferation measured
         three days later (lower panel). The peritoneal macrophages
         took up the three sets of Listeria and presented them equally
         well. In contrast, the alveolar macrophages took up only
         antibody-coated Listeria, which were the only ones presented
         (reproduced from Weinberg and Unanue, 1981, with permission).

comparable inhibition of antigen presentation (23) (Figure 4). The
presentation was not inhibited when macrophages were treated with
drugs following internalization and catabolism. We concluded that,
in order for the Listeria immunogen to be presented to T cells, the
bacteria had to be internalized and somehow processed after which
the immunogenic moiety could appear on the membrane. It is noteworthy
that the Listeria- containing phagosome bears Ia antigens. Whether
these derived from an internal pool or represent a portion of the

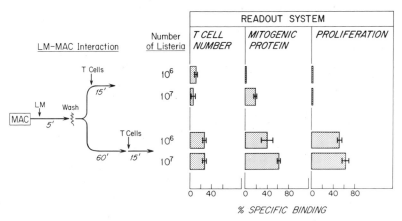

Fig. 2.   Experimental system for determining the optimal time for the
          binding of T cells to macrophages exposed to Listeria mono-
          cytogenes. Listeria were spun onto monolayers of macrophages
          (MAC) for five minutes, after which the non-bound bacteria
          were washed and removed. The macrophages were incubated with
          the bacteria five or sixty minutes at 37°C, after which the
          T cells were spun onto them, incubated for fifteen minutes,
          and then the non-adherent bacteria were removed and bioas-
          sayed. The readout system consisted of an estimation of T
          cells recovered as non-adherent cells (T cell number) or
          production of lymphocyte activating factor (interleukin I)
          assayed as a thymocyte mitogen, or T cell proliferation after
          four days of culture. T cells were bound best by macrophages
          that handled Listeria for sixty minutes (from Ziegler and
          Unanue, 1981, with permission).

plasma membrane, and their function, is not known (E. R. U., unpublished, mentioned in 1) (Figure 5). More recently, Paul Allen in
our laboratory has shown similar effects on the presentation of soluble Listeria proteins, suggesting that the chloroquine-ammonia-
sensitive step may apply also to some soluble proteins. Still to
be explained is whether small peptide antigens require this internalization step or can bypass it altogether. This is an important point
which bears on the very many studies indicating that metabolizability
of an antigen is a necessary step for immunogenicity. Certainly,
with complex structures as Listeria, it would appear to be an essential step in order to bring about specific immunity.

The nature of the membrane-bound immunogen is only partially
known. We have unequivocably identified Listeria peptides on isolated

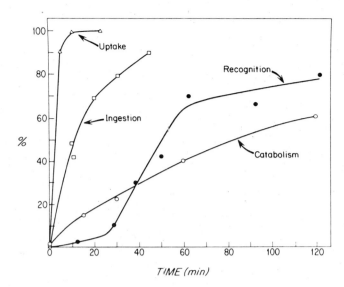

Fig. 3.   The different steps in the handling of <u>Listeria monocytogenes</u>
          by macrophages as a function of time. The binding of T cells
          (recognition) best takes place from thirty to sixty minutes
          after uptake (from Ziegler and Unanue, 1981, with permis-
          sion).

macrophage plasma membranes shortly after the uptake of the radiolabeled bacteria. It is our thinking that some of the peptides may
represent the immunogenic moiety recognized by T cells. The relationship between the membrane peptides and Ia antigens is being analyzed
biochemically with the expectation of finding a possible association.
Other investigators have claimed the existence of such complexes on
uncharacterized material released by macrophages (24,25). Postulating
an association between the peptide antigens and the I region product
best explains the I-region-restricted presentation. The eventual
characterization of the T cell receptor(s) will explain on a molecular
level the requirements for presentation of two distinct molecules by
the macrophage.

     Aside from the Ia-restricted immunogen, the macrophage is a
source of partially digested or undigested molecules. These molecules
have the potential for being taken up by B cells or other antigen-
presenting cells (26,27). In our early studies, using protein antigens, we called attention to membrane-bound molecules that were slowly
internalized which could be identified by antibody molecules (26,27).
Subsequently, Doctor Calderon and I reported that small peptides

resulting from incomplete lysosomal digestion of protein antigens
were invariably released from macrophages (28). These released peptides can be immunogenic when added to lymphoid cells. Table 1 is a
summary of our views, integrating our early studies with the experiments using an antigen-restricted system described earlier.

REGULATION OF MACROPHAGE Ia

    As might be expected for any important molecule, the expression
of Ia antigens by the macrophage is under very sophisticated regulation. It depends on the state of maturation and activation of the
macrophage, on the tissues where they localize, on the presence or
absence of T cell activity, and on the concentration and kinds of
prostaglandins.

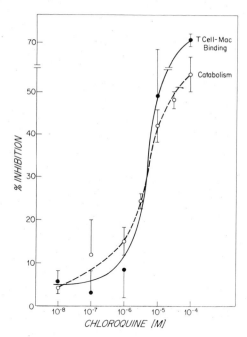

Fig. 4. Relationship between catabolism and presentation of antigen
         to T cells. In this experiment, macrophages were cultured
         with various concentrations of chloroquine during handling
         of Listeria monocytogenes. The catabolism, measured by the
         release of I^{125}-labeled tyrosine was inhibited by chloroquine
         and so was the presentation to T cells (from Ziegler and
         Unanue, 1982, with permission).

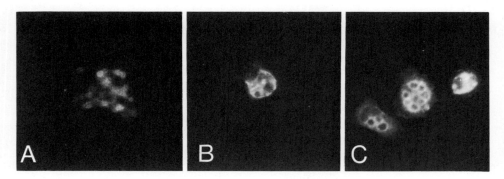

Fig. 5.   Three photomicrographs showing macrophages ingesting <u>Listeria</u>
          <u>monocytogenes</u> (left panel) or opsonized red cells (two panels
          at right).   The cells were treated with acetone, washed, and
          stained with fluorescent reagents specific for I-A^k deter-
          minants.   Note that the wall of phagosome is positive for
          I-A (from Unanue, 1981, with permission).

        Examination of macrophages in different tissues discloses two
clearly distinct populations based on the absence or presence of Ia
molecules (29-34).   The percentage and number in different tissues
is consistent unless an inflammatory process alters the normal physiology (Table 2).   It is noteworthy that the basal level of Ia-positive
macrophages in tissues is the same in nude athymic mice as in euthymic
mice, indicating that obvious T cell stimulation is not required (35).
The function of these two sets of macrophages were evaluated and were
found to differ only with regard to the antigen presentation to immune
T cells (Table 3).   Ia-negative macrophages did not present antigen,
a function limited only to Ia-positive macrophages.   Of great importance in our understanding of Ia expression by the macrophages are
the studies done with Dr. David I. Beller, which uncovered a regulation of Ia expression by T cells (34).   We found that non-immune
inflammatory agents injected intraperitoneally induced exudates rich
in Ia-negative macrophages, while, in contrast, antigenic molecules
induced exudates rich in Ia-positive macrophages (Table 2).   The
antigenic molecules were inducing Ia-positive macrophages by stimulating T cells to elaborate a lymphokine which was ultimately responsible for recruiting the Ia-positive macrophages (36,37).   This
lymphokine was termed Macrophage Ia-Positive Recruiting Factor, or
MIRF.   Thus, immune T cells transferred intraperitoneally with antigen
induced by two to three days an exudate rich in Ia-positive macrophages (34).   A culture supernatant from a mixture of T cells, macrophages, and antigen, injected intraperitoneally, reproduced this
finding.   Further studies have shown that MIRF was not acting <u>in vivo</u>

Table 1.   Two Forms of Antigen Presentation

	I-Region-Restricted	MHC-Independent
Assay	Antigen-pulsed macrophages added to purified T cells	Antigen-pulsed macrophages added to unfractionated T cells or to mixtures of T and B cells (with macrophages)
Effects of Antibody	Usually does not block interaction	Effective in blocking interaction
Effects of Anti-Ia Antibodies	Blocks the interaction	Not tested
Presentation by Allogeneic Macrophages	Present poorly	Can present
Presentation by Ia-bearing macrophages	Strictly required	Probably not required
Localization in Macrophages	Membrane. Obligate intracellular handling step?	Membrane-bound and soluble secreted molecules
Major Role	MHC-restrictive interactions with T cells	Source of antigen available for B cells and/or Ia-bearing macrophages

Complete explanations for each of these points can be found in Reference 1.

on the resident macrophage but on younger cells that could be derived from the bone marrow (37).   Thus, it was ineffective when injected intraperitoneally into X-irradiated mice which contained the radioresistant phagocytes; its injection into irradiated mice reconstituted with bone marrow cells, however, resulted in high Ia-positive exudates.   Interestingly, the analysis of X-irradiated mice showed that the expression of Ia by the macrophages in vivo was brief.   By twenty-four to seventy-two hours, the percent of Ia macrophages was reduced by about three-quarters of the initial level (37).   The nature of

Table 2.   Distribution of Ia-Positive Macrophages in Tissues

Tissue		% Ia$^+$	Reference
Thymus·		75	29
Spleen:	red pulp	30-50	30
	marginal zone	0	
Liver		30-50	32
Lung		5-10	15
Glomeruli		50	33
Peritoneal cavity		10	34
	after oil, peptone	10 or <	34
	after T cell stimulation	> 50	34

MIRF and its relationship to other T cell mediators is only partially known.   Beller's studies indicate that it is a protein of about 50,000 daltons.   A number of his ongoing studies suggest that it is distinct from colony-stimulating factor and from T cell growth factor. The relationship to the classical macrophage-activating factor is not known.

An understanding of the in vivo events regulating macrophage Ia has come about by studying cultured macrophages. We found that the time in which a macrophage synthesized Ia was brief- for about twenty-four to forty-eight hours, after which it ceased Ia expression and became, by definition, an Ia-negative cell.   Two conditions modulated this expression of Ia.   Addition of particles to the macrophage at the time of active synthesis of Ia potentiated and extended the amounts of time of synthesis (13).   Moreover, addition of T cell lymphokines to Ia-negative macrophages induced Ia synthesis and sur-face expression (38,39).   Further studies of David Beller have now shown that in culture there is a clear difference in response to the T cell lymphokine, depending on the age of the macrophage and its state of activation: monocytes and inflammatory macrophages re-sponded rapidly, expressing Ia by two to three days of culture with the factor, while, in contrast, resident macrophages were weak re-sponders.   This result confirms the in vivo studies described above (37) showing that the best target for MIRF was a young phagocyte.

Table 3. Comparison of Ia-Positive and Ia-Negative Macrophages

Property	Macrophages	
	Ia Positive	Ia Negative
Ia antigen expression	+++	---
Fc receptor	++	+++
C3 receptor	++	+++
Bacterial binding	++	++
Latex phagocytosis	++	++
Secretion of lymphocyte activating factor		
after endotoxin	++	++
after T cell-macrophage interaction	++	---
Ia-restricted T cell binding	++	---

An important culture study established that all bone marrow macrophage progenitors have the potential to express Ia (40). Thus, colonies of macrophages were grown from bone marrow stem cells using L cell-conditioned medium which is known to contain a growth-promoting molecule. The cells were then exposed to a T cell-conditioned medium known to contain MIRF. All bone-marrow-derived colonies contained Ia-positive macrophages. In the absence of T cell stimuli, very few were positive (less than 10%). Thus, it appears that there is only a single line of phagocytes that responds to Ia expression and not two distinct differentiation lineages.

In essence, the results of _in vivo_ and _in vitro_ experiments indicate that the phagocytes settle into the different tissues and there express Ia by virtue of unknown local stimuli. This expression of Ia is brief, to the point that the maintenance of the Ia-positive macrophage population depends on a continuous source of new cells. Thus, ablation of stem cell activity results in loss of antigen pre-

sentation function, clearly a radiosensitive process (37). Finally,
one definite regulatory mechanism that leads to an increment in Ia-
positive macrophages and the potential to likewise increase antigen
presentation is by way of T cell activity. The whole cycle of macro-
phage-T cell symbiosis becomes clearly apparent- an Ia-positive macrophage presents antigen to T cells which elaborate a protein that, in
turn, regulates Ia-positive expression by the macrophage.

DOWN REGULATION OF MACROPHAGE Ia

We discovered a suppressor pathway for Ia expression in studies
of neonatal mice. During the first two weeks of postnatal life,
Doctor Lu found that the percentage of Ia-positive macrophages was
very low; and consequently, antigen presentation to T cells was impaired (41). Of great interest is the fact that the thymus was the
only tissue with a normal content of Ia-positive macrophages from the
time of birth (42). We have speculated that, in the neonate, the
function of the thymus macrophage (29) is to regulate intrathymic
differentiation (43) rather than to have an antigen-presentation
function and that the limited potential for antigen presentation of
the macrophage of the secondary lymphoid organs may serve to insure
self-tolerance- by not providing a vehicle for T cell stimulation.

Where is the defect in the neonate that restricts Ia expression?
The macrophages of the neonate can release lymphostimulatory molecules
like LAF (IL-1) if stimulated with endotoxin, and can also be activated to a cytocidal state by lymphokines (44). Thus, the major
problem centers on their lack of Ia synthesis. Unpublished studies
of David S. Snyder in our laboratory have shown that these macrophages cultured with T cell lymphokines can be made to synthesize Ia.
The problem with the neonate relates to the production of prostaglandins that limit Ia expression (45). Doctors Snyder and Lu discovered that neonatal spleen cells, when transferred into adult mice
together with MIRF, inhibited the induction of Ia-positive macrophages. The neonatal cells acted by releasing a soluble dialyzable
inhibitor which was not produced in the presence of indomethacin,
a drug that blocks prostaglandin synthesis. David Snyder has now
brought forth definite proof of the inhibitory nature of prostaglandins by using purified compounds kindly provided by the Upjohn
Company. He found that prostaglandins E_1 and E_2 in vitro inhibited
the effect of MIRF at molar concentrations of 10^{-9} to 10^{-10}M (!); an
in vivo effect was also found when the drugs were injected with MIRF.

The cell in the neonate producing the prostaglandins belongs to
the phagocytic series and appears to be a young, immature monocyte.
The inhibitory cell is poorly adherent when freshly harvested from
the spleen and lacks Ig and Thy 1; when cultured, it develops into
typical macrophages (45). Further studies indicate that the macrophage line in the adult also regulates Ia expression via prostaglan-

din production. Thus, normal peritoneal cells, which are known to
be good prostaglandin producers, when added several times to mice
injected with MIRF depressed the induction of Ia-positive macrophages
(45). An inhibitory cell is also found in bone marrow from adults
but not in spleen and thymus. Clearly, the phagocyte line self-
regulates Ia expression by releasing the prostaglandins.

     We conclude that an important component that controls the amount
of Ia-positive macrophages in each tissue is the local concentration
of prostaglandin. In support of our thesis are recent unpublished
experiments also of David Snyder showing that the intraperitoneal
injection of aspirin or indomethacin to normal mice increases the
percent of Ia-positive macrophages by about threefold. Thus, the
peritoneal cavity seems to have a normally suppressive influence that
controls the level of Ia-positive macrophages.

CONCLUSION

     The studies on regulation of Ia in the macrophage- and therefore
of antigen presentation- indicate to us the importance of this macrophage protein and the need to establish important controlling influences. One would predict that overproduction of Ia-positive macrophages might be deleterious, resulting in uncontrolled stimulation
of the T cells to any environmental antigenic stimuli. In contrast,
a reduction would be incompatible with immune stimulation- a situation
that appears to occur in the neonate.

     Finally, macrophages are not the only Ia-antigen-presenting
cells. Langerhans cells of the skin have been shown to present
antigen (reviewed in 1 and 46). Dendritic cells of the spleen are
excellent stimulators of allogeneic interaction and perhaps may also
present antigen (47). More recently, B cells have been implicated
in antigen presentation (48,49) and so have endothelial cells (50).
The characteristic of each cell for antigen presentation, however,
needs to be critically evaluated. The macrophages, by virtue of their
number and ubiquitous presence in tissues, by their capacity to take
up virtually all antigens, by their ability to express I-region-linked
products, and by having the armamentarium to secrete biologically
active molecules, are, in our view, the most important of all the
presenting cells. We have speculated that perhaps it's time to
return to Aschoff's RES concept, having Ia as the common property-
the Ia-antigen-presenting system encompasses all cells with the
potential to handle antigen and initiate immune stimulation (1).
Clearly, the immune system is built so that an antigen-processing
event controlled by the I region is the essential first interaction.

Acknowledgement

     In this presentation, I have reviewed mostly the work from our

laboratory in which various individuals have participated. I thank
all of them for their many contributions. Our work is supported by
grants from the National Institutes of Health and from the Council
for Tobacco Research. I dedicate this presentation to my father,
Dr. Emilio P. Unanue, in celebration of his eightieth birthday- may
his love, teaching, and support continue for many more years.

REFERENCES

1.   Unanue, E. R., Adv. Immunol. 31:1, 1981.
2.   Shevach, E. M., and Rosenthal, A. S., J. Exp. Med. 138:1213, 1973.
3.   Rosenthal, A. S., New England J. Med. 303:1153, 1980.
4.   Benacerraf, B., Science 212:1229, 1981.
5.   Farr, A. G., Dorf, M. E., and Unanue, E. R., Proc. Natl. Acad.
       Sci. USA 72:3542, 1977.
6.   Farr, A. G., Kiely, J.-M., and Unanue, E. R., J. Immunol. 122:
       2395, 1979.
7.   Gery, I., Gershon, R. K., and Waksman, B. H., J. Exp. Med. 136:
       128, 1972.
8.   Gery, I., and Waksman, B. H., J. Exp. Med. 136:143, 1972.
9.   Calderon, J., and Unanue, E. R., Nature 253:359, 1975.
10. Calderon, J., Kiely, J.-M., Lefko, J. L., and Unanue, E. R., J.
       Exp. Med. 142:151, 1975.
11. Unanue, E. R., Kiely, J.-M., and Calderon, J., J. Exp. Med. 144:
       155, 1976.
12. Unanue, E. R., and Kiely, J.-M., J. Immunol. 119:925, 1977.
13. Beller, D. I., and Unanue, E. R., J. Immunol. 126:263, 1981.
14. Ziegler, K., and Unanue, E. R., J. Immunol. 127:1869, 1981.
15. Weinberg, D. S., and Unanue, E. R., J. Immunol. 126:794, 1981.
16. Katz, D. H., and Unanue, E. R., J. Exp. Med. 137:967, 1973.
17. Ziegler, K., and Unanue, E. R., J. Exp. Med. 150:1143, 1979.
18. Lipsky, P. E., and Rosenthal, A. S., J. Exp. Med. 141:138, 1975.
19. Werdelin, O., and Shevach, E. M., J. Immunol. 123:2779, 1979.
20. Ohkuma, S., and Poole, B., Proc. Natl. Acad. Sci. USA 75:3327,
       1978.
21. Gordon, A. H., D'Arcy Hart, P., and Young, M. R., Nature 286:79,
       1980.
22. Tietze, C., Schlesinger, P., and Stahl, P., Biochem. Biophys.
       Res. Commun. 93:1, 1980.
23. Ziegler, K., and Unanue, E. R., Proc. Natl. Acad. Sci. USA 79:
       175, 1982.
24. Erb, P., Meier, B., and Feldmann, M., J. Immunol. 122:1916, 1979.
25. Lonai, P., Puri, J., and Hammerling, G., Proc. Natl. Acad. Sci.
       USA 78:549, 1981.
26. Unanue, E. R., and Cerottini, J.-C., J. Exp. Med. 131:711, 1970.
27. Unanue, E. R., Immunological Rev. 40:15, 1978.
28. Calderon, J., and Unanue, E. R., J. Immunol. 112:1804, 1974.
29. Beller, D. I., and Unanue, E. R., J. Immunol. 124:1433, 1980.
30. Cowing, C., Schwartz, B. C., and Dickler, H. B., J. Immunol.
       120:378, 1979.

31. Humphrey, J. H., and Grennan, D., Eur. J. Immunol. 11:221, 1981.
32. Richman, L. K., Klingenstein, R. J., Richman, J. A., Strober, W.
     S., and Berzofsky, J. A., J. Immunol. 123:2602, 1979.
33. Schreiner, G. F., Kiely, J.-M., Cotran, R.S., and Unanue, E. R.,
     J. Clin. Invest. 68:920, 1981.
34. Beller, D. I., Kiely, J.-M., and Unanue, E. R., J. Immunol. 124:
     1426, 1980.
35. Lu, C. Y., Peters, E., and Unanue, E. R., J. Immunol. 126:2496,
     1981.
36. Scher, M. G., Beller, D. I., and Unanue, E. R., J. Exp. Med.
     152:1684, 1980.
37. Scher, M. G., Unanue, E. R., and Beller, D. I., J. Immunol. 128:
     447, 1982.
38. Steeg, P., Moore, R., and Oppenheim, J. J., J. Exp. Med. 152:
     1734, 1980.
39. Steinman, R., Noqueira, N., Witmer, M., Tydings, J., and Mellman,
     I., J. Exp. Med. 152:1248, 1980.
40. Calamai, E. G., Beller, D. I., and Unanue, E. R., J. Immunol.
     128, 1982, in press.
41. Lu, C. Y., Calamai, E. G., and Unanue, E. R., Nature 282:327,
     1979.
42. Lu, C. Y., Beller, D. I., and Unanue, E. R., Proc. Natl. Acad.
     Sci. USA 77:1597, 1980.
43. Beller, D. I., and Unanue, E. R., J. Immunol. 121:1861, 1978.
44. Lu, C. Y., and Unanue, E. R., Infect. Immunity, 1982, in press.
45. Snyder, D.S., Lu, C. U., and Unanue, E. R., J. Immunol. 128,
     1982, in press.
46. Stingl, G., Katz, S. I., Shevach, E. M., Rosenthal, A. S., and
     Green, I., J. Invest. Dermat. 71:59, 1978,
47. Steinman, R. M., and Witmer, M., Proc. Natl. Acad. Sci. USA 75:
     5132, 1978.
48. Kammer, G. M., and Unanue, E. R., Clin. Immunol. Immunopathol.
     15:434, 1980.
49. Chestnut, R. W., and Grey, H. M., J. Exp. Med. 126:1075, 1981.
50. Hirschberg, H., Bergh, O. J., and Thorsby, E., J. Exp. Med. 152:
     249s, 1980.

MACROPHAGES AND CANCER METASTASIS

Isaiah J. Fidler and George Poste

Cancer Metastasis and Treatment Laboratory
NCI-Frederick Cancer Research Facility
Frederick, Maryland   21701 USA

Smith Kline and French Laboratories
1500 Spring Garden Street
P.O. Box 7929
Philadelphia, Pennsylvania   19101 USA

INTRODUCTION: THE BIOLOGICAL DIVERSITY OF MALIGNANT NEOPLASMS

Metastasis is the principal cause of failure in the treatment of cancer. There are several reasons for the failure to treat metastasis. First, at time of diagnosis and excision of primary tumors, metastasis may have already occurred, but the lesions are often too small to be detected. Moreover, widespread dissemination of metastases frequently takes place before symptoms of disease occur. Second, the anatomic location of many metastases may limit the effective dose of therapeutic agents that can be delivered to the lesions without being toxic to normal tissues or the host. The most formidable problem, however, is the heterogeneous nature of malignant neoplasms which leads to the rapid emergence of metastases that are resistant to conventional therapy (1,2).

The cellular heterogeneity of neoplasms has been appreciated since the 1800's when histologic observations of malignant neoplasms characterized them as pleomorphic. More recently, however, cells obtained from individual animal and human neoplasms have been shown to be phenotypically diverse with regard to growth rate, antigenic and/or immunogenic properties, cell surface receptors and/or products, response to a variety of cytotoxic agents, invasion, and metastatic potential (1-4).

Recent work from our laboratory (5,6) and many others (reviewed

65

in Ref. 1,2) using animal tumors of diverse histologic origins has revealed significant variations in the metastatic capabilities of sub-populations of cells isolated from a single neoplasm. These data indicate that metastases arise from the nonrandom spread of special-ized subpopulations of cells that preexist within the primary tumor and that the responsiveness of the proliferating metastatic popula-tions to therapeutic agents may differ from that of the nonmetastatic tumor cells that make up the major fraction of the primary neoplasm. Moreover, even within the same patient, different metastases have different susceptibilities to chemotherapeutic agents (7) because the lesions rapidly become biologically diverse (8). Immunologic hetero-geneity among tumor cells populating a primary neoplasm and between a primary neoplasm and its metastases can also pose serious problems in the treatment of metastases by specific immunotherapy (9-12). Thus, both chemotherapy- and/or specific immunotherapy-resistant vari-ants preexisting within the parent tumor population can proliferate unchecked after the sensitive populations are destroyed.

Collectively, the emergence of metastases that are resistant to conventional therapy could be the major reason for the failure to treat cancer metastasis. The data demonstrating that metastases may result from the proliferation of a minor subpopulation of cells within the primary tumor and that tumors are heterogeneous with regard to many phenotypic characteristics, including metastatic potential, provide a conceptual basis for explaining the emergence of such relentless tumor deposits. These studies imply that the successful approach to total eradication of metastases will be one that circumvents tumor cell heterogeneity and also against which resistance is unlikely to develop.

THE ROLE OF MACROPHAGES IN THE PATHOGENESIS OF METASTASIS

Although it is tempting to describe the importance of macro-phages in controlling neoplasia, their primary functions must be considered first. Cells of the macrophage-histiocyte series are im-portant components of the host system responsible for maintenance of homeostasis. The primary function of macrophages is the phago-cytosis and disposal of effete cells, such as aged red blood cells, cellular debris, and serum protein. The removal of effete red blood cells is a continuous process. To accomplish it macrophages must distinguish old from young cells as well as damaged from healthy cells. The highly phagocytic nature of macrophages in vivo can be used to advantage by methods discussed below that enhance their activity. In addition to phagocytosis, macrophages are involved in the controlled metabolism of lipids and iron, in host response to injury (inflammation), and in the defense against microbial infections and parasitic infestations. Macrophages are also an important com-ponent of both the afferent and efferent arms of the immune system,

and finally these cells are important in the defense against neoplasms.

     Hibbs et al. (13) suggested that macrophages may provide a sur-
veillance system for the detection and destruction of nascent transformed neoplastic cells. Studies by Norbury and Kripke (14) supported
this suggestion. In the latter studies (14) the effect of treatments
with macrophage toxins (silica or trypan blue) or a stimulant (pyran
copolymer) on skin carcinogenesis induced by ultraviolet radiation
(UV) was examined in mice. When the dose of the carcinogen (UV) was
not overwhelming, pyran treatment prolonged the latent period of
tumor development and protected against carcinogenesis by reducing the
incidence and number of the resulting skin neoplasms. Conversely,
treatment of mice with macrophage toxins shortened the latent period
of skin cancer induction. Similarly, in studies with transplantable
neoplasms, systemic impairment of macrophages by carrageenan or
silica increased the incidence of spontaneous (15,16) and experimental (17) metastasis. There are also several published reports regarding the efficacy of macrophages in the inhibition of metastasis.
Syngeneic mouse macrophages activated in vitro and then injected
intravenously reduced the formation of B16 melanoma metastases, and
the intravenous injection of nonspecifically activated macrophages
prevented the formation of spontaneous fibrosarcoma metastases (18-
20). Activated mouse macrophages also were shown to inhibit the
growth of tumors at primary sites (21). Differences in the cytotoxic
activity of macrophages isolated from metastasizing and nonmetastasizing tumors have been reported (22). Macrophages isolated from a
nonmetastasizing sarcoma were cytotoxic in vitro. In contrast, macrophages isolated from a weakly immunogenic, metastasizing variant were
not (22). Similar findings have been reported for progressing and
regressing mouse sarcomas (23).

MACROPHAGE CONTENT OF MALIGNANT NEOPLASMS

     The exact role of macrophages that infiltrate primary neoplasms
has been controversial. In rats, the macrophage content of six car-
cinogen-induced fibrosarcomas was correlated directly with their
immunogenicity and inversely with their metastatic potential (24,
25), suggesting that some tumors do not produce metastases because
they contain many macrophages. However, this is not generally the
case. We recently examined the macrophage content of 16 different
rodent tumors and did not find a correlation between the extent of
macrophage infiltration into neoplasms and the metastatic behavior
of the tumors (26). We also did not find a correlation between the
macrophage content of UV-induced murine fibrosarcomas growing in normal or immunosuppressed syngenic mice and the immunogenic potential
of the tumors. There are several factors that influence the extent
of macrophage infiltration into tumors. One of these factors, tumor
cell immunogenicity, was not correlated with macrophage content in
our study. This observation is in agreement with studies by Evans

and Lawler (27) and Evans and Eidlen (28), who examined the macrophage
content of 33 different methylcholanthrene-induced murine fibrosarcomas and rhabdomyosarcomas and concluded that there was no relationship between macrophage content and the immunogenicity of the tumors.
Clearly, the role of the mononuclear phagocyte system in metastasis
varies for different tumors and does not correlate with tumor cell
immunogenicity and/or metastatic properties. In some tumors, numerous
infiltrating macrophages can inhibit metastasis, but the absence of
macrophages in a benign neoplasm will not lead to metastasis. The
absence of macrophages is unlikely to compensate for the inability
of tumor cells to invade the host stroma and enter the circulation
to produce distant growths. Therefore, neoplasms with low macrophage content may or may not be metastatic, as demonstrated in
studies in which nonmetastatic clones isolated from a highly metastatic neoplasm also exhibited low macrophage content when growing
subcutaneously (29).

Although the factors influencing macrophage infiltration of
tumors are complex and poorly understood, it is clear that, in addition to specific recruitment generated by in situ immune reactivity,
nonimmune factors can also regulate the extent of macrophage infiltration (28). In fact, under conditions of progressive growth, many
tumors appear to be nonimmunogenic (30) and macrophage infiltration
into these tumors may be more dependent upon nonimmunological factors,
such as inflammatory conditions or tumor necrosis (31). The mere
presence of macrophages within tumors is not sufficient to bring
about tumor regression. Rather, the degree to which tumoricidal
activity of macrophages is generated in situ appears to determine
whether tumors will progress or regress (23). This finding may explain why progressively growing spontaneous metastases often contain
as many or more macrophages than their parent tumors (32). Two
recent independent studies also reported that the macrophage content
of pooled metastases was similar to that of primary tumors (33) and
that the metastases were not resistant to macrophage-mediated lysis
(34).

THE INTERACTION OF MACROPHAGES WITH HETEROGENEOUS NEOPLASMS

There is now increasing evidence that the activation of cells
of the macrophage-histiocyte series to the bactericidal and/or tumoricidal state enhances host defense against infectious diseases and/or
neoplasms. There are two major pathways to achieve macrophage activation in vivo. Frequently, macrophages are activated as a consequence
of their interaction with microorganisms and/or their products, for
example, endotoxins, the bacteria cell wall skeleton, and small components of the bacteria cell wall skeleton such as muramyl dipeptide
(MDP) (35,36). In vivo activation of macrophages can also take place
after their interaction with soluble mediators released by sensitized lymphocytes. The soluble lymphokine that induces macrophage

activation is referred to as macrophage-activating factor (MAF). MAF
is able to act across species barriers (review 37). Tumoricidal macrophages acquire the ability to recognize and destroy neoplastic cells
both in vitro and in vivo, while leaving nonneoplastic cells unharmed
(37-39), by a nonimmunological mechanism that requires cell-to-cell
contact (39).

The ability of tumoricidal macrophages to discriminate between
tumorigenic and normal cells has been demonstrated with syngeneic
and allogeneic tumors of mice, syngeneic rat tumors, and syngeneic
guinea pig tumors (review 38). Thus, although tumor cell populations can exhibit heterogeneity with respect to many phenotypes, they
are all susceptible to lysis mediated by activated macrophages (38).
Macrophage-mediated cytolysis of tumorigenic cells, at least in vitro,
occurs independently of such tumor cell characteristics as antigenicity, drug sensitivity, invasiveness, and metastatic potential
(4,11,38). For example, B16 melanoma variant lines that have a low
or high metastatic potential and that are either susceptible or resistant to syngeneic T-cell-mediated lysis are all lysed in vitro by
MAF-activated macrophages (38). Similarly, several cell lines isolated from a UV-2237 fibrosarcoma originating in a C3H mouse which
vary greatly in their invasive and metastatic potential in vivo (40),
their immunogenicity, or their susceptibility to NK-cell-mediated
lysis (41) are all susceptible to destruction in vitro by tumoricidal
macrophages.

Recently we attempted to select in vitro tumor cell variant lines
that exhibit a phenotype resistant to macrophage-mediated lysis. We
used techniques similar to those used previously to successfully
select a B16 melanoma tumor cell line resistant to lysis by cytotoxic T lymphocytes (42) or a UV-2237 fibrosarcoma resistant to NK-
cell-mediated lysis (41). Despite using repeated selections in vitro,
we failed to isolate tumor cells with increased resistance to macrophage-mediated lysis. For example, cells from either the B16 melanoma or the UV-2237 fibrosarcoma were incubated with the selective
killer cells (cytotoxic T cells, NK cells, or tumoricidal macrophages). The surviving target cells were grown in monolayers and then
incubated with the effector cells. After 4-5 sequential selections,
tumor variants resistant to cytolysis mediated by T cells (B16) or
NK cells (UV-2237) were obtained. In contrast, variants resistant to
macrophage-mediated lysis could not be obtained even though the
selection pressure (>90% kill) exceeded that of the T or NK cells.
Moreover, as already stated, despite their resistance to T-cell-
mediated lysis (B16-F1^R) or NK-cell-mediated lysis (UV-2237-NK^R), both
variants were still susceptible to destruction by tumoricidal macrophages.

Taken together, these data indicate that, at least in vitro,
tumoricidal macrophages discriminate between neoplastic and nonneoplastic cells by a process that is independent of transplantation

antigens, species-specific antigens, tumor-specific antigens, cell cycle time, or various phenotypes associated with "transformation". Although the exact mechanism(s) by which macrophages recognize and lyse tumor cells is still unclear, it is probably regulated by a cell characteristic that is linked with the tumorigenic capacity of cancer cells (43).

IN SITU STIMULATION OF MACROPHAGES TO BECOME TUMORICIDAL

Because macrophages from tumor-bearing animals can respond to activating stimuli and become tumoricidal (44), finding a means to activate macrophages in situ becomes desirable. Early attempts to activate macrophages by the systemic administration of crude preparations of lymphokines have not been successful. Although intratumoral inoculation of lymphokine preparations containing MAF has been shown to induce regression of skin tumors and cutaneous metastases (45,46), the administration of lymphokines to achieve systemic activation of macrophages has not been feasible. After injection into the circulation, lymphokines have a short half life because serum proteins rapidly inactivate them (47). The most serious obstacle to therapeutic activation of macrophages in situ by crude lymphokines may be the fact that macrophages are susceptible to lymphokine activation for only 3-4 days after their emigration from the circulation into tissues (47). Once activated, macrophages are tumoricidal for only 3-4 days, and, with the decay of their tumoricidal properties, they are refractory to reactivation by soluble MAF (47). Therefore, adoptive therapy with macrophages, although theoretically sound and therapeutically effective in syngeneic animal tumor systems, may be hampered by the difficulty of finding capable immune donors. Passive therapy with systemically injected lymphokines aimed at the in situ stimulation of the reticuloendothelial system may be hampered by the inability to achieve effective levels at the tumor bed.

The therapeutic use of water-soluble synthetic MDP has also been hindered because after parenteral administration, this agent is rapidly cleared (<60 min) from the body and excreted in the urine (48). This brief period of exposure is insufficient to render macrophages tumoricidal even under ideal conditions (49).

Recent advances in liposome technology have provided a mechanism for activating macrophages in situ with soluble MAF and/or MDP. Liposomes can be used to carry agents to cells of the reticuloendothelial system since these cells are responsible for the rapid clearance of particulate material from the circulation. There are several advantages to using liposome-encapsulated materials to activate cells of the macrophage-histiocyte series in vivo. Many macrophage-activating agents such as bacterial products or lymphokines can be antigenic, and repeated systemic administrations can lead to adverse reactions. Liposomes are nonimmunogenic, and thus elicitation of allergic re-

actions commonly associated with the systemic administration of other
immune adjuvants may be avoided (50).

We have recently shown that lymphokines and MDP encapsulated
within liposomes are most efficient in their ability to activate
macrophages to the tumoricidal state in vitro (47, 49, 51, 52) and
in vivo (53-55). Moreover, unlike activation by free (unencapsulated)
MAF, which requires binding of MAF to a fucoglycolipid receptor on
the macrophage surface (56), liposome-encapsulated MAF can activate
macrophages lacking functional receptors for MAF (57, 58). Moreover,
MAF encapsulated within liposomes can induce activation of subpopulations of tissue macrophages and intratumoral macrophages that are
completely refractory to activation by free MAF (47, 57, 58).

TREATMENT OF METASTASES BY SYSTEMICALLY ADMINISTERED IMMUNOMODULATORS
ENCAPSULATED IN LIPOSOMES

These data raised the possibility that macrophages could be
activated in situ to the tumoricidal state by systemically administered immunomudulators encapsulated in liposomes and that this could
provide a potential therapeutic modality for enhancing host destruction of metastasis. To test this possibility, we treated mice bearing
spontaneous metastases with intravenous injections of multilamellar
vesicle liposomes consisting of phosphatidylserine and phosphatidylcholine and containing entrapped MAF (54) or MDP (59). C57BL/6 mice
were each given an intrafootpad injection of syngeneic B16-BL6 melanoma cells. Four weeks later, when the implants had reached a size
of 10-12 mm, the leg bearing the tumor, including the popliteal lymph
node, was amputated. Three days later, animals were injected intravenously with 5 µmoles of liposomes containing immunomodulators or
placebo preparations. Both test and control groups were treated twice
weekly for 4 weeks. Two weeks after the final treatment, the animals
were killed and necropsied. The presence of metastases was determined
under a microscope and all suspected lesions were confirmed histolog-
ically. Spontaneous pulmonary and lymph node metastases were well
established in the animals at the time liposome therapy was started;
several individual lung metastases were visible macroscopically.
Without therapy these tumor foci rapidly developed into lesions ex-
ceeding 2-3 mm in diameter. Most mice (>70%) treated with liposome-
encapsulated MAF or MDP had no macroscopically or microscopically
detectable metastases. Moreover, even in animals with metastases,
the median number of metastases was significantly smaller than in
the other treatment groups (54, 57). Comparison of the same treatment protocols in survival assays revealed that mice treated with
liposome-encapsulated immunomodulators survived significantly longer
than control mice or mice injected with free MDP or liposomes containing phosphate-buffered saline suspended in free MDP. In both
series of experiments (liposome-MDP and/or liposome-MAF), the regression of lymph node and pulmonary metastases always was associated with

the induction of tumoricidal activity in alveolar macrophages. In
control studies, in which systemically administered liposomes containing control substances failed to activate macrophages, tumor
regression also failed to occur.

Although compelling, the evidence that macrophages destroyed
metastases in mice injected systemically with liposomes containing
immunomodulators is at best circumstantial. For this reason we
have performed recently several experiments designed to demonstrate that macrophages are the essential host cell responsible for
the destruction of metastases in mice treated with either MAF or MDP
entrapped in liposomes (60). Results from three separate types of
experiments support this conclusion. First, when macrophage-activating agents such as lymphokines or MDP were delivered in liposomes
that were not efficiently retained in the lung, little or no activation of lung macrophages was observed and growth of metastases was
unaltered. Second, when tumor-bearing animals were treated with
agents that impaired macrophage function (e.g., silica, carrageenan,
hyperchlorinated drinking water) before systemic therapy with liposome-encapsulated lymphokines or liposome-encapsulated MDP, metastases
were not destroyed. Third, when macrophages were activated in vitro
by liposome-encapsulated MDP and then injected intravenously into mice
bearing experimental lung metastases, the growth of lung metastases
was significantly inhibited. These results suggest that the augmented
host response against pulmonary and lymph node metastases generated
by the systemic administration of liposome-encapsulated lymphokines
or MDP is mediated via activated cytotoxic macrophages (60).

For all therapy experiments, we used multilamellar vesicles consisting of two natural phospholipids. We chose this particular class
of liposomes because studies of the distribution of liposomes of
different size and phopholipid composition throughout the body demonstrated that localization and retention of liposomes in the lung could
be achieved with these negatively charged liposomes (55). Moreover
toxicity studies in which these liposomes containing encapsulated MAF
were injected intravenously into mice (10 µmoles phospholipid) or
beagle dogs (5-60 mg phospholipid/kg) failed to reveal any adverse
reactions in recipient animals even after repeated injections (61).

Our results indicate that the multiple intravenous injections
of liposomes containing macrophage-activating agents eradicated spontaneous pulmonary and lymph node metastases (arising from B16-BL6
melanoma or K-1735 melanoma which were excised before therapy) in
mice. In mice bearing the B16-BL6 tumor, the tumor burden in lung
and lymph node metastases at the start of therapy was probably in
excess of 10^7 cells. Yet, seventy percent of mice treated with
liposome-encapsulated MDP survived for 200 days after intrafootpad
implantion of the tumor. In this tumor system, the median life span
of mice inoculated with as few as 10 viable cells has been shown to
be 40-50 days (62). Therefore, the tumor burden in surviving mice

must have been reduced to fewer than 10 viable cells, because the mice survived longer than required to be classified as disease free.

The optimal conditions for systemic therapy with liposome-encapsulated immunomodulators and the efficacy of this modality, alone or in combination with others, in treating large metastatic tumor burdens has not been defined. Although the initial results reported here are encouraging, it is unlikely that this therapeutic approach could serve as a single modality for treatment of metastatic disease. Potential therapeutic regimens designed to stimulate host immunity must be used in combination with other conventional treatment modalities such as chemotherapy in order to first reduce the tumor burden to a level at which activated macrophages can kill tumor cells which survive or are resistant to other agents.

SUMMARY

Activated macrophages appear to be able to recognize and destroy neoplastic cells without regard to their phenotypic diversity, and macrophage-mediated cytotoxicity appears invulnerable to the problem of cellular resistance to killing which is routinely encountered in efforts to destroy tumor cells by cytotoxic drugs. However, macrophage-mediated destruction of large tumor burdens may not be feasible. In many tumors the number of macrophages is too low to destroy all tumor cells, even if the macrophages are activated to the optimal tumoricidal state. For this reason, systemically administered immunomodulators encapsulated in liposomes should be used to activate macrophages to destroy those few tumor cells resistant to other means of therapy.

REFERENCES

1. Hart, I. R. and Fidler, I. J., Biochim. Biophys. Acta 651:37, 1981.
2. Poste, G., and Fidler, I. J., Nature 283:139, 1980.
3. Fidler, I. J., Cancer Res. 38:2651, 1978.
4. Fidler, I. J., and Poste, G., in "Tumor Cell Heterogeneity" (A. Owens, ed.), Fourth Bristol-Myers Symposium, in press.
5. Fidler, I. J., and Kripke, M. L., Science 197:893, 1977.
6. Kripke, M. L., Gruys, E., and Fidler, I. J., Cancer Res. 38:2962, 1978.
7. Tsuruo, T., and Fidler, I. J., Cancer Res. 41:3058, 1981.
8. Cifone, M. A., and Fidler, I. J., Proc. Natl. Acad. Sci. USA 78:6949, 1981.
9. Prehn, R. T., J. Natl. Cancer Inst. 45:1039, 1970.
10. Hepner, G. H., in "Commentaries on Research in Breast Disease" (R. D. Bulbrook, and D. J. Taylor, eds.), Alan R. Liss, Inc., New York, 1979.

11. Kerbel, R. S., Nature 280:358, 1979.
12. Fidler, I. J., and Kripke, M. L., Cancer Immunol. Immunother.
        7:201, 1980.
13. Hibbs, J. B. Jr., Lambert L. H. Jr., and Remington, J. S.,
        Science 177:998, 1972.
14. Norbury, K., and Kripke, M. L., J. Reticuloendothel. Soc. 26:
        827, 1979.
15. Sadler, T. E., Jones, P. D. E., and Castro, J. E., in "The
        Macrophage and Cancer" (K. James, B. McBride, and A. Stuart,
        eds.), pp. 155-163, Econoprint, Edinburgh, 1977.
16. Jones, P. D. E., and Castro, J. E., Br. J. Cancer 35:519, 1977.
17. Montovani, A., Giavazzi, R., Polentarutti, N., Spreafico, F.,
        and Gavattini, S., Int. J. Cancer 25:617, 1980.
18. Fidler, I. J., Cancer Res. 34:1074, 1974.
19. Liotta, L. A., Gatozzi, C., Kleinerman, J., and Saidel, G.,
        Br. J. Cancer 36:639, 1977.
20. Fidler, I. J., Fogler, W. E., Connor, J., in: "Immunobiology and
        Immunotherapy of Cancer" (W. D. Terry, and Y. Yamamura, eds.),
        pp. 361-375, Elsevier, New York, 1979.
21. Den Otter, W., Dullens Hub, F. J., Van Lovern, H., and Pels, E.,
        in: "The Macrophage and Cancer", (K. James, B. McBride, and A.
        Stuart, eds.), pp. 119-141, Econoprint, Edinburgh, 1977.
22. Mantovani, A., Int. J. Cancer 22:741, 1978.
23. Russell, S. W. and McIntosh, A. T., Nature 268:69, 1977.
24. Eccles, S. A., in: "Immunological Aspects of Cancer" (J. E.
        Castro, ed.), pp. 123-154, MTP Press, Lancaster, England,
        1978.
25. Eccles, S. A., and Alexander, P., Nature 250:667, 1974.
26. Talmadge, J. E., Key, M., and Fidler, I. J., J. Immunol., in
        press.
27. Evans, R., and Lawler, E. M., Int. J. Cancer 26:831, 1980.
28. Evans, R., and Eidlen, D. M., J. Reticuloendothel. Soc. 30:
        425, 1981.
29. Kerbel, R. S., and Twiddy, R. R., in: "Contemporary Topics in
        Immunology" (I. P. Witz, and M. G. Hanna, Jr. eds.), pp. 239-
        254, Plenum Press, New York, 1980.
30. Hewitt, H. B., Blake, E. R., and Walder, A. S., Br. J. Cancer
        33:241, 1976.
31. Dvorak, H. R., Dickersin, G. R., Dvorak, A. M., Manseau, E. J.,
        and Pyne, K., J. Natl. Cancer Inst. 67:335, 1981.
32. Key, M., Talmade, J. E., and Fidler, I. J., J. Immunol., sub-
        mitted.
33. Nash, J. R. G., Price, J. E., and Tarin, D., Br. J. Cancer 39:
        478, 1981.
34. Mantovani, A., Int. J. Cancer 27:221, 1981.
35. Lederer, E., J. Med. Chem. 23:819, 1980.
36. Chedid, L., Carelli, L., and Audibert, F., J. Reticuloendothel.
        Soc. 26:631, 1979.
37. Fidler, I. J., and Raz, A., in: "Lymphokines" Vol. 3 (E. Pick,
        ed.), pp. 345-363, Academic Press, New York, 1981.

38. Fidler, I. J., Isr. J. Med. 14:177, 1978.
39. Hibbs, J. B. Jr., J. Natl. Cancer Inst. 53:1287, 1974.
40. Fidler, I. J., and Cifone, M. A., Am. J. Pathol. 97:633, 1979.
41. Hanna, N., and Fidler, I. J., J.Natl. Cancer Inst. 66:1183, 1981.
42. Fidler, I. J., Gersten, D. M., and Budmen, M., Cancer Res. 36:
    3608, 1976.
43. Fidler, I. J., Roblin, R. O., and Poste, G., Cell. Immunol. 38:
    131, 1978.
44. Sone, S., and Fidler, I. J., Cancer Res. 41:2401, 1981.
45. Papermaster, B. W., Holtermann, O. A., Rosner, D., Klein, E.,
    Dae, T., and Djerassi, I., Res. Commun. Chem. Pathol. Pharma-
    col. 8:413, 1974.
46. Salvin, S. B., Youngner, J. S., Nishio, J., and Neta, R.,
    J. Natl. Cancer Inst. 55:1233, 1975.
47. Poste, G., and Kirsh, R., Cancer Res. 39:2582, 1979.
48. Parant, M., Parant, F., Chedid, L.,Yapo, A., Petit, J. R., and
    Lederer, E. Int. J. Immunopharmacol. 1:35, 1979.
49. Sone, S., and Fidler, I. J., Cell.Immunol. 57:42, 1981.
50. Allison, A. C., J. Reticuloendothel. Soc. 16:619, 1979.
51. Poste, G., Kirsh, R., Fogler, W., and Fidler, I. J., Cancer Res.
    39:881, 1979.
52. Sone, S., and Fidler, I. J., J. Immunol.125:2454, 1980.
53. Fogler, W. E., Raz, A., and Fidler, I. J., Cell. Immunol. 53:
    214, 1980.
54. Fidler, I. J., Science 208:1469, 1980.
55. Fidler, I. J., Raz, A., Fogler, W. E., Kirsh, R., Bugelski, P.,
    and Poste, G., Cancer Res. 40:4460, 1980,
56. Poste, G., Kirsh, R., and Fidler, I. J., Cell. Immunol. 44:71,
    1979.
57. Poste, G., Am. J. Pathol. 96:595, 1979.
58. Poste, G., Kirsh, R., Fogler, W. E., and Fidler, I. J. Cancer
    Res. 39:881, 1979.
59. Fidler, I. J., Sone, S., Fogler, W. E., and Barnes, Z. L., Proc.
    Natl. Acad. Sci. USA 78:1680, 1981.
60. Fidler, I. J., Barnes, Z., Fogler, W. E., Kirsh, R., Bugelski,
    P., and Poste, G., Cancer Res. in press.
61. Hart, I. R. Fogler, W. E., Poste, G., and Fidler, I. J., Cancer
    Immunol. Immunother. 10:157, 1981.
62. Griswold, D. P. Jr., Cancer Chemother. Rep. 3:315, 1972.

IMMUNOLOGICAL SURVEILLANCE VERSUS IMMUNO-
LOGICAL STIMULATION OF ONCOGENESIS- A FORMAL
PROOF OF THE STIMULATION HYPOTHESIS

Richmond T. Prehn

Institute for Medical Research
751 South Bascom Avenue
San Jose, California   95128 USA

When I was first introduced to cancer research in 1949, the immune mechanism was characterized by the attributes of specificity and memory, and the function of the thymus was unknown. We now have some knowledge of the role of the thymus and the definition of an immune reaction seems to have undergone a radical change. For example, a cell of the lymphoid family that has little specificity and is apparently unassociated with any form of memory, the so-called NK cell, is now promoted as the possible agent of immunological surveillance. Apparently, any function of any leucocyte is now deemed to be immunologic regardless of questions of specificity or memory.

My own work has largely concerned the role of a thymic-dependent acquired immunity in the biology of 3-methylcholanthrene(MCA)-induced sarcomas in mice. Indeed, it was this system that first convinced the majority of investigators that the study of immune mechanisms was a legitimate subject of cancer research, and, while these tumors are probably poor models for most types of human neoplasms, their study may nonetheless provide some information of general validity.

One of the major drawbacks of many experimental studies, including many of those using MCA-induced sarcomas, is the use of transplanted tumors. Much valuable information can be obtained from the study of such tumors, but their limitations must be understood. For example, the demonstration of tumor antigenicity by immunization-challenge techniques in secondary or tertiary syngeneic hosts was a relatively easy matter (1), but it was difficult to induce or demonstrate acquired resistance to tumor growth in the autochthonous animal. George Klein did succeed, but it required hyperimmunization (2); it is clear that the autochthonous mouse reacts very differently

from a secondary host. The reason for this difference is not only
the presence of MCA in the autocthonous animal (which can also be
provided in a secondary host) but also with the manner of antigen
presentation (3).

The reactions of the autochthonous animal can be simulated, at
least to some extent, in a secondary host if the tumor inoculum is
exceedingly small. The small inoculum presumably is analogous to
growth of the de novo tumor from a very small nidus of transformed
cells and results in the "sneaking through" phenomenon (4,5). The
immune reaction in the primary host is usually smaller than that in
a secondary host unless the inoculum is very small. There is evidence
that the partial tolerance induced by a very small inoculum or by the
nidus of incipient tumor in the autochthonous host is due to the
generation of suppressor cells (6).

There is another way in which the MCA-induced tumor probably
fails as a model of human cancer: induction is usually with a concentration of chemical that is unlikely to be encountered by a human.
Even if many human cancers are due to environmental carcinogens,
itself a debatable point, it is quite clear that the exposure of the
human is usually to low chronic levels, a condition that is seldom
simulated under experimental conditions in the laboratory. The amount
of inducing chemical oncogen has a profound influence on the charac--
teristics of the resulting tumors; in general, tumors induced with
high concentrations are often more rapidly growing, more malignant,
and tend to be more immunogenic when tested by immunization-challenge
type tests (7).

The discovery that the growth of tumor implants in syngeneic mice
could be inhibited by specific immunization led quite naturally to
the concept of immunological surveillance (8) despite the dangers of
extrapolating from a secondary host to the original tumor bearer;
the role of the immune mechanism in resistance to bacteria and parasites had been appreciated since early in the century and if tumors
are antigenic and recognized by the host as foreign, what could be
more natural than to suppose that this same mechanism might play a
key role in anti-tumor defense? The immunity postulated to play the
surveillance role was, of course, the type that had been demonstrated
by the immunization-challenge tests, namely a thymic dependent, anti-
homograft type of reaction involving a high degree of tumor specificity as well as immunological memory. In fact, the concept was
carried to the extreme of postulating that anti-tumor defense was
the raison-dêtre for this type of immunity; defense against bacteria
was a happy by-product (9).

A principal prediction of the surveillance hypothesis is that
cancer should be more prevalent and easier to induce under conditions
in which the immune capacities of the subjects are compromised.
Numerous studies were soon reported that seemed to conform to the

prediction. In particular, those tumors caused by viral oncogens such as polyoma virus were much easier to induce in immuno-crippled animals (10,11); chemical oncogenesis proceeded faster in mice that had been thymectomized shortly after birth (12); and even in humans, who were immunodepressed for purposes of kidney grafting, an elevated rate of tumor formation was reported (13).

Despite these confirmations of prediction, there was room for skepticism. It was possible that the successes in the animal virus systems were due, at least in part, to unusually immunogenic tumors and perhaps to anti-viral rather than anti-tumor immunity; the potentiation of chemical oncogensis was of small magnitude and numerous failures were experienced (14); and the increases in cancer among kidney transplant recipients were, contrary to expectation, only in lympho-reticular and possibly in skin cancers rather than in the whole spectrum of types. After the first enthusiasm, a disenchantment occurred which was crowned by the observations, in several laboratories, that athymic nude mice were not more subject to spontaneous or chemically induced neoplasms than were their virtually normal heterozygous littermates (15,16). This was true even when the animals were kept under germ-free or gnotobiotic conditions, so that they lived a full life-span. It became clear that the homograft type of thymic dependent anti-tumor immunity is seldom, if ever, a very effective surveillance mechanism.

In 1971, in a keynote address to the annual meeting of the Reticuloendothelial Society, I proposed a modification of the surveillance theory which contained, as its center piece, the proposition that the immune reaction, of the thymic type I have been discussing, is a two-edged sword; that its predominant effect when present in modest degree is to stimulate rather than inhibit tumor growth (17). Since then much data congenial with this hypothesis has been assembled (18,19) and, at this time, almost formal proof of its basic validity is available. I will now outline the decisive argument.

I begin with the observation, made by several investigators, that there is a relationship between the immunogenicities of chemically induced tumors, as judged by the excision-challenge type of in vivo test performed in secondary or tertiary recipients (20) and the latencies of the primary tumors in the animals of origin (21,22,4). There is a marked tendency for the tumors that were of shorter latency in the original host to be more immunogenic when tested in an early transplant generation. (I am using the term immunogenic rather than antigenic, because the test does not measure tumor antigens directly; it measures instead the reaction process that involves not only antigen, but the responses of the tumor to the variable and complex reactions of the host.) Although the tumors that arise early after carcinogen exposure, tend to test as more immunogenic, some relatively non-immunogenic tumors are found at all latencies.

What are the reasons for this relationship between latency in the primary host and tumor immunogenicity? One possibility might be that, even among animals of a single inbred strain, some might metabolize the chemical oncogen more efficiently than others; therefore some would be more affected than others and consequently produce faster growing tumors. These same tumors might also tend to be more immunogenic if tumor antigens were induced by the direct action of the chemical metabolites.

This explanation, based upon possible variations among the inbred animals in their metabolism of a carcinogen, is supported by the observation that such variations may indeed exist (23). It is also known that antigenicity is a direct function of the chemical oncogen inasmuch as spontaneous tumors are almost always non-immunogenic (24,25,1) and tumors that arise in the immune-free environs of tissue culture or diffusion chambers are non-immunogenic unless induced by a chemical or viral agent (26,27,28).

Although the necessary preconditions are thus met, the hypothesis that the relationship between the immunogenicity of the transplanted tumors and the latency in the primary host is due directly to variations in the metabolism of the chemical oncogen is nonetheless false. Tumors induced by MCA in the immune-free environs of diffusion chambers do not show any relationship between latency and immunogenicity (22). Furthermore, the variations in susceptibility to oncogenesis, demonstrated among animals of an inbred strain, vanish when the mice are immunocrippled (29). Heavily radiated and thymectomized inbred mice are not demonstrably heterogeneous in their susceptibility to chemical oncogensis; there is still a spread in latent periods, but this is due to change rather than to differences among the mice (29). From these observations, one must conclude that an immune reaction by the primary host is a necessary condition for the relationship between tumor latency and immunogenicity. The relationship is thus mediated by the immune response; i.e., it is due to immunoselection in the original, primary host.

Having concluded that the relationship is due to immunoselection, there are two possibilities that are not mutually exclusive. The first is the surveillance hypothesis: perhaps the faster growing tumors of shorter latency can literally outgrow the inhibitory immune response and consequently even the highly immunogenic ones, if of intrinsically short latency, may escape surveillance; on the other hand, any highly immunogenic tumors that are not intrinsically fast growing would be eliminated and thus only tumors of low immunogenicity would be found among those of longer latency. The alternative to the surveillance hypothesis is the immunostimulation hypothesis which is based on the premise that the more immunogenic tumors are actually stimulated rather than inhibited by the immune reaction in the primary host and, for that reason, have shorter latencies.

The two hypotheses differ in the predictions they make about the details of the relationship between tumor immunogenicity and latency. The surveillance hypothesis predicts that, on average, the fastest growing tumors, i.e., those of shortest latency, being least affected by the immune reaction, would be most likely to have among their number tumors of highest immunogenicity. The immunostimulation hypothesis makes a somewhat different prediction: since, according to this theory, it is a low or intermediate level of immune response that is optimal for tumor growth, the tumors of shortest latency would, on average, be found to be tumors of intermediate immunogenicity.

With these two disparate predictions in mind, the relationship between latency and immunogenicity was reassessed in some previously published data involving 90 sarcomas that were induced with MCA in BALB/cAn female mice (30). With this number of tumors, some details of the relationship that were not apparent in smaller studies (4,22) became clear. The tumors of shortest latencies had intermediate rather than high levels of immunogenicity; the highest immunogenicities occurred among the second quartile of tumors to arise, rather than among the first (Table 1). The data thus conform to the predictions of the immunostimulation hypothesis; they are incompatible with the surveillance theory.

Could the surveillance theory be rescued by postulating that immunodepression by the MCA is a significant factor; if the maximum immunodepression were to occur at a time after the first tumors had appeared, is it possible that the period of maximum immunodepression might coincide with the latent periods associated with the highest tumor immunogenicities? In other words, if surveillance were not weakened by the MCA until after fully one quarter of the tumors had already appeared, the failure of the fastest growing tumors to be the most immunogenic might be compatible with the surveillance explanation. The time course of immunodepression by MCA, when the chemical is administered in the form of paraffin wafers, as is the routine practice in my laboratory, has not been systematically investigated. There is information that no depression is observable until after the first four weeks (31). Be that as it may, there is evidence in the existent data that argues conclusively against the adequacy of the surveillance hypothesis.

I have already pointed out that when inbred animals are given a uniform dosage of MCA, some are predisposed to more rapid oncogenesis than are others, and furthermore, that these differences among the animals are due to variations among them in their immunological capacities (29). This means that the animals that exhibit the shortest tumor latencies do so because, and only because, their immunological reactions are optimal for rapid tumor formation; but these same animals tend to develop tumors of only intermediate immunogenicity. Therefore, the level of immune reactivity that is optimal

Table 1. Relationship Between Latency and Immuno-
          genicity in 90 MCA Induced Sarcomas in
          BALB/cAN Mice

		Quart	ile	
Determination	I	II	III	IV
Number of tumors	22	23	24	21
Latencies (weeks)	8-12	13-15	16-23	24-40
Average Immunogenicity	2.24	5.32	2.05	1.40
		└ P <0.05 ┘		
Immunogenicities over 5	1	8	0	0
Percent over 5	5	33	0	0

Each of the 90 tumors was tested individually by the
excision-challenge type of test. Immunogenicity is
the average tumor size in approximately 20 normal con-
trols/average tumor size in approximately 20 immunized
experimentals. See (20). The raw data has been published (30). P value determined by Mann-Whitney U.
test (two-tailed).

for rapid tumor production is not the level that leads to either
highest or lowest immunogenicities. This is contrary to the surveillance hypothesis. Given these facts, the time course of immunodepression by the MCA becomes irrelevant; no matter how one looks
at these data, the surveillance hypothesis does not fit and the
immunostimulation hypothesis does.

The discussion, to this point, can be summarized as follows:
the fact that the maximal tumor immunogenicities failed to be found
among tumors that had been induced with the shortest latencies constitutes virtually formal proof that the surveillance hypothesis is
inadequate and that the immunostimulation hypothesis is a sufficient
explanation for the peculiar relationship between latency and immunogenicity in the MCA-sarcoma system. This conclusion is compatible
with a vast literature; an immune response can stimulate hyperplastic
growth in both normal and neoplastic target tissues (18,19). The
fact that an immune reaction is necessary for optimal tumor growth

explains the perplexing observation that immunogenic tumors usually retain their immunogenicity despite transplantation to many successive immuno-competent hosts (4).

The discussion thus leads to the following hypothesis: Perhaps the major effect of any type of oncogen is to produce a heritable antigenicity in somatic cells that can arouse a tissue-stimulating immunity. Carried one step further, could a tumor be defined as a hyperplasia caused by an immune response to abnormal cell-surface moieties that are heritable in a clone of somatic cells?

There are obstacles to such an immunological hypothesis of oncogenesis, but they may not be sufficient to force its immediate rejection. The most obvious obstacle is the fact that tumors vary greatly in their immunogenicities and, as far as can be ascertained, most so-called spontaneous tumors are non-immunogenic (1,24,25). However, they are non-immunogenic only by the immunization-challenge type of test; they may arouse a leucocytic reaction (an immune reaction in the broadest sense of the word!) while being negative as judged by that test. For example, it has been shown by Evans that MCA-induced sarcomas attract an infiltrate of macrophages, the number of which, although varying from tumor to tumor, remains quite characteristic and stable for each individual tumor. The percentage of macrophages in the tumor mass may constitute, in some cases, over 40% of the total. The interesting point is that this percentage is unrelated to the immunogenicities of the tumors as judged by immunization-challenge type tests (32). There is also evidence that if the tumor is deprived of the macrophage infiltrate by experimental manipulation, the tumor grows relatively poorly (33,34). That macrophages can be stimulatory to tumors and to certain normal cells is well established (34) and leads me to postulate that perhaps any immune effector, whether T-cell, natural killer cell, macrophage, lymphotoxin, or antibody may, in proper titer, be stimulatory. Considerable direct evidence to support this postulate has been published (19).

It is thus evident that the demonstration that specific thymic dependent immunity can be stimulatory in the case of immunogenic MCA-induced sarcomas is important as an example of what may be a broader principle. Many human tumors, especially early in their evolution, have leucocytic infiltrates and I have postulated that perhaps tumors are often (always?) dependent upon such infiltrates for varying periods in their early progression (35).

Perhaps the greatest problem with the thesis I have been advancing is how to relate it to the more general problems of growth control and tissue regulation. Put more crudely, why would nature have invented a mechanism to promote tumor growth? It seems obvious that this mechanism must convey some benefits to the host and tumor promotion must be only an unfortunate by-product.

It has been bruited since the time of Carrel (36) that leucocytes, especially lymphocytes, may stimulate tissue growth. More
recently, Burch and Burwell have set forth an elaborate theory in
which the lymphoid tissues serve as the yardstick in growth regulation (37,38). That leucocytes play a role in wound repair is widely
recognized, and the widespread relationship between lymphoid cell
accumulations and normal tissue hyperplasia has been repeatedly
noted (35). One must, I think, postulate that normal tissues can,
when injured or under certain other conditions, attract growth facil-
itating leucocytes and that this is a physiologic process that is
corrupted in the process of necplasia.

Acknowledgement

    Supported by Public Health Service (PHS) grants CA31837 and
CA31836 from the National Cancer Institute and an American Cancer
Society, Inc., grant.

REFERENCES

1. Prehn, R. T., and Main, J. M., J. Nat. Cancer Inst. 18:769, 1957.
2. Klein, G., Sjögren, H. O., Klein, E., and Hellström, K. E.,
        Cancer Res. 20:1561, 1960.
3. Basombrio, M. A., and Prehn, R. T., Cancer Res. 32:2545, 1972.
4. Old, L. J., Boyse, E. A., Clarke, D. A., and Carswell, E. A.,
        Ann. N. Y. Acad. of Sci. 101:80, 1962.
5. Marchant, J., Brit. J. Cancer 23:383, 1969.
6. Bonmassar E., Menconi, E., Goldin, A., and Cudkowicz, G., J. Nat.
        Cancer Inst. 53:475, 1974.
7. Prehn, R. T., J. Nat. Cancer Inst. 55:189, 1975.
8. Burnet, F. M., Prog. Exp. Tumor Res. 13:1, 1970.
9. Thomas, L., in "Cellular and Humoral Aspects of the Hypersensi-
        tive States" (H. S. Lawrence, ed.), pp. 529-532, Hoeber, New
        York, 1959.
10. Law, L. W., Cancer Res. 26:551, 1966.
11. Law, L. W., Cancer Res. 26:1121, 1966.
12. Grant, G., Roe, F. J., and Pike, M. C., Nature 210:603, 1966.
13. Penn, I., and Starzl, T. E., Transplantation 14:407, 1972.
14. Weston, B. J., in "Contemp. Topics Immunobiol." (A. J. S. Davies
        and R. L. Carter, eds.), pp. 237-252, Plenum Press, New York,
        1973, Vol. 2.
15. Outzen, H. C., Custer, R. P., Eaton, G. T., and Prehn, R. T.,
        J. Reticuloendothelial Soc. 17:1, 1975.
16. Stutman, O., Science 183:534, 1974.
17. Prehn, R. T., J. Reticuloendothelial Soc. 10:1, 1971.
18. Prehn, R. T., and Lappé, M. A., Transplantation Rev. 7:26, 1971.
19. Prehn, R. T., and Outzen, H. C., in "Prog. in Immunology IV,
        Immunology '80". (M. Fougereau and J. Dausset, eds.), pp.
        651-658), Academic Press, London, 1980.

20. Lawler, E. M., Outzen, H. C., and Prehn, R. T., Cancer Imm. and Immunotherapy 11:87, 1981.
21. Johnson, S., Brit. J. Cancer 22:93, 1968.
22. Bartlett, G. L., J. Nat. Cancer Inst. 49:493, 1972.
23. Nebert, D. W., Robinson, J. R., Niwa, A., Kumaki, K, and Poland, A. P., J. Cell. Physiol. 85:393, 1975.
24. Baldwin, R. W., and Embleton, M. J., Int. J. Cancer 4:430, 1969.
25. Hewitt, H. B., Blake, E. R., and Walder, A. S., Brit. J. Cancer 33:241, 1976.
26. Embleton, M. J., and Heidelberger, C. L., Int. J. Cancer 9:8, 1972.
27. Prehn, R. T., in "Immune Surveillance" (R. T. Smith and M. Landy, eds.), pp. 451-462, Academic Press, New York, 1970.
28. Parmiani, G., Carbone, G., and Prehn, R. T., J. Nat. Cancer Inst. 46:261, 1971.
29. Prehn, R. T., Int. J. Cancer 24:789, 1979.
30. Prehn, R. T., Ann. N. Y. Acad. Sci. 164:449, 1969.
31. Prehn, R. T., J. Nat. Cancer Inst. 31:791, 1963.
32. Evans, R., and Lawler, E. M., Int. J. Cancer 26:831, 1980.
33. Evans, R., Brit. J. Cancer 35:557, 1977.
34. Mantovani, A., Int. J. Cancer 22:741, 1978.
35. Prehn, R. T., J. Nat. Cancer Inst. 59:1043, 1977.
36. Carrel, A., J. Exp. Med. 36:385, 1922.
37. Burwell, R. G., Lancet 2:69, 1963.
38. Burch, P. R. J., Nature 225:512, 1970.

SECTION 2

CHEMOTAXIS AND PHAGOCYTOSIS

PHYSICAL AND CHEMICAL DETER-

MINANTS OF LEUCOCYTE LOCOMOTION

P. C. Wilkinson

Bacteriology and Immunology Department
University of Glasgow (Western Infirmary)
Glasgow, G11 6NT, Scotland

INTRODUCTION

It is now generally accepted that chemotaxis is an important determinant of leucocyte accumulation in sites of inflammation. A wide range of chemotactic factors has been identified and, more recently, progress has been made in understanding how they bind to the surface of the leucocyte and how they evoke a directional locomotor response. Nearly all of this information has been obtained using neutrophils, since these are the easiest of the leucocytes to work with. Cells of the mononuclear phagocyte system certainly also show chemotaxis, but we are much further from understanding their response in detail. Mononuclear phagocytes, whether as blood monocytes or as exudate macrophages, are more difficult to purify than neutrophils and, although methods have improved of late, even the most purified populations contain lymphocytes and cells of uncertain identity. Probably more of a problem is the fact that mononuclear phagocytes are capable of considerable differentiation and, as yet, we know very little of the effect of differentiation on their locomotor capacity, or about the effect of macrophage heterogeneity on chemotaxis. This subject is discussed by Dr. Leonard in this volume.

The assay systems we use _in vitro_ to study chemotaxis are deliberately simplified so that we can study leucocyte locomotion towards an attractant in isolation from other influences on locomotion which must be present _in vivo_. However, there are a number of important factors other than chemotaxis that may lead to a non-random distribution of cells and that may have a part to play in inflammation. Some of these factors have been studied using cells which are rather distant from the interests of leucocyte workers. In this paper, I should like to draw to the attention of people working with

macrophages and other leucocytes some of these locomotor responses,
since they may contribute to, or hinder, leucocyte accumulation. I
shall do this in the first part of the paper. In the second part I
shall present some of our own findings on chemotaxis and other locomotor responses in mononuclear phagocytes using visual assays such
as time-lapse cinematography.

CHEMICAL DETERMINANTS OF LOCOMOTORY BEHAVIOUR

     For many years a distinction between chemotaxis (Richtungs-
bewegung) (1) and chemokinesis (2) has been accepted. This distinction came rather late to the leucocyte field, chiefly because the
assay methods in use in that field were non-visual and did not permit
the distinction to be made operationally. The proposals for nomen-
clature put forward by Keller et al. (3) drew attention to this distinction, and the definitions attempted in that paper have met with
rapid acceptance. One can summarize them by stating that chemotaxis
is a reaction by which a chemical substance determines the direction
of cell locomotion. Chemokinesis may take two forms (a) orthokinesis
in which a chemical substance determines the speed of locomotion;
and (b) klinokinesis in which the chemical substance determines the
frequency of turning of the cell. We need not consider klinokinesis
further at present. It is a very important locomotor determinant
in, for instance, bacteria, but is not known to play an important
part in the locomotion of leucocytes. Chemical substances can certainly modify the speed of leucocytes, and it is probable that many
leucocyte chemotactic factors have a dual effect. In a gradient of
such factors, leucocytes move up the gradient by chemotaxis. However, even in the absence of a gradient, leucocytes may vary their
speed of movement depending on the concentration of the attractant,
which, in these conditions, is acting as a chemokinetic factor. The
result will be a concentration-dependent acceleration or deceleration
of movement, the direction of which is random. Can chemokinesis
(orthokinesis) alone cause a non-random distribution of cells? The
quick answer to this question is yes, if the chemokinetic factor is
present at different concentrations in different parts of the system.
For example, cells in areas of high concentration will move fast,
those in areas of low concentration will move slowly. The net result
will be that cells will spend more time in areas of low concentration
than of high. However, this is unlikely to be a good mechanism for
cell accumulation on its own unless some other factor such as an
'adhesive trap' is present. Of course any factor that simultaneously
increased cell speed and induced directional locomotion up a gradient
would produce marked cell accumulation.

     The evidence is now good enough to say that many, if not most,
chemotactic responses are generated by binding of chemotactic molecules to receptors on the leucocyte surface, using the term 'receptor'
in an open sense; and that the ligand-receptor interaction results

in signal transduction and a directional motor response. The mechanisms for generating directionality are considered by Zigmond (4),
Zigmond and Sullivan (5) and Dunn (6). It is still totally unknown
whether purely chemokinetic molecules exist that generate a locomotor
response by interactions with leucocyte surface receptors.

PHYSICAL DETERMINANTS OF LOCOMOTORY BEHAVIOUR

     Adhesion. Leucocytes move by contact with surfaces so that the
strength of contact may determine the speed of locomotion. If adhesion is too strong, cells may become tethered. If it is too weak,
they may not grip sufficiently well to move. Neutrophil adhesion
to surfaces is modified in a dose- and time-dependent way by chemotactic factors (7,8); however, it is not certain that these adhesive
changes have much effect on neutrophil locomotion. It is possible
that neutrophil locomotor capacity is not highly sensitive to small
changes in cell adhesion. In the case of the mononuclear phagocytes,
we know nothing at present about locomotion and adhesion. Lymphocytes
are well known to be poorly adhesive, and this raises the question
of how they generate sufficient traction to move in contact with
surfaces. In our studies (9), we found that lymphocytes failed to
grip and moved very inefficiently or not at all on plane protein-
coated surfaces (glass or plastic), but that they moved very well
within the matrix of three-dimensional networks such as collagen
gels. In such gels, they may be able to gain purchase without making
strong adhesions, for example, by inserting a pseudopod through a
gap in the gel lattice, expanding the pseudopod on the opposite side
and using it as a purchase point for further movement in any direction. In a different type of experiment, when lymphocytes are added
to cultures of tissue cells (especially lymph node reticular cells)
they penetrate the tissue cell layer and move vigorously while under-
lapping the cells (moving between the tissue cells and the substratum). This again is movement in a non-planar environment (10).
There is therefore not necessarily a discrepancy between the active
locomotory capacity of lymphocytes and their failure to adhere, since
the lymphocyte may be using a rather unique mechanism for locomotion
that involves the ability to make non-adhesive attachments to non-
planar surfaces.

     Contact Guidance. The shape of the substratum may determine the
direction of cell locomotion. For example, cells may move in grooves,
or, when on a curved surface such as a cylindrical glass fiber, they
may align in the axis of least curvature. Thus fibroblasts on such
fibers align along the long axis of the fiber rather than aligning
with their long axes around its circumference (11). Similar experiments can be done using more physiological surfaces than glass fibers.
For example, collagen gels can be aligned easily by gentle tension,
and fibroblasts show orientation in the axis of alignment of the
fibers in such 3-D collagen gels (12). These responses to the shape

of the environment have not until lately been considered in relation
to leucocytes. However, we have recently shown guidance responses
in human neutrophil leucocytes (13). Neutrophils on the grid of a
serum-coated glass hemacytometer show orientation in the grooves
of the grid and move up and down the grooves. As time proceeds, they
tend to accumulate on the grooves (show a preference for the grooves
rather than for plane glass). Likewise in aligned 3-D gels of collagen or fibrin, the direction of neutrophil locomotion shows a bias
towards the axis of alignment. Cells prefer to move up and down the
collagen fibers rather than across them, though this is a bidirectional locomotor response (the cells moving in one axis, but in both
directions in that axis) unlike chemotaxis which is a unidirectional
response. Many connective tissues show alignment in vivo, and it is
clear that the interaction between contact guidance and chemotaxis
needs to be studied in relation to leucocyte emigration into sites
of inflammation.

Certain other preferences for surfaces are shown by cells.
Macrophages accumulate on roughened surfaces rather than on smooth
surfaces in vitro, a phenomenon named 'rugophilia' by Rich and Harris
(14). The same authors also showed that macrophages accumulate on
hydrophobic rather than hydrophilic surfaces, a preference that is
opposite to what fibroblasts show. Another possible response to a
surface is haptotaxis or migration up an adhesion gradient (15,16),
but this has not been studied in leucocytes or macrophages.

VISUAL STUDIES OF LOCOMOTION OF MONONUCLEAR PHAGOCYTES

Time-lapse cinematography has been used very successfully to
measure the locomotor behaviour and chemotactic responses of neutrophil leucocytes, but, surprisingly, there have been very few similar
studies of mononuclear phagocytes. The latter cells have been studied
chiefly using filter assays which give much useful information on
many aspects of chemotaxis, but which are uninformative about the
details of cell behaviour. In fact, the evidence that monocytes and
macrophages really do show chemotaxis sensu stricto is still quite
insubstantial. We have studied various populations of mononuclear
phagocytes including human blood monocytes, and mouse peritoneal
macrophages, both resident and thioglycollate-elicited, as well as
a mouse macrophage cell line, J774B10, for chemotactic activity in
visual assays. The assay used was described by Allan and Wilkinson
(17) and employs spores of Candida albicans in fresh normal serum as
point sources of chemotactic gradients.

A summary of the results is shown in Table 1. Human blood monocytes from the upper layer of a density-sedimentation tube (Ficoll-
Hypaque, Pharmacia) which were washed, but not further separated from
non-adherent lymphocytes, migrated as vigorously as human neutrophils
and responded accurately to gradient sources (18). However, if the

Table 1. Chemotaxis of Mononuclear Phagocytes to Candida albicans in Visual Assays

Determination	Mouse Macrophages			Human Blood Leucocytes	
	Resident peritoneal	Thioglycollate elicited peritoneal	J774B10 cell line	Monocytes	Neutrophils
Number analysed	107	60	63	40	28
Locomotor forms	18%	72%	62%	70%	71%
Chemotactic ratio	0.9 ± 0.03	0.88 ± 0.02	0.9 ± 0.02	0.89 ± 0.02	0.96 ± 0.01
Speed (μm/min)	2.0 ± 0.17	3.5 ± 0.15	3.9 ± 0.15	14.2 ± 0.9	14.4 ± 1.2

In all cases, 25% fresh homologous serum was used to generate chemotactic activity at the surface of the Candida spores. Chemotactic ratio and speed are reported as mean $\pm$ standard error.

Chemotactic ratio = $\dfrac{\text{Straight-line distance to gradient source}}{\text{Total distance travelled by cell to gradient source}}$

monocytes were further purified by an adhesion step (not shown in
table), the speed of locomotion of the adherent monocytes was considerably decreased. In the latter experiments, the monocytes were
separated from lymphocytes by the procedure of Ackerman and Douglas
(19) which involves an adhesion of the monocytes to substrata which
were formerly coated with BHK fibroblasts, the fibroblasts having
been removed to leave a conditioned surface. The mean migration speed
of these cells was 3.5µm per minute, as opposed to 14.2µm per minute
for unpurified monocytes. The adhesion step may therefore slow
locomotion quite considerably, though the adherent monocytes still
showed accurate chemotactic responses to Candida spores.

When mouse macrophages were examined, their speed of locomotion
was found to be slow compared to neutrophils or monocytes from human
blood (20). Also there was a pronounced difference between resident
and thioglycollate-elicited cells. Few of the resident macrophages
showed a chemotactic response, even to nearby spores (Table 1) and
few of them were motile. Thioglycollate-elicited macrophages responded better, with a higher proportion of motile cells in the
population and a high proportion responding to Candida (Fig. 1).
Many of these thioglycollate-elicited macrophages were very large,
spread cells. Some responded to Candida by typical chemotactic locomotion. Others responded by formation of an extensive anterior hyaline
veil which could stretch forward 20-30 µm to ingest the nearest Candida without displacement of the cell body. This directional response
without translocation might be regarded as chemotropism rather than
chemotaxis, since the latter response involves translocation. The
word chemotropism has been used by botanists to describe turning of
an organism (without translocation) towards a source of a stimulus.
Cells of the J774B10 line showed chemotactic responses to Candida
without much spreading, and moved towards Candida spores at about
the same speed as the thioglycollate-elicited cells (3-4 µm per
minute). This is consistent with findings (21) using filter assays
in which chemotactic responsiveness of macrophages of a J774 line
was also demonstrated.

These results thus demonstrate an unequivocal chemotactic response in several types of mononuclear phagocyte. The speed of locomotion of human blood monocytes, unseparated by an adhesion step,
was much higher than that of the other cells studied. This is not
surprising since the monocyte is the cell that must leave the blood
in response to stimuli at sites of inflammation. Macrophages in tissues may be more slow-moving, though one should be cautious about
extrapolating from movement on a protein-coated surface in vitro to
movement through tissues in vivo.

One should also ask the question whether macrophages respond
to the shape or curvature of the surfaces along which they move i.e.
if they show contact guidance. If mouse thioglycollate-elicited
macrophages are placed on Neubauer counting chambers coated with mouse

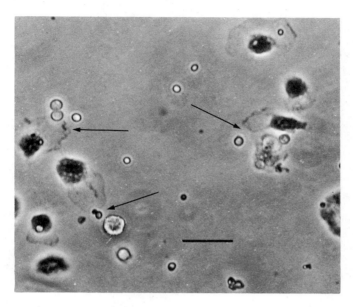

Fig. 1.   Thioglycollate-elicited mouse peritoneal macrophages in homologous serum respond chemotactically to spores of Candida
albicans by protrusion of an extensive anterior hyaline veil.
Three cells, each about to ingest Candida spores, are indicated by an arrow.   Bar = 20 µm.

serum, they can be shown within 1 to 2 hours, to align along the
grooves in the glass (Figure 2), and, if these cells are studied over
a period of 72 hours, they remain aligned in the grooves.   During the
first few hours, they move along the grooves but eventually, they
become immotile.   This may be a form of contact guidance.   On the
other hand, it is possible that grooves engraved with a diamond, as
these were, are rougher than the polished glass surface of the slide,
and that the macrophages are showing 'rugophilia' as in the experiments of Rich and Harris (14).   A more discriminating test of contact
guidance might be to study macrophage locomotion in aligned collagen
gels, as we have done using human neutrophils (13).   However, up
to now, we have not succeeded in getting these rather large cells
to migrate through the matrix of a collagen gel.

     In visual studies then, at least two forms of directional loco-
motion can be distinguished in mononuclear phagocytes and neutrophils.
One is chemotaxis, a well-studied response to chemical signals from
the environment.   The other is contact guidance (or a very similar

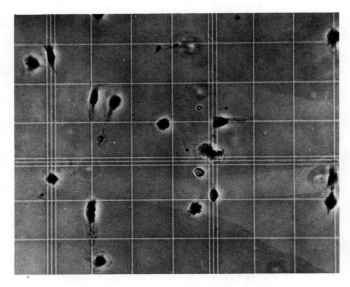

Fig. 2.   Response of thioglycollate-elicited mouse peritoneal macro-
          phages to grooved glass surfaces. The cells were in 25%
          homologous serum. Many cells have elongated with their long
          axes in the line of the grooves and their anterior lamellae
          are formed preferentially along that line. The distance
          between parallel grooves on the counting chamber is 50 µm.

response) directed by physical features of the environment. Chemotaxis is a unidirectional response; guidance is bidirectional. Obviously the influence of guidance responses on chemotaxis needs to
be considered, especially since connective tissue is often aligned
in vivo, and it will be necessary to devise experiments to study these
interactions. The complex architecture of tissues and organs in the
body may be an important determinant of the effectiveness of cellular
mobilization and thus of the ability to mount an efficient inflammatory response. The study of leucocyte locomotion in relation to
tissue structure has hardly begun.

Acknowledgement

        The work of the author is supported by the Medical Research
Council.

REFERENCES

1. Pfeffer, W., Untersuchungen aus dem Botanischen Institut zu
      Tübingen 1:363, 1884.
2. Rothert, W., Flora 88:371, 1901.
3. Keller, H. U., Wilkinson, P. C., Abercrombie, M., Becker, E. L.,
      Hirsch, J. G., Miller, M. E., Ramsey, W. S., and Zigmond, S.
      H., Clin. Exp. Immunol. 27:377, 1977.
4. Zigmond, S. H., Nature 249:450, 1974.
5. Zigmond, S. H., and Sullivan, S. J., J. Cell Biol.82:517, 1979.
6. Dunn, G. A., in "Biology of the Chemotactic Response" (J. M.
      Lackie and P. C. Wilkinson, eds.), pp. 1-26, Cambridge Univer-
      sity Press, Cambridge, 1982.
7. Smith, R. P. C., Lackie, J. M., and Wilkinson, P. C., Exp. Cell
      Res. 122:169, 1979.
8. Smith, C. W., Hollers, J. C., Patrick, R. A., and Hassett, C.,
      J. Clin. Invest. 63:221, 1979.
9. Haston, W. S., Shields, J. M., and Wilkinson, P. C., J. Cell
      Biol. 92, 1982, in press.
10. Haston, W. S., Cell. Immunol. 45:74, 1979.
11. Dunn, G. A., and Heath, J. P., Exp. Cell Res. 101:1, 1976.
12. Dunn, G. A., and Ebendal, T., Exp. Cell Res. 111:475, 1978.
13. Wilkinson, P. C., Shields, J. M., and Haston, W. S., Exp. Cell
      Res., 1982, in press.
14. Rich, A., and Harris, A. K., J. Cell Sci. 50:1, 1981.
15. Carter, S. B., Nature 213:256, 1967.
16. Harris, A. K., Exp. Cell Res. 77:285, 1973.
17. Allan, R. B., and Wilkinson, P. C., Exp. Cell Res. 111:191,
      1978.
18, Wilkinson, P. C., and Allan, R. B., in "Mononuclear Phagocytes,
      Functional Aspects" (R. van Furth, ed.), pp. 475-500, Mar-
      tinus Nijhoff, The Hague, 1980.
19. Ackerman, S. K., and Douglas, S. D., J. Immunol. 120:1372, 1978.
20. Wilkinson, P. C., Immunobiology, 1982, in press.
21. Snyderman, R., Pike, M. C., Fischer, D. G., and Koren, H. S.,
      J. Immunol. 119:2060, 1977.

DISCUSSION

SILVERSTEIN: What distinguishes the movement of lymphocytes from
that of monocytes or PMNs in the same collagen gel?

WILKINSON: But I'm looking at two different sorts of collagen gels.
With the lymphocytes, I am looking at a non-aligned gel, whereas in
the case of the PMNs, I am looking at an aligned gel.

SILVERSTEIN: But in a non-aligned gel, what distinguishes the movement
of the lymphocytes from that of the PMNs?

WILKINSON: When PMNs move through a 3 dimensional gel, the changes
in their shape are very different from those observed during movement
on a 2 dimensional surface. With the latter, it's far more difficult
to predict the front from the back. One of the problems is that the
PMNs may put out two pseudods and the PMN has to choose between them.
Obviously in an aligned gel, there is more freedom for the cell to
move into the pseudopod which is in the direction of alignment. Thus,
the shapes of these cells moving through 3 dimensional gels show more
variation than when the cell moves on the classical 2 dimensional
surface.

SILVERSTEIN: Couldn't it be that the two dimensional surface that
we all work with and that you described for the lymphocyte is the
way the cells move through the connective tissue in real life?

WILKINSON: In the body, the cells usually move in a three dimensional
matrix. I suppose there are times when they may be moving in two
dimensions as well, for instance when they're moving across a serous
surface, or across the capsule of an organ.

UNKNOWN: When the lymphocyte comes to the end of the groove, does
it step out or would it rather reverse its direction?

WILKINSON: We've never really observed them move to the end of the
line. However, what we have observed is that they tend to persist
in the direction in which they start; they tend not to reverse directions easily. If they do reverse directions, they round up, stop
for a period of time, and then move in the opposite direction. I
would like to emphasize that movement on a glass groove may be a
very different phenomenon than movement in an aligned tissue. These
may be two quite distinct mechanisms.

LEONARD: We have seen this striking alignment on glass tissue culture
plates. Occasionally there is a micro-defect in the manufacture of
the plate and the monocytes and macrophages line up along that line.
So, it need not relate to the material of the glass.

WILKINSON: Well it may. It depends on how the scratch was made.
The microstructure of the scratch may be different from the microstructure of the glass. Albert Harris has recently described a
phenomenon called rugophilia, by which he means a preference by macrophages for rough surfaces. He's roughened plastic and showed that
the macrophages accumulate on the rough surface. You could explain
that phenomenon by preference for movement on the smooth surface.
If the cells were moving on the smooth surface and got stuck on the
rough surface, they'd all accumulate on the rough surface. It could
be a difference in rugosity of the surface or it could be a response
to the 3-dimensional shape of the groove.

LEONARD: In the old days, one could make beautiful actin-myosin threads by taking solutions of actin and myosin in appropriate ionic strength and squirting them through a narrow tube into a medium of different composition. The results were lots of threads which might have a lot of alignment.

G. KAPLAN: Do the PMNs move into the gel similar to lymphocytes?

WILKINSON: Yes. They'll invade the gel. They actually move better in aligned collagen than in non-aligned collagen.

G. KAPLAN: They don't stay on the surface of the collagen?

WILKINSON: Some do, particularly on aligned collagen gels. Here they tend to move up and down the top surface. However, if you flatten the gel out so that it's two dimensional, they don't move on it anymore.

UNKNOWN: Do lymphocytes not need adhesion in order to be able to push out pseudopods?

WILKINSON: A cell can form pseudopods quite readily in suspension. Pseudopod formation is not dependent on adhesion. Cells can change shape then they're in suspension.

FUNCTIONALLY DISTINCT SUBPOPULATIONS OF HUMAN
MONOCYTES: RECEPTORS FOR F-MET-LEU-PHE ARE EX-
PRESSED ONLY ON THE CHEMOTACTICALLY RESPONSIVE CELLS

Werner Falk, Liana Harvath and Edward J. Leonard

National Cancer Institute
Bethesda, Maryland   20205 USA

In 1971 Horwitz and Garrett introduced a modification of the
Boyden chamber method that made it possible to quantify chemotaxis
of mononuclear phagocytes (1). The innovation was replacement of
the 150 um thick cellulose membrane separating cell and attractant
chambers by a 10 um thick polycarbonate membrane in which 5 um diameter holes were randomly distributed. Macrophages migrated through
these holes toward attractant and remained adherent to the attractant
side of the membrane. One of the advantages of this method was that
since all the migrated cells were in one optical plane, the number
migrated could be readily counted. The authors showed that macrophages migrated toward E. coli filtrate, and that this was presumably
not a chemokinetic effect, since migration was greatly reduced when
equal concentrations of attractant were placed in cell and attractant
wells. The method was also applied to blood monocytes (2) and many
studies followed that showed decreases in the number of chemotaxin-
responsive monocytes from patients with cancer and other diseases
(3,4,5).

One fact not addressed in any of the above studies was the low
percentage of monocytes from normal subjects that migrated to chemotaxins. In our experience, the percentage of input monocytes that
migrated rarely exceeded 40% and could be as low as 10-20%. We therefore began studies to determine why a large fraction of normal human
monocytes did not migrate to attractants. The results of our previous work can be summarized under 3 headings.

CONDITIONS OF THE ASSAY

We tested many variables that might account for suboptimal migration, including chemoattractant concentration, duration of the

101

gradient, time for migration, diameter of pores in the membrane, pore number/mm^2 of membrane, duration of cell capacity to respond, cell-cell interaction, and dropping of migrated cells from filter surface into the attractant well. Since none of these factors accounted for the limited number of migrating monocytes, we concluded that monocytes capable of migrating to chemotaxins represent a distinct subpopulation of the total blood monocytes (6).

DESENSITIZATION STUDIES

Human monocytes are excellent for study of specific desensitization, since in contrast to neutrophils they survive various manipulations without aggregation or loss of viability. Using 3 different attractants, we showed that incubation of monocytes with desensitizing concentrations of one attractant abolished or markedly diminished the subsequent chemotactic response to the desensitizing attractant, whereas responses to other attractants remained intact. Therefore, monocytes have functionally distinct receptors for the 3 different attractants. We could then ask whether the low percentage of total monocytes migrating to a single attractant was due to the existence of subpopulations of monocytes with different attractant specificities.

FUNCTIONAL RECEPTOR ANALYSIS

When various chemotaxins and chemotaxin combinations were tested, we found that a single chemotaxin attracted almost as many cells as combinations of 2 or 3 chemotaxins. Analysis showed that at least 75% of chemotaxin-responsive monocytes had receptors to all 3 of the attractants tested. This was confirmed qualitatively by showing that populations selected by their capacity to migrate to one attractant could migrate to a different attractant. If at least 75% of the migrating cells were a uniform population with receptors for all 3 of the chemotaxins tested, what were the characteristics of the non-migrators, which comprised 60-80% of the total monocyte population? Did they lack chemotaxin receptors, or did they have receptors but lack the capacity to migrate because of a block subsequent to chemotaxin-receptor interaction?

Our experimental approach was to allow human blood monocytes to migrate toward optimal concentrations of C5a, then separate migrated from non-migrating cells and determine binding of f-Met-Leu-(^{3}H)Phe to each cell population (7). We designed a separation chamber by modifying our 48-well chamber (8) so that there were 4 rectangular upper and lower compartments separated by a 25x80 mm polycarbonate membrane. After the bottom compartments were filled with an optimal concentration of C5a, the chamber was assembled and to each of the 4 top compartments we added 2 ml of a suspension containing 2x10^6 monocytes/ml. The chamber was incubated at 37°C for 70 min. By the

end of this time, attractant-responsive monocytes had migrated to
the bottom surface of the membrane where they remained attached without falling into the bottom compartment. The non-migrated cells on
the top of the membrane were collected by resuspending with a pipet.
The membrane was then removed and the attached migrated cells were
washed into a polypropylene tube. Virtually all the migrated cells
were monocytes. Total yield of migrating and non-migrating monocytes
was 75-90% of the input number in a series of experiments. Only
2-8% of the non-migrating monocytes responded to attractant when
tested a second time.

     The migrating and non-migrating cell pools were suspended in
Gey's medium containing 2% BSA and also 1 mM azide to inhibit possible
ligand internalization. Monocyte number was adjusted to 5×10^5/ml,
and 1 ml aliquots were incubated for 30 min at 24°C with 10^{-9}M to
10^{-7}M radioactive peptide, with or without a 100-fold excess of non-
radioactive peptide. After incubation, the cells were washed free
of fluid phase radioactivity, mixed with Ultrafluor scintillation
fluid (National Diagnostics, Sommerville, NJ), equilibrated for at
least 3 hrs and counted. Since non-specific binding in the presence
of a 100-fold excess of non-radioactive peptide was less than 5%
of the total, data are presented without correction.

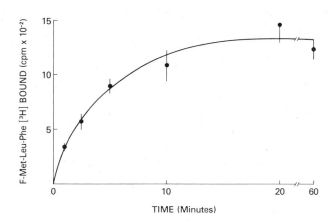

Fig. 1.   Time course of binding of f-Met-Leu-(^{3}H)Phe to monocytes
          separated by chemotaxis to C5a.

     Figure 1 shows that binding of optimal concentrations of f-Met-
Leu-(^{3}H)Phe to migrated monocytes occurred rapidly. A plateau was
reached after about 20 min at room temperature. The amount bound
at 60 min was always slightly less.

Figure 2 shows a typical dose-response curve for binding of f-Met-Leu-(^{3}H)Phe to the monocytes that had migrated to C5a. Binding was saturable, reaching a plateau at a chemotaxin concentration of 5×10^{-8}M. Fifty percent of maximal binding occurred at 10^{-8}M, the concentration that induces the maximal chemotactic response. In 8 different experiments, the saturation level for 5×10^5 monocytes was 1300-2000 cpm, which represents 28-43 fmol peptide per 10^5 monocytes. In contrast, the non-migrating pool of monocytes from the top of the collection chamber bound only 100-150 cpm per 5×10^5 monocytes.

Since binding to non-migrating monocytes was negligible and these cells represent more than half of the total blood monocytes, we expected that f-Met-Leu-(^{3}H)Phe bound to unseparated monocytes would be considerably lower than for the migrating population. Figure 3 shows that the cpm bound at saturation for 5×10^5 unseparated monocytes is about half that for 5×10^5 migrated monocytes.

In this experiment, approximately 40% of the input monocytes migrated toward chemoattractant.

These studies show that the non-responding monocytes fail to bind f-Met-Leu-Phe and that functionally these cells lack receptors for this chemotaxin. Additional deficiencies in subsequent stages of the stimulus-response pathway have not been ruled out. We plan to determine whether the functional receptor deficiency is peculiar to the peptide or whether similar results will be found with C5a. It will also be of great interest to determine whether the lack of receptor represents a maturational stage or cyclic variation in blood monocytes of a single lineage, or whether the migrators and non-migrators represent different lineages. We have found that the number of chemotaxin-responsive monocytes that repopulate the circulation

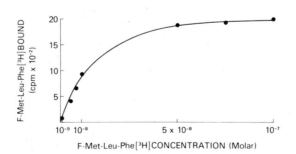

Fig. 2.   Dose-response curve for binding of f-Met-Leu-(^{3}H)Phe to migrated monocytes.

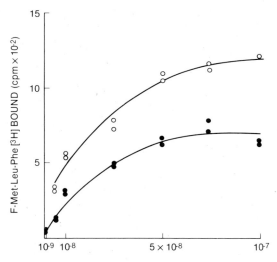

Fig. 3. Dose-response curves for binding of f-Met-Leu-(^{3}H)Phe to migrated (open circles) and non-separated (closed circles) monocytes.

after a 2 hour leukopheresis in which 10^9 monocytes were removed is about 60% of the control (9). This suggests that the capacity to respond to chemoattractants may depend on state of maturation. Receptor studies on these monocytes are in progress.

REFERENCES

1. Horwitz, D. A., and Garrett, M. A., J. Immunol. 106:649, 1971.
2. Snyderman, R., Altman, L. C., Hausman, M. S., and Mergenhagen, S. E., J. Immunol. 108:857, 1972.
3. Boetcher, D. A., and Leonard, E. J., J. Natl. Cancer Inst. 52: 1091, 1974.
4. Hausman, M. S., Brosman, S., Snyderman, R., Mickey, M. R., and Fahey, J., J. Natl. Cancer Inst. 55:1047, 1975.
5. Gallin, J. I., Annals Int. Med. 92:520, 1980.
6. Falk, W., and Leonard, E. J., Infect. Immun. 29:953, 1980.
7. Falk, W., Harvath, L., and Leonard, E. J., Infect. Immun., in press.
8. Falk, W., Goodwin, R. H. Jr., and Leonard, E. J., J. Immunol. Methods, 33:239, 1980.
9. Alteri, E., and Leonard, E. J., Clin. Research, in press.

DISCUSSION

SILVERSTEIN: If you let the cells mature under non-adherent conditions, will they develop receptors for f-Met-Leu-Phe?

LEONARD: That's a hard question for us to answer. We're not entirely
satisfied with our culture conditions but at 24 hours we still have
non-migrating cells. Our other approach was to look at monocyte
donors before and after a 2 hour leukophoresis. At the end of 2
hours, about 2×10^9 monocytes will have been collected and the donor
will have lost about one circulating blood volume worth of monocytes.
We've looked at chemotaxis before, at the end of the procedure, and
then 2 or 3 hours later. Chemotactic responses at the end of leukophoresis averaged 60% of control. We're doing peptide binding studies
on those cells now. If you look 3 hours after leukophoresis, they're
well on their way back to normal.

SILVERSTEIN: Those monocytes that just came out of the bone marrow,
or appeared after leukophoresis, which receptors do they have?

LEONARD: We only have one actual binding study. But we have 8
experiments in which we've measured the chemotactic responses of newly
released cells and it is only 60% of the control. For example, if
30% of the control monocytes migrated to f-Met-Leu-Phe, then one
would have 15% at the end of the procedure and it would be back up to
30% 3 hours later.

OLIVER: So you would predict that there are fewer receptors on immature monocytes?

LEONARD: Our experiments on immature monocyte receptor number are in
progress.

WILKINSON: Synderman's work on the monocyte line suggests that there
may be fewer. His monocyte line had less receptors before he stimulated with lymphokine than after.

LEONARD: Yes, but that may be another question. It may not be
maturation so much as a response to whatever goes on with lymphokine.

GINSBERG: What is the evidence that f-Met-Leu-Phe acts as a chemoattractant in vivo?

LEONARD: With E. coli filtrates, there is cross desensitization
between products of the E. coli and the peptide.

MODULATION OF GRANULOCYTE RESPONSE

TO THE CHEMOATTRACTANT F-MET-LEU-PHE

Claes Dahlgren and Olle Stendahl

Department of Medical Microbiology
Linköping University Medical School
S-581 85 Linköping, Sweden

Cells involved in host defense, including macrophages and granulocytes, change their surface and functional properties during maturation (1,2,3). In addition, mature granulocytes are able to change their surface structure and functional characteristics in response to mediators of the inflammatory process (4,5,6,7).

In the present investigation we studied alterations in the physiochemical surface properties of mature granulocytes in vitro, in relation to the chemiluminescent response induced by f-Met-Leu-Phe, and as a result of f-Met-Leu-Phe binding.

MATERIALS AND METHODS

Polymorphonuclear leukocytes (PMNL) were isolated at $4^{\circ}C$ from peripheral human EDTA blood according to Bøyum (8).

The chemotactic peptide formyl-methionyl-leucyl-phenylalanine (f-Met-Leu-Phe), phorbol myristate acetate and luminol were obtained from Sigma (Sigma Chemical Co., St. Louis, Mo.). Heat-killed Saccharomyces cerevisiae opsonized with rabbit anti-yeast IgG as described by Hed (9), was used as particulate stimulus.

Physicochemical surface changes were assayed in aqueous two-polymer phase systems of dextran and polyethylene glycol (PEG) with addition of charged bis-tri-methylamino-PEG (TMA^+-PEG) or hydrophobic palmitoyl-PEG (P-PEG) (10).

Luminol enhanced chemiluminescence was assayed in a liquid scintillation counter kept at $22^{\circ}C$ (11).

RESULTS AND DISCUSSION

    Human PMNL were prepared and kept at 4°C, then transferred to
22°C and immediately assayed for chemiluminescent response using
different stimuli. Addition of phorbol ester or the particulate
stimulus IgG-yeast gave rise to a pronounced response. By contrast,
the peptide f-Met-Leu-Phe did not induce any chemiluminescent response. Storage of the PMNL at 22°C for as long as 240 min before
adding the stimulus, resulted in a gradual increase in the chemiluminescent response to f-Met-Leu-Phe, whereas no such increase occurred
when phorbol myristate acetate or IgG-yeast were used (Figure 1).
This increased response to f-Met-Leu-Phe developed more rapidly with
conditioning at 37°C than at 22°C.

    Mature peripheral blood PMNL show initial metabolic unrespon-
siveness to the chemotactic peptide f-Met-Leu-Phe. Since the response
to IgG-yeast particles and to the potent soluble stimulus phrobol
ester were maximal in similarly treated cells, the data suggest that
the lack of response is due to an insufficient number of receptors

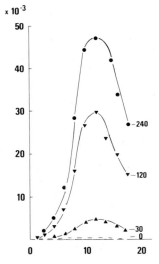

Fig. 1.   Time course of chemiluminescent response of human polymor-
          phonuclear leukocytes (PMNL) (10^5 cells) induced by 10^{-7}M
          f-Met-Leu-Phe. The PMNL samples were obtained at various
          time intervals (0-240 min after preparation) from a cell-
          suspension (10^7 PMNL/ml) kept at 22°C. Abscissa: Time of
          study. Ordinate: Light emission in CPM. (From Dahlgren
          et al. Ref. 11, with permission).

for f-Met-Leu-Phe. Since up to 65% of the receptors to chemotactic peptides have been shown to be hidden (12), exposure of these receptors as a result of degranulation (4) or receptor recycling (13) could be one mechanism by which PMNL regulate their activity.

Using the chemiluminescent response as a measure of f-Met-Leu-Phe binding to its receptors, we applied Scatchard plots for the different storage times at $22^{\circ}C$ (Table 1). A comparison of the values derived from the intercept on the abscissa of the best-fit lines obtained by linear regression analysis (r >0.98), indicates that cell conditioning is accompanied by an increase in the number of functional receptors for f-Met-Leu-Phe. This treatment also changes the binding constant K_a, as calculated from the slope of the lines in the Scatchard plot analysis.

Cell surface properties of PMNL are changed concomitantly with the development of f-Met-Leu-Phe receptors. Changes in charge and hydrophobicity were measured by cell partitioning in aqueous two-phase systems of dextran and PEG with part of the PEG exchanged for either positively charged TMA^+-PEG or hydrophobic P-PEG. By adding TMA^+-PEG to the system, the charged PEG will interact with negatively charged surface components of the cells, and thereby change their

Table 1. Effect of Storage of Human Peripheral Blood Leukocytes on the f-Met-Leu-Phe Binding Constant and the Number of Binding Sites

Duration of Storage $22^{\circ}C$	Binding Constant K_a μM^{-1}	Number of Binding Sites
5	2.45	2,300
30	7.56	10,000
120	18.7	41,000
240	14.2	78,000

Cells were stored at $22^{\circ}C$ for variable periods before addition of f-Met-Leu-Phe. The binding constant and the number of binding sites (reported in counts per minute) were calculated from Scatchard plot analysis of chemiluminescence data.

partition. Accordingly, addition of hydrophobic P-PEG forms a hydrophobic top phase that pulls cells interacting with the ligand into
the PEG-rich top phase. In phase systems containing TMA$^+$-PEG or
P-PEG, both PMNL with few receptors and cells with many receptors
were collected in the PEG rich top phase. However it was shown that
the magnitude of the redistribution of the PMNL conditioned at 37°C
(many receptors) was greater than that of the PMNL used directly after
preparation (few receptors) (Figure 2).

Binding of f-Met-Leu-Phe to the receptors also resulted in PMNL
surface property changes. The net negative surface charge decreased,
and the effect was most pronounced in the PMNL population conditioned
at 37°C prior to peptide binding. These cells also showed a decrease
in surface hydrophobicity as a result of peptide binding. The surface
property changes induced by f-Met-Leu-Phe cannot be explained by a
PMNL response to the chemotactic factor, which has been shown to
induce cell surface property changes (14,15) since in our experiments
the temperature for the binding was kept at 4°C.

Attempts have been made to characterize chemotactic factor receptors on PMNL (16,17,18), but due to the structural complexity of
cell surfaces, a precise description - in molecular terms - of relevant attributes in recognition is still remote. Since the development
of functional receptors to f-Met-Leu-Phe is associated with an in-

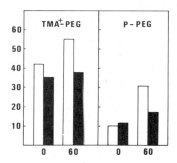

Fig. 2.   Partition of human polymorphonuclear leukocytes (PMNL) in
          aqueous biphasic systems containing covalently bound ligands.
          Two PMNL populations were studied, one used directly after
          separation and one used after incubating at 37°C for 60 min.
          The cells were cooled and incubated for 15 min at 4°C without
          (open bars) or with (filled bars) 5×10^{-8}M f-Met-Leu-Phe.
          Figures represent % increase in the top phase in the pres-
          ence of charged bis-tri-methlamine-polyethylene glycol
          TMA$^+$-PEG) or hydrophobic palmitoyl-polyethylene glycol
          (P-PEG).

creased negative surface charge and an increased tendency to hydrophobic interaction, and since binding of f-Met-Leu-Phe to PMNL is
associated with a masking of charged and hydrophobic structures on
the PMNL surface, we suggest that these properties are important in
the interaction between PMNL and the chemotactic peptide f-Met-Leu-
Phe.

REFERENCES

1. Altman, A. J., and Stossel, T. P., Br. J. Hematol. 24:241, 1974.
2. Stendahl, O., Dahlgren, C., and Hed, J., Second European Confer-
    ence on Phagocytic Leukocytes. Trieste, Sept. 1980, Plenum
    Press, in press.
3. Lichtman, M. A., and Weed, R. I., Blood 39:301, 1972.
4. Fletcher, M. P., and Gallin, J. I., J. Immunol. 124:1585, 1980.
5. Kay, A. B., Glass, E. J., and Salted, D., Clin. Exp. Immunol. 38:
    294, 1979.
6. Sullivan, S. J., and Zigmond, S. H., J. Cell Biol. 85:703, 1980.
7. Zigmond, S. H., J. Cell Biol. 88:644, 1981.
8. Bøyum, A., Scand. J. Clin. Lab. Invest. Suppl. 97:77, 1968.
9. Hed, J., FEMS Lett. 1:357, 1977.
10. Dahlgren, C., Kihlström, E., Magnusson, K.-E., Stendahl, O., and
    Tagesson, C., Exp. Cell Res. 108:175, 1977.
11. Dahlgren, C., Magnusson, K.-E., Stendahl, O., and Sundqvist, T.,
    Int. Arch. Allergy Appl. Immunol. 67 No 6 (Karger, Basel
    1982).
12. Liao, C. S., and Freer, R. J., Biochem. Biophys. Res. Commun.
    93:566, 1980.
13. Weitzman, S. A., Desmond, M. C., and Stossel, T. P., J. Clin.
    Invest. 64:321, 1979.
14. Gallin, J. I., Durocher, J. R., and Kaplan, A. P., J. Clin.
    Invest. 55:967, 1975.
15. Schaak, T. M., Takeuchi, A., Spilberg, I., and Persellin, R.H.,
    Inflammation 4:37, 1980.
16. Goetzel, E. J., and Hoe, K. Y., Immunology 37:407, 1979.
17. Schiffman, E., Aswanikumar, S., Venkatasubramanian, K., Corcoran,
    B. A., Pert, C. B., Brown, J., Gross, E., Day, A. R., Freer,
    R. J., Showell, A. H., and Becker, E. J., FEBS Lett. 117:1,
    1980.
18. Snyderman, R., and Goetzel, E. J., Science 213:830, 1981.

DISCUSSION

OLIVER: I'm concerned that you'd get similar data if, instead of an
absence of peptide binding, there was absence of oxidase activation.
Do you find the same phenomena if you look at another parameter of
f-Met-Leu-Phe, like perhaps a change in ion transport.

DAHLGREN: I don't know. We haven't done the experiments.

UNKNOWN: You used $22^\circ C$ for measuring chemilluminescence. Why did you use this unphysiological temperature?

DAHLGREN: I don't think it matters if you measure at 22° or $35^\circ C$. You do get a faster response if you measure at $37^\circ C$.

UNKNOWN: How do you account for the change in the affinity constant?

DAHLGREN: If there is expression of new receptors on the surface as a result of storage, these new receptors could have a different binding constant than the receptors that are there from the beginning.

CYTOSKELETON-MEMBRANE INTERACTION AND THE REMODELING OF THE CELL SURFACE DURING PHAGOCYTOSIS AND CHEMOTAXIS

J. M. Oliver and R. D. Berlin

Departments of Physiology and Pathology
University of Connecticut Health Center
Farmington, Connecticut 06032 USA

INTRODUCTION

Chemotaxis and phagocytosis are membrane processes. They begin with the interaction of particles or chemoattractants with cell surface receptors. Their earliest manifestations include changes in ion fluxes (1), the activation of membrane oxidases (2) and the physical reorganization of the membrane and underlying cytoskeleton to form new structures: pseudopods during phagocytosis, uropods and lamellipodia during chemotaxis (3). In this paper we review evidence for a remarkable topographical reorganization of membrane structural determinants and of surface functions during chemotaxis and phagocytosis. It will be shown that membrane receptors may segregate to specific regions of the surface. After ligand binding to these receptors, the resulting complexes may be translocated to other specific regions. Functions such as endocytosis and membrane transport are also restricted to specified areas of the surface. The mechanisms that initiate and sustain these topographical asymmetries will be explored. Recognition of these membrane events may clarify several characteristic properties of polymorphonuclear leukocytes (PMN) and macrophages, for example their ability to maintain unidirectional movements in very shallow chemotactic gradients or to preserve the integrity of membrane transport systems during the removal of cell surface by phagocytosis.

CELL SHAPE AND CYTOSKELETAL ORGANIZATION DURING PHAGOCYTOSIS AND CHEMOTAXIS

Unstimulated PMN and macrophages maintain a more or less rounded shape with an irregularly ruffled surface and a rather uniform sub-

membranous meshwork of microfilaments. Phagocytosis begins with the
recruitment of microfilaments to sites of particle binding, forming
pseudopodia that enclose and eventually engulf suitable particles
(Figure 1). Once internalization has occurred, phagocytic vacuoles
rapidly lose their coat of microfilaments, enabling the approach and
fusion of lysosomes. The typical shapes of cells exposed to chemoattractants are illustrated by scanning electron microscopy (SEM)
in Figure 2 and by transmission electron microscopy (TEM) in Figure
3. The broad, ruffled anterior lamellipodium contains a dense meshwork of microfilaments. The posterior uropod is also supported by
microfilaments, often surrounding a central core of intermediate
(100 Å) filaments. An indented region of the surface that separates
highly ruffled from smoother membrane is characteristically seen by
SEM (Figure 2, arrow). By TEM this indentation is less readily identified (an arrow in Figure 3 indicates its probable location).

Microtubules are rarely observed near the membrane of phagocytizing or oriented cells. In fact microtubules are specifically
absent from the microfilament-rich pseudopods, lamellipodia and uropods. Nevertheless, microtubule assembly from centrioles is increased
following the binding of phagocytic particles or chemotactic agents
to membrane receptors (reviewed in Ref. 4). In oriented PMN, these
microtubules typically extend in parallel bundles from centrioles
located behind the lamellipodia and within the nuclear lobes, towards
the uropod (Figure 3).

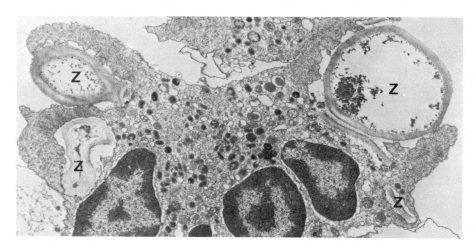

Fig. 1. A human peripheral blood PMN fixed after 30 seconds exposure
         to complement-opsonized zymosan (Z), demonstrating the forma-
         tion of microfilament-rich pseudopodia at points of particle
         contact.

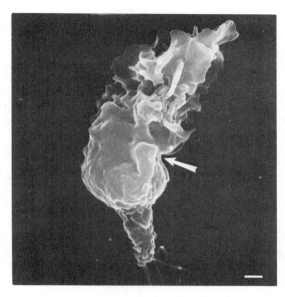

Fig. 2.   Scanning electron microscopic view of a human peripheral
          blood PMN polarized by 20 minutes incubation with 5 x 10^{-7}
          M f-Met-Leu-Phe. The cell shows a distinctive bipolar mor-
          phology with a veil-like anterior lamellipodium and posterior
          uropod. The arrow indicates a characteristic surface inden-
          tation that separates highly ruffled from smooth membrane.
          Bar = 1 μm.

SURFACE TOPOGRAPHY DURING PHAGOCYTOSIS AND CHEMOTAXIS

     The formation of pseudopodia and removal of surface membrane
during phagocytosis is associated with the development of remarkable
topographical asymmetries of both structural and functional determinants of the cell surface. Some of these are summarized in Table 1.

     The first clear evidence that membrane determinants may assume
non random distributions on cell surfaces was obtained during studies
of phagocytosis. Tsan and Berlin (5) reported in 1971 that the
activities of the membrane transport systems for adenine and lysine
are completely unaltered after removal of up to 40% of the rabbit
PMN or alveolar macrophage surface by phagocytosis of latex. The
most likely explanation was the segregation of transport carriers
out of membrane that engulfs phagocytic particles and is internalized.
Subsequently, we (6,7) reported the segregation of some Con A recep-

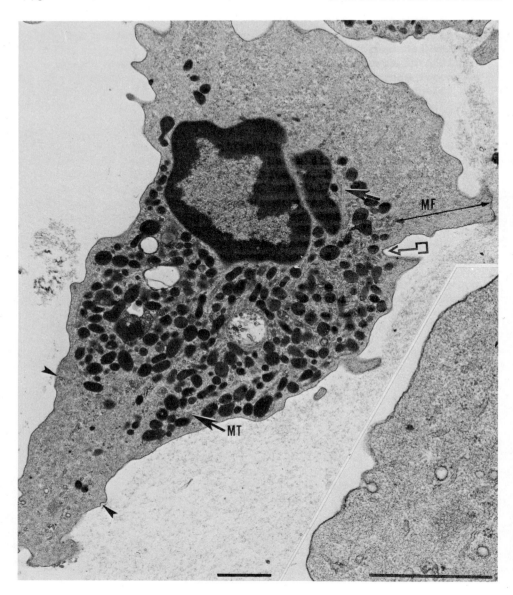

Fig. 3.   Human peripheral blood PMN polarized by 20 min. incubation
          with 10^{-7}M f-Met-Leu-Phe. Broad lamellipodium (top right)
          is packed with microfilaments (MF) and excludes granules.
          Uropod (lower left) is similarly occupied by a meshwork of
          MF. Microtubules (MT) extend towards, but not into, both
          cell poles. The arrow heads and inset show the localizations
          of coated pits and vesicles. Large arrow, upper right, indi-
          cates probable location of surface indentation. Bar = 1 µm.

Table 1.   Patterns of Membrane Asymmetry During Phagocytosis
           and Chemotaxis

Components and Functions that Segregate away from Uropod
and/or Pseudopod Membrane:

       Transport carriers
       Fc receptors
       C3b receptors

Components and Functions that Distribute into Uropod and/
or Pseudopod Membrane:

       Con A-receptor complexes
       Fc-receptor complexes
       C3b-receptor complexes
       Fluid pinocytosis
       Adsorptive pinocytosis
       Phagocytosis
       Coated pits

Components and Functions that Remain Partially or Fully
Dispersed

       Con A receptors

---

tors and the bulk of Con A-receptor complexes into pseudopod membrane
(Figure 4c); and Silverstein (8) established that C_3b receptors are
preserved on the surface of mouse macrophages during the uptake of
IgG-opsonized particles or spreading on IgG-coated surfaces.   The
recent demonstration of Aggeler and colleagues (9) that clathrin
baskets are specifically associated with pseudopod membrane raises
the possibility that macromolecular complexes may also show segrega-
tive movement during phagocytosis.

   Cells oriented in a chemotactic gradient or simply incubated in
suspension with a chemoattractant show an analogous segregation of
membrane components.   The topographical expression of Fc receptors
on PMN has been inferred from the distribution of antibody coated
erythrocytes (EA) during brief (30 sec.) labeling at 37° C.   These
probes for inherent Fc receptor topography bind to the lamellipodium
and not the uropod (10,11).   Ig aggregates, which also bind Fc recep-

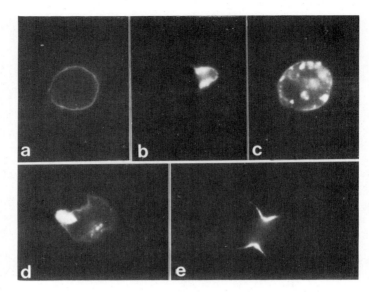

Fig. 4.   The distribution of fluorescein-Con A (10 μg/ml) on variously
          treated cells incubated for 5 minutes at 37° C with lectin.
          The cells are respectively (a), untreated; (b), microtubule-
          depleted (protuberant), (c), phagocytizing (carboxylated
          polystyrene particles that do not bind Con A); (d), oriented
          (by incubation for 20 minutes in F-Met-Leu-Phe) and (e),
          dividing. All cells are human PMN except for the mitotic
          cell, which is a Chinese Hamster Ovary cell.

tors, show the same anterior distribution (12; Figure 5). Wilkinson
and colleagues (10) have observed further that Fc receptor- EA complexes migrate from the lamellipodium to the uropod if incubation
at 37° C is continued for 5-10 min. This presumably reflects a ligand-
induced migration of Fc receptors. Similarly Con A receptors may
show an essentially uniform distribution on oriented cells (11,13:
see Figure 7, first frame), but Con A-receptor complexes move rapidly
to the uropod during incubation of labeled cells at 37°C (Figure 4D).

    Perhaps most importantly, we have recently demonstrated that
topographical asymmetries of both membrane macromolecular complexes
and membrane functions occur on locomoting cells (12). In particular,
oriented cells accumulate coated pits exclusively in the uropod
region. This result, obtained by morphometric analysis of pits and
vesicle distribution in PMN oriented by f-Met-Leu-Phe, is illustrated
in Figure 6. Accompanying this strict segregation of coated pits,

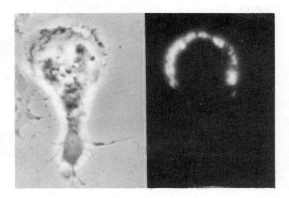

Fig. 5. The distribution of Fc receptors on oriented PMN. Cells
         were oriented in a gradient of $10^{-7}M$ to $10^{-6}M$ f–Met–Leu–
         Phe then exposed to Ig aggregates (30 seconds at $37^{\circ}C$).
         Binding is restricted to the lamellipodium.

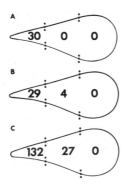

Fig. 6. The topography of coated pits and coated vesicles during
         chemotaxis in human PMN. The numbers of coated pits (A,B)
         or pits and coated vesicles (C) occupying the anterior, cen-
         tral or posterior thirds of the total cell perimeter were
         determined in human PMN polarized by 20 min. incubation in
         a gradient (A,C) or in suspension (B) with F–Met–Leu–Phe
         $(10^{-7}M)$. The results in A were obtained by analysis of 10
         cells from 2 separate experiments; B is from 16 cells from
         2 experiments; and C analyzes 17 cells from 3 experiments.

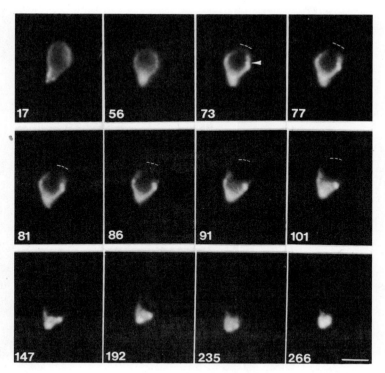

Fig. 7.   The redistribution of Ccn A on locomoting cells. Monolayers
           of human PMN on glass coverslips were oriented by incubation
           at 37°C for 15 min with 5x10⁻⁷M f-Met-Leu-Phe, then labeled
           at 10°C for 30 sec. with 100 µg/ml fluorescein-Con A and
           observed by epiillumination through the 63X neofluor objec-
           tive of a Zeiss Universal Fluorescence microscope. A polar-
           ized cell was selected for observation at a low light level
           using a Venus DV3 video intensification camera in series with
           a time generator (TEL) and an IVC 1010 tape recorder. The
           stage was warmed to 25°C at 45 sec. after labeling. The re-
           distribution of fluorescence with time was photographed on
           Polaroid 611 film with a Tektronix C-28 camera during play-
           back from the recorder to a Quantex digital image processor
           (set in the 16 frame averaging mode) and Tektronix 634 moni-
           tor. The dashed line indicates the position of the lamel-
           lipodial tips. The arrowhead shows the initial position of
           the surface indentation. The microscope stage was moved
           between 101 and 147 seconds and 192 - 235 seconds to keep
           the locomoting cell within the same high power field. Time
           is given in seconds. Bar = 10 µm.

the uptake of fluorescein-dextran and of fluorescein-Con A is confined
to the uropod. This indicates the topographical restriction of both
fluid and adsorptive pinocytosis on oriented cells.

SURFACE TOPOGRAPHY DURING OTHER MEMBRANE EVENTS

The membrane asymmetries that occur on phagocytizing or oriented
cells are not unique. Precisely analogous asymmetries occur on macrophages and other leukocytes under conditions that permit membrane
capping of surface-bound ligands (reviewed in 4 and 21). In general,
a small proportion of leukocytes exhibits an asymmetric shape with
a microfilament-rich bulge or protuberance at one pole. The proportion of such cells is increased from fewer than 10% to over 70% by
incubation with drugs that disassemble microtubules. The protuberance
of colchicine-treated leukocytes characteristically excludes unoccupied Fc and C_3b receptors but shows a time and temperature dependent
accumulation of complexes of Fc and C_3b receptors with appropriate
particles. Con A-receptor complexes also accumulate in the protuberance, forming familiar cap structures (Figure 4B). Finally, coated
pits are concentrated and phagocytosis, fluid pinocytosis and adsorptive pinocytosis are all restricted to the protuberance. The recruitment of microfilaments to form the contractile ring of dividing cells
provides an analogous focus for membrane segregation: cleavage furrow
membrane accumulates ligand-receptor complexes (Figure 4E) and coated
pits (reviewed in 14,15). Endocytosis, which ceases during mitosis,
is re-initiated at the region of cleavage as cells emerge into the
G1 phase of the cell cycle.

THE PROCESS OF LIGAND-RECEPTOR REDISTRIBUTION DURING CHEMOTAXIS: DIFFUSION VS. FLOW

How do ligand-receptor complexes redistribute into pseudopods,
uropods and analogous structures? The simplest hypothesis is the
free diffusion of ligand-receptor complexes and their accumulation
by crosslinking or by immobilization in "trap" regions, eg. the pseudopods of phagocytizing cells; the uropod of locomoting cells. In
this case rather slow rates of redistribution would be predicted:
measured diffusion constants, D, in animal cells usually range from
$5 \times 10^{-9} cm^2/sec$ to $10^{-11} cm^2/sec$ (or slower) for proteins or ligand-
protein receptor complexes (16). Alternatively ligand-receptor complexes may redistribute by a process of membrane flow, where the force
for complex migration is provided by cytoskeletal-membrane interaction
or other active mechanisms. Finally, some investigators have suggested
that ligand-receptor complexes may accumulate passively at the uropod
of motile cells as a consequence of cell movement in which unlabeled
membrane migrates away from cross-linked receptors. These various
possibilities are discussed in (14). We have approached the mechanism
of ligand-receptor redistribution during chemotaxis, using fluorescent

ligands whose movement is monitored by video intensification microscopy and quantified after digitization by computer analysis. Analogous studies during cytokinesis and capping are described elsewhere
(15,17).

In the video sequence shown in Figure 7, Con A was bound more
or less uniformly to the surface of an f-Met-Leu-Phe-treated human
PMN during 30 sec. labeling at 10°C. The temperature of the microscope stage was increased to 25°C at 45 sec. on this recording. Warming was followed by an immediate contraction of the uropod (between
50-60 sec.) and subsequent extension of the lamellipodium (beginning
at around 70 sec.). Con A was displaced first from the lamellipodium
and subsequently from the central section of the cell. Essentially
all the Con A was in the uropod region by 100 sec. A coordinate
change in cell shape accompanied the redistribution of Con A. In
particular, a surface indentation and a posterior bulge formed at
the base of the lamellipodium and were displaced progressively towards
the uropod. The Con A was further concentrated in the uropod during
the subsequent cycles of tail retraction and lamellipodial extension
that occurred between 100 and 300 sec.

To determine the velocity of this Con A movement (backward) and
of cell movement (forward), selected time frames between 50 and 100
sec. were transferred from tape to the memory of a Quantex digital
image processor interfaced to an LM^2 computer. A grid of 15 parallel
lines, 1.5 μm apart, was drawn over these frames (Figure 8). Since
the grid is fixed in space, it nay be calculated that the cell is
displaced forward by approximately 6 μm in 45 sec., or around 8 μm/
min. It is also clear from Figure 8 that the indented region and its
associated bulge remain relatively fixed in space as the cell moves
forward. The fluorescence intensities across each line (approximately
80 locations on the cell per line) corrected for background, were
next transferred to computer memory where they were summed. Figure
9A shows the summed fluorescence intensities with time across equivalent lines running through the lamellipodial tip (0 μm) and through
the midpoint of the uropod (15 μm). Con A is almost completely displaced from the lamellipodium between 70-80 sec. It accumulates
continuously at the uropod between 70-100 sec. Figure 9B illustrates
the summed fluorescence intensities with time across lines drawn 1.5
and 4.5 μm from the lamellipodial tip. The loss of Con A-receptor
complexes from the 1.5 μm line occurs at a maximum rate between 70-
80 sec. The loss of complexes from the 4.5 μm line occurs in parallel
but is displaced in time by around 7 sec. A similar displacement
(not shown) was measured for the loss of Con A from the 7.5 μm line
as compared to the 4.5 μm line. Based on this displacement of approximately 7 sec. over a 3 μm distance, we calculate that Con A moved
off the lamellipodium at about 26 μm/min.

To account for this rate by a mechanism of diffusion with trapping would require an effective diffusion coefficient approaching

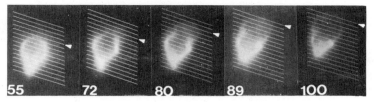

Fig. 8.   The velocity of cell movement and Con A-receptor redistribution. Five time frames of the same locomoting cell illustrated in Figure 7 were selected for analysis. A grid composed of 15 lines at 1.5 μm intervals was superimposed cn
each frame. Fluorescence intensity across each line was
transferred to the memory of an LM^2 computer for further
analysis. Since the grid is fixed in space, this series
enables calculation of the rate of cell migration. The arrow
heads indicate the line across the lamellipodial tip (0 μm)
for each frame. Note that the apparent differences in Con
A fluorescence between Figure 8 and 9 reflect use of the
image processor's nonlinear transform mode in Figure 9 in
order to photograph both cells and grid. The total fluorescence intensity summed between lines 1 and 15 was essentially
unchanged from frames 55 to 100, indicating relatively little
loss of fluorescence due to bleaching, self adsorption, saturation or other sources within this experiment (data not
shown).

10^{-8} cm^2/sec. This value approaches the measured diffusion coefficient
of lipids and exceeds by 1 or 2 orders of magnitude the usual diffusion coefficients of proteins in biological membranes. Thus a mechanism of ligand-receptor redistribution based on diffusion seems unlikely. Similarly the hypothesis that ligand or particles may be
simply left behind at the uropod as the cell moves forward (10,18)
is not well supported: the rate of Con-A receptor complexes movement
off the lamellipodium is at least 3 times faster than the rate of
net forward cell locomotion. These data are thus most consistent with
ligand-receptor redistribution by a process of membrane flow.

MECHANISMS OF MEMBRANE FLOW

     Various investigators have suggested that direct or indirect
receptor-microfilament interactions cause the long range surface
movements (flow) of ligand-receptor complexes into caps and equivalent
structures (19,20: reviewed in 14). The increased density of microfilaments under membrane that accumulates ligand-receptor complexes

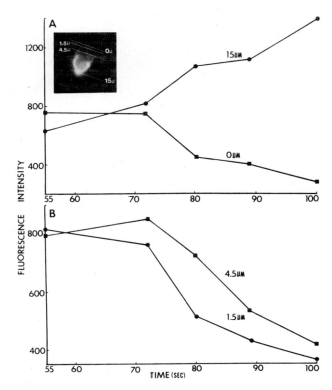

Fig. 9.   Summed fluorescence intensities with time of selected lines
          across a Con A-labelled, locomoting PMN. In A, fluorescence
          intensity was summed across lines at 0 μm (lamellipodial
          tip) and 15 μm (uropod).   In B, fluorescence intensity was
          summed across lines at 1.5 μm and 4.5 μm behind the lamel-
          lipodium.   The inset shows the position of these 4 lines
          at one time point (89 seconds).   For the other time frames,
          the position of the lines may be calculated from Figure 8
          as: 0 μm, arrowed line: 15 μm, arrow plus 9 lines; 1.5 μm,
          arrow plus 1 line; 4.5 μm, arrow plus 3 lines.

and the inhibition of ligand-receptor redistribution processes by
cytochalasins encourage this view. In Figure 7, the changes in cell
shape accompanying the migration of Con A to the uropod are also suggestive of dynamic interactions of microfilaments with the membrane
that may cause lectin-receptor redistribution. However, the varying
patterns in which membrane constituents are segregated during processes like chemotaxis and phagocytosis (Table 1) would lead to an

extraordinary criss-cross traffic of microfilaments if these were
attached to the different membrane components. In addition, not all
regions of microfilament concentration are regions of ligand-receptor
accumulation. In particular the anterior lamellipodia of moving
neutrophils are supported by a dense meshwork of microfilaments
(Figure 3) but ligand-receptor complexes redistribute to the posterior
or uropod region.

Radically different flow models, also based on capping, have been
proposed by Bretscher (21) and Harris (22). In these models membrane
lipid or whole membrane is postulated to flow continuously from the
front of the cell to the rear where it is removed and recycled via
endocytosis to the front. This flow causes the movement of ligand-
receptor complexes whereas unoccupied receptors escape flow by back-
diffusion. A molecular filter excludes ligand-receptor complexes
from internalization and recycling. The net result is cap formation
at the site of filtration. This model relieves cytoplasmic microfilaments of an obligatory direct role in topographical control.
However, the filter is not defined and the exclusion of ligand-receptor complexes from the cytoplasmic milieu seems inconsistent with the
rapid internalization of complexes that usually follows capping of
multivalent ligands.

We have thus proposed a new model for membrane flow in which wave
motion provides the force for the selective movement of surface constituents by a process resembling surfboarding (14). The model is
based in part on the theoretical work of Hewitt (23).

In our view, the initial event which makes possible ligand movement is the formation of a uropod, pseudopod or analogous structure
by local recruitment of microfilaments. The filaments in turn orga-
nize into bundles or meshworks that interact with the membrane, generating tension directed inward and along the membrane. During phagocytosis this tension is generated by the extension of microfilament-
rich pseudopodia. During chemotaxis, tension may be initiated by the
contraction of the uropod. The membrane indentation that subsequently
appears at the base of the lamellipodium and is displaced to the
uropod (24-28; Figures 2,7) may add to and propagate this tension.
Such a role is implied by previous investigators who described this
propagated indentation as a 'constricting ring'. The asymmetric
application of tension to the membrane initiates movements perpendicular to the surface with the peak and crest displaced laterally with
time, i.e., mechanical waves. The precise point of origin and the
orientation of these waves is determined by the relationship of the
regions of tension to cell geometry.

We postulate that Con A-receptor complexes are among a class
of membrane determinants that respond essentially as surfboards to
the passage of these membrane waves. These components associate with
waves such that they attain a maximally energetically favorable inter-

action at some point on the surface of the wave (illustrated in Figure
10). Then they move along with the wave in order to preserve this
maximal interaction, i.e., any displacement from this point requires
energy. This hypothesis gains experimental support from the sequential rise and fall in fluorescence intensity observed on individual
lines across the lamellipodial region of the locomoting cell analyzed
in Figures 7-9 (note especially Figure 8B).

We emphasize that not all membrane determinants will show the
same response to the passage of waves. Some components (lectin
receptors for example) may behave as corks on water: as the wave
passes they may undergo changes in orientation without significant
net movement (Figure 10). In addition, some components (Fc receptors
and transport carriers for example) are excluded from uropod and
pseudopod membranes. To explain this, we suggest that entrainment
is influenced not only by the relation of membrane determinants to
wave geometry but also by 'drag' due to the viscosity of the lipid
bilayer, to interactions with other membrane components and to interaction of cytoskeletal elements with the membrane bilayer structure
(discussed in 14). Factors contributing to 'drag' may antagonize
entrainment. We also note that entrainment in waves may not be

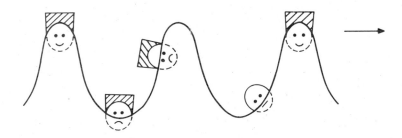

 = Ligand-Receptor Complex

Fig. 10:   Postulated responses of membrane determinants to the pas-
           sage of waves. Certain receptors sense a change in orien-
           tation but no large change in environment as the waves
           progress. These occupy random positions on the cell sur-
           face. Other receptors and ligand-receptor complexes are
           preferentially associated with particular regions of the
           waves (the crests in this diagram) and retain this distri-
           bution. These move with the waves to the region of force
           generation.

restricted to ligand-receptor complexes but may extend to a macromolecular complex, the coated pit.

The immediate challenge is to identify the structural nature of
the proposed waves involved in ligand-receptor redistribution during
chemotaxis, phagocytosis and analogous processes. Since waves are
postulated to flow from the lamellipodial tip towards the uropod on
locomoting cells, closer examination of membrane events at the lamellipodium may clarify this issue. In addition, the relationship of
the propagated membrane indentation ('constricting ring') and its
lateral bulge to ligand-receptor redistribution on locomoting cells
requires closer analysis. The lateral bulge is probably not itself
the wave that engages membrane determinants: it develops before
extensive displacement of bound Con A occurs (see Figure 7, 73 sec.);
and the bulk of the Con A moves past this region onto the uropod
(Figure 3,91 and 101 sec.). Our postulate above that the indentation
and bulge are involved in the generation of tension implies a region
of local microfilament-membrane interaction. Evidence for changes
in the density or orientation of microfilaments underlying the indented region is required to support this hypothesis.

THE SIGNIFICANCE OF MEMBRANE ASYMMETRY DURING PHAGOCYTOSIS AND
CHEMOTAXIS

The exclusion of unoccupied receptors and membrane transport
carriers from pseudopod membrane may preserve independent membrane
activities during phagocytosis. Conversely, the inclusion of other
membrane proteins and lipids may enable lysosomes to recognize and
fuse with incoming phagocytic vesicles but not with the bulk plasma
membrane. Consistent with this latter hypothesis, the segregation
of membrane component is altered during phagocytosis in colchicine
treated cells (4): lysosomal degranulation and PMN bactericidal
activity are also impaired in these cells.

During chemotaxis, the segregation of membrane receptors may
facilitate the recognition and binding of phagocytic particles and
other ligands. In addition, it may play a significant role in the
maintenance of unidirectional locomotion. PMN can sense extremely
shallow chemotactic gradients and can migrate in a particular direction for extended periods of time even in the absence of a gradient.
To explain these properties, we propose simply that the asymmetric
membrane topography of PMN includes either chemotactic receptors or
membrane components that determine the direction of movement following
peptide-receptor interaction. Once established, an anterior localization of such determinants would effectively project a shallow gradient
in the medium to a steep gradient across a cell. This amplification
of the transcellular gradient would tend to sustain migration in one
direction.

Acknowledgement

        Supported in part by NIH grant CA-15544 and grant BC-179 from
the American Cancer Society. J. M. Oliver holds an ACS Faculty
Research Award. We thank Ms. C. B. Pearson and Ms. J. O'Brien for
their invaluable technical assistance.

REFERENCES

1.   Naccache, P. H., Showell, H. J., Becker, E. L., and Sha'afi,
        R. I., J. Cell Biol. 73:428, 1972.
2.   Becker, E. L., Sigman, M., and Oliver, J. M., Am. J. Pathol. 95:
        81, 1979.
3.   Bessis, M., and DeBoisfleury, A., Blood Cells 2:365, 1976.
4.   Oliver, J. M., and Berlin, R. D., Symp. Soc. Exp. Biol. 33:227,
        1979.
5.   Tsan, M. F., and Berlin, R. D., J. Exp. Med. 134:1016, 1971.
6.   Oliver, J. M., Ukena, T. E., and Berlin, R. D., Proc. Natl. Acad.
        Sci. USA 71:394, 1974.
7.   Berlin, R. D., and Oliver, J. M., J. Cell Biol. 77:789, 1978.
8.   Michl, J., Pieczonka, M. M., Unkeless, J. C., and Silverstein,
        S. C., J. Exp. Med. 150:607, 1979.
9.   Aggeler, J., Heuser, J. R., and Werb, Z., J. Cell Biol, 91:2649,
        1981.
10. Wilkinson, P. C., Michl, J., and Silverstein, S. C., Cell Biol.
        Int. Rep. 4:736, 1980.
11. Walter, R. J., Berlin, R. D., and Oliver, J. M., Nature 286:
        724, 1980.
12. Davis, B. H., Walter, R. J., Pearson, C. B., Becker, E. L., and
        Oliver, J., M., submitted for publication.
13. Oliver, J. M., Krawiec, J. A., and Becker, E. L., J. Reticulo-
        endothelial Soc. 24:697, 1978.
14. Oliver, J. M., and Berlin, R. D., Int. Rev. Cytol. 74:55, 1982.
15. Koppel, D. E., Oliver, J. M., and Berlin, R. D., J. Cell Biol.,
        in press, 1982.
16. Cherry, R. J., Biochim. Biophy. Acta 559:289, 1979.
17. Berlin, R. D., and Oliver, J. M., J. Theor. Biol. in press.
18. Schreiner, G. F., and Unanue, E. R., Adv. Immunol. 24:38, 1976.
19. Edelman, G. M., Science 192:218, 1976.
20. DePetris, S., in "Dynamic Aspects of Cell Surface Organization"
        (G. Poste and G. L. Nicolson, eds.), p. 644, Elsevier, Amster-
        dam, 1977.
21. Bretscher, M. S., Nature 260:21, 1976.
22. Harris, A. K., Nature 263:781, 1976.
23. Hewitt, J. A., J. Theor. Biol. 80:115, 1979.
24. Ramsey, W. S., Exp. Cell Res. 72:489, 1972.
25. Senda, N., Tamura, H., Shibata, N., Yoshitake, J., Kendo, K.,
        and Tanaka, K., Exp. Cell Res. 91:393, 1975.

26. Senda, N., Shibata, N., Tamora, H., and Yoshitake, J., _Meth. Achiev. Exp. Pathol._ 9:169, 1979.
27. Englander, L. L., _J. Cell Biol._ 87:89a, 1980.
28. Keller, H. U., and Cottier, H., _Cell Biol. Int. Rep._ 5:3, 1981.

DISCUSSION

WILKINSON: Your fluorescence studies throw light on the movement of particles or receptors to the back of the cell but they don't shed much light on what I consider to be the most interesting redistribution: namely, the redistribution of f-Met-Leu-Phe or Fc receptors to the front of the cell. Could you discuss the mechanism of that type of redistribution?

OLIVER: We realize that phenomena other than geometric interactions must play a role in stabilizing this form of receptor asymmetry. Regional changes in membrane viscosity, membrane-membrane inter-actions, and membrane-cytoskeletal interactions could be involved. It has also been suggested that counter-current movements of membrane determinants occur. But in the sense of having any real information, we don't.

LEONARD: Redistribution of the f-Met-Leu-Phe receptor is a very interesting idea. If such redistribution did take place, it might explain desensitization. It would be very interesting to know if the f-Met-Leu-Phe receptors are redistributed in response to f-Met-Leu-Phe binding.

OLIVER: At the present time, the asymmetric localization and ligand-induced redistribution of the f-Met-Leu-Phe receptor has not been demonstrated. A cycle of anterior binding, redistribution and re-cycling is an hypothesis. A similar concept has been advanced by Sally Zigmond to explain why cells don't turn their lamellipodia into uropods.

WILKINSON: Actually, they can turn the uropod into a lamellipod. There's a paper by Gerish and Keller (J. of Cell Science, 52:1, 1982) using micropipettes, which shows this phenomenon. And we've seen it on films too. They can use both mechanisms.

SNYDERMAN: We have used f-Met-Leu-Phe coupled to fluoresceinated latex beads and have seen an asymmetrical distribution of the ligand-receptor complex. However, we haven't seen the beads located in the lamellipod. We've seen them almost exclusively bound to the uropod, which would go along with the concept that there's almost instanta-neous sweeping of receptors to the uropod upon binding.

OLIVER: I think the cell would have to be well fixed because we

were shocked to see that Con A redistribution happens in some ten
seconds.

SNYDERMAN: I think the best we could do is about 30 seconds, and
that's probably not fast enough. It's all in the uropod by that
time. My guess is that it began at the lamellipod but was swept
almost instantaneously upon binding.

OLIVER: Will f-Met-Leu-Phe coupled to fluoresceinated latex beads
bind to fixed cells?

SNYDERMAN: Yes

OLIVER: That should work, because I think even at 4°C the movement
is so fast that the inherent receptor distribution would be hard to
observe.

WILKINSON: The f-Met-Leu-Phe receptor redistribution from the lamellipod is obviously crucial to the chemotactic response.

UNKNOWN: Does Con-A receptor movement correlate with microtubule
assembly or disassembly? And where do you find myosin during this
process.

OLIVER: Con A consistently accumulates at regions of cell shape
asymmetry and microfilament accumulation. One way to create such a
region is by microtubule disassembly: up to 90% of cells will develop
a protuberant shape and will cap Con A to the microfilament-rich
protuberance. However chemotactic peptide induces an analogous cell
shape change, with Con A redistribution to the uropod in the presence
of microtubules. In short, the place where the microfilaments accumulate and generate tension determines where the ligand-receptor
complex will end up. In our view, microtubules control some aspect
of microfilament distribution.

     I suspect that wherever there's endocytic activity, you'll find
myosin. It accumulates at places like caps, pseudopodia and cleavage
furrows that also accumulate receptor complexes. I predict that
the lamellipodium has microfilaments and all sorts of other microfilament associated proteins but probably not myosin.

SILVERSTEIN: If the receptors that occupied the anterior end are
going posteriorly, then there must be additional receptors being
inserted. There is receptor down regulation and there's receptor
recovery. So, there must be a counter-circulation or insertion of
membrane receptors into the cell surface. Is there any evidence
for reinsertion of Con A sites at the front end after they have been
swept to the rear. You describe it as the top half of the cell, but
of course the whole cell membrane is being cleared of Con A both
top and bottom. If the membrane is to be preserved, then other things

must be going anteriorly at the same time or new membrane vesicles
have to be inserted in the front end.

OLIVER: Many people have suggested that that's what happens. However, the lamellipodium has incredibly dense microfilaments and so
it's hard to imagine membrane vesicles getting through. Sylvia
Hoffstein has studied the addition of membrane during chemotaxis and
I think her data showed that there is a large expansion of the membrane during chemotaxis, which was essentially complete in the first
10 seconds of f-Met-Leu-Phe stimulation. Of course, our wave model
allows for the selective redistribution of ligand-receptor complexes
without wholesale membrane recycling.

SILVERSTEIN: How long after the removal of all the Con A ligand
binding sites can you see Con A binding again?

OLIVER: If you add Con A to saturation, these cells round up and
they cease to do anything that looks like chemotaxis.

UNKNOWN: You spoke about rapid displacement of Con A receptors. Is
there any idea of how much these receptors take out of the plasma
membrane in this short period?

OLIVER: We think that the ligand receptor complex is moving through
the membrane, so I don't think the whole membrane is necessarily
consumed. The Con A receptor is a major protein of the membrane but
we're working at subsaturating concentrations. So, probably not a
vast amount of membrane protein is actually being displaced to the
rear of the cell.

ENHANCEMENT OF PHAGOCYTOSIS BY NEUROTENSIN,

A NEWLY FOUND BIOLOGICAL ACTIVITY OF THE NEUROPEPTIDE

R. Goldman,[1] Z. Bar-Shavit,[1] E. Shezen,[1] S. Terry,[2] and S. Blumberg[3]

[1]Department of Membrane Research, The Weizmann
Institute of Science 76100 Rehovot, Israel
[2]The Israel Institute of Biological Research
Ness Ziona, Israel
[3]Department of Biophysics, The Weizmann Institute
of Science 76100 Rehovot, Israel

INTRODUCTION

Neurotensin is a tridecapeptide isolated originally from calf hypothalamus and subsequently from a variety of other tissues and species (1,2). It exhibits a broad spectrum of pharmacological effects both in the central nervous system and in peripheral tissues (reviewed in Ref. 2-4).

Recently, we showed that the undecapeptide substance P stimulates phagocytosis by mouse peritoneal macrophages and human polymorphonuclear leukocytes, and that this activity resides in its N-terminal tetrapeptide sequence (5). The phagocytosis stimulating activity was comparable to that obtained with the tetrapeptide tuftsin (5-8) and both substance P and its N-terminal tetrapeptide were shown to compete with tuftsin for its binding sites (5).

In the present study, we show that neurotensin binds specifically to thioglycollate elicited macrophages and to bone marrow derived mononuclear phagocytes and that the binding has a biphasic effect on the phagocytic capacity of the cells. The characteristics of the binding and the effect on phagocytosis were explored using partial sequences of neurotensin, as well as substance P and tuftsin.

MATERIALS AND METHODS

Neuropeptides and their Partial Sequences: Neurotensin (NT)

133

and (^3H)-neurotensin, $(^3H)NT$ (61 Ci/mmol) were purchased from Vega
Biochemicals (Tucson, Az.) and New England Nuclear (Boston, Mass.),
respectively. The N-terminal decapeptide fragment of NT (NT 1-10)
was a gift from Drs. F. Rioux, S. St. Pierre and R. Quirion (University of Sherbrooke, Sherbrooke, Que., Canada). The C-terminal tri-
peptide (NT 11-13) was prepared by stepwise synthesis and the C-
terminal hexa- and octa-peptide fragments, NT (8-13) and NT 6-13)
were prepared by the solid phase method (9). Substance P and its
partial sequences were synthesized by conventional methods (5).
Tuftsin and (^3H)-tuftsin (24 Ci/mmol) were gifts of P. Gottlieb and
Dr. M. Fridkin.

    Cells: Thioglycollate-elicited macrophages (TG-macrophages)
were collected from BALB/c male mice (6-8 weeks old), 4 days post-
intraperitoneal injection of thioglycollate, and cultivated in 24-
well tissue culture plates (Costar, Cambridge, Mass.) in Dulbecco's
modified Eagle's medium (DMEM) supplemented with antibiotics and
10% heat-inactivated fetal calf serum as described (7). Bone marrow
cells were collected and cultivated for 7 days in DMEM containing
antibiotics, 10% horse serum and 15% L-cell conditioned medium as
described (10). $5x10^4$ and $2.5x10^5$ cells were cultivated in wells
of Costar plates and 35x10 mm Falcon tissue culture plates for phagocytosis assays and binding assays, respectively.

    Phagocytosis Assay: (^3H)-Zymosan particles were prepared by
oxidation of zymosan particles with sodium metaperiodate at pH 4.5
and reduction of the generated aldehyde groups with borotritiated
hydride (11). The specific activity of the particles was $3.2x10^4$
dpm/10^5 particles.

    Phagocyte monolayers were washed twice in phosphate buffered
saline (PBS) and incubated with 0.4 ml of PBS containing the specified
peptide for 5 min at $37^\circ C$. ^3H-Zymosan particles were then added
(100 μl containing $2-3x10^6$ particles) and incubation continued for
30 min at $37^\circ C$. The monolayers were then washed thoroughly in PBS
(5 times) and dissolved in 5% sodium dodecyl sulfate. The radioactivity in dpm was assessed in a Prias liquid scintillation counter.

    Binding Assay: Cell monolayers were washed in PBS and incubated
with $(^3H)NT$ and any of the additional specified peptides at the specified concentration in PBS (0.5 ml/well) for 30 min at $22^\circ C$. The
non-specific component of the binding was assessed by a parallel
incubation of $(^3H)NT$ in the presence of $5x10^{-6}M$ NT. Specific binding
was defined as the total amount of $(^3H)NT$ bound minus the corresponding value of non-specific binding.

RESULTS

    Thioglycollate elicited peritoneal macrophages responded to in-
creasing concentrations of NT in a biphasic manner (Figure 1). At

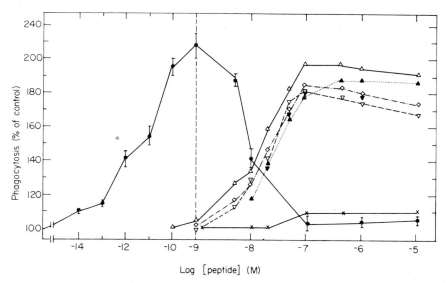

Fig. 1.   Effect of neurotoxin, its partial sequences, substance P,
           and tuftsin on the phagocytic capability of thioglycollate
           elicited macrophages. Symbols used are as follows:

● neurotoxin (NT)   ✕ NT (11-13)

△ NT (1-10)   ▲ substance P

▽ NT (8-13)   ▼ tuftsin

◇ NT (6-13)

The data given for neurotoxin are averages of 5 full experiments ± SEM. Each experiment was carried out in quadrupli-
cates with SEM within 5%. Each of the other peptides was
tested in 2 or 3 experiments. The average of the value for
phagocytosis in 5 control experiments (e.g. 100%) was 41336
± 2237 dpm.

concentrations of 10^{-14}M to 10^{-9}M NT, a dose-dependent augmentation
of phagocytosis occurred. A further increase in NT concentration
in the assay mixture led to a decreased phagocytic response relative
to the maximal augmentation observed at 10^{-9}M, reaching the basal
response (in the absence of NT) at $>10^{-7}$M. Three partial sequences
of NT, NT (1-10), NT (6-13) and NT (8-13), also interacted with the

macrophages to augment their phagocytic capability. The maximal
enhancing effect by these peptides, however, was reached at about
$10^{-7}M$ of the peptides and no inhibitory phase was involved up to a
peptide concentration of $10^{-5}M$ (Figure 1). The dose response curve
of the three reactive peptide fragments was essentially that obtained
for tuftsin and substance P (Figure 1).

     The interaction of NT with resident peritoneal macrophages, as
well as with macrophages differentiated in vitro from bone marrow
cells gave essentially a similar biphasic dose response curve of
phagocytosis.

     (^{3}H)NT bound specifically to both thioglycollate elicited macro-
phages and bone marrow derived macrophages. The binding increased
with time and reached a plateau at about 40-60 min of incubation
at 22° C. Specific binding was found to be concentration dependent and
a Scatchard analysis of both the binding to thioglycollate elicited
macrophages (not shown) and to bone marrow derived macrophages (Figure
2) suggested the existence of two populations of binding sites: a
major population of relatively low affinity binding sites and a small
population of high affinity binding sites (Table 1). The values given
for the high affinity binding sites are probably an approximation of
the actual values because of experimental limitations at these low
concentrations of (^{3}H)NT.

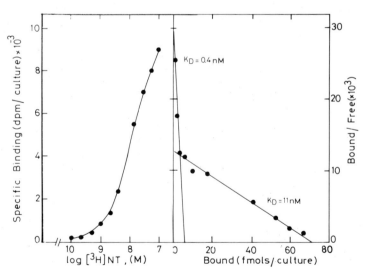

Fig. 2.   Specific binding of tritiated neurotoxin, (^{3}H)NT, to bone
          marrow derived macrophages (left) and its Scatchard analysis
          (right).

The concentration of neurotoxin required to achieve 50% of the maximal enhancement of phagocytosis (Kp) was 10-25 times less than that needed for 50% saturation of the high affinity sites (Table 1). Thus, only a small fraction of those sites had to be occupied in order to trigger the biological response. The concentration at which 50% inhibition of the maximal enhancement was achieved (Kp) was in the range at which a significant proportion of the low affinity sites were occupied (Table 1).

To further explore the characteristics of the binding sites for (^{3}H)NT and their possible interaction with substance P and tuftsin, we carried out binding competition studies which are summarized in Table 2. Of all the peptides tested, only NT was effective in inhibiting 100% of the binding of (^{3}H)NT. The rest of the peptides appeared to be (a) as effective as NT in inhibiting about 85-90% of the binding (which would imply competition for the binding at the low affinity sites) or (b) essentially non-competitive inhibitors of

Table 1. Comparison of the Interaction of Neurotensin with Bone Marrow Derived Macrophages and Thioglycollate Elicited Peritoneal Macrophages

Measurement	Bone Marrow Macrophages		Peritoneal Macrophages	
	High-affinity sites	Low-affinity sites	High-affinity sites	Low-affinity sites
K_D	0.4 nM	11 nM	0.9 nM	28 nM
Binding sites/cell	15,800	178,000	4,800	33,500
K_P	0.016 nM (↑)	3.1 nM (↓)	0.01 nM (↑)	10 nM (↓)

Thioglycollate elicited macrophages were cultured at 7×10^5/well, 71 µg protein/10^6 cells. Bone marrow derived macrophages had 38 µg protein/well on the day of assay, (estimated to be 2.4×10^5 cells/well). An independent estimate of protein content of bone marrow macrophages was carried out for suspended cells yielding 157 µg protein/10^6 cells. K_p equals the concentration of peptide that produces 50% of the maximal enhancement of phagocytosis (↑) or 50% of reduction of the maximal enhancement (↓).

Table 2.   Competition for (^{3}H)neurotensin Binding and Effect
           on Phagocytosis by Thioglycol
           of Neurotensin (NT), Partial Sequences of NT, Sub-
           stance P, and Tuftsin

Compound	Binding Assay[a] K_I (nM)	Phagocytosis[b] K_P (nM)
NT	12	0.01 (↑); 10 (↓)
NT (8-13)	13	18 (↑)
NT (6-13)	13	18 (↑)
NT (1-10)	>>10^5 [c]	16 (↑)
NT (11-13)	>>10^5 [c]	N.S.
Substance P	20	26 (↑); 16 (↑)[d]
Substance P (1-4)	17	14 (↑)[d]
Substance P (5-11)	>>10^5 [c]	N.S.
Tuftsin	16	22 (↑)

[a] $K_- = (I)_{0.5}\left(\dfrac{K_D}{K_D + {^3}HNT}\right)$, where $(I)_{0.5}$ is the concentration
of the unlabeled peptide that produced 50% inhibition of (^{3}H)NT
binding and K_D is the dissociation constant of (^{3}H)NT.   The K_D
derived from Scatchard analysis of binding in the range of
2x10^{-9}M-10^{-6}M was 28 nM.
[b] K_P equals the concentrations of peptide that produced 50% of
the maximal enhancement of phagocytosis (↑) or 50% of reduction
of the maximal enhancement (↓).   N.S. = no significant effect
on phagocytosis.
[c] About 20-30% inhibition of (^{3}H)NT binding was observed at
10^{-4}M of the peptides.
[d] Data taken from phagocytosis of heat-killed yeast cells (5).

(^{3}H)NT binding. Of the peptides that did not compete for the binding, most notable is the N-terminal fragment (NT (1-10)) which showed a biological activity comparable to that of the C-terminal fragments, NT (8-13) and NT (6-13).

A comparison of K_P values derived from Figure 1 with the K_I values of the respective peptides led to the conclusion that progressive saturation of the low affinity sites with either of the effective peptides (except for NT and NT (1-10)) enhances the phagocytic response. NT (1-10) did not bind to the low affinity sites and apparently showed also a much lower affinity than NT to the high affinity sites, but the high affinity sites were supposedly the site of its interaction with macrophages. NT is unique in that it exerted inhibitory effects on the augmentation of phagocytosis when bound to the lower affinity sites. It is worthwhile mentioning at this point that various tuftsin analogs have been shown to have inhibitory activity on the enhancement of phagocytosis by tuftsin (12).

DISCUSSION

Specific NT binding sites have been characterized on rat brain synaptic membranes and on extraneural receptors (a cell line derived from a human colon carcinoma). Values for K_D in the range of 0.9-2 nM were obtained for both the membrane preparation and the cells (13,14) and substance P was not able to compete with NT for its binding sites. These findings are in accord with the present observations regarding the high affinity binding sites.

The biological significance of NT-phagocyte interaction is not clear. NT was found to exhibit chemotactic activity towards human leukocytes, and high levels of NT-related antigens have been found in synovial fluids from rheumatoid arthritics (2). The level of radioimmunoassayable NT in the plasma of several species is about 10-50 f·mols/ml and is further augmented after fat intake (2,4). This level of NT (~10^{-11}M) was shown in our study to be within the range of 50% augmentation of phagocytosis of thioglycollate elicited macrophages and bone marrow derived macrophages. Preliminary results suggest that NT interacts also with human polymorphonuclear leukocytes to augment their phagocytic response.

NT is a potent vasodilator (2) and we suggest that it may also modulate the function of the phagocytes that migrate through the blood vessels into adjacent tissues and possibly also at other sites of peptide-phagocyte interaction.

SUMMARY

Specific binding of neurotensin (NT) to mouse peritoneal thio-

glycollate-elicited macrophages and macrophages differentiated in
vitro from bone marrow cells was demonstrated and characterized.
NT binding to these phagocytes modulated their phagocytic capacity
in a biphasic manner. At concentrations of 10^{-14} to 10^{-9}M NT, a dose-
dependent augmentation of phagocytosis (up to 2-fold) was observed.
Further increases in the concentration of NT resulted in a gradual
decrease of the augmented response until the basal phagocytic activity
(in the absence of NT) was reached. Three partial sequences of NT,
NT (8-13), NT (6-13) and NT (1-10), were also effective in augmenting
the phagocytic response of thioglycollate elicited macrophages, but
the maximal effect was attained at about 10^{-7}M and stayed at that
level up to a concentration of 10^{-5}M. The activity of the three NT
partial sequences was comparable to that of substance P and tuftsin.
Scatchard analysis of (^{3}H)NT binding to macrophages suggested the
existence of two populations of binding sites, a major population of
relatively low affinity binding sites and a small population of high
affinity binding sites. NT (8-13), NT (6-13), substance P and tuftsin
competed with (^{3}H)NT binding to the low affinity sites with a comparable K_I to that of NT. NT (1-10) did not compete for the binding at
the low affinity sites.

It is suggested that NT binding to the high affinity sites leads
to enhancement of phagocytosis, whereas its binding to the low affinity sites leads to inhibition of the augmented response. However,
the low affinity sites are the sites of interaction of NT (8-13),
NT (6-13), substance P and tuftsin with the phagocytes and their
saturation with the peptides leads to augmentation of phagocytosis.

Acknowledgement

We thank the Israel Academy of Sciences and Humanities-Basic
Research Foundation and the Mitchell Fund for Higher Education for
the support of this study. S. B. is an incumbent of the William and
Mildred Levine Career Development Chair.

REFERENCES

1. Carraway, R., and Leeman, S. E., J. Biol. Chem. 248:6854, 1973.
2. Leeman, S. E., Mroz, E. A., and Carraway, R. E., in "Peptides in
       Neurobiology" (H. Gainer, ed.), pp. 99-144, Plenum Press, New
       York, 1977.
3. Bissette, G., Manberg, P., Nemeroff, C. B., and Prange, Jr., A J.,
       Life Sci. 23:2173, 1978.
4. Miller, R. J., Med. Biol. 59:65, 1981.
5. Bar-Shavit, Z., Goldman, R., Stabinsky, Y., Gottlieb, P., Fridkin,
       M., Teichberg, V. I., and Blumberg, S., Biochem. Biophys. Res.
       Commun. 94:1445, 1980.
6. Najjar, V. A., Cytobios 6:97, 1972.

7. Bar-Shavit, Z., Stabinsky, Y., Fridkin, M., and Goldman, R., J.
       Cell. Physiol. 100:55, 1979.
8. Stabinsky, Y., Bar-Shavit, Z., Fridkin, M., and Goldman, R., Mol.
       Cell. Biochem. 30:71, 1980.
9. Merrifield, R. B., J. Am. Chem. Soc. 85:2149, 1963.
10. Bar-Shavit, Z., Bursuker, I., and Goldman, R., Mol. Cell. Biochem.
       30:151, 1980.
11. van Lenten, L., and Ashwell, G., in "Methods in Enzymology" (V.
       Ginsburg, ed.), 28, pp. 209-211, Academic Press, New York,
       London, 1972.
12. Fridkin, M., Stabinsky, Y., Zakuth, V., and Spirer, Z., Biochim.
       Biophys. Acta 496:203, 1977.
13. Kitabgi, P., Carraway, R., van Rietschoten, J., Granier, C.,
       Morgat, J. L., Menez, A., Leeman, S., and Freychet, P., Proc.
       Natl. Acad. Sci. USA 74:1846, 1977.
14. Kitabgi, P., Poustis, C., Granier, C., van Rietschoten, J.,
       Rivier, J., Morgat, J.-L., and Freychet, P., Mol. Pharm. 18:
       11, 1980.

DISCUSSION

FELDMAN: When you use neurotensin at a concentration which inhibits
phagocytosis, can this effect be reversed by tuftsin?

GOLDMAN: When neurotensin is inhibitory at 10^{-7}M, you can not stimulate the macrophages by any other agent: not with substance P nor
with tuftsin.

UNKNOWN: Does neurotensin stimulate receptor mediated endocytosis
in any other type of cell?

GOLDMAN: Such studies are just beginning. We have observed this
effect on bone marrow derived macrophages and on the macrophage cell
line P388D1. We also think we can detect it in PMNs. With respect
to tuftsin, we have demonstrated that tuftsin receptors are restricted
to macrophages, PMNs, and monocytes. They can not be demonstrated
on the surfaces of lymphocytes, red blood cells or fibroblasts. High
affinity neurotensin binding sites are found on synaptosomal membranes
and on a cell line derived from a colon carcinoma (HT 29).

PICK: Is this stimulation of phagocytosis in any way specific for
certain particles? You used zymosan as your test particle. Was it
opsonized and did you try any other particles?

GOLDMAN: The zymosan was not opsonized. We see the same effect with
heat killed yeast. With substance P, we have also shown that IgG
coated erythrocytes are taken up better.

A MEMBRANE LECTIN ON MACROPHAGES:
LOCALIZATION AND DEMONSTRATION
OF RECEPTOR RECYCLING

Hubert Kolb, Victoria Kolb-Bachofen, Dieter Vogt
and Jutta Schlepper-Schäfer

Diabetes Research Institute and Institute for
Biophysics and Electron Microscopy
University of Düsseldorf
D 4000 Düsseldorf F.R.G.

It has been observed repeatedly that erythrocytes after neur-
aminidase-treatment do not stay in circulation but adhere to Kupffer
cells and thus are rapidly trapped in the liver (1). Spontaneous
binding of neuraminidase-treated erythrocytes to Kupffer cells also
occurs under serum free conditions in vitro (2,3). We have recently
presented evidence that in vitro as well as in vivo cell contacts
are due to a membrane lectin on Kupffer cells binding terminal D-
galactosyl residues exposed on desialylated erythrocytes (3-6). We
now have localized the lectin on the macrophage cell surface. In
studying the function of the receptor we have found that the lectin
mediates endocytosis of particulate ligands followed by recycling
of the receptor to the cell surface. A preliminary report of these
findings has been published elsewhere (7).

MATERIALS AND METHODS

Kupffer cells: Livers from male Wistar rats weighing 170-200 g
were perfused with collagenase (Boehringer, Mannheim, FRG) as des-
scribed previously (4). Kupffer cells were enriched by differential
centrifugation (4).

Electron Microscopy: Procedures for coupling glycoproteins to
colloidal gold particles and for processing of cells for electron
microscopy were the same as described previously (4,8).

Cell Adhesion Test: Kupffer cells (10^5 cells in 50 µl) were
mixed with an equal volume of erythrocytes (5×10^6 cells), centrifuged

143

for 5 min at 100 g and resuspended after incubation at $4^{\circ}C$ for 1 hour. Kupffer cells binding 3 or more erythrocytes were counted in a hemocytometer (4).

RESULTS

When Kupffer cells were incubated with asialofetuin-gold particles (ASF-gold) for 15 min at $4^{\circ}C$, binding of ligand to the plasma membrane was observed in discrete clusters (Figure 1). Binding of ligand was reduced by 98% in the presence of 25 mM N-acetyl-D-galactosamine but not after addition of N-acetyl-D-glucosamine or D-mannose, demonstrating that ASF-gold particles interacted with the N-acetyl-D-galactosamine D-galactose-specific lectin on Kupffer cells.

When the reaction mixture was warmed to $37^{\circ}C$, uptake of ASF-gold was seen (Figure 2). Endocytosis occurred via the coated pit/coated vesicle pathway. The final stage was the accumulation of ligands in secondary lysosomes.

Evidence for reutilization/recycling of receptors was as follows: Within 15 minutes at $37^{\circ}C$ Kupffer cells took up many times the initial number of particles bound and still expressed receptors on the cell surface (Figure 2). In another type of experiment, membrane lectins were destroyed by treating the cells with trypsin (Figure 3). It was found that lectins were replaced by new receptors within 3 minutes at $37^{\circ}C$ (Figure 4).

The rapid reappearance of membrane lectins after trypsin treatment was also seen when, instead of ASF-gold binding, lectin-dependent cell contact formation was determined. Kupffer cells were mixed with desialylated erythrocytes and the formation of cell aggregates was followed. Cell contacts were specifically blocked by N-acetyl-D-galactosamine or D-galactose and thus were lectin mediated. Trypsin treatment of the macrophages destroyed their capacity to bind desialylated cells. Lectin-mediated cell contacts, however, were fully restored after briefly warming the cells to $37^{\circ}C$ in the absence of protein biosynthesis (Table 1).

DISCUSSION

The experiments described here show that the N-acetyl-D-galactosamine/D-galactose specific lectin on Kupffer cells is arranged in small discrete clusters which are randomly distributed over the cell surface. After interaction of receptors with particulate ligands, endocytosis occurs via the coated pit/coated vesicle pathway. Ligands are finally accumulated in secondary lysosomes. After consumption of membrane lectins, new receptors are expressed on the

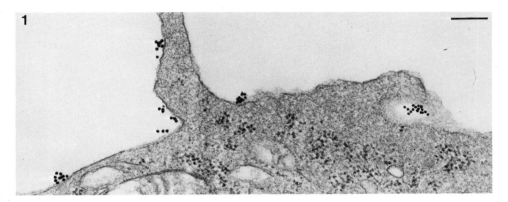

Fig. 1.   Incubation of Kupffer cells with ASF-gold at 4°C for 15 min
results in clustered binding to the cell membrane. The bar
represents 0.2 µm.

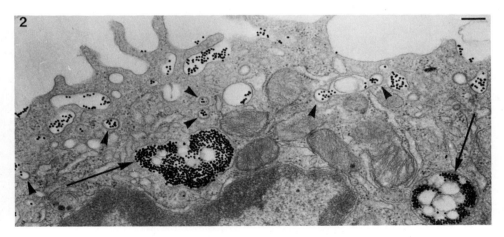

Fig. 2.   Incubation of Kupffer cells with ASF-gold at 37°C for 15
minutes leads to endocytosis via the coated pit/coated
vesicle pathway (arrowheads point to coated membrane structures), and particle accumulation in large vacuoles (arrows).
The bar represents 0.2 µm.

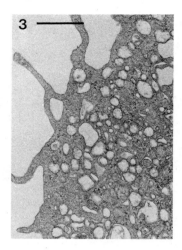

Fig. 3. Mild trypsin-treatment of Kupffer cells completely abolishes
         ASF-gold binding sites. No binding is observed after an
         incubation of 20 minutes at 4°C. The bar represents 1 μm.

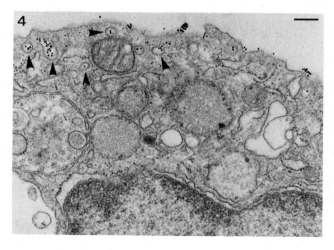

Fig. 4. Trypsin treated Kupffer cells reexpress ASF-gold binding
         sites within 3 minutes at 37°C in the presence of 50 μM
         cycloheximide. The bar represents 0.2 μm.

Table 1. Rapid Reappearance of Kupffer Cell Surface Lectin
          after Trypsin Treatment

Treatment	% Kupffer Cells Expressing Lectin Activity
Before trypsin	89±5
After trypsin	
0 minutes	16±2
15 minutes no cycloheximide	91±4
with cycloheximide	87±3

Lectin activity was determined in the cell adhesion test (see
Materials and Methods). Kupffer cells were incubated with
trypsin (2 mg/ml) for 5 minutes at 0°C, washed 3 times and
subsequently incubated at 37°C with or without the presence of
50 µM cycloheximide. Data are reported as mean values from 3-
5 experiments with standard deviation.

cell surface in the absence of protein biosynthesis. This finding
indicates receptor recycling and the presence of an internal pool
of lectin receptors. Using the same experimental system as described
here, we have also analyzed the distribution of D-galactose receptors
on hepatocytes (8). In marked contrast to our present observation
of receptor clusters on Kupffer cells, it was found that hepatocyte
lectins are not aggregated (8). This finding may explain why hepatocytes are able to take up soluble asialoglycoproteins (9) whereas
Kupffer cells only ingest particulate ligands. The interaction of
single ASF molecules with lectin clusters on Kupffer cells apparently
is not a sufficient trigger for endocytosis, whereas the reaction
with a multivalent ASF-gold particle leads to a physiological response. We have reported preliminary evidence that in vivo the Kupffer cell lectin may be involved in the sequestration of aged erythrocytes (10).

Acknowledgement

We thank U. Lammersen, A. Schloemer and B. Schmidt for expert
technical assistance and Prof. C. Bauer for a gift of asialofetuin.

This work was supported by the Deutsche Forschungsgemeinschaft and by the Minister für Forschung and Wissenschaft des Landes Nordrhein-Westfalen.

REFERENCES

1. Jancik, J. M., Schauer, R., Andres, K. H., and von Düring, M., Cell Tissue Res. 186:209, 1978.
2. Aminoff, D., vor der Bruegge, W. F., Bell, W. C., Sarpolis, K., and Williams, R., Proc. Natl. Acad. Sci. USA 74:1521, 1977.
3. Kolb, H., and Kolb-Bachofen, V., Biochem. Biophys. Res. Commun. 85:678, 1978.
4. Schlepper-Schäfer, J., Kolb-Bachofen, V., and Kolb, H., Biochem. J. 186:827, 1980.
5. Kolb, H., Friedrich, E., and Süss, R., Hoppe-Seyler's Z. Physiol. Chem. 362:1609, 1981.
6. Müller, E., Franco, M. W., and Schauer, R., Hoppe-Seyler's Z. Physiol. Chem. 362:1615, 1981.
7. Kolb, H., Vogt, D., and Kolb-Bachofen, V., Biochem. J. 200:445, 1981.
8. Kolb-Bachofen, V., Biochim. Biophys. Acta 645:293, 1981.
9. Ashwell, G., and Morell, A. G., Adv. Enzymol. 41:99, 1974.
10. Schlepper-Schäfer, J., Kolb-Bachofen, V., Friedrich, E., and Kolb, H., in "Glycoconjugates" (T. Yamakawa, T. Osawa, S. Handa, eds.), pp. 386-387, 1981.

DISCUSSION

KOLB: The question was asked if there's identity between this D-galactose specific lectin on Kupffer cells and the one on hepatocytes. With regard to specificity of binding, we don't see any difference. Right now we are raising antibodies in order to compare the two receptors. This is still an open question but they are very much alike.

UNKNOWN: Is the receptor also present on the subcellular membranes of the Kupffer cell in addition to the plasma membrane?

KOLB: It is present on the rough endoplasmic reticulum and also on the nuclear envelope.

PICK: Does the killing mechanism of senescent cells by the Kupffer cell involve an oxidative burst? Do you induce an oxidative burst in either a Kupffer cell or a macrophage whose surface lectin has bound D-galactose?

KOLB: I cannot yet tell you.

WILKINSON: Why are you postulating that this receptor is really one for aged erythrocytes or other aged cells? Why couldn't its function involve microorganisms?

KOLB: I would presume that every substance with exposed terminal D-galactose residues might be bound by this receptor and sequestrated. Aged erythrocytes may be one of these and a series of bacteria strains might be also.

UNKNOWN: If you do your trypsinization experiment in the presence of cyclohexamide, do you then get reexpression of new receptor?

KOLB: No.

UNKNOWN: How were your cells isolated?

KOLB: Our experiments used Kupffer cells isolated by a collagenase perfusion technique. For electron microscopy, you don't need pure Kupffer cells, because you can differentiate between Kupffer cells and hepatocytes. However, it is relatively easy to isolate pure Kupffer cells by differential centrifugation.

CATIONIC POLYELECTROLYTES AND LEUKOCYTE FACTORS FUNCTION AS OPSONINS, TRIGGERS OF CHEMILUMINESCENCE AND ACTIVATORS OF AUTOLYTIC ENZYMES IN BACTERIA: MODULATION BY ANIONIC POLYELECTROLYTES IN RELATION TO INFLAMMATION

Isaac Ginsburg[1], Meir Lahav[1], Mina Ferne[2], and Sybille Müller[3]

[1]Department of Oral Biology, Hebrew University-Hadassah School of Dental Medicine
[2]The Streptococcus Reference Laboratory, Government Laboratories, Ministry of Health, Jerusalem, Israel
[3]The Robert Koch Institute, Berlin, West Germany

INTRODUCTION

Both antibodies and complement components are essential for successful phagocytosis of many virulent microorganisms (1,2). Although the mechanisms by which opsonins promote particle uptake are not fully understood, it has been suggested that both electrostatic and hydrophobic forces act in concert with specific receptors for Fc and C3b to facilitate interiorization of particles (2,3). In the case of group A streptococci, opsonization by immunoglobulins abolishes the anti-phagocytic properties of the M-antigen (4,5). Since one mechanism by which opsonins may act is to decrease repulsion forces between negative charges present on the surface of the particle and phagocyte, cationic ligands may function as effective opsonins (6-11). In addition, cationic substances may participate in bacteriolysis. We recently suggested (11) that the breakdown of bacterial cells following phagocytosis is mediated indirectly by leukocyte cationic proteins and phospholipases which activate autolytic enzymes and not by lysosomal enzymes directly.

The present communication further describes the role played by cationic and anionic macromolecular substances as modulators of phagocytosis and chemiluminescence by polymorphonuclear neutrophils (PMNs) and macrophages.

MATERIALS AND METHODS

Resident peritoneal macrophages were obtained from Swiss mice

by lavage using Hanks balanced salt solution (HBSS) buffered with
Hepes pH 7.4.   Human granulocytes (PMNs) were separated from fresh
heparinized blood by sedimentation with dextran (12).   Macrophages
and PMNs (10^6/ml) in HBSS supplemented with 10% fetal calf serum
(FCS) were allowed to adhere for 45 min at 37°C to 22 x 22 mm coverslips in 35 mm Falcon dishes.   The monolayers were washed in HBSS
and covered with 1 ml HBSS containing 1% FCS. 10^7 group A type 2
M+ streptococci or 10^7 Staph. aureus, which had been preopsonized in
vitro either with 1% rabbit antiserum supplemented with 10% fresh
guinea pig serum or with a variety of cationic agents were added to
the monolayers.   After 60 min at 37°C the monolayers were washed
three times with HBSS.   The coverslips were then immersed for 5 sec
in a hypotonic solution containing 1 volume saline, 2 volumes distilled water and 1% FCS to swell the phagocytes which allowed a better
assessment of phagocytosis.   Both the percentage phagocytosis and
the phagocytic index (number of cocci per cell) were determined on
200 cells.   The effect of anionic polyelectrolytes on macrophage
phagocytosis was assayed by first exposing the cells for 30 min at
37°C to the anions, followed by washing and the addition of opsonized
bacteria.   Mouse fibroblasts (L-cells) and human epithelial cells
were cultivated in Medium 199 supplemented with 10% FCS to achieve
semiconfluency (11).   The monolayers were rinsed in HBSS and allowed
to phagocytose serum or histone (Type IIA)-opsonized Candida albicans
(heat-killed) at a ratio of 5 fungi per fibroblast or epithelial cell.

Chemiluminescence was measured using 10^{-5}M Luminol as described
(13).   Bacteriolysis was assayed using Staph. aureus prelabeled with
^{14}C-glucose as described (14) and expressed as the percentage radioactivity solubilized from a standard labeled staphylococcal suspension
in 0.1M acetate buffer pH 5.0.

RESULTS

Studies on Phagocytosis

Table 1 summarizes the data on the phagocytosis by mouse macrophages of Staph. aureus.   It appears that cationic ligands like
histone, RNase and leukocyte extracts function as very effective
opsonins.   Similar results were obtained with human PMNs or when
group A M+ streptococci were employed except that the uptake of unopsonized bacteria was very low.   Table 2 shows results of double
opsonization of streptococci by antisera and other non-specific
opsonins.   Streptococci, which have been first exposed to antiserum,
then washed and re-exposed to a variety of cationic ligands and leukocyte extracts, were phagocytosed to a much higher extent than those
preopsonized with antiserum alone, or when first exposed to the
various ligands then to antiserum.   It appeared that double opsonization introduced more binding sites, which enabled more streptococci
to gain access to the macrophages.

Table 1.   Opsonization of Staphlococcus aureus by Polyelectrolytes for Phagocytosis by Mouse Macrophages

Opsonizing agent	% Phagocytosis	Phagocytic Index
None	74	2.43
Antiserum + complement	89	28.25
Histone type II A	99	15.00
Ribonuclease	100	21.00
Lysozyme	100	16.30
Leukocyte extract	99	13.10
Leukocyte extract – FC	100	42.00
Inflammatory exudate	100	14.50

Opsonization was performed by preincubating 1 mg protein/ml
with 10^8/ml Staph. aureus in PBS.   The bacteria were then
washed.   Human leukocyte extracts were prepared by freeze-
thawing or by a fluorocarbon (FC) extract of whole blood
leukocytes.   Data are the average of 3 experiments performed
in triplicate.

We also tested the uptake of opsonized streptococci in the peritoneal cavity of mice.   Streptococci, which had been preopsonized
with antiserum or with cationic polyelectrolytes in vitro, were injected intraperitoneally into Swiss mice.   Thirty minutes later the
macrophages were removed by lavage, plated in Falcon petri dishes
in the presence of 10% FCS, and the percentage phagocytosis and phagocytic index determined.   Whereas unopsonized streptococci were phagocytosed to a very small extent (5-7% and a phagocytic index of about
1.0), streptococci opsonized either with antiserum plus complement
or with histone, leukocyte and platelet extracts and with RNase were
readily taken up (75-95% phagocytosis and a phagocytic index of 10-
20).   To prove that phagocytosis really occurred in vivo and not in
the petri dishes, we added 25-50 µg/ml of liquoid to the cultures
which completely inhibited the phagocytosis of opsonized streptococci
by the macrophages.

Table 2.   Multiple Opsonization of Type 2 Streptococci
           for Phagocytosis by Peritoneal Macrophages

Streptococci pretreated by	Followed by	% Phagocytosis	Phagocytic Index
None	–	9.6	0.75
Antiserum (1%)	–	62.0	6.25
Antiserum	Histone MX	73.2	11.70
Histone	Antiserum	44.0	4.10
Histone	–	62.7	8.17
Antiserum	RNAse	72.3	12.50
RNase	Antiserum	57.5	7.70
RNase	–	44.0	5.03
Antiserum	Leukocyte Ext.	80.0	13.60
Leukocyte Ext.	Antiserum	63.5	6.80
Leukocyte Ext.	–	53.0	5.10
Antiserum	Platelet Ext.	76.5	16.75
Platelet Ext.	Antiserum	49.5	5.35
Platelet Ext.	–	54.5	4.85
Antiserum	Histone Arg.	88.0	12.88
Histone Arg.	Antiserum	69.0	6.9
Histone Arg.	–	26.0	2.9
Antiserum	Synovial Fluid	77.0	15.5
Synovial Fluid	Antiserum	45.0	4.4
Synovial Fluid	–	21.0	1.7

Type 2 streptococci 10^7 /ml were opsonized for 30 minutes with the various agents (1 mg/ml). Some of the
opsonized cells were again exposed to certain agents
(double opsonization). The data are the average of 4
experiments performed in duplicates.

    Since cationic ligands were effective opsonins for professional
phagocytes, we determined their role in the phagocytosis of particles
by fibroblasts. Figures 1 and 2 show phagocytosis of histone-coated
Candida albicans by mouse fibroblasts in culture. Similar results

were obtained with human epithelial cells, HeLa cells and beating
rat heart cells in culture. Non-opsonized Candida or Candida opsonized by fresh human serum were not taken up by any of these cells.

Inhibition of Phagocytosis by Polyelectrolytes

Table 3 shows that phagocytosis of antiserum-opsonized streptococci was strongly inhibited if the macrophages were pre-exposed
for 30 minutes with DNA, hyaluronic acid and liquoid. Similar results
were obtained with human saliva, concanavalin A and phytohemagglutinin. Macrophages pre-exposed to liquoid, concanavalin A or phytohemagglutinin appeared extremely vacuolated.

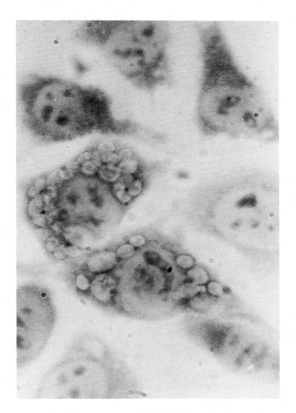

Fig. 1.   Phagocytosis by mouse fibroblasts in culture of Candida
          albicans pre-opsonized with nuclear histone. Giemsa x 680.

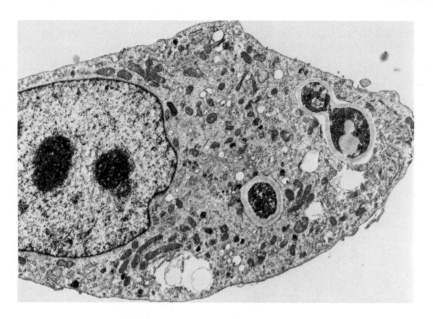

Fig. 2.   Mouse fibroblast which has engulfed 3 Candida albicans.
Electron micrograph x9000.

Studies on Chemiluminescence

Bacteria preopsonized with antibodies trigger a strong chemiluminescent reaction in PMNs and macrophages (15). Figure 3 shows
that staphylococci coated with nuclear histones can also trigger
chemiluminescent reactions in macrophages and PMNs (not shown). This
reaction was faster and steeper than that induced by antibody-opso-
nized bacteria. Similar results were also obtained with a variety
of other cationic substances, as well as with leukocyte and platelet
extracts suggesting that the membrane of the phagocyte cannot differentiate between a specific trigger induced through Fc and C3b
receptors and a non-specific perturbation induced by a cationic
ligand. It was therefore surprising when we found that liquoid, which
strongly inhibited phagocytosis (Table 3), proved to be a very potent
trigger of the chemiluminescent reaction in both macrophages and
PMNs. Under the same experimental conditions none of the other
anionic substances tested (heparin, chondroitin sulfate, etc.) were
capable of inducing a chemiluminescent reaction. The mechanism by
which liquoid induces membrane perturbation is not understood.

Table 3.   Inhibition by Anionic Polyelectrolytes of the
           Phagocytosis of Group A Streptococci by Mouse
           Peritoneal Macrophages

Macrophages pretreated with	µg/ml	% Phagocytosis	Phagocytic Index
None		80	15.6
Hyaluronic acid	10	80	13.7
	50	57	9.2
	100	53	8.1
	500	50	7.6
	1000	38	4.3
Deoxyribonucleic acid	10	46	5.2
	50	40	4.9
	100	35	3.1
	500	27	2.2
	1000	19	1.5
Liquoid	5	49	5.3
	10	38	4.1
	15	16	1.7
	20	14	1.5
	25	14	1.3
	30	9	0.78
	40	8	0.29

Macrophage monolayers (10^6/plate) were exposed to the
polyelectrolytes for 45 minutes at 37°C and then washed
in HBSS and allowed to phagocytose type 2 streptococci
preopsonized with 1% rabbit antiserum + 10% guinea pig
serum. The data are the average of 4 experiments performed in duplicates.

Studies on Bacteriolysis

     Since cationic macromolecular substances like histone, RNase,
and lysozyme trigger activation of autolytic enzymes in staphylococci
(14), it was of interest to determine the smallest cationic substance
which could replace these high-molecular weight substances. We tested
a series of amines and amino acids and found that whereas arginine,

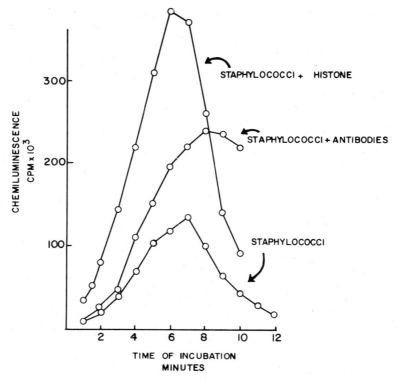

Fig. 3. Chemiluminescence induced in mouse peritoneal macrophages
         by staphylocci opsonized with antiserum or with nuclear
         histone.

lysine and histidine were ineffective in triggering bacteriolysis,
spermine and spermidine were excellent lytic agents for staphylococci.
We also found that human lysozyme is more active on a weight basis
than egg-white lysozyme in triggering bacteriolysis. Therefore,
perturbation of bacterial membranes with cationic ligands appeared
sufficient to trigger activation of autolytic enzymes and the subsequent degradation of their cellular constituents. Bacteriolysis
was abolished by boiling the bacterial cells presumably destroying
their autolytic systems.

DISCUSSION AND CONCLUSIONS

     The data presented suggest that a variety of cationic substances
and leukocyte and platelet extracts function as effective opsonins
for phagocytosis of streptococci by professional phagocytic cells

both in vitro and in vivo. Furthermore, streptococci which have been opsonized with cationic ligands are capable of triggering a strong chemiluminescent reaction in both PMNs and macrophages. It is very likely that the cationic ligands function to minimize the repulsion forces between negative charges present on both the phagocyte surface and the bacterial cell (3,8). The data showing either enhanced or depressed phagocytosis following treatments of streptococci with "sandwiches" comprised of antibodies and leukocyte extracts or cationic ligands showed that the sequence in which a particle encounters opsonizing agents is important to the outcome of bacteria-leukocyte interactions.

Pretreatment of macrophages with anionic polyelectrolytes strongly modulates phagocytosis (Table 3) and chemotaxis (see Ref. 12). It appears that polyelectrolytes act both on the particle and on the phagocytic cells to inhibit or enhance phagocytosis. Cationic ligands also modulate chemiluminescence suggesting that cell membrane does not differentiate between stimulation mediated through a specific receptor and that induced by a non-specific cationic ligand.

The opsonization by cationic ligands of Candida albicans for phagocytosis by "non-professional" phagocytes might lead to persistence of fungi in cells poor in digestive enzymes. Such a phenomenon may exist in fibroblasts in vivo.

In conclusion, cationic and anionic polyelectrolytes are important to leukocyte function. The concentration and chemical nature of the cationic and anionic substances may determine whether or not bacterial cellular constituents are degraded by bacteriolysis, or whether non-biodegradable components of bacteria persist in macrophages or in tissues to trigger chronic inflammation.

Acknowledgement

This study was supported by research grants obtained from Dr. Samuel Robbins of Cleveland, Ohio.

REFERENCES

1. Rabinowitch, M., in "Mononuclear Phagocytes" (R. van Furth, ed.), pp. 299-313, Blackwell Scientific Publications, Oxford, 1970.
2. Ehlenberger, A. G., and Nussenzweig, V., J. Exp. Med. 145:357, 1977.
3. van Oss, C. J., and Gillman, C. F., J. Reticuloendothelial Soc. 12:497, 1972.
4. Peterson, P. K., Schmeling, D., Clearg, P. P., Wilkinson, B. J., and Quie, P. G., J. Infect. Dis. 139:575, 1979.
5. Fox, E. N., Bacteriol. Rev. 38:57, 1976.
6. De Vries, A., Salgo, J., Matoth, Y., Nevo, A., and Katchalski, E., Arch. Int. Pharmacodyn. 104:1, 1955.

7. Stossel, T. P., J. Cell Biol. 58:346, 1973.
8. Pruzanski, W., and Saito, S., Exp. Cell Res. 117:1, 1978.
9. Deierkauf, F. A., Benkerg, H., Deierkauf, M., and Riemersma, C.,
        J. Cell. Physiol. 92:169, 1977.
10. Westwood, F. R., and Longstaff, E., Br. J. Cancer 33:392, 1976,
11. Ginsburg, I., Sela, M. N., Morag, A., Ravid, Z., Duchan, Z.,
        Ferha, M., Inflammation 5:301, 1981.
12. Ginsburg, I., and Quie, P. G., Inflammation 4:301, 1980.
13. Muller, S., Falkenberg, S., Fromtling, R. A., Fromtling, A. M.,
        and Klimetzek, V., in "Bioluminescence and Chemiluminescence"
        (M. A. De Luca and W. D. McElroy, eds.), pp. 721-727,
        Academic Press, Inc., New York, 1981.
14. Lahav, M., and Ginsburg, I., Inflammation 2:165, 1977.
15. Allan, R. C., Biochem. Biophys. Res. Commun. 47:679, 1980.

DISCUSSION

UNKNOWN: What happens if you expose your cells to Candida albicans
in the presence of polylysine or ribonuclease?

GINSBERG: Actually the question is whether you can opsonize both a
macrophage and the particle. The answer is both. If you opsonize
the cells with amounts of polylysine insufficient to kill them, you
can enhance phagocytosis of non-opsonized particles by about 1 1/2
to 2 fold, not more. On the other hand, you can enhance by 50 fold
if you put the ligand on the surface of the particle.

GOLDMAN: Several years ago we opsonized yeast with Concanavalin A,
and we're pretty sure that the Con A caused a tremendous increase in
phagocytosis by the macrophages. Also, fibroblasts can phagocytose
such Con A opsonized particles. If we treated the macrophages and
not the yeast with Con A, we also got enhanced phagocytosis.

UNKNOWN: How do you think substances such as heparin sulfate might
function in vivo in the removal and killing of bacteria?

GINSBERG: When macrophages pinocytose such negatively charged polyelectrolytes as dextran sulfate, carrageenan sulfate or heparin sulfate, it switches off the digestive systems. I think there is a very,
very subtle balance between cationic activators and anionic inhibitors
and the balance between them will determine whether a particle will
be engulfed, whether it will be killed, and whether it will be
digested or extruded from the macrophage. I think we have to look
into the contents of the milieu in which the macrophages are living
in order to understand what may happen in vivo.

PHAGOCYTIC ENHANCEMENT OF GRANULOCYTES
AND MONOCYTES BY CATIONIC PROTEINS FROM
HUMAN POLYMORPHONUCLEAR LEUKOCYTES

W. Pruzanski, N.S. Ranadive and S. Saito

Department of Medicine and Pathology
University of Toronto
Toronto, Ontario, Canada

INTRODUCTION

Defense against infection includes both humoral and phagocyte-mediated mechanisms. Close cooperation between these two mechanisms has been well documented. Certain substances that belong to the humoral compartment also serve as opsonins and chemotactic attractants. On the other hand, phagocytes produce and secrete several proteins which participate in bactericidal and bacteriolytic activities including complement and lysozyme. We described a few years ago another complex composed of nuclear and cytoplasmic cationic proteins which directly affects microorganisms and also enhances phagocytosis (1). In this paper we provide evidence that cationic proteins enhance phagocytosis in man.

MATERIALS AND METHODS

Human polymorphonuclear leukocytes (PMN) and mononuclear phagocytes were purified from fresh blood of healthy volunteers or patients with polycythemia. Viability was greater than 98% while purity was 98% for PMN and 88-93% for monocytes. Phagocytic assays and intracellular bactericidal activity (ICBA) were tested as described (1). Viable Staphylococcus aureus and the smooth strain of E. coli were used and the ICBA Index estimated by relating the number of residual microorganisms to the phagocytic rate and dividing the results by those of controls.

Cationic fractions were obtained from sonicated human PMN by acid extraction using either Sephadex G-50 or G-75 column chromatography. In each fraction the following were estimated: total protein, histamine-release capacity, neutral protease, peroxidase, lysozyme, glucuronidase, chymotrypsin, cationic leukemia antigen, immuno-globulins, complement components and properdin.

RESULTS

    Cationic lysosomal proteins from human PMN were divided ini-tially into 3 fractions- A, B and C and later into 5 fractions I and II corresponding roughly to A, III corresponding to B and IV and V ccrresponding to C. Fraction I had 4 electrophoretic bands of slower mobility than lysozyme (LZM); fraction II had 5 or 6 bands slower than LZM; fraction III had at least 7 bands slower and 2 bands faster than LZM; fraction IV contained LZM, 2 bands faster and a few faint bands slower than LZM; fractioa V was composed of almost pure LZM. Partial characterization of the fractions showed presence of neutral protease in fractions I, II, III and IV, chymotrypsin in fraction III and lysozyme in fractions IV and V.

    Fractions A and B markedly enhanced the phagocytic index of human PMN using Gram positive or Gram negative microorganisms (Table 1).

Table 1. Influence of Cationic Fractions on Phagocytic Index and Intracellular Bactericidal Activity (ICBA) of Polymorphonuclear Cells

| | Phagocytic Index | | |
Fraction	Control	Presence of Cationic Proteins	ICBA Index
A	2.6	5.3	40
B	3.1	6.8	91
C	3.7	3.3	6
I	2.4	3.2	49
II	2.6	3.1	33
III	2.4	4.5	100
IV	2.4	5.7	100
V	2.6	5.1	45

Fractions A to C were tested at 100 mg/ml while frac-tions I-V were tested at 50 mg/ml. An ICBA index less than 80 represents a significant enhancement of bacteri-cidal activity.

Phagocytic activity of monocytes was enhanced by all 3 fractions. Intracellular bactericidal activity (ICBA) was markedly enhanced by fractions A and C. Fractions I to V were tested at a lesser concentration than fractions A to C (50 mg/ml vs 100 mg/ml). Enhancement of the phagocytic index was observed with all fractions except fraction II. ICBA was markedly enhanced by fractions I, II and V. Addition of DNA or cytochalasin B inhibited or abolished the enhancing activity of cationic fractions on phagocytosis while their influence on ICBA was less pronounced with the exception of fraction II.

The data suggest that some cationic lysosomal fractions from human PMN enhance phagocytosis and intracellular bactericidal activity of human PMN as well as the phagocytic activity of monocytes.

Acknowledgement

Supported by grant-in-aid from the Medical Research Council of Canada.

REFERENCES

1. Pruzanski, W., and Saito, S., Exper. Cell Research 117:1, 1978.

DISCUSSION

GINSBERG: Understanding the role of cationic proteins is very important not only as modulators of cell surfaces but also as enhancers of intracellular killing. Bacterial susceptibility to cationic proteins varies tremendously. This has been shown in vitro but we don't know if it exists in vivo. We also don't know the exact nature of the modifiers which enhance phagocytosis. They may be C5a or lysates of bacteria.

PRUZANSKI: We have just started our chemotactic experiments using cationic proteins. I'm afraid that the results may be conflicting between your group, my group and the Gordon group* which recently reported that human lysozyme and egg white lysozyme do not enhance chemotactic activity. The latter result was quite puzzling. I understand you used egg white lysozyme and obtained results which were quite different from Gordon's group. Our results with human lysozyme are also different from Gordon's group. Now, we don't know whether differences in these results are caused by different techniques or because we are all using different lysozyme preparations. In the meantime, we have done some preliminary experiments which show that cationic fractions probably are not strong enhancers of chemotaxis. We theorized from a teleological view that if you have an abcess or an inflammatory focus, nature is not interested in attracting ad

infinitum polymorphonuclear leukocytes and macrophages to the focus
because in this way the abcess will expand enormously. On the other
hand, the PMNs and some monocytes already present in the abcess die
and release cationic substances which are capable of enhancing the
phagocytic activity of the surviving cells. But this is just a
hypothesis. In the meantime, preliminary experiments have shown that
there is no enhancement or no marked enhancement of chemotaxis by
cationic fractions.

WILKINSON: You have to be careful with interpreting effects of
cationic substance in terms of cell locomotion. Most of them increase
cell adhesion considerably and stop cell movement. So, any effect
you observe can't be interpreted in terms of chemotaxis because the
effects on locomotion itself are so marked.

PRUZANSKI: By which method?

WILKINSON: If you place cells on a cationic surface such as a polylysine coated surface, the cells get stuck.

PRUZANSKI: We measured the distance migrated under agarose so as to
avoid problems of adherence. In this system, there was no inhibition
of migration by cationic proteins.

PICK: Do you claim now an alternative mechanism which is clearly not
receptor mediated? If you look closely at Dr. Gordon's results compared to yours, you will see a receptor mediated event which is not
very efficient (two-fold increase). In contrast, your cationic protein
was enormously efficient. We have been used to thinking it's a
receptor initiated event. Now you are coming up with a proposal which
is radically different.

---

*Gordon, L. I., Douglas, S. D., Kay, N. E., Osserman, E. F., Jacob,
    H. S., Modulation of neutrophil function by lysozyme. Poten-
    tial negative feedback system of inflammation, J. Clin. In-
    vest. 64:226-232, 1979.

INHIBITION OF COMPLEMENT DEPENDENT PHAGOCYTOSIS BY NORMAL HUMAN MONOCYTES FOLLOWING PREINCUBATION WITH IMMUNE COMPLEXES

Anthony J. Pinching

Departments of Medicine and Immunology
Royal Postgraduate Medical School
London W12 OHS   U.K.

INTRODUCTION

Interaction between immune complexes and cells of the monocyte/ macrophage lineage is important not only for the normal removal of immune complexes by the reticuloendothelial system but also in relation to defects found in putative immune complex diseases. Previous studies (1,2,3) with soluble complexes or immunoglobulin G aggregates have shown inhibition of Fc dependent monocyte or macrophage function, using relatively high concentrations of complexes. These studies used experimental designs in which the mononuclear phagocytes that had been preincubated with complexes did not encounter plasma, since preopsonised targets (usually erythrocytes) were used. In the course of studying phagocyte function in patients with a variety of auto-allergic diseases, it was noted that patients with active immune complex disease (eg. systemic lupus erythematosis) had a defect of phagocytosis when tested using Candida guilliermondii targets in normal or autologous plasma (4). The effect of preincubating normal human monocytes with soluble heat-aggregated human IgG (AHG) or BSA anti BSA complexes was therefore investigated. Further studies were also performed using plasma from patients with immune complex diseases.

MATERIALS AND METHODS

Mononuclear cells in suspension (2×10^6/ml) were obtained by Ficoll Hypaque separation and preincubated in siliconized tubes with varying concentrations of AHG (30-35S) or BSA anti BSA complexes at equivalence (20S) or in 10:1 antigen excess (10S). The monocytes were then washed before phagocytic activity was measured with Candida guilliermondii and normal plasma. For preincubation, immune complexes

165

(AHG, BSA anti BSA 10:1) were placed in medium alone or in plasma/ serum (all types of immune complexes) at 4°C or 37°C for 1 hour with mixing (to avoid adherence). Candida assays were performed for 1 hour at 37°C using 200 μl cells, 100 μl normal plasma and 60 μl of a Candida suspension at 2×10^7 organisms per ml. In some experiments, monocytes preincubated with AHG or BSA anti BSA complexes were incubated with normal or modified plasma for 15 minutes at 4°C or 37°C, washed and assayed with preopsonised Candida in the absence of plasma.

Phagocytosis was determined by light-microscopic examination of cytocentrifuge slide preparations stained with non-specific esterase. Monocytes ingesting one or more Candida were expressed as a percentage of total monocytes (% phagocytosis). Inhibition of phagocytosis was expressed as the percent reduction from the value for cells preincubated in medium or plasma alone (100%). Opsonin-specific phagocytosis was calculated as follows: Fc dependent: heat-inactivated plasma minus heat-inactivated gnotobiotic (colostrum-free) calf serum (or heat-inactivated human hypogamma-globulinemic plasma); C3 dependent: gnotobiotic calf serum (or hypogamma-globulinemic plasma) minus heat-inactivated gnotobiotic calf serum (or heat inactivated hypogamma-globulinemic plasma). These results were expressed as percentage inhibition as above.

RESULTS

Following preincubation of monocytes with immune complexes, inhibition of phagocytosis occurred in a dose-dependent fashion (Figure 1). Inhibition was observed with AHG in a concentration range of 5-80 μg/ml, with BSA anti BSA (10:1) at 1.5 - 24 μg/ml or with BSA anti BSA (equivalence) at 3 - 50 μg/ml. Inhibition of phagocytosis was equal between preincubation in plasma/serum or in medium alone, either at 4°C or 37°C. Inhibition of phagocytosis was not seen using normal human IgG or anti-BSA alone. Phagocytosis by neutrophils was not altered by similar treatment. Experiments with ^{125}I labelled immune complexes showed that about 2% of the complexes bound to monocytes at all concentrations. Prolonged washing (1-3 hours) at 4°C did not reduce phagocytic inhibition, but at 37°C it was reduced by about 50% after 1 hour, with no further reduction after 3 hours.

Although phagocytosis of Candida did occur to a minor extent without complement or antibody, both factors were required for maximal phagocytosis. Complement alone or specific antibody alone allowed about 50% of maximal phagocytosis. In the concentration ranges used and for all immune complexes tested, the inhibition of phagocytosis was due to decreased C3 dependent phagocytosis (Figure 1). Inhibition of Fc dependent phagocytosis only occurred when AHG was used at ten times the highest concentration used generally in these experiments.

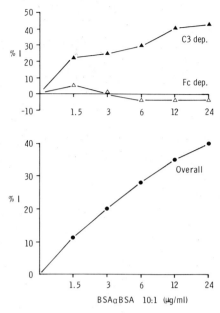

Fig. 1.   Inhibition of phagocytosis (% I) of normal human monocytes
          by preincubation at 4°C for one hour with various concentra-
          tions of BSA anti BSA immune complexes at 10:1 antigen excess
          in medium without plasma.   Bottom: overall inhibition when
          assayed in normal plasma.   Top: the effect on C3 and Fc de-
          pendent phagocytosis. While Fc dependent phagocytosis is
          unaffected, there is a dose-dependent inhibition of C3 de-
          pendent phagocytosis.

Reduction in C3 dependent phagocytosis was not due to fluid phase
C3 consumption and subsequent failure of opsonization, because supernatants from assays in which inhibition had occurred still opsonized
fresh Candida for normal monocytes.

    The inhibition of phagocytosis resulted from an interaction
between the monocytes which had been exposed to immune complexes and
plasma, because the same monocytes tested with preopsonized targets
and without plasma showed no inhibition. If such monocytes were incubated with plasma at 37°C, washed and then tested with preopsonized
targets, inhibition of phagocytosis occurred. In contrast, if the
intermediate incubation with plasma was performed at 4°C, there was
no inhibition. Further, the concentration of plasma was critical:
inhibition occurred between 10% and 30% plasma, was reduced at 4% or
2% plasma and was abolished at less than 1% plasma.

C2 deficient, C4 depleted and Mg-EGTA plasma did not sustain inhibition. If C2 deficient plasma was reconstituted with purified C2, inhibition occurred. This was confirmed with intermediate incubation experiments. Using such intermediate incubation procedures, plasma depleted of factor B or factor D of the alternative pathway did not cause inhibition. Inhibition did occur when the depleted plasmas were reconstituted with purified factor B or D respectively.

Using a similar preincubation procedure at 37°C, plasma containing cryoglobulins (IgM-IgG or IgG-IgG) obtained from two patients with mixed essential cryoglobulinemia (Type II) were studied. The cryoglobulins were redissolved by warming to 37°C just before use. Preincubation with cryoglobulins caused a substantial inhibition of phagocytosis (Figure 2). At low concentrations, this was only due to decreased C3 dependent phagocytosis, but at higher concentrations it was due to a decrease in both Fc and C3 dependent phagocytosis.

Patients with active immune complex disease had detectable circulating complexes in plasma and monocytes which showed defective phagocytic function at the time of testing. Preincubation of normal monocytes with such plasma caused inhibition of phagocytosis comparable to that seen with the patient's monocytes. The inhibition was due to reduced C3 dependent phagocytosis. The reduced phagocytosis seen with the monocytes of patients with active immune complex disease is also generally due to decreased C3 dependent phagocytosis, although reduced Fc dependent phagocytosis may coexist.

DISCUSSION

Exposure of monocytes to soluble immune complexes leads to a defect in C3 dependent phagocytosis. This defect occurs at a concentration range of immune complexes that is equivalent to that seen in human immune complex disease (confirmed by rheumatoid factor binding assays performed on preincubation mixtures as well as on patients' samples). Further, the concentration is much less than that needed to cause a decrease in Fc dependent function.

The defect in phagocytosis is dependent upon the interaction of monocytes which had been exposed to immune complexes and plasma. This interaction requires >2% plasma, and both the classical and alternative pathways of complement must be functionally intact. The attachment of immune complexes appears to be by Fc receptors, but at too low a density to affect Fc dependent function. The defect of complement dependent phagocytosis could arise by the local generation of C3b on the monocyte surface by the immune complexes. This C3b, either because of high local concentration or by remaining attached to immune complexes, could then compete with C3b opsonized Candida for attachment to C3b receptors or for C3 dependent ingestion. Alternatively the complexes could cause alteration in or movement of the

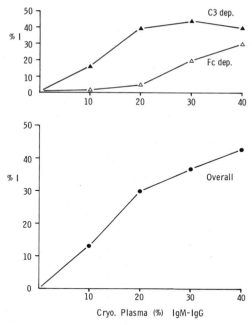

Fig. 2.   Inhibition of phagocytosis (1% I) in normal human monocytes
preincubated at 37°C for one hour with various concentrations
of cryoglobulinemic (IgM-IgG) plasma (original cryoglobulin
concentration about 4.5 g/L).   Bottom: overall inhibition
assayed in normal plasma.   Top: the effect on C3 and Fc
dependent phagocytosis.   At 10% and 20% plasma concentrations, only inhibition of C3 dependent phagocytosis is seen;
at higher plasma concentrations, Fc dependent phagocytosis
is also inhibited.

C3b receptors rendering them less available for attachment or ingestion
of Candida.

The relevance of the phenomenon to human immune complex disease
is indicated by the fact that preincubation of normal monocytes with
immune complexes or with plasma from patients with active immune complex disease and defective monocyte function causes a similar defect
to that seen in the patients' cells both in degree and type.   These
findings highlight the limitations inherent in the use of preopsonized
targets and the value of microbial targets in assessing opsonin-
specific defects.

This paper appears to be the first report of the induction of a C3 dependent monocyte functional defect by soluble immune complexes. The defect appears to require complement interaction with surface bound immune complexes on the monocytes. Two previous reports have shown that immobilized immune complexes can alter C3 dependent monocyte (5) or macrophage (6) function but only if complement is provided, lending support to the present observations using a quite different experimental system.

Acknowledgement

I am grateful to the Wellcome Trust, the National Kidney Research Fund and the Royal Postgraduate Medical School for their financial support, and to O. Moshtael and Miss L. A. Smith for technical assistance. I also thank Dr. I. E. Addison, Dr. B. D. Williams, Dr. J. Schifferli, Miss N. Yousaf, W. Kilgallon and L. Thomas for generously donating reagents. It is a pleasure to acknowledge the encouragement and advice of Professor J. H. Humphrey and Professor D. K. Peters.

REFERENCES

1. Kurlander, R. J., J. Clin. Invest. 66:773, 1980.
2. Griffin, F. M. Jr., J. Exp. Med. 152:905, 1980.
3. Griffin, F. M. Jr., Proc. Nat. Acad. Sci. USA 78:3853, 1981.
4. Pinching, A. J., unpublished observations.
5. Arend, W. P., and Massoni, R. J., Immunology 44:717, 1981.
6. Michl, J., Pieczonka, M. M., Unkeless, J. C., and Silverstein, S. C., J. Exp. Med. 150:607, 1979.

DISCUSSION

SILVERSTEIN: Could the immune complexes deposit complement on their surfaces? Such immune complex complement could then block selectively the complement receptor and, since C3 is bound to the immune complex, it could mask the Fc fragment.

PINCHING: That is one of the mechanisms that could occur.

SILVERSTEIN: Such a mechanism could explain why the immune complexes are not affecting Fc mediated phagocytosis.

PINCHING: At the concentrations used, there was no affect on Fc mediated phagocytosis. The reason that I think that the complexes are bound by the Fc receptors is that binding can occur in the absence of complement.

SILVERSTEIN: When the immune complexes bind to the cell receptors in the absence of complement, can they be displaced by the addition of serum as a complement source?

PINCHING: I don't yet know, for that experiment requires rather sensitive technology.

SECTION 3

PROMONOCYTES AND MONOCYTES

COMPARISON OF THE IN VIVO AND IN VITRO PROLIFERATION OF MONOBLASTS, PROMONOCYTES, AND THE MACROPHAGE CELL LINE J774

R. van Furth, Th. J. L. M. Goud, J. W. M. van der Meer,
A. Blussé van Oud Alblas, M. M. C. Diesselhoff-den Dulk,
and M. Schadewijk-Nieuwstad

Department of Infectious Diseases
University Hospital
Rijnsburgerweg 10
2333 AA Leiden
The Netherlands

This report concerns studies on the identification and characterization of monoblasts and promonocytes, and the macrophage cell line J774, and gives data on the in vivo and in vitro proliferative behavior of these cells and the kinetics of monocytes.

CHARACTERIZATION AND NOMENCLATURE

For modern research on the kinetics of cells, adequate definition of the cells under study is obligatory. Formerly, cells were characterized mainly on the basis of morphological characteristics identified by light microscopy. Electron microscopy showed that cells which had seemed to be similar were really different. Other differences were revealed by (immuno)cytochemical methods. The determination of membrane characteristics (e.g., presence of specific antigens or receptors) and functional features of cells have made it possible to define the mononuclear phagocytes more precisely. Criteria for the characterization of a cell as a mononuclear phagocyte are summarized in Table 1 (1).

On the basis of these criteria, mononuclear phagocytes can be identified in the bone marrow, peripheral blood, and tissues. These cells are called monoblasts, promonocytes, monocytes, and macrophages. The monoblasts and promonocytes are the dividing cells (see below) that reside in the bone marrow, the monocytes are non-dividing cells that are formed in the bone marrow and transported via the circulation to the tissues, where they become macrophages. The macrophages are

175

Table 1. Criteria for the Characterization of a Cell as a
          Mononuclear Phagocyte

Morphological characteristics

Cytochemical characteristics
    non-specific esterase
    peroxidase-positive or -negative granules
    lysozyme
    5'nucleotidase
    aminopeptidase

Membrane characteristics
    Fcγ receptors
    C3b receptors
    specific antigens

Functional characteristics
    phagocytosis
        opsonized bacteria
        IgG-coated red cells
        IgMC-coated red cells
        latex beads
    pinocytosis

---

the most mature cells of this cell line. According to the tissue
where they are located, macrophages have different names, for instance
histiocyte in the connective tissue, Kupffer cell in the liver,
alveolar macrophage in the lung, pleural and peritoneal macrophage
in the serous cavities, type A cell in the synovial lining, and
osteoclast in bone tissue. All of these cell types belong to the
same cell line, called Mononuclear Phagocyte System (MPS)- (1,2).

If the criteria listed in Table 1 are applied to the cells of
the MPS, it is clear that these characteristics help to identify
mononuclear phagocytes. An example of such characterization is given
for murine mononuclear phagocytes in Table 2.

IN VIVO PROLIFERATION OF MONOBLASTS AND PROMONOCYTES

The bone marrow origin of the monocyte was firmly established
many years ago (3), and its direct precursors, i.e., the promonocyte
and the monoblast, were characterized more recently in murine and

Table 2. Functional and Cytochemical Characteristics of Murine Mononuclear Phagocytes

	Monoblasts[a] %	Promonocytes[b] %	Bone marrow Monocytes[b] %	Resident peritoneal macrophages[b] %	Cell line J774 %
Cytochemical characteristics					
Esterase	91	91	97	99	100
Peroxidase	43	93	87	0	0
Lysozyme	0	100	100	100	100
5' nucleotidase	1	ND	ND	69	87
Aminopeptidase		ND	ND	14	0
Membrane receptors					
Fc receptor	94	93	93	100	97
C receptor	16	80	80	100	100
Phagocytosis					
Opsonized bacteria	30	77	74	98	49
IgG-coated red cells	96	93	93	91	59
IgM-coated red cells	0	23	23	13	90
Pinocytosis	21	91	91	99	99

a 96 hour culture
b 24 hour culture

human bone marrow (4-10). Morphological, cytochemical, and functional studies have shown that murine monoblasts and promonocytes
have many characteristics in common with monocytes (Table 2).

   In vivo labeling studies have shown that under steady-state
conditions the cell-cycle time of the promonocyte is 16.2 hours (11).
During an inflammatory response, however, the cell-cycle time of
these cells is initially shorter, mainly due to reduction of the
DNA-synthesis time (Table 3). Later this value returns to the steady-
state level as the number of promonocytes in the bone marrow increases
(11). Under these conditions, the production of monocytes during
an inflammation can increase rapidly and remain at the higher level
for some time.

   The proliferative behavior of monoblasts cannot be studied in
vivo, because the techniques now available do not permit identification of these cells in smears or short-term cultures of bone marrow
and we therefore work provisionally with the cell-cycle times obtained
in vitro (Table 3) (7). The estimated number of monoblasts in bone
marrow is 1 per 10^3 nucleated cells (8), which is about half the
number found for promonocytes (i.e., per mouse 5×10^5 per 2.5×10^8
nucleated bone marrow cells (11)). On the basis of these data, the
most likely model of division in vivo is as follows (Figure 1): the
monoblast derives from a (committed) stem cell which divides once
and gives rise to two monocytes. Thus, from monoblast to monocyte
there is a four-fold increase in the number of cells.

   Monocytes do not divide (3,8), and leave the bone marrow randomly
within 24 hours. These cells remain relatively long in the circulation (mouse: half life 17.4 hours; man: half life 71.0 hours)
compared with granulocytes. They leave this compartment randomly
(12), and become macrophages in tissues and serous cavities (1,2).
The distribution of the monocytes to the various tissues and cavities
has been calculated from data collected in a number of studies done
in rodents: at any given time, about 56% of the monocytes migrate
to the liver (13), about 16% to the lungs (14), about 8% to the
peritoneal cavity (3), and about 5% to the intestines (15).

   Tissue macrophages obtained from non-inflamed sites are non-
dividing cells. The small percentage (less than 5%) of macrophages
that label in vitro with ^{3}H-thymidine are mononuclear phagocytes
that have very recently (less than 24-48 hours before harvesting)
arrived in the tissue from the bone marrow and do not belong to the
resident population of macrophages. Most probably these cells divide
only once. A schematic representation of the bone marrow origin of
tissue macrophages and the dividing mononuclear phagocytes in the
tissues, which are immature cells arising from the bone marrow, is
given in Figure 2. We developed a mathematical approach to the analysis of macrophage kinetics in a tissue compartment, taking into
account the influx of monocytes, the local division of mononuclear

Table 3. Cell-cycle Times of Murine Monoblasts and Promocytes, and the Macrophage Cell Line J774

| Measurement | Monoblast | Promonocyte | | | Macrophage Cell Line J774 |
| | | Steady State | Inflammation | | |
			at 12 hr	at 24 hr	
Cell-cycle time	11.9	16.2	10.8	16.7	18.5
DNA-synthesis time	5.7	11.8	7.6	12.8	9.5

Data are reported in hours. Monoblasts were determined on day 4 in bone marrow cultures (7) while promonocytes were determined *in vivo* (11).

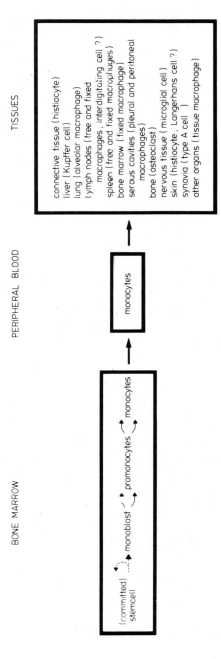

Fig. 1. Cells included in the Mononuclear Phagocyte System. These cells occur in the bone marrow, circulation, various tissues, and serous cavities.

phagocytes, and the efflux of macrophages. When this approach was
applied to the data on pulmonary macrophages from mice in the normal
steady state, the calculations showed that about 75% of this macrophage population is maintained by monocyte influx and about 25% by
a single local division of immature mononuclear phagocytes originating
from the bone marrow (16).

IN VITRO PROLIFERATION OF MONOBLASTS AND PROMONOCYTES

     Bone marrow cells cultured in the presence of colony-stimulating
factor proliferate in vitro (6,17). In principle, this property
makes the investigation of precursor cells possible, but in practice
the substrate, i.e., agar or methyl cellulose, used for cell support
hampers the study of the characteristics of these cells. Therefore,
the original method was modified to permit culture of the leukocyte
colonies in a liquid medium (6,18). In these cultures mononuclear
phagocytes adhere to the glass surface of a coverslip, which allows
study of their morphology, cytochemistry, function, and proliferation
(6,7,19).

     In these liquid bone marrow cultures two kinds of colonies are
found, i.e., mononuclear phagocyte colonies and granulocyte colonies;
mixed colonies are never observed (6,19). Granulocyte colonies can
be recognized easily, because they grow in a very compact form.
They are in the minority during the first 7 days of culture, and
then gradually disappear. After 10 days of culture, only mononuclear
phagocyte colonies remain (19). The mononuclear phagocyte colonies
grow in a single layer with some crowding of rounded cells in the
center and stretched cells in the periphery. Colony formation does
not occur when cells are cultured on a Teflon surface (20,21). The
rate of proliferation of mononuclear phagocytes is very high initially -- between day 0 and day 14 the progeny of one monoblast
amounts to 5×10^3 mononuclear phagocytes -- but levels off after
two weeks of culture (19). After 3 or 4 weeks of culture without
replating, the cells round up, the nucleus becomes pycnotic, and the
cells detach from the glass (19). Replating of the cells made it
possible to maintain proliferation for more than 100 days. Calculation showed that in these cultures one monoblast had a progeny of
more than 1×10^5 mononuclear phagocytes by day 100 (19).

     Three types of cells can be distinguished in the mononuclear
phagocyte colonies: the monoblast, the promonocyte, and the macrophage (6,19), whose relative numbers on the various days of incubation
are given in Table 4. The number of monoblasts remains rather constant, whereas the number of promonocytes and macrophages increases
strongly during the culture period. The monoblasts and promonocytes
are the most actively proliferating cells; however, within the population of macrophages there is also some proliferation. These macrophages were tentatively called young macrophages.

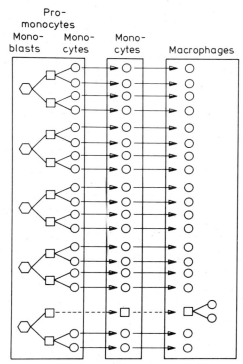

Fig. 2.   Current concept of the origin and kinetics of mononuclear
           phagocytes. Monoblasts and promonocytes divide in the bone
           marrow, are transported via the circulation to the tissues,
           where they become macrophages. The small percentage of
           mononuclear phagocytes that divide in the tissues, and have
           recently originated in the bone marrow, are also included.

       The cell-cycle times of monoblasts and promonocytes in these
liquid cultures were determined using four independent methods (7).
The results were 11.9 hours for the monoblast (Table 3) and 11.4
hours for the promonocyte. The cell-cycle time of the promonocytes
in vitro is shorter than that obtained under steady-state conditions
in vivo, but is about equal to the value found during the early phase
of an inflammatory response (Table 3).

MACROPHAGE CELL LINES

       Macrophage cell lines are transformed mononuclear phagocytes

Table 4.   Total Number of Monoblasts, Promonocytes and Macrophages
in Liquid Bone Marrow Cultures

Day	Monoblasts x 10^3	Promonocytes x 10^3	Monocytes and Macrophages x 10^3	Total x 10^3
0	0.001			0.001
4	0.03	0.03	0.07	0.13
7	0.2	0.4	1.1	1.7
10	0.2	1.1	4.0	5.3
14	0.4	2.0	5.3	7.7
17	0.3	2.5	5.9	8.7
21	0.3	3.8	8.9	13.0

Values are means of normalized data of 4 experiments, each done in
triplicate (19); the initial number of monoblasts is taken as one
cell at day 0.

that proliferate continuously in vitro. This raises the question
as to which stage of development and state of activation these dividing
mononuclear phagocytes represent. In other words, are these cells
in one respect comparable to monoblasts or promonocytes and in other
respects similar to resident or activated macrophages? This point
cannot be settled before the cells in question have been characterized
according to the criteria shown in Table 1.

Such a characterization was carried out for cell line J774, and
showed the following morphological characteristics. The cells are
round to oval and lack pseudopods; the nucleus is round without inden-
tations, and the nucleus-to-cytoplasm ratio is about 1. These data
suggest a monocyte precursor. The other characteristics are summarized in Table 2. With respect to cytochemical characteristics (e.g.,
esterase and lysozyme) these cells are true mononuclear phagocytes,
and their high 5'nucleotidase and low aminopeptidase activity indicates that these cells are not activated and correspond with resident

macrophages. With respect to the presence of membrane receptors,
these cells are true mononuclear phagocytes. The level of phagocytosis of opsonized bacteria or red cells is rather low, whereas
complement-coated red cells are endocytized by the great majority
of cells. In relation to the latter, the cells are in an activated
state (22). The 1 hour labeling index amounts to 33%, which indicates
that these cells proliferate. Further studies on the proliferative
behavior of the J774 cells were performed with three methods: determination of the course of ^{3}H-thymidine-labeled mitosis, determination
of the labeling index in the presence of continuous labeling with
^{3}H-thymidine, and determination of the stathmokinetic index under
cholchicine or vinblastine. Preliminary analysis of these results
(performed together with T. Rytömaa and H. Toivonen, Helsinki) showed
that the cell-cycle time of this line amounts to 18.5 hours. However,
the doubling time of cell line J774 amounts to 27.5 hours. The
discrepancy between these values can only be explained by the death of
an appreciable proportion of the cells during proliferation. It is
of interest that the cell-cycle time of these cells is about similar
to the cell-cycle time of promonocytes in the steady state. In that
respect the cells of this cell line resemble monocyte precursors.

SUMMARY

     Mononuclear phagocytes are localized in the bone marrow compart-
ment (monoblasts, promonocytes and macrophages), the circulation
(monocytes), and the tissues and serous cavities (macrophages).

     In vivo and in vitro studies done in murine bone marrow have
shown that monoblasts and promonocytes are the most immature, dividing
cells of the mononuclear phagocyte cell line; monocytes and resident
macrophages do not divide.

     A very small percentage of the mononuclear phagocytes in the
tissues, which are bone-marrow derived and have already the morphology
of macrophages, divide once in vivo. The progeny of these dividing
cells contribute to the maintenance of the tissue population of macrophages under steady-state conditions. In vitro an appreciable percentage of (young) macrophages divide, in all probability due to the
influence of colony-stimulating factor. The cells of macrophage cell
lines are transformed cells that proliferate continuously.

     The morphological, cytochemical and functional characteristics
of all these kinds of cells, as well as their proliferative behavior
in vivo and in vitro, show great similarity, although there are also
distinct differences.

REFERENCES

1. Furth, R. van, in "Mononuclear Phagocytes. Functional Aspects"
    (R. van Furth, ed.) p. 1, Martinus Nijhoff Publishers, The
    Hague, 1980.
2. Furth, R. van, Cohn, Z. A., Hirsch, J. G., Humphrey, J. H.,
    Spector, W. G., and Langevoort, H. L., Bull. WHO 46:845, 1972.
3. Furth, R. van, and Cohn, Z. A., J. Exp. Med. 128:415, 1968.
4. Furth, R. van, and Diesselhoff-den Dulk, M. M. C., J. Exp. Med.
    132:813, 1970.
5. Furth, R. van, Hirsch, J. G., and Fedorko, M. E., J. Exp. Med.
    132:794, 1970.
6. Goud, Th. J. L. M., Schotte, C., and Furth, R. van., J. Exp. Med.
    142:1180, 1975.
7. Goud, Th. J. L. M., and Furth, R. van, J. Exp. Med. 142:1200,
    1975.
8. Furth, R. van, Raeburn, J. A., and Zwet, Th. L. van, Blood 54:
    485, 1979.
9. Meer, J. W. M. van der, Beelen, R. H. J., Fluitsma, D. M., and
    Furth, R. van, J. Exp. Med. 149:17, 1979.
10. Meer, J. W. M. van der, Gevel, J. S. van de, Beelen, R. H. J.,
    Fluitsma, D. M., and Furth, R. van, submitted for publication.
11. Furth, R. van, Diesselhoff-den Dulk, M. M. C., and Mattie, H.,
    J. Exp. Med. 138:1314, 1973.
12. Furth, R. van, Diesselhoff-den Dulk, M. M. C., Raeburn, J. A.,
    Zwet, Th. L. van, Crofton, R. W., and Blussé van Oud Alblas,
    A., in "Mononuclear Phagocytes. Functional Aspects" (R. van
    Furth, ed.), p. 279, Martinus Nijhoff Publishers, The Hague,
    1980.
13. Crofton, R. W., Diesselhoff-den Dulk, M. M. C., and Furth, R.
    van, J. Exp. Med. 148:1, 1978.
14. Blussé van Oud Alblas, A., and Furth, R. van, J. Exp. Med. 149:
    1504, 1979.
15. MacPherson, G. G., and Steer, H. W., in "Mononuclear Phagocytes.
    Functional Aspects" (R. van Furth, ed.), p. 299, Martinus
    Nijhoff Publishers, The Hague, 1980.
16. Blussé van Oud Alblas, A., Mattie, H., and Furth, R. van, sub-
    mitted for publication.
17. Meer, J. W. M. van der, in "The Reticuloendothelial System. A
    Comprehensive Treatise I. Morphology" (I. Carr, and W. Th.
    Daems, eds.), p. 735, Plenum Press, New York, London, 1980.
18. Meer, J. W. M., van der, Gevel, J. S. van de, and Furth, R. van,
    in "Methods for Studying Mononuclear Phagocytes" (P. J.
    Edelson, H. Koren, and D. O. Adams, eds.), p. 121, Academic
    Press, New York, 1981.
19. Meer, J. W. M. van der, Gevel, J. S. van de, and Furth, R. van,
    submitted for publication.

20. Meer, J. W. M. van der, Bulterman, D., Zwet, Th. L. van, Elzenga-
         Claasen, I., and Furth, R. van, J. Exp. Med. 147:271, 1978.
21. Meer, J. W. M. van der, Gevel, J. S. van de, Elizenga-Claasen,
         I., and Furth, R. van, Cell. Immunol. 42:208, 1979.
22. Bianco, C., Griffin, F. M. Jr., and Silverstein, S. C., J. Exp.
         Med. 141:1278, 1975.

DISCUSSION

METCALF: I'd like to ask van Furth what he thinks is the difference
between his monoblasts in vivo that generate four progeny and his
monoblasts in culture which suddenly become stem cells capable of
generating infinite numbers of monocytes? Is the in vitro system
abnormal or is the in vivo situation limited by the steady state?

VAN FURTH: This puzzled us too and we thought initially that the
monoblasts in vitro would resemble monoblasts in vivo. From all
calculations, they did not. I think that you are right in suggesting
that in vivo we have steady state conditions whereas in vitro the
cells are responding to growth factors in the conditioned media and
continue to divide as stem cells. The same analysis applies to the
macrophage. Macrophages divide in vitro but do so to only a limited
extent in vivo. I think the answer lies in the nature of the stimulus
provided by the growth factors in conditioned media.

MILLER: Would van Furth elaborate on his statement that some alveolar
macrophages arise from in situ precursors capable of division?

VAN FURTH: Under steady state conditions, there is a small percentage
of cells which will incorporate tritiated thymidine by pulse labeling.
These cells divide only once and are recruited from the bone marrow.
In non-steady state conditions, such as after BCG injection intravenously or by aerosol, this percentage of labeled cells will increase. However, the increase is temporary and will return to normal
levels shortly after the inflammatory response subsides.

UNKNOWN: Are monoblasts capable of moving directly to the tissues
and dividing there?

VAN FURTH: We know that some sort of precursor cell enters the tissues from the circulation and is presumably derived from the bone
marrow. Until we have good markers to differentiate between monocytes, promonocytes, and monoblasts - such as monoclonal antibodies -
we can't determine if these cells are promonocytes or monoblasts.
However, we can conclude that these cells are recruited into the

tissues because with corticosteroids and other situations in which you get a monocytopenia, they do not arrive in the tissues.

VALENTINE: How many dividing cells are present in human peripheral blood?

VAN FURTH: The number of dividing cells in the circulation is very small - maybe 1% or lower both in rodents and in humans.

PROPERTIES OF THE FACTOR INCREASING MONO-

CYTOPOIESIS (FIM) IN RABBIT PERIPHERAL BLOOD

W. Sluiter, I. Elzenga-Claasen, E. Hulsing-Hesselink
and R. van Furth

Department of Infectious Diseases
University Hospital
Leiden
The Netherlands

INTRODUCTION

Monocyte production in mice is stimulated during the initial
phase of acute inflammation by a humoral factor called the factor
increasing monocytopoiesis (FIM) (1). This factor, a protein of
about 20,000 daltons without detectable carbohydrates at its active
site, induces monocytosis by stimulating proliferation of promonocytes
in the bone marrow (2). The number of granulocytes and lymphocytes
is not affected.

The monocytosis-inducing activity can be assayed in vivo by
intravenous injection of FIM into normal animals and determining
the increase in the number of blood monocytes. For this assay, the
animals must be in the steady state. The latter is characterized
by a fairly constant number of monocytes during a certain period
before the experiments are carried out and by a reproducible increase
in the number of blood monocytes after induction of an experimental
inflammation (3).

The occurrence of FIM is not restricted to mice, since indications exist of its presence in rabbits (4). The aim of the present
study was to determine the chemical and biological characteristics
of this factor as found in the peripheral blood of rabbits.

RESULTS AND DISCUSSION

The Rabbit Model

We first determined if it was possible to select male Chinchilla rabbits in the steady state condition from a conventionally bred and housed population. Accordingly, blood monocyte levels were monitored during five consecutive days in a number of animals. When the number of monocytes on the first and second days fell within one standard deviation of the mean of the total population, the number of blood monocytes remained fairly constant during the next three days (5). Since blood monocytes could be followed in the same animal, a paired design could be used to test the significance of a difference between two sample means. This approach greatly decreased the effect of variation.

In rabbits fulfilling these criteria, an acute inflammation was evoked by the injection of polystyrene latex particles into the peritoneal cavity. After injection, the number of macrophages in the peritoneal exudate increased, reached a maximum between 18 and 48 hours, and then decreased to a normal level (4). Concomittantly, monocytosis developed in the peripheral blood, with a maximum at 48 hours. It was concluded that such rabbits fulfilled the conditions for studies on the mechanisms regulating monocyte production during inflammation.

If monocyte production and release during an inflammatory reaction is regulated in rabbits by a mechanism similar to that found in mice (1), the factor increasing monocytopoiesis (FIM) would be expected to appear in the circulation during the initial phase of inflammation. Accordingly, the monocytosis-inducing activity of rabbit plasma and serum were tested at various time-points during the inflammatory reaction (4,5).

Effect of Rabbit FIM on Leucocyte Production

Monocytosis could not be induced in rabbits with normal rabbit plasma or serum. As early as 12 hr after an intraperitoneal injection of latex, however, monocytosis-inducing activity was demonstrable in the plasma and serum. This activity increased rapidly to a peak at 18 hours, then subsided, and was no longer detectable from 48 hours. The numbers of blood granulocytes and lymphocytes remained in the normal range.

The effect on monocyte production in the bone marrow, as expressed by the number of labeled blood monocytes, was studied after intravenous injection of ^{3}H-thymidine simultaneously with active serum collected 18 hr after intraperitoneal injection of latex or normal rabbit serum. The labeling index of blood monocytes after injection of serum from animals with inflammation was about 1.6 times higher than the control value. The labeling index of blood granulocytes and lymphocytes was unaffected. On the basis of these results, it was concluded that rabbit serum obtained during an inflammatory response contained a factor (rabbit FIM) that regulates the number of circulating monocytes by increasing specifically the production of monocytes in the bone marrow.

Characteristics of Rabbit FIM

The characteristics of rabbit FIM are summarized in Table 1. A linear relationship between the administered dose of FIM and the resulting monocytosis was found. However, the slopes of the regression lines formed between monocytosis and concentration of serum and plasma differed significantly, possibly due to interference by some plasma factor. The FIM activity was readily abolished in plasma during incubation in vitro at 37°C. This loss of activity might be due to the action of proteolytic enzymes, since the addition of soybean trypsin inhibitor, a serine protease inhibitor, stabilized the factor considerably.

The molecular weight of rabbit FIM lies between 10,000 and 25,000 daltons, as determined with ultrafiltration membranes. Rabbit FIM was not sensitive to treatment with a mixture of glycosidases, but after addition of proteases the FIM activity of plasma was rapidly lost. These observations indicate that the active site of FIM is protein in nature and lacks carbohydrate moieties.

Table 1. Characteristics of the Factor Increasing Monocytopoiesis (FIM) from Rabbits

Biological activity	specific stimulation of monocyte production; no effect on granulocyte and lymphocyte production
Dose – effect relationship	positive, linear
Stability in vitro (37°C)	less than 15 min
Molecular weight	between 10,000 and 25,000 daltons
Chemical nature	protein, no carbohydrates at its active site
General aspects	no relation to complement factors, clotting factors, or colony-stimulating factor

In <u>vitro</u> activation of complement in normal rabbit serum with zymosan did not generate monocytosis inducing activity. This observation makes it unlikely that rabbit FIM is an intermediate product formed during complement activation.

Since both plasma and serum of rabbits given an intraperitoneal injection of latex have monocytosis-inducing activity, it is unlikely that this activity is generated during the clotting process. Rabbit sera containing FIM could not be substituted for colony-stimulating factor (CSF) in the bone marrow colony assay (6), which indicates that rabbit FIM is not identical with CSF.

Comparison of rabbit FIM with murine FIM (2) showed a striking resemblance. Both factors seem to lack species specificity, since it was possible to evoke moderate monocytosis in mice with rabbit FIM (5) and moderate monocytosis in rabbits with murine FIM.

It was concluded that in rabbits and mice a similar mechanism regulates monocytopoiesis during an acute inflammation.

Acknowledgement

This study was supported by the Foundation for Medical Research FUNGO, which is subsidized by the Netherlands Organization for the Advancement of Pure Research (ZWO).

The authors are greatly indebted to Mr. Th. Stijnen for the computer analysis of the data.

REFERENCES

1. Waarde, D. van, Hulsing-Hesselink, E., Sandkuyl, L. A., and Furth, R. van, <u>Blood</u> 50:141, 1977.
2. Waarde, D. van, Hulsing-Hesselink, E., and Furth, R. van, <u>Blood</u> 50:727, 1977.
3. Waarde, D. van, Bakker, S., Vliet, J. van, Angulo, A. F., and Furth, R. van, <u>J. Reticuloendothelial Soc</u>. 24:197, 1978.
4. Sluiter, W., Waarde, D. van, Hulsing-Hesselink, E., Elzenga-Claasen, I., and Furth, R. van, <u>in</u> "Mononuclear Phagocytes-Functional Aspects" (R. van Furth, ed.), p. 325, Martinus Nijhoff Publishers, The Hague, Boston, London, 1980.
5. Sluiter, W., Elzenga-Claasen, I., Hulsing-Hesselink, E., and Furth, R. van, submitted for publication.
6. Worton, R. G., McCullock, E. A., Till, J. E., <u>J. Cell. Physiol</u>. 74:171, 1969.

DISCUSSION

METCALF: By what crystal ball did you decide that the macrophage was the cell of origin of your monocytopoiesis factor?

SLUITER: FIM is released by peritoneal macrophages after stimulation with latex beads. We found no indication that normal macrophages release this factor spontaneously.

METCALF: How many cell types did you test?

SLUITER: To date, we have tested granulocytes and lymphocytes and neither cell produced the factor.

WEINER: Does Dr. Sluiter have any data or work in progress concerning the identification of FIM by a monoclonal antibody? Such antibodies could be used to study the function of FIM by either eliminating or blocking its activity.

SLUITER: The production of monoclonal antibodies against FIM is in progress.

FRIEDLANDER: Since the serum half life of FIM is only 20 minutes, is it possible to assay its colony stimulating activity in vitro either on marrow cells or on exudate cells? CSF has to be present continuously to be effective.

SLUITER: I agree that colony stimulating factor has to be present continuously to stimulate colony formation. However, we can stabilize our factor with soybean trypsin inhibitor, but even then we do not find colonies produced in vitro or an enhancement of colony production with optimal concentrations of CSF. I might add that colony stimulating factor is known to be degraded at 65°C whereas FIM is degraded at 37°C. This indicates some differences between the factors.

FRIEDLANDER: Does it work on exudate macrophages?

SLUITER: It is produced by resident peritoneal macrophages and it exerts its action on promonocytes and probably monoblasts. I didn't check if it had any effect on peritoneal exudate cells.

FRIEDLANDER: Have you used anti-CSF1 antibody to see if it will block the effect of FIM?

SLUITER: No.

MONOCYTE PRODUCTION AND KINETICS IN RESPONSE TO LISTERIOSIS IN RESISTANT AND SUSCEPTIBLE MURINE HOSTS

P. A. L. Kongshavn,[1] C. Punjabi (née Sadarangani),[1] and S. Galsworthy[2]

[1]Department of Physiology and Montreal General Hospital Research Institute, McGill University, Montreal, Quebec
[2]Department of Microbiology and Immunology, University of Western Ontario, London, Ontario, Canada

INTRODUCTION

Infection of mice with the facultative, intracellular, bacterial parasite, Listeria monocytogenes, has been established as a model of an acute bacterial infection in which host resistance is brought about by a cellular form of immunity (1). From our investigations (4-6) and those of others (5,6) it is evident that host resistance to this infection is controlled genetically by a single, autosomal, dominant, non-H-2-linked gene designated Lr (for Listeria resistance) (6). Mouse strains bearing the resistant (Lr^r) allele (C57BL/6 and substrains; NZB, SJL) are able to control bacterial proliferation following a challenge inoculum which is lethal to strains bearing the susceptible (Lr^s) allele (A, BALB/c, DBA/2, CBA) (2,4,6). This effect is seen during the early (innate) phase of the response, before the development of acquired (T-cell mediated) immunity. The Lr gene is expressed in the response of the mononuclear phagocyte system rather than that of the T-cell (7).

In earlier findings, we obtained indirect evidence that resistance in Lr^r mouse strains was associated with a prompt accumulation at infective foci of young (radiosensitive) mononuclear phagocytes, which provided the crucial anti-bacterial protection (8). More recently, we have shown that the macrophage inflammatory response (for example, to sterile irritants introduced into the peritoneal cavity) is controlled by a trait which is linked genetically to Lr (9) and which, therefore, is probably an expression of the same gene. The Lr gene product is apparent not an autonomous property of the macrophage per se (10). Rather, it seems to influence the environ-

ment in which the macrophages differentiate from their bone marrow
precursors. In order to investigate where, in the macrophage differentiation pathway, the Lr gene product might be exerting its influence, we have compared monocyte production and kinetics in Lr^r and
Lr^s mice following infection with Listeria. As well, we measured
these parameters after stimulation with an aqueous extract of listerial cell wall, which contains a monocytosis-producing activity
(11,12).

MATERIALS AND METHODS

     A (Lr^s) and B10.A (Lr^r) strain mice used in these experiments
were bred in one laboratory. Bacteria, infection of mice and bacterial enumeration have all been described previously (1). Extracts
of L. monocytogenes containing monocytosis promoting activity were
prepared as described previously (11) and the saline-extractable
material used. Cytokinetic determinations were carried out as described previously (12,13) unless stated otherwise. The generation
time (T_G) of promonocytes was determined using the method of van
Furth and Diesselhoff-den-Dulk (14). The half-time ($T_{1/2}$) of monocytes in the blood was estimated using the method of Volkman and
Collins (15). For T_G determinations, cells with four or more grains
over the nucleus were considered to be labeled. For $T_{1/2}$ determinations, the cohort of cells with 20-50 grains per cell was considered
to be labeled. At each point, at least 200 promonocytes were scored
from triplicate bone marrow cell cultures from each of four mice.
Labeling   on peripheral blood monocytes were ascertained from
at least 100 monocytes from triplicate blood smears of each of four
animals at any time point.

RESULTS

Promonocyte Generation Time Following Listerial Infection or Treatment with Listerial Extract

     Groups of A (Lr^s) and B10.A (Lr^r) strain mice were infected with
10^4 colony forming units (CFU) L. monocytogenes or injected with
phosphate buffered saline (PBS) as a control and the average generation time (T_G) of promonocytes estimated by autoradiographic determination over the first few days of infection (14). Other groups
of mice were injected intraperitoneally with monocytosis promoting
extracts (1 mg in PBS) or PBS alone, and the average T_G of promonocytes determined 8 hours later using the same method.

     Under steady state conditions, there was no significant dif-
ference in the T_G of promonocytes between the two strains. However,
following infection with L. monocytogenes, T_G shortened significantly
in B10.A strain mice whereas no change was seen for at least two days

in A strain mice (Figure 1). In a similar fashion, 8 hours following
treatment with extract, T_G fell from 17.2 to 13.8 hours in B10.A
strain mice whereas the values were 18.8 and 19.1 hours respectively
in A strain mice (PBS control values were 17.3 hours (B10.A) and 18.8
hours (A)). P was <0.025 for B10.A extract-treated vs all other
groups.

Half-Time of Blood Monocytes Following Listerial Infection or Treatment with Listerial Extract

Mice were either infected with 10^4 CFU Listeria or treated with
monocytosis promoting extracts, or given PBS as a control, and the
half-time ($T_{1/2}$) of blood monocytes estimated by autoradiographic
determination (Table 1). It was shown from analysis of the regression
line fitted by the method of least squares that the $T_{1/2}$ of blood
monocytes in B10.A ($\underline{Lr^r}$) strain mice was significantly lower following
infection with Listeria (P<0.05) or after treatment with MPA listerial
extracts (P <0.05). No such effect was seen in A ($\underline{Lr^s}$) strain mice.

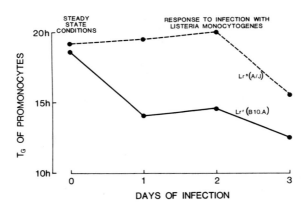

Fig. 1.   Promonocyte generation time (T_G) in susceptible and resistant mouse strains to infection with Listeria monocytogenes.
Resident B10.A strain mice (solid line) and susceptible
A/J mice (dashed line) each received 10^4 CFU of Listeria.
Values for control groups of mice receiving phosphate buffered saline (not shown) did not differ significantly from
non-treated values. At each time point, mice were injected
with tritiated thymidine either once or twice within a 2
hour interval. Bone marrow was harvested 2 hours after the
final injection, cultured for 6 hours, and subjected to autoradiography. T_G was calculated as described (14). P value
is <0.025 for B10.A infected vs. B10.A control and A infected
groups, for each time point.

Table 1.   Half-Time $(T_{1/2})$ of Blood Monocytes following
           Infection with Listeria or Treatment with Extracts
           Containing Monocytosis Promoting Activity (MPA)

Mouse Strain	Treatment[a]	Half-Time[b] (hours)
B10.A	Non-Infected	30.0 (27.7 - 32.6)
B10.A	Listeria-Infected	21.1 (19.6 - 23.4)
A	Non-Infected	32.0 (28.6 - 34.7)
A	Listeria-Infected	30.4 (24.1 - 40.6)
B10.A	PBS-Treated	27.4 (26.0 - 30.2)
B10.A	MPA-Treated	19.2 (18.1 - 20.3)
A	PBS-Treated	28.9 (26.8 - 31.4)
A	MPA-Treated	28.2 (26.0 - 30.3)

[a]Mice were infected intravenously with 10^4 CFU Listeria. Non-infected mice received an equal volume of phosphate buffered saline (PBS). MPA-treated mice received an intraperitoneal injection of 1 mg MPA in 0.2 ml PBS; mice in the control group received an equal volume of PBS.
[b]Mice were given a single injection of (^{3}H)TdR 8 h after injection with MPA or 24 h after infection. Mice were bled at intervals after injection of isotope. Blood smears were fixed, saved until the end of the experiment, and subjected to autoradiography. The labeling index of blood monocytes at each time point was determined, and the half-time calculated as described (15). Values in parentheses represent 95% confidence limits.

CONCLUSIONS

The above cytokinetic determinations show that within hours following infection with Listeria or treatment with listerial cell wall extract monocyte production is increased in resistant, but not in susceptible, mice. As well, monocytes leave the circulation more rapidly in resistant hosts, as measured by $T_{1/2}$, and this presumably reflects an increased accumulation of young, bactericidal macrophages in infective foci. The Lr gene thus appears to be controlling the production and emigration of mononuclear phagocytes shortly following infection, this in turn determining the degree of susceptibility of the host to listeriosis.

Acknowledgement

This work was supported by MRC grants # 5448 and 5090.

REFERENCES

1. Mackaness, G. B., J. Exp. Med. 116:381, 1962.
2. Skamene, E., and Kongshavn, P. A. L., Infect. Immun. 25:345, 1979.
3. Skamene, E., Kongshavn, P. A. L., and Sachs, D., J. Infect. Dis. 139:228, 1979.
4. Kongshavn, P. A. L., Sadarangani, C., and Skamene, E., in "Genetic Control of Natural Resistance to Infection and Malignancy" (E. Skamene, P. A. L. Kongshavn, and M. Landy, eds.), pp. 149-163, Academic Press, New York, 1980.
5. Robson, H. G., and Vas, S. I., J. Infect. Dis. 126:378, 1972.
6. Cheers, C., and McKenzie, I. F. C., Infect. Immun. 19:755, 1978.
7. Kongshavn, P. A. L., and Skamene, E., J. Reticuloendothelial Soc. 24:51a, 1978.
8. Sadarangani, C., Skamene, E., and Kongshavn, P. A. L., Infect. Immun. 28:381, 1980.
9. Stevenson, M. M., Kongshavn, P. A. L., and Skamene, E., J. Immunol. 127:402, 1981.
10. Kongshavn, P. A. L., Sadarangani, C., and Skamene, E., Cell. Immunol. 53:341, 1980.
11. Galsworthy, S. B., Gurofsky, S. M., and Murray, R. G. E., Infect. Immun. 15:500, 1977.
12. Shum, D. T., and Galsworthy, S. B., Can. J. Microbiol. 25:698, 1979.
13. Otokunefar, T. V., Shum, D.T., and Galsworthy, S. B., Can. J. Microbiol. 25:706, 1979.
14. van Furth, R., and Diesselhoff-den-Dulk, M. M. C., J. Exp. Med. 132:813, 1970.
15. Volkman, A., and Collins, F. M., J. Exp. Med. 139:264, 1974.

DISCUSSION

MACPHERSON: Are the differences in monocyte production between the two mouse strains specific for Listeria?

KONGSHAVN: They are probably not specific for Listeria.   We have recently identified a gene which controls inflammatory responsiveness in general and in our example it manifests itself in enhanced Listeria resistance.   The crucial point is that this gene allows the host to mobilize the macrophages to the site of infection as rapidly as possible.

VAN FURTH: What type of chimera did you use and at what time were your experiments conducted?

KONGSHAVN: They were bone marrow chimeras examined after 9 weeks.

MG-1, IDENTIFICATION OF AN EARLY
DIFFERENTIATION MARKER FOR BONE MARROW
CELLS IN A MONOCYTIC/GRANULOCYTIC LINEAGE

B. J. Mathieson,[1] F.-W. Shen,[2] A. Brooks,[3] and
W. M. Leiserson[3]

[1]Basel Institute for Immunology, Basel, Switzerland
[2]Memorial Sloan-Kettering Cancer Center, New York, New
York USA
[3]National Institute of Allergy and Infectious Diseases
Bethesda, Maryland USA

Several reagents have been developed for use in different species
that reportedly detect macrophages and subsets of bone marrow cells
such as monocytes and granulocytes (1-5). At least two of these
reagents are known to detect both mature macrophages and a subset of
bone marrow cells or myelocytic cells of mice (2,5). One of these
reagents, 2.4G2, detects Fc receptors (FcR) on several different
types of cells in mice (5). The other reagent defines another specificity known as Mac-1 that is different from FcR by molecular weight
and tissue distribution. With the exception of the alloantiserum
to Mph-1 (1), the other reagents (monoclonal antibodies) have been
produced across xenotropic barriers and do not recognize genetic
strain differences (2-5).

This report will describe experiments that identify and characterize a subpopulation of bone marrow cells detected with a monoclonal
antibody which we will tentatively call anti-MG-1. This reagent
apparently detects neither FcR nor the same cell populations as those
defined by the reagent Mac-1. Furthermore, the strain distribution
is not colinear with that previously reported for Mph-1 antigen (1).

MATERIALS AND METHODS

Mice used for these studies were obtained from the Jackson Laboratories, Bar Harbor, ME or the Animal Production Unit, NIH, Bethesda,

MD.   The source is indicated by J or N.   B6-Ly congenic mice were
produced by Corbel Laboratories, Rockville, MD on contract to NIAID.

Cell suspensions were obtained by washing bone marrow cells from
the marrow cavities of femora and tibiae of 2-3 mice with Hanks'
balanced salt solution (HBSS).   Cells were further suspended by
pipetting and then filtered through nylon mesh to remove cell aggregates.

Bone marrow cells were separated by unit gravity velocity sedimentation as described elsewhere (6).   In each separation, 1×10^8
cells were separated for a total of 4 hours in a 20cm diameter chamber
with a sucrose gradient (1-2.5%).

Fluorescent staining of cells was performed in a 2-step immunofluorescence procedure that has been reported elsewhere for thymocytes
(7).   Reduced sample size (from 1×10^6 to 5×10^5 cells) was required
for sedimentation velocity fractions with low cell yield.   Cells
were stained by initial incubation with anti-MG-1 at 1/100 dilution
of ascites from tumor bearing mice.   The secondary incubation was
done with fluoresceinated Fab fraction of affinity purified goat anti-
IgG2 (Fl-GAMIgG2) provided for these studies by B. J. Fowlkes (NIAID).
Unstained cells were monitored in HBSS with 0.1% sodium azide and 0.1%
bovine serum albumin at 4°C.

Single and dual parameter flow microfluorometry analysis carried
out on a FACS II with minor modifications from the type of analyses
described elsewhere (8).   That is, fluorescence data were collected
and displayed on a log scale in 512 channels.   The contour plots
collected in a 64 x 64 channel matrix, are presented with contour
lines at doubling levels of the percent of cells per channel.

Antibody dependent, complement-mediated cytotoxicity (CTX) was
performed by incubation of unfractionated bone marrow cells suspended
in 1/100 dilution of anti-MG-1 (titer 1/5000) for 30 minutes at 37°C,
removal of antibody followed by 2 sequential 30 minute treatments
of selected non-toxic rabbit serum as a complement source, also at
37°C.   This dual treatment was required for complete elimination of
the antigen positive cells in the bone marrow as monitored by flow
microfluorometric analysis.

Colony forming units in the spleen (CFU-S) were assayed by
counting colonies in fixed spleens as reported by Till and McCulloch
(9).   Bone marrow cells used in these assays were divided into 4
equal aliquots, each aliquot was then treated by anti-MG-1 only,
anti-MG-1 + complement, complement only or not treated.   The cells
were injected at fixed doses relative to the initial number of untreated cells, i.e., cell counts were not adjusted after CTX.

RESULTS

A cytotoxic (IgG2) monoclonal antibody (anti-MG-1) against bone
marrow cells was produced as a hybridoma of a myeloma fused with
spleen cells of ASW mice immunized with SJL cells. This antibody can
lyse in CTX assays 40-70% of bone marrow cells, about 5% of spleen
cells and <1% of thymocytes or lymph node cells. Thus, this reagent
appeared to have no tissue correlation with any previously defined
reagent for lymphoid cells. Screening bone marrow from several
strains of mice by immunofluorescence showed that 40-60% of bone
marrow cells from the following strains were stained: SJL/J, C57BL/6
(B6), and several B6-Ly congenics, SWR/J, DBA/2J, NZB/BLN, NZW/BLN,
NFS/N. Bone marrow cells from ASW, A/J, BALB/c/AnN and C3H/HeJ mice
were negative. Therefore, this reagent may detect an allelic difference between different strains of mice that makes it useful for
bone marrow repopulation studies.

To further characterize which bone marrow cells were binding
this reagent, we initially chose to analyze bone marrow subpopulations
selectively enriched for size differences. Subpopulations obtained
by sedimentation at unit gravity were analyzed by flow microfluorometric analysis for immunofluorescence. In parallel, the samples
were monitored by differential counts on cytocentrifuge preparations
that were stained with Wright-Giemsa reagents. On sedimentation at
unit gravity, 3 major bands of cells were observed. The top band,
(fractions 7-11), containing red cells and lymphocytes, had less than
10% cells that stained with anti-MG-1 + Fl-GAMIgG2. The lowest band,
(fractions 25-30), contained a mixture of cell types that included
lymphoblasts, monocytes, and cells in the granulocytic series.
The proportion of positive cells in these fractions approximated the
sum of the monocytic cells and the granulocytic cells. A fraction
from the middle band of the gradient (fraction 20) that was enriched
to 92% for "ring" forms of segmented neutrophils, with 2% polymorphonuclear leukocytes and 6% lymphoid blast cells, contained 94%
positive cells (Figure 1-B).

In Figure 1, the immunofluorescence on the unfractionated bone
marrow cells (A) is compared with the immunofluorescence on cells
from fraction 20 (B) of the gradient. A strong enrichment for positive cells is seen in the fraction. When the cells were examined in
parallel for relative cell size by forward angle light scatter, a
major enrichment for the larger sized cells was seen in bone marrow
from fraction 20 (D) relative to the 3 peaks observed in unfractionated bone marrow (C). These 3 peaks approximately correspond to
red cells and dead cells up to channel 300, lymphocytes, channels
300-400, and larger cells, channels 400-1000. Because it was clear
that not all of the larger cells from the gradient were binding the
monoclonal anti-MG-1 antibody, we attempted to further characterize
the positive subset of cells by CTX elimination experiments.

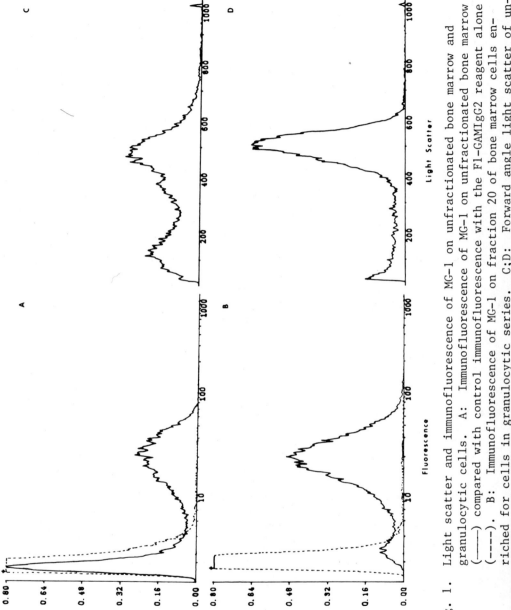

Fig. 1.   Light scatter and immunofluorescence of MG-1 on unfractionated bone marrow and granulocytic cells. A: Immunofluorescence of MG-1 on unfractionated bone marrow (———) compared with control immunofluorescence with the Fl-GAMIgG2 reagent alone (----). B: Immunofluorescence of MG-1 on fraction 20 of bone marrow cells enriched for cells in granulocytic series. C;D: Forward angle light scatter of un-

Anti-MG-1 treated cells were compared with cells treated with anti-MG-1 plus complement. Dual parameter analysis for immunofluorescence and light scatter of such cells is seen in Figure 2. These panels show that a larger sized subpopulation of cells (greater than channel 28 on this scale) seen in Figure 2-A, treated with anti-MG-1 antibody only, is eliminated by treatment with anti-MG-1 antibody plus complement (Figure 2-B).

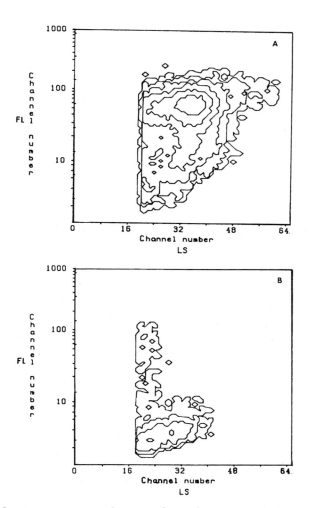

Fig. 2.   Dual parameter analysis of MG-1 immunofluorescence and light scatter. A: Anti-MG-1 on unfractionated B6 bone marrow cells. B: Analysis of remaining bone marrow cells after anti-MG-1 plus complement treatment.

When the light scatter of the viable cells (channels 300-1000)
was compared between populations treated with antibody alone or antibody plus complement (Figure 3) a depletion of the largest sized cells
is seen. The relative proportion of medium sized cells and smaller
cells was enriched after the lytic treatment.

In 3 experiments, we further analyzed the CFU-S of bone marrow
cells treated with anti-MG-1. When animals were injected with 8-10
CFU-S/10^5 bone marrow cells, we observed no significant reduction
in the total number of CFU-S when antibody plus complement treated
cells were compared with antibody treated, complement treated or
untreated bone marrow cells. This inability to eliminate CFU-S was
seen despite the fact that over 50% of the cells were lysed by antibody plus complement treatment.

Preliminary experiments with D. Pluznik indicate that colony
formation in agar of bone marrow cells treated with anti-MG-1 antibody plus complement is inhibited. Additional studies with immunofluorescence indicate that the antibody binds to some, but not all,
plastic adherent cells from spleens but not to phagocytic cells from
peritoneal exudates.

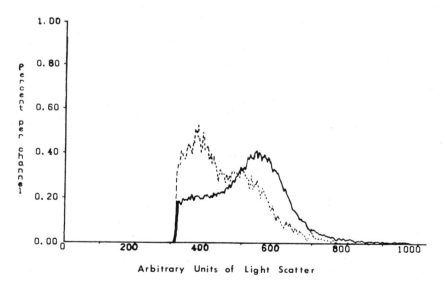

Fig. 3.   Comparison of relative cell size after antibody dependent
          complement lysis with anti-MG-1 light scatter of viable cells
          from B6 bone marrow treated with monoclonal anti-MG-1 only
          (————). Light scatter of viable cells after treatment with
          anti-MG-1 plus complement (----).

Previous studies on bone marrow cells have indicated that cells of a myelocytic or granulocytic lineage include cells committed to a monocytic lineage (10,11). The anti-MG-1 antibody that we have described here apparently detects this subpopulation. Therefore, it appears that the antibody binds to committed but possibly not mature cells of the monocytic/granulocytic lineage.

REFERENCES

1. Archer, J. R., and Davies, D. A. L., J. Immunogenet. 1:113, 1974.
2. Springer, T., Galfré, G., Secher, D. S., and Milstein, C., Eur. J. Immunol. 9:301, 1979.
3. Breard, J., Reinherz, E. L., Kung, P. C., Goldstein, G., and Schlossman, S. F., J. Immunol. 124:1943, 1980.
4. Rumpold, H., Swetby, P., Boltz, G., and Förster, O., in "Hetero-geneity of Mononuclear Phagocytes" (O. Förster and M. Landy, eds.), pp. 47-52, Academic Press, London, 1981.
5. Unkeless, J. C., Mellman, I. S., McGettigan, M., and Plutner, H., in "Heterogeneity of Mononuclear Phagocytes" (O. Förster and M. Landy, eds.), pp. 91-96, Academic Press, London, 1981.
6. Peterson, E. A., and Evans, W. H., Nature 214:824, 1967.
7. Mathieson, B. J., Sharrow, S. O., Campbell, P. S., and Asofsky, R., Nature 277:478, 1979.
8. Fowlkes, B. J., Waxdal, M. J., Sharrow, S. O., Thomas, C. A. III, Asofsky, R., and Mathieson, B. J., J. Immunol. 125:623, 1980.
9. Till, J. E., and McCulloch, E. A., Rad. Res. 14:213, 1961.
10. Pluznik, D. H., and Sachs, L., Exp. Cell Res. 43:553, 1966.
11. Bradley, T. R., and Metcalf, D., Aust. J. Biol. Med. Sci. 44: 287, 1966.

ISOLATION AND CHARACTERIZATION OF

HUMAN MONOCYTE SUBSETS

Roy S. Weiner

Division of Medical Oncology
University of Florida
Gainesville, Florida 32610 USA

Human peripheral blood monocytes manifest several functions in vitro that are thought to be important to host defenses. Monocytes and/or their more differentiated forms exhibit chemotaxis, phagocytosis, bacterial killing, tumor cytotoxicity, and modulation of immunity and hematopoiesis. Peripheral blood monocytes, for all their demonstrable functions, are thought to be cells in transit, i.e., merely passengers in the blood stream traveling from their origin in the bone marrow to their destiny in tissues where they function as macrophages (1-3). Tissue macrophages in animals demonstrate a staggering heterogeneity wherein function differs depending upon the site from which they are obtained, the stimulus which induced their accummulation, and the interval between that stimulus and their collection and testing (4-9).

While populations of cells can manifest heterogeneity of function and structure, they tell us little about the functional restrictions of any given cell or the mechanisms by which a cell has differentiated such that it is committed to a limited spectrum of functions. For instance, is a cell which produces colony stimulating factor -- a glycoprotein which stimulates the proliferation and differentiation of granulocyte and monocyte progenitor cells (10,11)-- also capable of responding to lymphokine and becoming cytotoxic to tumor cells (12,13)? The study of monocyte ontogeny and macrophage differentiation requires the collection and purification of cells that can be characterized structurally and functionally, and studied in systems which permit controlled manipulation of their maturation and differentiation. Only recently has attention been directed at studies of human peripheral blood monocytes. Techniques have been developed which permit the collection of large numbers of monocytes, their isolation from other nucleated cells in the blood, and their

209

separation into subpopulations with an expanding cohort of definable characteristics.

COLLECTION

Monocytes constitute less than 5% of the nucleated cells in normal peripheral blood, and thus each milliliter of blood potentially yields but $2.5 - 5 \times 10^5$ monocytes. A series of structural and functional assays using 5×10^7 monocytes, for example, would require phlebotomy of 250ml of whole blood and needless waste of red cells and plasma. Leukopheresis, however, can provide more than 5×10^8 monocytes in a 100ml buffy coat while returning to the donor virtually all the red cells, granulocytes, and plasma (14-16). Semi-continuous flow centrifugation using the Haemonetics Model 30 blood cell separator has become a standard blood banking procedure in the collection of platelets. By modifying the platelet collection protocol so that a 50ml buffy coat is collected at a flow rate of 30ml/min, 80% of the monocytes in a unit of peripheral blood is obtained. Four units can be pheresed in two hours. Other investigators have used continous flow centrifugation for leukopheresis with similar results (17). The buffy coat is processed by isopyknic centrifugation over Ficoll Hypaque to purge the preparation of residual erythrocytes and granulocytes (18). The resulting mononuclear cell preparation yields greater than 90% of the monocytes in a purity that ranges between 25-40% (14-16,19).

ISOLATION

Isolation of peripheral blood monocytes has been dependent on their property of adherence to glass or plastic (20-23) or on their density (24-27). Isolation by adherence is unsatisfactory for the study of monocyte biology because it selects for cells that adhere under a given set of experimental conditions. Yields of 50-90% are reported and purity varies between 80-95%. In our experience and the experience of others, the presence of non-specific esterase or myeloperoxidase positive cells in the non-adherent population is variable but often significant (28,29). Further, adherence causes marked physical and functional changes in the monocytes recovered. Attachment for as little as 2 hours results in a decrease in non-specific esterase stain and Fc receptor function when the cells are recovered (29) and prolonged incubation of the adherent cells induces functional changes in both complement and Fc receptor mediated phagocytosis (17,30,31). Separation by density gradient centrifugation or sedimentation at unit gravity can result in yields of up to 80% and purity of 60-80%. These methods have the advantage of inducing fewer perturbations in structure and function, but they have inherent limitations in the quantity of monocytes obtainable. Unit gravity sedimentation has the potential of subfractionating semi-purified monocytes on the basis of small size differences, especially when only a few cells are needed to study surface or kinetic characteristics. Adaptations of standard systems are being explored to accommodate large numbers of cells but there appears to be little advantage of this technique for bulk separation over new centrifugal methods (25).

Sanderson et al. (32) showed that monocytes could be separated from lymphocytes in isotonic media by counter flow centrifugal elutriation (CCE). The monocytes were obtained in high yield and purity though the quantities were limited. We adapted the technique of CCE elutriation to purify monocytes from the Ficoll-Hypaque separated mononuclear cells from pheresis specimens (15). The optimum conditions for separation using a Beckman JE-6 rotor in a Beckman J-21C centrifuge were a rotor speed of 2500 rpm (625 x G), a rotor temperature of $10^{\circ}C$, and a suspension media of Hanks Balanced Salt Solution without calcium and magnesium, but with 100mg of EDTA per liter. A counter flow rate of 17ml/min was found to purge all the lymphocytes present and a flow rate of 22ml/min yielded 70% of the monocytes with a purity of 95%. Monocytes as defined by myeloperoxidase positive cells were present in the lymphocyte rich fraction eluted at a counter flow rate of 17ml/min. These cells represented up to 5% of the lymphocyte rich population and up to 25% of the total monocyte population. The large monocytes represent a homogeneous population of cells with a modal volume of approximately $380\mu^3$. The populations of cells contaminating the lymphocytes have recently been enriched and found to be slightly smaller, having a modal volume of approximately $345\mu^3$ (33). At 17ml/min counter flow rate, the first 100ml elutriated volume contains most of the lymphocytes and less than 5% monocytes. The modal volume is approximately $170\mu^3$. The second 100ml, however, contains approximately 60% monocytes and 40% lymphocytes. The size profile of this population is bimodal with a peak at $170\mu^3$ and a second peak at $335-350\mu^3$. These smaller monocytes were characterized in an enriched population without the problems of selection and perturbation inherent in adherence procedures. Similar results have recently been reported by Yasaka et al. (34), but they characterized their subpopulations of monocytes after adherence. Distinct subpopulations of monocytes have been detected by several authors using CCE (15,33-38). The population of smaller monocytes, however, are always a minor contaminant of the lymphocyte rich population and require purification by adherence or other procedures for further study.

Isolation of monocytes by density gradient centrifugation in Percoll has been studied by several authors (26,39-41). This nontoxic suspension of heterogeneous sized silica particles establishes a density gradient upon centrifugation at high G force and permits cells to come to equilibrium as a function of their density. The gradients can be calibrated with colored beads of known density and thus the separation can be standardized and the recovered cells characterized. However, conditions which permit high yield of monocytes tend to sacrifice purity and vice versa. By and large, one can achieve a 50% yield of monocytes with a purity of 70% when the starting population is a Ficoll-Hypaque separated mononuclear cell preparation. We have used an in situ generated Percoll gradient to purify the monocytes obtained from CCE (41). Two to eight x 10^7 cells in 14ml Percoll at a specific gravity of 1.070 is centrifuged at 30,000 x G for 40 minutes. The population of smaller monocytes can be separated from the lymphocytes with a purity of >80% and a yield of 90%. The Percoll separation procedure is complimentary to the CCE since the latter is influenced more by size and the former more by density. In fact, we found that while the modal volumes of the two populations of monocytes separated by CCE differ by approximately $40\mu^3$ in volume, they have identical densities in the Percoll separation. Each population retains its characteristic size distribution after Percoll separation. Adherence causes a notable increase in size of recovered monocytes and this is reflected in an observed decrease in their density.

In summary, monocytes can be obtained in large numbers by leukopheresis. Two subsets can be purified without the need for prior adherence by CCE followed by density gradient centrifugation in Percoll. A two hour leukopheresis and five hours of laboratory manipulation can result in the purification of 5 x 10^8 large monocytes and 5 x 10^7 small monocytes.

CHARACTERIZATION

The monocyte heterogeneity in peripheral blood may represent different stages in maturation or stable subpopulations, each subject to maturational or differentiating stimuli in vivo. Since isolation itself depends on a definable monocyte characteristic such as size, density, cell surface receptors or cell surface antigens, or a given function such as adherence or chemotaxis, the isolated population(s) require characterization by independent parameters. Functional assessment of monocytes separated and isolated on the basis of a physical characteristic provides a baseline from which to evaluate changes occurring with maturation as well as to distinguish among subsets which may have differentiated such that their functional spectrum is restricted. Enzyme content, surface receptors for Fc fragments of immunoglobulins, complement, or surface antigens defined by specific antibodies would serve as convenient markers with which to distinguish among functionally defined subsets or functionally defined stages in differentiation or maturation. Functional characterization and characterization by structural attributes, however, are fraught with difficulties stemming from variability in methods of isolating the cells, variability in assay techniques, and the current lack of standardization of reagents used to define cell surface antigens.

We have characterized monocyte subpopulations that were purified

on the basis of size and/or density with respect to hexose monophosphate shunt (HMS) stimulation by zymosan, phagocytosis, chemotaxis,
antibody dependent cellular cytotoxicity, and native cytotoxicity.
The CCE purified large population of monocytes has a significantly
greater stimulation of HMS in response to zymosan than does either
the smaller monocytes or the monocytes in the unseparated Ficoll-
Hypaque preparation. The large population of monocytes consistently
displayed 125-150% of the HMS activity in the unseparated Ficoll-
Hypaque preparation and 300-400% of the activity of the small monocyte
population. These results are analogous to those described by Yasaka
et al. (34) for superoxide release after exposure to zymosan. With
respect to monocyte mediated antibody dependent cellular cytotoxicity,
we observed that both populations of monocytes are equally active
against human O Rh+ erythrocytes sensitized with anti-D (33). Our
data is in disagreement with that reported by Norris et al. (35) who
found increased ADCC activity in the larger monocytes. We have shown,
however, that the demonstration of ADCC activity is dependent upon
the concentration of effector cells in the reaction mixture. The
ADCC activity of both large and small monocytes can be suppressed
by the addition of either lymphocytes or competing, unlabelled and
unsensitized erythrocytes if the proportion of effector cells is
reduced below 10%. Thus, the CCE derived "small monocyte fraction"
in which monocytes comprise only 5-10% of the total cells is inadequate to accurately assess ADCC function. Enrichment of the small
monocytes by fractionated CCE, reversible adherence to microexudate
coated tissue culture plates, or density gradient centrifugation on
Percoll reveals that the small monocytes bear Fc receptors and are
as active in ADCC as the large monocytes. Chemotactic response of
monocytes is thought to be dependent upon a subset comprising 60%
of the monocytes which respond to multiple chemoattractants and may
be related to a receptor for chemotactic peptides (42). Both large
and small monocytes respond to both lymphokine and AB serum as chemoattractants indicating that subsets of monocytes with adequate receptors exist in both large and small populations. Native cytotoxicity
for P815 murine mastocytoma cells is a property of human peripheral
blood monocytes (16). The cytotoxicity is increased in response to
both lymphokine and lipopolysaccharide. Subsequently, we found that
this native cytotoxicity is a property of the smaller, but not the
larger, monocyte population (13). An intriguing observation currently under study in our laboratory is the ability of dimethyl-
sulfoxide (DMSO) to induce native cytotoxicity in the population of
large monocytes. DMSO is known to induce differentiation in a variety of cells including the differentiation to monocytes in the pro-
granulocytic leukemic cell line, HL60. In this instance, if the DMSO
were promoting differentiation, it would imply that the smaller rather
than the larger monocyte is the more differentiated cell at least
with respect to native cytotoxicity. On the other hand, native or
"natural" killing has been ascribed to bone marrow promonocytes, and
DMSO in this instance may well be inducing a property that the more
mature, large monocyte had lost in the process of differentiation
(43).

Monocytes have long been characterized by enzymatic parameters, and enzyme profiles have been correlated with differentiation (44). More recently, esterase polymorphism as determined by iso-electric focusing patterns has been used to characterize monocytes that give rise to peritoneal macrophages (45). This may represent a useful refinement of histochemical techniques that could distinguish between differentiating populations of monocytes and functionally distinct subsets.

Clearly the ability to identify and quantitate functionally significant surface receptors offers the possibility of characterizing monocyte subsets in a meaningful manner. The Fc receptor for IgG is characteristic of monocytes and associated with binding, phago-cytosis, and ADCC. It appears that several types of Fc receptors exist on the monocyte. Virtually all monocytes have Fc receptors for rabbit IgG and about 65% of monocytes have, in addition, recep-tors for human IgG (46). In a recent study, it was shown that ap-proximately 40% of peripheral blood monocytes have Fc receptors for IgA and about 12% of the monocytes were shown to have receptors for both IgA and IgG. While cells having receptors for IgG were found to have ADCC activity, the subpopulation having Fc receptors for IgA did not appear to lyse antibody coated cells (47). Moreover, approximately 20% of human peripheral blood monocytes have Fc recep-tors for IgE (48). While the Fc receptors provide attractive markers for monocyte characterization, they represent cell surface components which are known to change dramatically with adherence (17,31) and are subject to both reversible and irreversible loss as a consequence of interaction with aggregated or immobilized IgG (49).

Yet another defined antigen on the surface of monocytes is an Ia like antigen detectable on 70-80% of peripheral blood monocytes. The expression of Ia antigen is associated with the population of monocytes active in modulating proliferative responses of lymphocytes to antigens. The expression of the Ia like antigen on human monocytes has recently been shown to be reduced by exposure to endotoxin or zymosan activated serum (50). The Ia like antigen may serve as a convenient marker by which to characterize the sensitivity of sub-populations of cells to endotoxin. In the modulation of hemato-poiesis, lactoferrin appears to react with an Ia antigen positive subpopulation of monocytes to inhibit the production of colony stimu-lating factor (51). Endotoxin may exert its stimulatory effect on hematopoiesis by decreasing the expression of both Ia antigen and lactoferrin receptor on the same population of cells, thus neutral-izing an inhibitor of CSF production. Combinations of functionally defined receptors such as Ia, lactoferrin, and Fc provide an op-portunity to study differentiation under defined experimental condi-tions.

Numerous antigens on the surface of monocytes are being detected by hybridoma derived heterologous monoclonal antibodies. Since the description by Breard et al. of a monoclonal antibody that reacted with human peripheral blood monocytes (52), several antibodies have been described with specificities restricted to monocytes or sub-populations thereof (53-55). While OKM1 (52) detects antigens shared by peripheral blood monocytes, granulocytes and null cells, MAC120 recognizes only 30% of the peripheral blood monocytes and apparently detects a subpopulation of Ia like antigen positive monocytes required for antigen induced T cell proliferation (53). Data presented at this session shows that surface antigens may reflect stages of dif-ferentiation (56) and provide a means of defining a specific precursor population (57). New monoclonal antibody probes are being prepared and used to isolate the antigens they recognize so that they in turn can be chemically defined and serve as an entrée to the study of molec-ular control of antigen expression, differentiation, and functional integrity. These reagents, used in conjunction with sophisticated cell sorters, provide an opportunity to isolate highly restricted subpopulations in order to define their functional correlates and monitor their differentiation.

The development of techniques capable of collecting large numbers of human peripheral blood monocytes and subsequently isolating them from lymphocytes and granulocytes has facilitated study of their functional and structural heterogeneity. It appears that significant advances in the detection of cell surface antigens and the ability to sort cells on the basis of antigenicity and physical character-istics presages significant breakthroughs in our knowledge of monocyte ontogeny.

Acknowledgement

We acknowledge the expert editorial assistance of Leslie Rigg in the preparation of this manuscript and the expert technical assis-tance of Virendra Shah, Richard Mason, and Lisa Anderson. Supported by NIH Grant RO1-CA-29266.

REFERENCES

1. Goud, T. J. L. M., Schoote, C., and vanFurth, R., J. Exp. Med. 142:1180, 1975.
2. vanFurth, R., and Cohn, Z. A., J. Exp. Med. 128:415, 1968.
3. Volkman, A., J. Reticuloendothelial Soc. 19:249, 1976.
4. Rhodes, J., J. Immun. 114:976, 1975.
5. Hibbs, J. B., Taintor, R. R., Chapman, H. A., and Weinberg, J. B., Science, 197:279, 1977.

6. Lee, K.-C., and Berry, D., J. Immun. 118:1530, 1977.
7. Normann, S. J., Sorkin, E., and Schardt, M., J. Natl. Cancer Inst. 63:825, 1979.
8. Rice, S. G., and Fishman, M., Cell. Immunol. 11:130, 1974.
9. Walker, W. S., Immunology 26:1025, 1974.
10. Golde, D. W., Finely, T. N., and Cline, M. J., Lancet ii:1397, 1972.
11. Opitz, H. G., Neithammer, D., Jackson, R. C., Lemke, H., Huget, R., and Flad, H. D., Cell. Immunol. 18:70, 1975.
12. Fidler, I. J., Israel J. Med. Scis. 14:177, 1978.
13. Normann, S. J., and Weiner, R. S., unpublished observations.
14. Weiner, R. S., Richman, C. M., and Yankee, R. A., Blood 49:391, 1977.
15. Weiner, R. S., and Shah, V. O., J. Immunol. Methods 36:89, 1980.
16. Weiner, R. S., and Normann, S. J., J. Natl. Cancer Inst. 66:255, 1981.
17. Stevenson, J. C., Katz, P., Wright, D. G., Contreras, T. J., Jemionek, J. F., Hartwig, V. M., Flor, W. J., and Fauci, A. S., Scand. J. Immunol., 14:243, 1981.
18. Boyum, A., Scand. J. Clin. Invest. 21(Suppl. 97):1, 1968.
19. Soman, S. and Kaplow, L. S., J. Immunol. Methods 32:215, 1980.
20. Rabinowitz, Y., Blood 23:811, 1964.
21. Shortman, K., Williams, N., Jackson, H., Russel, P., Byrt, P., and Diener, E., J. Cell Biol. 48:566, 1971.
22. Ackerman, S. K., and Douglas, S. D., J. Immunol. 120:1372, 1979.
23. Zanella, A., Mantovani, A., Mariani, M., Silvani, C., Peri, G., and Tedesco, F., J. Immunol. Methods 41:279, 1981.
24. Bennett, W. E., and Cohn, Z. A., J. Exp. Med. 123:145, 1966.
25. Bont, W. S., deVries, J. E., Geel, M., vanDongen, A., and Loos, H. A., J. Immunol. Methods 29:1, 1979.
26. Gmelig-Meyling, F., and Waldmann, T. A., J. Immunol. Methods 33:1, 1980.
27. deBoer, M., Reijneke, R., van de Griend, R. J., Loos, J. A., and Roos, D., J. Immunol. Methods 43:225, 1981.
28. Jerrells, T. R., Dean, J. H., Richardson, G. L., and Herberman, R. B., J. Immunol. Methods 32:11, 1980.
29. Treves, A. J., Yagoda, D., Haimovitz, A., Ramu, N., Rachmilewitz, D., and Fuks, Z., J. Immunol. Methods 39:71, 1980.
30. Newman, S. L., Musson, R. A., and Henson, P. M., J. Immunol. 125:2236, 1980.
31. Stevenson, H. C., Katz, P., and Fauci, A. S., Cell. Immunol. 53:94, 1980.
32. Sanderson, R. J., Shepperdson, F. T., Vatter, A. E., and Talmage, D. W., J. Immunol. 118:1409, 1977.
33. McCarley, D. L., Shah, V. O., and Weiner, R. S., in preparation.
34. Yasaka, T., Manitich, N. M., Boxer, L. A., and Baehner, R. L., J. Immunol. 127:1515:1981.

35. Norris, D. A., Morris, R. M., Sanderson, R. J., and Kohler, P.
       F., J. Immunol. 123:166, 1979.
36. Arenson, E. B. Jr., Epstein, M. B., and Seeger, R. C., J. Clin.
       Invest. 65:613, 1980.
37. deMulder, P. H. M., Wessels, J. M. C., Rosenbrand, D. A.,
       Smeulders, J. B. J. M., Wagener, D. J. Th., and Haanen, C.,
       J. Immunol. Methods 47:31, 1981.
38. Figdor, C. G., Bont, W. S., deVries, J. E., and vanEs, W. L.,
       J. Immunol. Methods 40:275, 1981.
39. Pertoft, H., Johnsson, A., Warmegard, B., and Seljelid, R., J.
       Immunol. Methods 33:221, 1980.
40. Fluks, A. J., J. Immunol. Methods 41:225, 1981.
41. Weiner, R. S., and Mason, R. R., Presented at 9th International
       RES Congress, February 1982.
42. Cianciolo, G. J., and Snyderman, R., J. Clin. Invest. 67:60,
       1981.
43. Lohmann-Matthes, M. L., Domzig, W., and Roder, J., J. Immunol.
       123:1883, 1979.
44. Bursuker, I., and Goldman, R., J. Reticuloendothelial Soc. 26:
       205, 1979.
45. Parwaresh, M. R., Radzun, H. J., and Dommes, M., Am. J. Path.
       102:209. 1981.
46. Barrett, S., Garratty, E., and Garratty, G., Brit. J. Haemet.
       43:575, 1979.
47. Fanger, M. W., Shen, L., Pugh, J., and Bernier, G. M., Proc.
       Natl. Acad. Sci. USA 77:3640, 1980.
48. Melewicz, F. M., and Spiegelberg, H. L., J. Immunol. 125:1026,
       1980.
49. Kurlander, R. J., J. Clin. Invest. 66:773, 1980.
50. Yem, A. W., and Parmely, M. J., J. Immunol. 127:2245, 1981.
51. Broxmeyer, H. E., J. Clin. Invest. 64:1717, 1979.
52. Breard, J., Reinherz, E. C., Kung, P. C., Goldstein, G., and
       Schlossman, S. F., J. Immunol. 124:1943, 1980.
53. Raff, H. A., Picker, L. J., and Stobo, S. D., J. Exp. Med. 152:
       581, 1980.
54. Todd, R. F., Nadler, L. M., and Schlossman, S. F., J. Immunol.
       126:1435, 1981.
55. Rosenberg, S. A., Liger, F. S., Ugolini, V., and Lipsky, P. E.,
       J. Immunol. 126:1473, 1981.
56. Treves, A. J., Haimovitz, A., and Fuks, Z., Presented at 9th
       International RES Congress, February 1982.
57. Mathieson, B. J., Shen, F.-W., Brooks, A., and Leiserson, W. M.,
       Presented at 9th International RES Congress, February 1982.

DISCUSSION

STEVENSON: You mentioned that larger monocytes conceivably are
activated as opposed to smaller monocytes. You mentioned that they

have increased phagocytosis, but less spontaneous cytotoxicity against the targets you tested. Do you have any idea what they might be activated for? Have you looked at monokine production, IL-1 production, interferon or any other function?

WEINER: The only indication we have for increased activity is their increased hexose monophosphate shunt activity in response to zymosan phagocytosis.

STEVENSON: In your elutriation system, do you have any idea what might be activating your monocytes for increased HMS activity?

WEINER: No.

STEVENSON: I noticed that you had an additive effect for cytotoxicity when you used lymphokine and LPS, but have you looked for endotoxin in your eluate? How do you sterilize your system?

WEINER: We have found that flushing the system with 70% alcohol essentially sterilizes it so that elutriated cells have been sterile in culture for 14 days. Studies of endotoxin in the eluate are currently underway.

STEVENSON: Do you use a variety of fractions and then pool them together?

WEINER: The first or small monocyte fraction was collected at 17 mls per minute, 625 g, eluted to exhaustion. Subsequently, the flow rate was increased to 22 mls per minute and all cells which elute at 22 mls per minute constituted the large monocyte fraction.

STEVENSON: We use a substantially different form of elutriation. I believe that the modal distribution of monocytes by size is relatively artificial. We collect fractions by increasing the flow rate at a half ml per minute, so in your system, we would go from 17 mls per minute to 17.5 to 18 to 18.5. Under these conditions, we can find many different sizes of human monocytes. There is no question that some are small monocytes that co-elute with lymphocytes. As you proceed to higher flow rates, you get relatively purified monocyte populations. When we do size analysis, the monocytes march their way up to 400 cubic microns. Using these different sized monocytes, we are in the process of looking at a variety of different functions from surface markers to antigen presentation.

WEINER: It would seem very reasonable that a continuum of sizes may exist. In our studies with half ml per minute flow rates between 17 and 22, we got about 10^5 cells per 50 ml of collection at each step of the way, so it was difficult to pin these down as separate mini subpopulations. Even so, we did get a major flush of large monocytes at 22 mls per minute.

NORMANN: I wish to comment on our experience using elutriation
centrifugation to isolate guinea pig peripheral blood monocytes.
By progressively raising the flow rate, we isolate monocytes with a
spectrum of sizes similar to the situation described by Stevenson for
the human. The advantage of selecting conditions that isolate the
two extremes, so to speak, is it concentrates the cells for functional
analysis. In this way, one can look at pools of cells with quite
different sizes and explore their functional differences.

STEVENSON: We've been focusing on elutriation as a separation procedure to produce large numbers of purified monocytes, but it also
is a valuable tool for those people interested in monocyte/lymphocyte
interactions. Elutriation is very good at producing highly purified
populations of monocyte depleted lymphocytes. We've looked at these
lymphocytes for their ability to respond in monocyte dependent systems, certain antigen dependent systems, and certain mitogen dependent
systems: we essentially get zero responses using these highly purified lymphocytes. But when we add back just a few monocytes they
respond well. So elutriation is a technique for purifying both
monocytes and lymphocytes.

SEPARATION AND CHARACTERIZATION OF GUINEA PIG MONOCYTE SUBSETS USING COUNTER-FLOW CENTRIFUGATION ELUTRIATION (CCE)

S. J. Noga, S. J. Normann, and R. Weiner

Department of Pathology and Division of Medical Oncology
University of Florida
Gainesville, Florida   32610   USA

INTRODUCTION

Counter-flow centrifugation elutriation (CCE) has been used to separate human monocytes into two populations that differ in size, Fc receptors (FcR), antibody-dependent cellular cytotoxicity (ADCC), and native or acquired tumoricidal activity (1-4). To determine whether a similar monocyte heterogeneity exists in species other than man, we examined guinea pig (GP) monocytes. Unlike mouse and rat, GP peripheral blood has a white cell composition similar to humans (5). Our studies suggest that three distinct GP monocyte fractions can be isolated by CCE that differ in adherence, phagocytosis, and non-specific esterase activity as well as modal volume (MV), FcR's, and acid phosphatase activity.

SEPARATION OF GUINEA PIG MONOCYTES BY CCE

Peripheral blood (150 ml) was obtained by cardiac puncture from 7 male Hartley strain guinea pigs (650-700 g) and the mononuclear cells isolated using Ficoll-Hypaque with a specific gravity of 1.101. This density was necessary in order to recover >95% of the monocytes. After two washings, contaminating RBC's were lysed with ammonium chloride and the mononuclear cell pellet washed and resuspended in 5-10 ml of elutriation media consisting of Hanks' balanced salt solution without Ca and Mg, but containing 100 mg EDTA/L and 5 mg/ml bovine serum albumin. The mononuclear cells (2×10^8) were injected into the Beckman JE-6B elutriator rotor system at a loading flow rate of 10 ml/min and a rotor speed of 3,000 RPM. Cell fractions were then collected at 1 ml/min incremental increases between 24 and 29 ml/min. Cells remaining in the rotor were collected by turning the

rotor off (R/O) and maintaining the flow. Cells were washed twice
before use. Modal volume measurements were determined by a pulse-
height analyzer.

CCE RECOVERY DATA

       Cell recovery after elutriation ranged from 91-99% with a via-
bility of >95%. Although up to six fractions were collected only the
24ml/min, 29 ml/min and R/O cells were analysed. The 24 ml/min fraction had a MV of 153 u^3 and contained 2% monocytes by morphology using
Wright's giemsa cytocentrifuge preparations. This fraction contained
15% of the total eluted monocytes. The 29 ml/min fraction had a MV
of 317 u^3, a monocyte purity of 70% and contained 57% of the total
monocytes. The R/O fraction contained 11% of the eluted monocytes
with 38% purity and MV of 354 u^3. Small lymphocytes were the major
cell found in the 24 ml/min fraction whereas large lymphocytes and
Kurloff cells were present in the monocyte-rich 29 ml/min fraction.
Granulocytes and their precursors, large Kurloff cells and a few
large lymphocytes contaminated the R/O fraction. Morphologically,
monocytes found in the 24 and 29 ml/min fractions appeared identical,
having an eccentric reniform nucleus, glassy sky blue cytoplasm, and
slight vacuolization. R/O monocytes were larger, possessed a cen-
trally located larger nucleus, and had a pyroninophilic cytoplasm with
heavy vacuolization.

ADHERENCE CHARACTERISTICS OF POST-ELUTRIATION FRACTIONS

       Post Ficoll-Hypaque preparations of GP monocytes suspended in
RPMI 1640 with 10% GP serum did not adhere after overnight incubation
in plastic tissue culture flasks. However, addition of gram negative
bacteria (Pseudomonas) or 5 ug/ml endotoxin caused virtually all monocytes to adhere within 2 hours. The microexudate remaining on plastic
tissue culture plates after the removal of the BHK-21 cell line has
been used to recover human monocytes in high yield and purity by
detaching adherent cells with a 15 minute exposure to 5 mM EDTA (6).
However, post-elutriation fractions of guinea pig monocytes (24 & 29
ml/min) incubated on microexudate plates showed no adherence in one
hour while R/O monocytes were only moderately adherent. Nylon wool
is often used to separate macrophages from lymphocytes and NK cells.
After a one hour incubation on nylon wool columns in the presence of
5% GP serum, only 18% of 24 ml/min monocytes adhered whereas 29 ml/
min and R/O monocytes were found to be considerably more adherent
(74 & 77% respectively).

PHAGOCYTOSIS OF OPSONIZED C. ALBICANS OR CARBON

       Monocytes readily phagocytized C. albicans only when opsonized

with fresh GPS, in agreement with other investigators who propose
a   C3b dependent opsonization of C. albicans   (7).   R/O monocytes
phagocytized the greatest number of organisms;29 ml/min monocytes were
intermediate and 24 ml/min monocytes the least phagocytic.   When
carbon particles were used as a measure of C3b independent phagocytic
ability, R/O monocytes displayed excellent uptake but the 24 or 29
ml/min monocytes ingested virtually no carbon.

HISTOCHEMICAL STAINING

     All GP monocytes stained for non-specific esterase with 24 ml/min
monocytes having a 1+, 29 ml/min a 2+ diffuse staining pattern and
R/O monocytes a 4+ staining with both diffuse and local reaction.
Acid phosphatase was graded as 2+ in the 29 ml/min and R/O fractions
and zero in monocytes of the 24 ml/min fraction.   Myeloperoxidase
was negative in all 3 fractions showing only a few cells with $\pm$ peri-
nuclear staining in the 29 ml/min and R/O fractions.

FcR DETERMINATION

     Guinea pig monocytes were analyzed for FcR using IgG sensitized
sheep RBC's.   The number of rosette-forming cells was identified
by the use of phase-microscopy and cytocentrifuge preparations.
100% of 24 ml/min and 91% of R/O monocytes formed rosettes with IgG-
coated SRBC's but only 15% of 29 ml/min monocytes formed rosettes.

CONCLUSIONS

     Counter-flow centrifugation elutriation (CCE) consistently sep-
arated guinea pig monocytes into 3 populations, more on the basis of
size and shape than density.   Use of continuous Percoll gradients
demonstrated that all GP monocytes concentrated in a band at specific
gravity 1.075.   Cells identified as monocytes either fit established
criteria for phagocytes as isolated or could be induced to meet these
criteria.   Our studies argue for the existence of a monocyte population having a size range with a Poisson distribution.   CCE allows the
study of monocytes far smaller (24 ml/min) and larger (R/O) than the
peak population (29 ml/min).   Analysis of these fractions indicated
that major differences existed between the small, intermediate and
large monocytes.

     We entertain three interpretations of these data.   (a) The
various sized monocytes simply represent the maturation spectrum of
monocytes within the peripheral blood.   Increasing histochemical
staining activity, phagocytic ability, and adherence properties correlated with the increased size of the monocytes.   Promonocytes just
released into the peripheral blood would not be expected to possess

large amounts of specific products or be fully functional. The small
monocytes identified in the 24 ml/min fraction fit this criteria.
The intermediate-sized monocytes (29 ml/min) might represent the
normal differentiation of the guinea pig monocyte. If so, the
majority of GP monocytes are less mature than their human counterparts. The macrophage-like properties of the R/O cells argue for
this cell being well differentiated. That small and large monocytes
possess FcR's while the majority of circulating monocytes (intermediate size) lack them has similarities to the macrophage Ia antigen which is lost and then resynthesized after phagocytosis (8).
(b) The various sized monocytes represent three distinct subsets differing quantitatively in their functional properties. Although the
FcR can be found in both small and large monocytes, the two populations are morphologically distinct. (c) The guinea pig may possess
only an FcR positive and an FcR negative monocyte population similar
to that described in man. The FcR+ monocytes may represent a rather
broad size distribution of the total monocyte size profile. These
monocytes would thus comprise all the monocytes at either end of
the Poisson distribution but only a relatively small percentage in
the intermediate monocyte range (15%). Super-imposed upon these
FcR + monocytes would be the FcR - subpopulation which comprises the
majority of GP monocytes of intermediate size range.

     Regardless of the correct interpretation for the 3 monocyte
fractions, their existence is a clear demonstration of monocyte heterogeneity and establishes the guinea pig as an excellent experimental
model for studying such heterogeneity.

Acknowledgement

     This work was supported by Grant CA 29266-02 from the National
Cancer Institute, USA. We thank Lisa Anderson, Janet Cornelius and
Julie Waltz for their expert technical assistance.

REFERENCES

1. Norris, A., Morris, M., Sanderson, J., and Kohler, P., J. Immunol.
        123:166, 1979.
2. Weiner, R., and Shah, V., J. Immunol. Methods 36:89, 1980.
3. Coutreras, T., Jemionek, J., Stevenson, H., Hartwig, V., and
        Fauci, A., Cell. Immunol. 54:215, 1980.
4. Normann, S., and Weiner, R., in: "Heterogeneity of Mononuclear
        Phagocytes" (O. Förster and M. Landy, ed.), pp. 496-503,
        Academic Press, New York, 1981.
5. P. Altman, D. Katz, ed., in: "Inbred and Genetically Defined
        Strains of Laboratory Animals", Part II, pp. 534-535, FASEB,
        Maryland, 1979.
6. Ackerman, S., and Douglas, S., J. Immunol. 120:1372, 1978.

7. Morrison, R., and Cutler, J., J. Reticuloendothelial Soc. 29:23,
        1981.
8. Beller, D., Unanue, E., J. Immunol. 126:263, 1981.

DISCUSSION

VAN FURTH: I wonder if the esterase stain is sufficient for the small
monocytes. With a weak reaction, you can't discern whether you have
a negative or weak positive response. I suggest using lysozyme instead of esterase because it's always positive in all kinds of monocytes.

NOGA: In the beginning, we had a hard time getting a non-specific
esterase stain to work. In order to do so, we raised the temperature
of incubation from $22^{o}C$ to $37^{o}C$.   At this temperature, we had a
nice reaction with T lymphocytes which stained with a punctuate pattern as expected. This indicated to me that the esterase stain was
working properly, even if it did not stain the small monocytes.

VAN FURTH: But in our experience you can occasionally get a positive
reaction with lymphocytes with a poor reactive stain if you just let
the reaction incubate for awhile. So again, I suggest lysozyme might
be better.

NOGA: Right now we are looking at a battery of enzymes to analyze
these cells and of course lysozyme is a major one that is being
considered.

VAN FURTH: You're in a position to label dividing cells in your
animals to determine which population of monocytes (smaller or larger)
emerges first from the bone marrow. It could well be that the small
cells after they have been born immediately leave the bone marrow or
the reverse could be true and the larger ones are the first to leave.

NOGA: Yes, that is a very exciting possibility. We have now established that monocytes exist in both large and small sizes. We've
further characterized them by separation on Percoll. The small monocytes are interesting because they are acid phosphatase negative and
that is the only monocyte population we know that is. As you suggested, it is now possible to see which are the newly emergent cells,
to determine their transit times in the circulation and possibly where
they go.

CHANGES IN SURFACE MARKERS OF HUMAN
MONOCYTES FOLLOWING THEIR IN VITRO
MATURATION TO MACROPHAGES

Abraham J. Treves, Adriana Haimovitz and Zvi Fuks

Department of Radiation and Clinical Oncology
Hadassah University Medical Center
Jerusalem 91120 Israel

INTRODUCTION

Peripheral blood monocytes are a transitory population between
bone marrow promonocytes and tissue macrophages (1,2). Blood monocytes have several characteristics which assign them to the macrophage
lineage. However, each of the macrophage markers is not always
expressed on the entire monocyte population (3,4) and some markers
and functions are altered during monocyte maturation or activation
(5,6,7). In vitro cultivation of human monocytes is accompanied by
a series of changes which results in their transition to macrophages
(8). In the present study, we analyzed membrane markers on freshly
obtained monocytes and on mature macrophages. These were the receptor for peanut agglutinin (PNA), which is associated with cell maturation in T lymphocytes (9) and the HLA-DR which is associated with some
immune functions (10-12).

MATERIALS AND METHODS

Adherent monocytes were obtained from human peripheral blood
mononuclear cells (PBM) as described (4). Part of the cultures were
incubated for 10-13 days and by the end of this period, the adherent
cells acquired the morphological appearance of large macrophages.
The adherent monocytes and the matured macrophages were tested for
the different markers.

Adherent cells were washed once with phosphate buffered saline
(PBS) and then either 0.1 ml of fluorescein isothiocyanate conjugated
Concanavalin A (F-Con A) 100 µg/ml or 0.1 ml of 50-125 µg/ml fluores-
cein isothiocyanate PNA (F-PNA) was added. The F-PNA was either

227

purchased from Vector Laboratories, Inc. (Burlingame, California) or
donated by Dr. N. Sharon of the Weizmann Institute of Science,
Rehovot, Israel. F-Con A was obtained from Miles-Yeda (Rehovot,
Israel). After incubation for 40 min at room temperature, the cells
were washed once with PBS and further incubated overnight at 4°C
in PBS: Glycerol (9:1). In some experiments, 0.1 M of D-galactose
(GAL) or 0.05 M methyl-α-D-manopyranoside (MAN) were added during
incubation with the fluorescein labeled lectins. The slides were
read under epi-illumination in a fluorescence microscope. In some
experiments, adherent cells were washed once with PBS and fresh medium
containing 10 μl of 500 U/ml neuraminidase (GIBCO) was added for 30
min and then washed twice with culture medium. The results are expressed as the mean percentage of fluorescent positive cells in
several repeated experiments $\pm$ standard error (S.E.).

For determination of HLA-DR positive cells, we used mouse anti-
HLA-DR monoclonal antibody (L243), provided by Dr. Ronald Levy,
Stanford University Medical School, Stanford, California (13).
Adherent cells were washed once with PBS and incubated with human
serum for 10 minutes at 37°C. After removing the serum, the cells
were washed with PBS, incubated with the L243 ascitic fluid (1:40 in
PBS) for 30 minutes at 37°C, washed again, incubated with fluorescein conjugated goat antimouse IgG (1:20 in PBS) for 30 minutes at
37°C, washed, covered with PBS:Glycerol (9:1) and kept at 4°C until
read the next day. Control cultures were stained with only the
fluorescent antisera and were always negative. Preincubation with
human serum was performed to block the monocyte Fc receptors. We
found that such pretreatment completely blocked Fc receptors on monocytes and macrophages as measured by the binding of antibody coated
sheep erythrocytes.

RESULTS

Binding of F-PNA to freshly obtained monocytes and to mature
macrophages is presented in Table 1. Most monocytes (85 $\pm$ 5%)
bound F-PNA whereas twelve days later, only 12 $\pm$ 6% of the monocyte
derived macrophages bound the lectin. Thus, in vitro maturation of
monocytes was accompanied by a sharp decrease in the proportion of
F-PNA binding cells. In contrast, F-Con A binding did not change
during monocyte maturation to macrophages. Practically all the cells
in both cell populations bound this lectin. The binding of both
lectins was specific since each lectin was inhibited only by its
specific sugar inhibitor and not by the other sugar. A brief treatment of monocytes and macrophages with neuraminidase, increased F-PNA
binding to 98-100% of the cells (14). These results indicated that
the decrease in F-PNA binding to mature macrophages was probably
due to masking of the receptors by sialic acid. A kinetic study of
F-PNA binding revealed that a sharp decrease occurred within 1-2 days
following monocyte isolation and plating (14).

Table 1. Peanut Agglutinin (PNA) and Con A Binding to
          Monocytes and Macrophages and their Inhibition
          by Specific Sugars

Treatment	Monocytes	Macrophages
PNA	85 ± 5	12 ± 6
PNA + GAL	0	0
PNA + MAN	67 ± 12	11 ± 5
Con A	100 ± 0	100 ± 0
Con A + GAL	100 ± 0	100 ± 0
Con A + MAN	0	0

Data are reported as mean percentage of fluorescent
positive cells ± 1 SE on 5 donors. The binding assay
was performed in the presence or absence of sugar
inhibitors: GAL = D-galactose and MAN = Methyl-α-D-
manopyranoside.

   We then tested the binding of antibodies against surface HLA-DR
antigens (Table 2). The results indicated that most of the monocytes
(87 ± 3%) expressed this antigen and this proportion was only slightly
reduced following their maturation to macrophages (66 ± 6%).

DISCUSSION

   The in vitro maturation of monocytes to macrophages is accompanied by several morphological and functional changes. In the present study, we investigated possible changes in surface markers. A
sharp decrease in F-PNA binding was observed after a short period of
monocyte cultivation and since this decrease was stable, only a small
proportion of mature macrophages expressed this activity. On the
other hand, the binding of Con A did not change following monocyte
maturation. The binding of each of the lectins was inhibited only by
its specific sugar inhibitor. Similar changes in PNA binding are
also characteristic for T cell maturation (9). The reduction in
PNA binding to macrophages was probably due to masking with sialic

Table 2.   Surface HLA-DR Antigens on Monocytes and
           Macrophages

Donor Number	Monocytes	Macrophages
1	76	66
2	69	65
3	98	90
4	85	95
5	79	50
6	100	100
7	100	55
8	100	54
9	90	81
10	75	32
11	90	65
12	80	39
Mean $\pm$ S.E.	87 $\pm$ 3	66 $\pm$ 6

Data are reported as percentage of fluorescent positive
cells after staining with the monoclonal L243 antihuman
HLA-DR and with fluorescent goat antimouse IgG.

acid (14). Aging of erythrocytes has been found to be accompanied
by changes in the content of membrane sialic acid (15). It is therefore possible that changes in membrane sialic acid may be part of the
regulatory mechanism leading to monocyte differentiation and migration from the peripheral blood to other organs.

The expression of membrane HLA-DR antigens was found to be different from the expression of PNA receptor, but also different from
the expression of the equivalent Ia antigen in mice. The proportion
of monocytes which expressed HLA-DR antigens was 87% and decreased to
66% following maturation to macrophages. The binding of the L243
reagent was specific since we blocked the Fc receptor of the cells by
IgG found in normal serum. The decrease in the proportion of HLA-
DR positive cells, although variable, was statistically significant
(P<0.001). It was reported earlier that the expression of Ia on mouse
macrophages decreased sharply following their in vitro incubation but
increased as a result of treatment with lymphokines (7). Complement
and Fc receptors on the surface of monocytes also increased following

maturation to macrophages (6). Some tissue macrophages also express Ia antigens: namely, Kupffer (16), skin Langerhans cells (17) and alveolar macrophages (18). In the mouse, splenic macrophages express the I-A subregion of Ia to a much higher degree than peritoneal macrophages (5).

These studies indicate that the expression of any surface marker on macrophages may represent a dynamic situation which is influenced not only by cell maturation but also by activation. It is also possible that heterogeneity in marker expression may be due to the occurrence of subpopulations of cells.

Acknowledgement

This work was supported by research grants from the USA-Israel Bi-National Science Foundation and the National Council for Research and Development, Israel and the D.K.F.Z. Heidelberg, Germany.

REFERENCES

1. VanFurth, R., and Cohn, Z. A., J. Exp. Med. 128:415, 1968.
2. Volkman, A., in "Immunobiology of the Macrophage" (D.S. Nelson, ed.), pp. 291-322, Academic Press, New York, 1976.
3. Sanderson, R. J., Shepperdson, F. T., Vatter, A. E., and Talmage, D. W., J. Immunol. 118:1409, 1977.
4. Treves, A. J., Yagoda, D., Haimovitz, A., Ramu, N., Rachmilewitz, D., and Fuks, Z., J. Immunol. Meth. 39:71, 1980.
5. Cowing, C., Schwartz, B. D., and Dickler, H.B., J. Immunol. 120: 378, 1978.
6. Newman, S. L., Musson, R. A., and Henson, P. M., J. Immunol. 125: 2236, 1980.
7. Steinman, R. M., Noguiera, N., Witmer, M. D., Tydings, J. D., and Mellman, I. S., J. Exp. Med. 152:1248, 1980.
8. Johnson, D. R. Jr., Mei, B. and Cohn, A. Z., J. Exp. Med. 146: 1613, 1977.
9. London, J., Berrih, S., and Bach, J. F., J. Immunol. 121:438, 1978.
10. Geha, R. S., Milgrom, H., Broff, M., Alpert, S., Martin, S., and Yunis, E. J., Proc. Natl. Acad. Sci. USA 76:4038, 1979.
11. Bergholtz, B. O., and Thorsby, E., Scand. J. Immunol. 6:779, 1977.
12. Rich, R. R., Abramson, S. L., Seldin, M. F., Puck, J. M., and Levy, R., J. Exp. Med. 152:218s, 1980.
13. Grumet, F. C., Charron, D. J., Fendly, B. M., Levy, R., and Ness, D. B., J. Immunol. 125:2785, 1980.
14. Haimovitz, A., Fuks, Z., Galili, N., and Treves, A. J., J. Reticuloendothel. Soc. 31:187, 1982.
15. Durocher, J. D., Payne, R. C., and Conrad, M. E., Blood 45:11, 1975.
16. Rogoff, T. M., and Lipsky, P. E., J. Immunol. 124:1740, 1980.

17. Silberger-Sinakin, I., Gighi, I., Bar, R. L., and Thorbecke,
        G. J., Immunol. Rev. 53:203, 1980.
18. Weinberg, D. S., and Unanue, E. R., J. Immunol. 126:794, 1981.

DISCUSSION

UNKNOWN: The existence of peanut agglutinin positive monocytes is
very puzzling. PNA positive cells in general are readily opsonized
by natural antibodies which are directed against the PNAs in the
serum. It is known that PNA positive cells, such as immature thymocytes, are readily trapped in the liver but this apparently does
not happen to monocytes. What is the explanation for this?

TREVES: These experiments can be done also in the absence of serum,
yet there might still be some serum proteins bound to the monocytes.
However, I do not believe that the PNA is reacting with cell bound
serum protein because the whole maturation of monocytes to macrophages
in the absence of serum yields macrophages which will stain very
nicely. It is a very small proportion but after treatment with neuraminidase, it will be again 100%.

MATHIESON: In my experience working with Bart Hanes using OKM1 and
4F2, it was clear that we brought down some T cells as well as monocytes with agglutination. I wondered if you might be seeing a subset
of T cells and not a monocyte with the PNA reagent?

TREVES: The strongly adherent cells are more than 95% phagocytic
cells, non-specific esterase positive and Fc receptor positive. The
PNA positivity was more than 75%, which is highly suggestive that we
are staining monocytes.

MATHIESON: My question really is have you ever seen T lymphocytes
staining with PNA?

TREVES: About 4% of adherent cells look like lymphocytes so T lymphocytes might be there.

CORRELATION BETWEEN TWO NEW METHODS FOR THE ISOLATION
OF HUMAN MONOCYTES: SPECIFIC REVERSIBLE AGGLUTINATION WITH
A CARBOHYDRATE FROM POKEWEED AND BY USE OF THE LECTIN PA-4

M. J. Waxdal, C. Kieda and C. Murre

Laboratory of Immunology
National Institute of Allergy and Infectious Diseases
National Institutes of Health
Bethesda, Maryland   20205 USA

INTRODUCTION

Monocytes play essential roles in the activation of both T and
B lymphocytes (1,2,3). To study the interactions of monocytes, as
well as other cells, with the responding lymphocytes, methods of
fractionating the different cell types are necessary. During our
studies on cell surface carbohydrates and carbohydrate receptors on
lymphoid cells, it became apparent that human monocytes strongly
bound both a complex carbohydrate, and rather specifically bound the
lectin Pa-4. These observations led to two new methods for the
fractionation of human monocytes from peripheral blood.

MATERIALS AND METHODS

Peripheral blood mononuclear cells were isolated by plateletpheresis of venous blood of healthy donors, followed by centrifugation
of the cell pellet on Ficoll-Hypaque. A high molecular weight, highly
branched arabinoxyloglucan (designated PWCH) was isolated from the
root of Phytolacca americana (4). The single polypeptide chain, non-
agglutinating, poly GlcNAc specific lectin, Pa-4, was isolated as
previously described (5). Pa-4 was labeled with fluorescein isothiocyanate in sodium carbonate buffer to produce Fl-Pa-4. Cyanogen
bromide activated Sepharose 6MB beads were coupled to Pa-4, or to
bovine serum albumin, in sodium borate buffer (6). The mouse monoclonal antibody, OKM1, which primarily stains human monocytes, was
obtained from Ortho Diagnostic Systems (Raritan, New Jersey, USA).

The non-specific esterase staining of monocytes was performed

233

by a modification of the method described by Tucker et al. (7).
Reagents for this stain were obtained from Technicon (Terrytown, New
York, USA). Fixed cells were incubated with substrate and dye for
8 minutes at 37°C. Microscopic examination of a typical suspension
of human peripheral blood leukocytes showed three populations; unstained (73%), pink to red (10%) and reddish brown (17%). All colored
cells were scored as esterase positive, but designated as lightly or
darkly stained.

Small angle light scatter and flow microfluorometry analyses (8)
were performed using a FACS II (Becton Dickinson, Mountain View, CA,
USA) as previously described for Fl-lectins (9). Fluorescence intensity and forward light scatter intensity (a measure of viability and
cell size), were measured simultaneously on 2.5×10^5 individual
cells. These data were stored as cell number in a 64 x 64 matrix of
light scatter (X axis) and fluorescence (Y axis). These dual parameter analyses are displayed in the figures as contour plots, where
succeeding contour lines represent double the cell number of the
preceeding contour line.

Agglutination of peripheral blood leukocytes by PWCH (4) was
carried out in buffered saline at 2.5×10^8 cells/ml and 500 µg
PWCH/ml. The best results were obtained if some autologous red cells
(2.5×10^7/ml) were included. After agglutination for 15 minutes
at room temperature, the suspension was layered on a two step density
gradient (5% and 10% BSA in buffered saline) and allowed to sediment
for 30 minutes at room temperature. Most of the agglutinated cells
were found at the bottom and most of the non-agglutinated cells were
recovered from the top and upper 1/2 of the 5% BSA layer. The agglutinated cells were disaggregated by treating for 5 minutes with
0.2 M α-methylmannoside in buffered saline. The few contaminating
erythrocytes were removed by hypotonic shock. Double E rosetting of
T cells with sheep erythrocytes was carried out as described by Fauci
et al. (10).

Absorption of peripheral blood leukocytes by Pa-4 coupled beads,
or by BSA-coupled beads, was carried out as previously described (6).
One milliliter of 5×10^6 cells in RPMI-1640 was mixed with 100 µl
of a 50% suspension of the derivatized beads, centrifuged to a pellet
and incubated at 37°C for 30 minutes. Unbound cells (Pa-4 −) were
removed by gentle suspensions and decantations with buffered saline.
The bound cells (Pa-4 +) were recovered from the beads by treatment
with buffered, Ca++ free, saline containing 25mM $(GlcNAc)_4$. The
molarity refers to the concentration of GlcNAc residues. In this
technique it is essential that each cell has contact with the Pa-4
beads.

Peripheral blood leukocytes were separated into adherent and
non-adherent fractions by incubating 5×10^7 cells in 5 ml RPMI-1640
for 1 hr at 37°C on a 10 cm plastic culture plate. Several gentle

washes with RPMI-1640 were used to remove the 'non-adherent' cells.
The plates were then treated with 2 mM EDTA in phosphate buffered
saline for 30 minutes at 37°C, after which the 'adherent' cells were
removed by decantation and washing.

RESULTS AND DISCUSSION

Flow microfluorometric analyses of Fl-Pa-4 stained unfractionated
peripheral blood leukocytes indicated that Pa-4 bound primarily to
monocytes (Figure 1A). The contour plot of light scatter vs fluores-
cence with Fl-Pa-4 is essentially identical to that obtained with
OKM1, a monoclonal antibody which primarily stains human monocytes.
This was also true for each of the remaining panels in Figure 1.
The contour plots for OKM1 are not shown here, but are presented in
Ref. 6.

Fractionation of leukocytes with PWCH resulted in two populations
of cells. The non-agglutinated population (PWCH) negative was almost
totally depleted of darkly stained esterase positive cells (<0.1%,
Table 1) and consisted primarily of small lymphocytes. This population varied from <0.1% to 2% in lightly esterase positive cells.
The agglutinated population (PWCH) positive was enriched in esterase
positive cells (Table 1) and also markedly enriched in large Fl-Pa-4
brightly fluorescent cells (predominantly monocytes, Figure 1B).
The smaller PWCH positive cells are mostly T cells which can be removed by E rosetting, recovered, and stained with Fl-Pa-4. However,
they are either unstained or weakly stained by Fl-Pa-4 (Figure 1C).
These T lymphocytes are also markedly depleted in esterase positive
cells (Table 1). The larger population of PWCH positive cells are
all esterase positive (Table 1) and predominantly brightly stained
with Fl-Pa-4 (Figure 1D). There is a small subpopulation of these
cells which are moderately stained with Fl-Pa-4, the number of which
correlates well with the number of lightly esterase positive cells.
All of the larger PWCH positive cells are adherent and appear to be
comprised predominantly of monocytes.

Although Fl-Pa-4 readily bound to subsets of human blood leukocytes, it did not agglutinate them even at high concentrations.
Consequently, Pa-4 was chemically coupled to Sepharose 6MB to yield
multivalent Pa-4 beads. These beads were used to specifically absorb
cells with receptors for Pa-4. The results are presented in Table 1
and show that Pa-4 beads removed almost all of the darkly stained
esterase positive cells from human blood leukocyte preparations, but
was less efficient in removing the lightly positive cells. During
incubation, contact between each cell and the Pa-4 beads was critical
for depletion. Variations in purity may be due primarily to this
factor. The Pa-4 bead bound cells could be readily released from
the beads and recovered by treatment with 25 mM $(GlcNAc)_4$ in buffered
saline. Control experiments, using bovine serum albumin derivatized

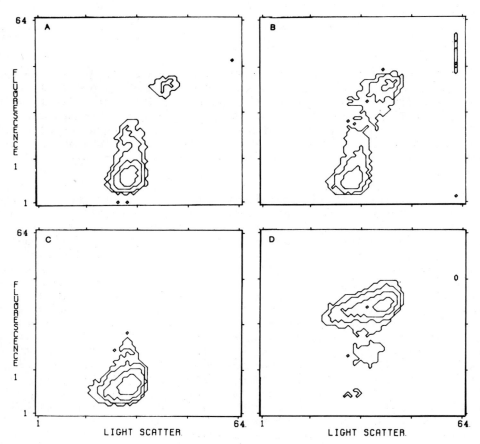

Fig. 1.   Correlated dual parameter flow cytometric analyses of Fl-Pa-4
          stained human peripheral blood leukocytes fractionated by
          PWCH agglutination and E rosetting.   (A) Unfractionated
          cells.   (B) PWCH positive cells (agglutinated and recovered).
          (C) PWCH positive T cells.   (D) PWCH positive monocytes.
          The X axis is forward light scatter intensity, a measure of
          cell size.   The Y axis is the logarithm of fluorescence
          intensity due to bound Fl-Pa-4.   Successive contour lines
          indicate two-fold increases in cell number.   The samples were
          stained with 50 µg/ml of Fl-Pa-4.   Light scatter gates were
          set to exclude erythrocytes and dead cells.

Table 1. Esterase Staining of Fractionated Cells

Fractionation Procedure	Predominant Cell Types	% Esterase Positive		
		Light Staining	Dark Staining	Total
Unfractionated[a]		10	17	27
PWCH negative	small lympho-cytes	1	<0.1	1
PWCH positive	mixed T cells & monocytes	20[b]	70	90
small cells	T cells	5	2	7
large cells	monocytes	15	85	100
Pa-4 negative		9	1	10
Pa-4 positive (eluted)		27	42	69
Non-adherent		15	3	18
Adherent		40	43	83

[a]Values reported are typical results from a single patient and vary less than ± 3% in repeated determinations on the same preparation of peripheral blood leukocytes.
[b]These cells were preferentially lost during the PWCH fractionation and may be found also at the 5% to 10% interface in a bovine serum albumin gradient.

beads, removed only about 10% of the esterase positive cells. Further Pa-4 bead binding experiments were performed with the PWCH fractionated populations of blood leukocytes. Greater than 99% of the larger PWCH positive population (all esterase positive, Table 1) were bound to the Pa-4 beads. On the other hand, less than 1% of the PWCH negative population (less than 1% esterase positive, Table 1) were bound.

These data suggest that the interactions between F1-Pa-4 and human esterase positive cells is specific and that it may be useful as a new marker for these cells. Pa-4 beads also may be used for the depletion or enrichment of human monocytes. The complex carbohydrate, PWCH, binds to, and agglutinates essentially all human monocytes and should be useful for their removal or purification. PWCH also binds and agglutinates a specific subpopulation of T cells. These T cells may be removed by E rosetting, either before or after PWCH treatment. Both of these new techniques are superior to adherence for the removal of esterase positive cells (Table 1) and provide an inexpensive, quick and simple procedure for the fractionation of human peripheral blood leukocytes.

Acknowledgement

We thank Steve Trost for technical assistance, Virginia Poke for preparation of the complex carbohydrate, Susan Sharrow, David Stephany for the flow microfluorometric analyses, and Dr. Michael Monsigny for the poly GlcNAc.

REFERENCES

1. deVries, J. E., Caviles, A. P., Bont, W. S., and Mendelsohn, J., J. Immunol. 122:1099, 1979.
2. Mosier, D. E., Science 158:1573, 1967.
3. Feldmann, M., J. Exp. Med. 135:1049, 1972.
4. Kieda, C., and Waxdal, J. J., submitted for review.
5. Waxdal, M. J., Biochemistry 13:3671, 1974.
6. Murre, C., Kieda, C., and Waxdal, M. J., submitted for review.
7. Tucker, S. B., Pierre, R. V., and Jordan, R. E., J. Immunol. Methods 14:267, 1977.
8. Loken, M. R., and Herzenberg, L. A., Ann. N. Y. Acad. Sci. 254: 163, 1975.
9. Fowlkes, B. J., Waxdal, M. J., Sharrow, S. O., Thomas, C. A., Asofsky, R., and Mathiescn, B. J., J. Immunol. 125:623, 1980.
10. Fauci, A. S., Pratt, K. R., and Whalen, G., J. Immunol. 117: 2100, 1976.

DISCUSSION

NORMANN: What was the yield of monocytes by your agglutination methodology and what was the purity?

WAXDAL: The yield has always been greater than 90%. Most of the cell loss occurs at the interface of the five and ten percent BSA boundaries. The actual purity of monocytes based on their cytoplasmic esterase reactivity is about 85%.

NORMANN: To what extent does agglutination change the functional characteristics of the cells?

<u>WAXDAL</u>: Probably not much.   The only thing we've done to date is to
take the agglutinated cells, remove the carbohydrate, reconstitute
the original mixtures, and then look for mitogen activation and for
immunoglobulin production in culture. We find essentially no differences between reconstituted mixtures and unfractionated ones. We
were rather surprised at the results because we thought we would
activate the macrophages but we did not. The most obvious guess is
that PWCH is binding to the mannose receptor, which is then displaced
by mannose. This could be confirmed by using a fluorescent probe.
I don't think it's internalized.

CHARACTERISTICS OF HUMAN MONOCYTE SUB-

POPULATIONS WITH THE USE OF FLOW CYTOMETER

Mária Kávai,[1] Edit Bodolai,[1] and János Szöllösi[2]

[1]IIIrd Department of Medicine, University Medical
School of Debrecen, Debrecen, Hungary
[2]Department of Biophysics, University Medical School
of Debrecen, Debrecen, Hungary

Monocytes isolated by counterflow centrifugation have a bimodal volume distribution with the large monocytes predominantly Fc receptor (FcR) positive and active in ADCC assays (1). Volume analysis of human monocytes with a multiparameter cell sorter has revealed three distinct populations with different chemotactic capabilities (2).

In this study, purified human monocytes were separated and analyzed in a fluorescence activated cell sorter (FACS III) in order to investigate their heterogeneity with respect to size, esterase activity, and Fc receptor number and association constant.

METHODS AND RESULTS

Human monocytes that were separated from peripheral blood by Ficoll-Hypaque and adherence to serum-coated dishes (93±5 % monocytes) (3) were analyzed with FACS III. The scattered light intensity distribution histograms of the cells are shown in Figure 1. The light scattering was proportional to the cross-sectional area of the cells (4). Three peaks were observed on the frequency distribution histograms. The characteristics of the cell fractions were further defined by (a) nonspecific esterase staining, (b) microscopic visualization of phagocytized latex (1.091 μ in diameter) and (c) the reaction of FcR with sensitized sheep red blood cells (SRBC). The first peak belonged to the red blood cells and debris; the second one to lymphocytes and small monocytes; the third one to large monocytes.

The monocyte suspension was stained with fluorescein diacetate

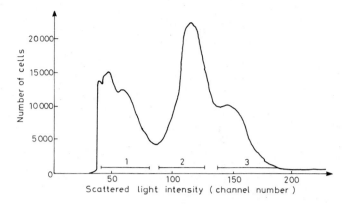

Fig. 1.   Scattered light intensity distribution of purified monocytes.
          Fraction 1: red blood cells and debris; fraction 2: lympho-
          cytes and small monocytes; fraction 3: large monocytes.

(FDA); 10 µl of FDA of 10^{-4} M were added to 1 ml of 2×10^6 cell/ml,
at 25°C. The cell distribution was examined according to size and
fluorescein content. The data analysis was triggered by fluorescence
intensity and only two peaks could be observed on the scattered light
distribution histograms and two on the fluorescence intensity distribution histograms (Figure 2). The first peak belonging to the
red blood cells and debris (Figure 1) disappeared because these cells
and debris did not accumulate fluorescein. In order to measure the
size distribution of monocytes with high fluorescence intensity
separately, data collection was triggered with discriminated fluorescence signals as indicated in Figure 2. Cells with low scattered
light intensities (small size ones) had low fluorescence, whereas the
large ones had high fluorescence intensities.

     During FDA hydrolysis the fluorescence intensity distribution
of cells changes with time. The average fluorescence intensity of
small monocytes and that of large ones increased at the beginning
of the reaction and reached different plateaus (Figure 3). The exact
fluorescein content of cells could be determined by comparing the
average fluorescence intensities of cells to that of chicken RBC as
a standard. This finding indicates two and a half times higher intracellular fluorescein concentraticn in large monocytes compared to
that of small ones, taking into account the difference in size.

     Monocytes (2×10^6/ml) were incubated with increasing amounts of
human monoclonal IgG1 labeled with fluorescein isothiocyanate (FITC)
(5) or with medium alone, at 25°C for an hour and analyzed for

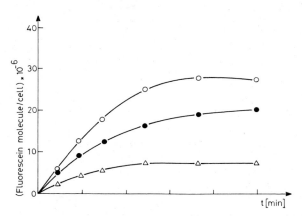

Fig. 2.   Distribution histograms of scattered light and fluorescence
           intensity of cells stained with fluorescein diacetate.
           Solid line: data collection triggered by light scattering;
           slanted lines: data collection triggered by the hatched
           region of fluorescence intensity.

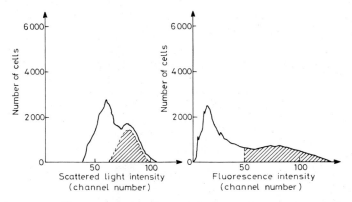

Fig. 3.   Average fluorescence intensities of monocytes during fluores-
           cein diacetate hydrolysis. Open circles o: large monocytes;
           closed circles •:   total monocytes; triangles: small mono-
           cytes.

fluorescence at 25°C within half an hour.   Cells lacking FITC-IgG1
had a sharp distribution with relatively low fluorescence intensity.
With increasing amounts of FITC-IgG1, the peak broadened and moved
to higher channel numbers.   The average fluorescence intensities
were calculated from the two peaks of the fluorescence intensity
distribution and were compared to the fluorescence intensity of
chicken RBC in order to determine the amount of bound IgG1 molecules
per cell.

Figure 4 shows the specific binding data displayed as Scatchard
plots.   Small and large monocytes had a different number of binding
sites (3.3×10^5 and 1×10^6,   respectively), but practically identical
association constants (1.28×10^5 M^{-1} and 1.17×10^5 M^{-1}).   The cell surface density of the FcR of large monocytes was 2 to 3 times higher
than that of the small ones, even after normalizing the differences
in size.   As controls, the binding of FITC-F(ab')2 and FITC-BSA to
monocytes was measured.   A very low association constant and very
high numbers of binding sites were observed in both experiments.

Summarizing the data of eight separate experiments, the average
association constant for the reaction of IgG1, was in the range of
$1-2 \times 10^5$ M^{-1} and the numbers of FcR of small and large monocytes were
within the range of $2-5 \times 10^5$ and $8-13 \times 10^5$, respectively.

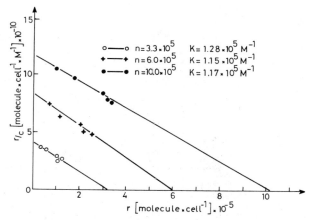

Fig. 4.   Scatchard plots demonstrating the variation in numbers of
          receptor sites of IgG1 (n) without significant difference
          in average association constants (K).   Solid circles •:
          large monocyte; open circles o: small monocyte; -x-:
          total monocyte.

These studies suggest that the activity of the cell surface receptors in terms of the relative adherence of IgG1 was related to the number of the receptor sites and not to their affinity. The different FcR numbers on monocyte subpopulations may be associated with the maturation of the mononuclear phagocytic cell line (6).

SUMMARY

Human monocytes separated from peripheral blood by Ficoll-Hypaque and by adherence to serum-coated dishes showed a bimodal volume distribution measured with a fluorescence activated cell sorter. The rate of the fluorescein diacetate hydrolysis by small monocytes was lower than that of large ones. The majority of large monocytes reacted with sensitized SRBC while only a minority of small monocytes bound sensitized SRBC. Scatchard plots on the binding of FITC-labeled human monoclonal IgG1 to the two subpopulations indicated similar association constants, $K=1.2\pm0.3\times10^5$ M^{-1}. The number of FcR was significantly different for the small ($3.3\pm0.6\times10^5$) and the large monocytes ($10\pm1\times10^5$).

REFERENCES

1. Norris, D. A., Norris, R. M., Sanderson, R. J., and Kohler, P. F., J. Immunol. 123:166, 1979.
2. Arenson, E. B., Epstein, M., and Seeger, R., J. Clin. Invest. 65: 613, 1980
3. Kumagai, K., Itoh, K., Hinuma, S., and Tada, M., J. Immunol. Methods 29:17, 1979.
4. Salzman, G. C., Mullaney, P. F., and Price, B. J., in "Flow Cytometry and Sorting" (M. R. Melamed, P. F. Mullaney and M.L. Mendelsohn, eds.), pp. 105-124, Wiley Medical Publication, New York, 1979.
5. de Petris, S., in "Methods in Membrane Biology" (E. D. Korn, ed.), 9:1-201, Plenum Press, New York, 1978.
6. van Furth, R., Raeburn, J. A., and van Zwet, T. L., Blood 54:485, 1979.

SECTION 4

GROWTH REGULATION BY MACROPHAGES

GROWTH REGULATION BY MACROPHAGES

Walker Wharton, Edwin Walker and Carleton C. Stewart

Experimental Pathology Group
Los Alamos National Laboratory
Los Alamos, New Mexico   87545 USA

INTRODUCTION

Macrophages play an important role in several aspects of immune regulation, as verified by the articles within this volume. In recent years it has become clear that macrophages play a pivotal regulatory role by influencing the rates of proliferation of multiple cell types. This article is a review focusing on the products secreted by macrophages which alter cell growth.

COLONY STIMULATING ACTIVITIES

Colony stimulating activities (CSA) which cause hematopoietic progenitor cells to form colonies in vitro have been obtained from a multitude of sources. The most extensively used demonstration system for these activities employs a soft agar culture system incorporating either human bone marrow or, more typically, murine bone marrow as the target cells. Table 1 lists several sources of human and murine colony stimulating activities and the predominant colony type which forms (1-20). Since different types of colonies form depending on the source of CSA, most investigators favor the notion that distinct colony stimulating factors exist which specifically stimulate the proliferation of either neutrophil, eosinophil, macrophage, megakaryocyte or erythrocyte progenitor cells. It is clear that either some types of CSA are produced by macrophages or that macrophages can induce other cells to produce CSA.

Of particular importance is the observation by Schreier and Iscove (21) that elaboration of some classes of CSAs are dependent on T-lymphocytes (T-cells) and syngeneic accessory cells. They showed

249

Table 1. Sources and Target Cells Responsive to Colony Stimulating
          Activities

Source of Colony Stimulating Activity	Colony Types Found
Human:	
peripheral blood leukocytes	granulocytic and macrophagic
embryonic kidney cells	granulocytic and macrophagic
spleen cells	granulocytic and macrophagic
vascular cells	granulocytic and macrophagic
lung cells	granulocytic and macrophagic
giant tumor cells	granulocytic and macrophagic
human placenta	granulocytic and macrophagic
human serum	granulocytic and macrophagic
Murine:	
mouse lung conditioned medium	granulocytic and macrophagic
endotoxin serum	granulocytic and macrophagic
pregnant mouse uterine extract	granulocytic and macrophagic
pokeweed mitogen spleen conditioned medium	granulocytic, macrophagic, megakaryocytic and erythoid
endotoxin spleen conditioned medium	granulocytic, macrophagic, magakaryocytic and erythoid
embryo fibroblast conditioned medium	macrophagic
L-cell conditioned medium	macrophagic
WEHI-3 conditioned medium	granulocytic and megakaryocytic

that macrophage, granulocyte and erythroid CSA was present in culture supernatants of interleukin-2 (IL-2) dependent cloned T-cells only when the specific antigen and accessory cells of the correct haplotype were also present. Little activity was found in medium conditioned constitutively.

When pokeweed mitogen (PWM) (14,15) or endotoxin (LPS) (16,22) was added to fractionated spleen cells, both T-cells and adherent cells were absolutely required for production of CSA; PWM stimulated spleen cells from nude mice did not produce activity unless normal T-cells were also added. It cannot be determined whether macrophages or T-cells were actually producing the CSAs even though both were required.

When LPS was added directly to macrophages and the one to four day conditioned medium was then added to nude mouse spleen cells, CSA was produced (16). Maximal macrophage derived activity was obtained after 24 hours of culture. When macrophages were derived from LPS-unresponsive C_3H/HeJ mice and cultured with LPS, the conditioned medium failed to induce nude mouse spleen cells to produce CSA. But when macrophages were derived from the responsive C_3H/eB mouse, CSA was produced. This data suggested that the T-cells which secrete IL-2 are different from those secreting CSA since nude mice cannot produce IL-2 (23).

Staber, et al., (24) tested pure preparations of the bacterial cell-wall components Lipid A, LPS and murein in cultures of spleen and lymph node cells. They also found that adherent cells were absolutely required for CSA release; mitogenicity, however, was not a prerequisite for release. Muramyl dipeptide and LPS degradation products also induced release of CSA. These data suggest that LPS added directly to cultures of adherent peritoneal cells induced CSA, i.e. activity was directly elaborated by macrophages. However, adherent lymphocytes which contaminate the adherent fraction may have been stimulated by an LPS induced macrophage factor.

Pelus, et al., (25) have shown that resident murine peritoneal "monocytoid" cells are capable of constitutive release of colony stimulating activity. Using velocity sedimentation, one population of secretory cells sedimented over a range of 5-8mm/hr and the other had a mean velocity of 4mm/hr. Lactoferrin inhibited the release from the 5-8mm/hr cells, although this inhibitory effect was reversed by addition of LPS. In addition, when LPS was added, a new cell population sedimenting at 1-3mm/hr began releasing CSA.

These apparent discrepancies between constitutive and stimulated release of CSAs need to be resolved. Quantitative differences in the isolation of target bone marrow cells, cell contamination differences in the semi-purified populations under test, and LPS contamination by the culture medium constituents are clearly intangible differences among laboratories which could lead to the conflicting interpretations. Even so it seems very clear that several different colony stimulating factors exist and that macrophages are involved in the production of some of them. Indeed, perhaps the only one they do not produce, or are not involved in producing, is the 70,000 dalton macrophage growth factor (CSF-1) derived from L-929 cells and embryo fibroblasts (17,18). Whether they produce the factors themselves or produce factors which act on lymphocytes which in turn secrete these colony stimulating activities is still unresolved. Until absolutely pure populations of each cell type can be obtained and co-cultured in a controlled manner, the resolution of this problem is likely to evade us.

LYMPHOCYTE STIMULATING ACTIVITIES

     In addition to colony stimulating activities, cells of the mono-nuclear phagocyte compartment also release lymphostimulatory molecules which modulate and profoundly influence the function of T and B lym-phocytes (reviewed in 26-28). Principal among these activities is the soluble factor, lymphocyte activating factor (LAF), more recently referred to as interleukin 1 (IL-1). The first experimental descrip-tion of this molecular activity came from the observations by Gery and Waksman (29,30) using adherent cells from human peripheral blood. In this early study and in subsequent work with IL-1, the discrimi-nating activity ascribed to this factor is the ability to stimulate thymocyte proliferation after 72-96 hours in culture. IL-1 acts either alone or synergistically with phytohemagglutinin (PHA) or concanavalin A (Con A) to trigger thymocyte proliferation. Recent work has confirmed the fact that the mononuclear phagocytes are the cellular source of IL-1. Thus, in experiments using murine peritoneal exudate cells, murine macrophage cell lines, and human peripheral blood monocytes (31-35), IL-1 was generated using a variety of induction protocols. These studies employed LPS, latex particles, viable Listeria organisms, immune complexes, dimethyl sulfoxide, and phorbol myristate acetate (PMA) to induce IL-1 secretion. Co-culture of murine peritoneal macrophages or the murine macrophage tumor cell line P388.D1 with activated T lymphocytes (36,37) also stimulated IL-1 secretion by the macrophage effector cells. Thus IL-1 secretion may be dependent upon the macrophage receiving an activating signal mediated by T lymphocytes.

     IL-1 from a variety of sources has been partially purified and characterized (reviewed in 26-28). Human derived IL-1 has a molecular weight of 12,000 to 14,000 daltons and an isoelectric point of pH 6.8 to 7.0. Peaks of activity with larger molecular weight character-istics have been described but not yet fully characterized (32,38,39). Studies using murine IL-1 show it to be a molecule of approximately 15,000 daltons (40), with an isoelectric point of pH 4.8 (41). Both human and murine IL-1 display charge heterogeneity when further sepa-rated by DEAE chromatography or by isoelectric focusing (39,41,42-44).

A recent report by Mizel describes the purification to homogeneity of IL-1 generated by the murine macrophage tumor cell line
P388.D1 under a "superinduction" protocol using PMA, cyclohexamide
and actinomycin D (39). However, the results of this study can be
questioned by virtue of the fact that recent unpublished data of
S. Shimizu, R. T. Smith, and V. C. Maino, University of Florida School
of Medicine, shows that PMA binds both to some undefined element in
fetal calf serum, and to purified soluble BSA such that a PMA-protein
complex shows potent stimulation of thymocyte proliferation by itself
and acts synergistically with either Con A or PHA to induce even
greater ^{3}H-Tdr incorporation by stimulated thymocytes. This PMA-
protein complex ("SLAF", synthetic lymphocyte activating factor)
migrates to a pI value of 4.5-5.0 after isoelectric focusing separation.

Although the effects of IL-1 on thymocytes are well documented,
the direct effect of this lymphoproliferative factor on B lymphocyte
effector target cells is controversial. Studies by several different
laboratories have shown that both human and murine macrophages produce
a soluble factor of approximately 13,000-15,000 daltons which augments
in vitro plaque-forming responses (PFC) by splenic lymphocytes from
nude mice or from B cell enriched (T cell depleted) cultures (31,42,
43). This enhancement of the B cell response (PFC response) is produced by conditions exactly comparable to those required for the
generation of IL-1 activity. Thus, the use of LPS induction protocols
or stimulated T cells would trigger macrophage production of the B
cell differentiating factor function as well as IL-1 activity (33,36,
42). This seemingly B cell-specific activity has been referred to as
B cell activating factor (BAF) by at least one group of workers (45).
There have been conflicting reports from different laboratories suggesting that the two factor activities, IL-1 and BAF, can be separated
from each other biochemically (38), and, conversely, by others that
the two activities co-purify and are probably properties of a single
molecular species which cannot be chemically resolved from one another
(42,43). To date there is no definitive analysis of the issue of
how the macrophage derived activities variously described as BAF
or IL-1 might act on B cells directly or how they might affect B
cell maturation by a synergistic interaction with T helper cell
activity. Experimental analysis of this question remains a current
area of intensive activity. Nevertheless, the data are clear that
macrophages produce activities which are involved in the regulation
of lymphocyte proliferation.

FIBROBLAST STIMULATING ACTIVITIES

The growth of nontransformed fibroblasts is strictly dependent
on soluble extracellular hormones and growth factors (46), which are
usually supplied in tissue culture by the addition of whole blood
serum to a chemically defined culture medium. Pledger, et al. (47,

48) showed that serum components could be separated into two functionally distinct groups. The initiation of cell cycle traverse
was controlled by platelet-derived growth factor (PDGF) which is
released from a sequestered state if platelets are present during
clot formation. PDGF was said to make quiescent cells "competent"
to respond to factors present in platelet-poor plasma (PPP). PPP,
which is made by removing the formed elements from whole blood before
clotting is allowed to occur, regulated the temporal traverse of
competent cells through the cell cycle and was thus said to contain
"progression" activity. Based on these data, Stiles et al. (49)
proposed a "dualistic" model of in vivo fibroblast proliferation.
This model postulates that cells in the body are bathed in an ultrafiltrate of plasma which contains all the factors necessary for
proliferation except those possessing competence activity. Fibroblast
growth in any particular tissue is postulated to be regulated by the
localized concentration of competence factors.

This model appears to accurately describe the serum-stimulated
in vitro growth of fibroblasts as well as the proliferation which
is observed in vivo during early stages of wound healing where PDGF
is locally released as platelets are lysed during clot formation.
There are, however, several examples of in vivo fibroblast proliferation which can be observed in the apparent absence of the accumulation
and lysis of platelets. Since several of these examples, such as
the proliferation commonly found in sites of chronic inflammation or
in tumors, are also characterized by a significant accumulation of
macrophages, it seems likely that cells in the mononuclear phagocyte
series might be a source of competence activity.

Leibovich and Ross (50) reported that in vivo wound healing
was impaired in guinea pigs with an experimentally depressed number
of macrophages. Later, the same authors reported that peritoneal
macrophages cultured in PPP produced a factor(s) that stimulated the
proliferation of fibroblasts (Leibovich and Ross, 51). However, in
a later publication, Leibovich (52) presented evidence that the
mitogenic substance was produced by a macrophage-dependent metabolism
of a plasma constituent, casting doubt on the role of this cell type
as a primary source of stimulatory activity. More recently Glenn and
Ross (53) reported that activated human peripheral monocytes cultured
in medium containing PPP and either Con A or endotoxin also produced
mitogenic activity for fibroblasts.

Over the past several years there has been a number of papers
showing a putative role for macrophages in the control of fibroblast
proliferation (54,55,56,57). These results were obtained using a
multitude of experimental systems asking, at least in principle, if
macrophages could produce activity which could act in any of several
ways to stimulate fibroblast proliferation. Although these data
are of potential interest from an in vitro point of view, if the model
of proliferation developed by Stiles, et al. (49) is correct, any

macrophage elaborated mitogens which are important regulators of
fibroblast proliferation in vivo, should act synergistically with
PPP in a manner similar to the PDGF-mediated effects. An important
question is whether macrophages elaborate a mitogenic activity which
acts as a competence factor.

As shown in Table 2, both resident peritoneal cells and thioglycollate-elicited peritoneal macrophages,which were maintained for
3 days in culture medium containing 5% PPP,elaborated mitogenic
activity for BALB/c-3T3 cells. Because of the limited number of cells
available in primary cultures and since there are multiple cell types
present in peritoneal washes, the ability of cloned murine macrophage-
like tumor cells to elaborate mitogenic activity was also investigated. As shown in Table 3, the macrophage-like tumor line $P388D_1$
also elaborated activity which stimulated the proliferation of fibroblasts. The amount of mitogenic activity at any concentration of
conditioned medium was dependent on the length of exposure to the
$P388D_1$ cells (data not shown). In addition, the macrophage-like
cells produced activity when they were maintained in serum-free
medium. Similar results were obtained using several of the common
macrophage-like cell lines, and indicated that the metabolism of a
plasma factor was not responsible for the production of the activity.

For a mitogen to act as a competence factor, it is necessary
that it act synergistically with PPP to stimulate proliferation.

Table 2. Mitogenic Activity Elaborated by Primary Murine Macrophages

Source of Mitogenic Activity	Percentage Labeled Nuclei			
	% Macrophage Medium			
	0	10	30	100
Resident Peritoneal Cells	4	6	16	20
Thioglycollate-Elicited Cells	6	17	47	67

Peritoneal cells were maintained for 72 hours in serum free DME.
The conditioned medium was then centrifuged and filtered, and the
indicated concentration added together with 5% platelet poor
plasma and 5 μCi/ml ^{3}H-thymidine to quiescent cultures of BALB/
c-3T3 cells. The number of cells initiating DNA synthesis during
a 36 hour period was determined by autoradiography.

Table 3. Elaboration of Mitogenic Activity by the Macrophage-like
          Tumor Cell Line P388D$_1$

Character of Conditioned Medium	Percentage Labeled Nuclei			
	% Macrophage Medium			
	0	10	30	100
Conditioned with PPP	4	41	82	100
Conditioned without PPP	5	32	62	84

P388D$_1$ cells were maintained for 72 hours in DME either with or without supplementation with 5% platelet poor plasma (PPP). The medium
was then harvested, centrifuged and sterile filtered, and added to
quiescent BALB/c-3T3 cells with the PPP concentration adjusted to 5%.
The number of cells initiating DNA synthesis was determined by autoradiography.

As shown in Table 4, this criteria was satisfied for medium conditioned by P388D$_1$ cells. The addition of only plasma-free medium
conditioned for 3 days by P388D$_1$ cells did not stimulate the proliferation of 3T3 cells. The addition of medium containing only PPP
was also not effective in stimulating growth. However, when medium
containing both the macrophage-elaborated products and PPP were added
together, a stimulation of DNA synthesis equivalent to that seen following serum stimulation was observed. The other criteria for a
competence factor proposed by Stiles, et al. (49), is that a transient
exposure to the mitogen is sufficient to stimulate one round of DNA
synthesis and that cells exposed to the factor alone remain 12 hours
from the initiation of DNA synthesis. These criteria have also been
shown for crude macrophage-conditioned medium (data not shown). The
activity which stimulates the proliferation of fibroblasts is not
identical to IL-1, since they can be separated on DEAE columns (data
not shown).

SUMMARY

     The evidence reviewed here indicates that macrophages, either
acting alone or in concert with other cells, influence the proliferation of multiple types of cells. Most of the data indicate that these

Table 4.   Synergism of P388D$_1$ Conditioned Medium with PPP

Concentration of PPP in Conditioned Medium	Percentage Labeled Nuclei % Macrophage Medium			
	0	10	30	100
0.25% PPP	2	4	7	5
1.00% PPP	4	17	21	26
2.50% PPP	4	27	39	58
5.00% PPP	3	62	87	98

Serum-free DME which had been exposed to P388D$_1$ cells for 72 hours was added with 5µCi/ml ^{3}H-thymidine and the indicated concentrations of platelet poor plasma (PPP) to quiescent cultures of BALB/c-3T3 cells.   Following a 36 hour incubation the fibroblasts were harvested and the percentage of labeled nuclei was determined by audioradiography.

effects are mediated by soluble macrophage-elaborated products (probably proteins) although the role of direct cell-to-cell contacts cannot be ruled out in all cases. A degree of success has been achieved on the biochemical characterization of these factors, but such work has been hampered by the factors low specific activity in conditioned medium and the lack of rapid, specific assays. It is our belief that understanding the growth-regulating potential of macrophages is an important and needed area of research.

Acknowledgement

        This work was performed under the auspices of the Department of Energy. This investigation was supported by Grant Number AI15563 from the National Institute of Health, Bethesda, MD.

REFERENCES

1. Shah, R. G., Caporals, L. H., and Moore, M. A. S., Blood 50:811, 1977.
2. Pike, B. L., and Robinson, W. A., J. Cell.Physiol. 76:77, 1970.

3. Brown, C. H., and Carbone, P. P., J. Natl. Cancer Inst. 46:989, 1971.

4. Paran, M., Sachs, L., Barak, Y., Resnitzky, P., Proc. Natl. Acad. Sci. USA 67:1542, 1970.

5. Knudtzon, S., and Mortenson, B. T., Blood 46:937, 1975.

6. Foja, S. S., Wu, M-C., Cross, M. A., and Yunis, A. A., Biochim. Biophys. Acta 494:92, 1977.

7. DiPersio, J. F., Brennan, J. K., Lichtman, M. A., and Speisser, B. L., Blood 51:507, 1978.

8. Nicola, N. A., Metcalf, D., Johnson, G. R., Burgess, A. W., Blood 54:614, 1979.

9. Furusawa, S., Komatsu, H., Saito, K., Enokihara, H., Hirosa, K., and Shishido, H., J. Lab. Clin. Med. 91:377, 1978.

10. Byrne, P., Heit, W., and Kubanek, B., Cell Tissue Kinet. 10:341, 1977.

11. Williams, N., and Burgess, A. W., J. Cell.Physiol. 102:287, 1980.

12. Bol, S., and Williams, N., J. Cell.Physiol. 102:233, 1980.

13. Staber, F. G., and Burgess, A. W., J. Cell.Physiol. 102:1, 1980.

14. Burgess, A. W., Metcalf, D., Russell, S. H. M., and Nicola, N. A., Biochem. J. 185:301, 1980.

15. Metcalf, D., and Johnson, G. R., J. Cell. Physiol. 96:31, 1978.

16. Apte, R. N., Hertogs, Ch.F., and Pluznik, D. H., J. Immunol. 124:1223, 1980.

17. Stewart, C. C., and Lin, H-L., J. Reticuloendothelial Soc. 23:269, 1978.

18. Stanley, E. R., in "The Lymphokines" (R. Stewart and J. Hadden, eds.), Humana Press, N.J., 1981.

19. Stanley, E. R., and Guilbert, L. J., J. Immunol. Methods 42:253, 1981.

20. Williams, N., Jackson, H., Ralph, P., and Nakoinz, I., Blood 57:157, 1981.

21. Schreier, M. H., and Iscone, N. N., Nature 287:228, 1980.

22. Apte, R. N., Hertogs, Ch.F., and Pluznik, D. H., J. Reticulo-endothelial Soc. 26:491, 1979.

23. Lipsick, J. S., and Kaplan, N. O., Proc. Natl. Acad. Sci.USA 78:2398, 1981.

24. Staber, F. G., Gisler, R. H., Schumann, G., Tarcsay, L., Schlafli, E., and Dukor, P., Cellular Immunol. 37:174, 1978.

25. Pelus, L. M., Broxmeyer, H. E., Desousa, M., and Moore, M. A. S., J. Immunol. 126:1016, 1981.

26. Rocklin, R. E., Bendtzen, K., and Greineder, D., Adv. Immunol. 29:56, 1980.

27. Cohen, S., Pick, E., and Oppenheim, J. J., in "Biology of the Lymphokines", Academic Press , New York, 1979.

28. Unanue, E. R., Adv. Immunol. 31:1, 1981.

29. Gery, I., and Waksman, B. H., J. Immunol. 107:1778, 1971.

30. Gery, I., and Waksman, B. H., J. Exp. Med. 136:143, 1972.

31. Calderon, J., Kiely, J. M., Lefke, J. L., and Unanue, E. R., J. Exp. Med. 142:151, 1975.

32. Blyden, G., and Handschumacher, R. E., J. Immunol. 118:1631, 1978.

33. Unanue, E. R., Kiely, J. M., and Calderon, J., J. Exp. Med. 144:
       155, 1976.
34. Lachman, L. B., Hacker, M. P., Blyden, G. T., and Handschumacher,
       R. E., Cellular Immunol. 34:416, 1977.
35. Lachman, L. B., Hacker, M. P., and Handschumacher, R. E., J.
       Immunol. 1:2019, 1977.
36. Unanue, E. R., Beller, D. J., Calderon, J., Kiely, J. M., and
       Stadecker, M. J., Am. J. Pathol. 85:465, 1976.
37. Mizel, S. B., Oppenheim, J. J., and Rosenstreich, D. L., J.
       Immunol. 120:1504, 1978.
38. Wood, D. D., Cameron, P. M., Poe, M. T., and Morris, C. A.,
       Cellular Immunol. 21:88, 1976.
39. Mizel, S. B., and Mizel, D., J. Immunol. 126:834, 1981.
40. Unanue, E., and Kiely, J. M., J. Immunol. 119:925, 1977.
41. Economu, J. S., and Shin, H. S., J. Immunol. 121:1446, 1978.
42. Koopman, W. J., Farrar, J. J., Oppenheim, J. J., Fuller-Bonar,
       J., and Dougherty, S., J. Immunol. 119:55, 1977.
43. Koopman, W. J., Farrar, J. J., and Fuller-Bonar, J., Cellular
       Immunol. 35:92, 1978.
44. Mizel, S. B., Rosenstreich, D. L., and Oppenheim, J. J., Cellular
       Immunol. 40:230, 1978.
45. Wood, D. D., and Cameron, P. M., J. Immunol. 121:53, 1978.
46. Holley, R. W., Nature 258:487, 1975.
47. Pledger, W. J., Stiles, C. D., Antoniades, H. N., and Scher,
       C. D., Proc. Natl. Acad. Sci. USA 74:4481, 1977.
48. Pledger, W. J., Stiles, C. D., Antoniades, H. N., and Scher,
       C. D., Proc. Natl. Acad. Sci. USA 75:3829, 1978.
49. Stiles, C. D., Capone, G. T., Scher, C. D., Antoniades, H. N.,
       Van Wyk, J. J., and Pledger, W. J., Proc. Natl. Acad. Sci.USA
       76:1279, 1979.
50. Leibovich, S. J., and Ross, R., Am. J. Pathol. 78:71, 1975.
51. Leibovich, S. J., and Ross, R., Am. J. Pathol. 84:501, 1976.
52. Leibovich, S. J., Exp. Cell Res. 113:47, 1978.
53. Glenn, K. C., and Ross, R., Cell 25:603, 1981.
54. Greenburg, G., and Hunt, T., J. Cell. Physiol. 97:353, 1978.
55. Wall, R. T., Harker, L. A., Quadracci, L. J., and Striker,
       G. E., J. Cell. Physiol. 96:203, 1978.
56. Jalkanen, M., Peltonen, J., and Kulonen, E., Acta Pathol. Micro-
       biol. Scand. C. 87:347, 1979.
57. Martin, B. M., Baldwin, W. M., Gimbrone, M. A., Unanue, E. R.,
       and Cotran, R. S., J. Cell Biol. 83:376a, 1979.

DISCUSSION

UNKNOWN: In myelo-proliferative syndromes, there are lots of fibers
in the marrow. Would you suggest that thrombocytosis, often seen in
these patients, is related to the fiber formation?

STEWART: I don't know whether to suggest that or not. Certainly if platelets are lysed, they will release platelet derived growth factors (PDGF) which could cause fibroblasts to proliferate.

UNKNOWN: Is the PDGF related to the factor which induces growth of smooth muscle cells in atherosclerosis and did you try your factor on smooth muscle cells?

STEWART: No, we have not. However, Martin and co-workers have shown that macrophages secrete a factor which causes growth of smooth muscle cells. Whether it's the same factor or not is still an open question.

UNKNOWN: Can you explain how calcium phosphate causes proliferation?

STEWART: Calcium phosphate precipitates may act by aggregation of receptors. Once receptors are aggregated, transmembrane signals may occur leading to proliferation.

    I might add that insulin-like growth factors and epidermal growth factors are both required for fibroblast proliferation. These factors are present in the plasma which normally baths cells in vivo. When purified epidermal growth factor and somatomedin C are added to fibroblasts, they will not proliferate unless either platelet derived growth factor or macrophage derived competence activity is added. You can think of fibroblasts as being bathed in everything they need to proliferate except competence activity. Pledger, et al. have recently shown that there is a 29,000 and a 35,000 dalton protein which is induced by platelet derived growth factor. These two induced proteins may be related to competence activity. Tumor cells express both of these proteins and that may be why they don't require these factors to proliferate.

REFERENCES

1. Martin, B. M., Gimbone, M. A., Unanue, E. R., and Cotran, R. F., Stimulation of non-lymphoid mesenchymal cell proliferation by a macrophage derived growth factor, J. Immunology 126:1510, 1981.
2. Pledger, W. J., Hart, C. A., Locotall, K. L., and Scher, C. D., Platelet derived growth factor modulated proteins: constitutive synthesis by a transformed cell line, Proc. Natl. Acad. Sci. USA 78:4358, 1981.

REGULATION OF BONE-MARROW MACROPHAGE

PROLIFERATION

David A. Hume and Siamon Gordon

Sir William Dunn School of Pathology
South Parks Road
Oxford OX1 3RE, U.K.

INTRODUCTION

Macrophages are derived from pluripotent bone marrow stem cells
(1). Their proliferation and differentiation can be induced in
vitro by the addition of "conditioned media" that contain colony-
stimulating factors (2-6). Colony-stimulating factors (CSF) are normally assayed by their ability to promote colony formation from stem
cells in semi-solid (soft agar or methylcellulose) medium. Two types
of colony-stimulating factors that influence macrophage growth have
been identified. CSF-1 (isolated from mouse L-cell conditioned medium
or human urine) gives rise to colonies entirely composed of macrophages while CSF-II (e.g. from WEHI-3 conditioned medium or endotoxin-
lung conditioned medium) yields mixed colonies of granulocytes and
macrophages (2-6).

The colony assay for CSF is not a convenient method for it is
tedious, time-consuming and relatively non-quantitative. As an alternative we have developed an automated ^{3}H-thymidine incorporation bioassay for CSF-1.

BIOASSAY FOR CSF-1

Our assay relies initially on the purification of a macrophage
population that has a high mitotic index and is dependent on CSF
for growth. Mouse femoral bone marrow cells are cultured in bulk
in Falcon T-200 flasks (2 femurs/flask) coated with gelatin in a
medium containing 20% mouse L-cell conditioned medium (LCM) (7).
After 4-7 days an adherent layer of macrophages develops. The non-
adherent cells (which at this stage no longer have a strong prolif-

261

erative response to LCM) are discarded; the adherent cells can be
detached using 10-15 mM Lidocaine or Trypsin-EDTA (the method used
for detachment does not influence subsequent proliferation). A
similar adherent cell layer forms in uncoated (plastic) flasks but
the cells are very resistant to detachment.

The detached adherent cells are transferred to 96 well plates
containing 200 μl of medium/well (usually Dulbecco's modified minimal
essential medium + 10% fetal bovine serum: adult sera contain varying amounts of CSF activity). After the desired time 6-^{3}H-thymidine
(20 μl/well, 0.5 μCi, 2 Ci/mmol) is added to each well and labeling
is continued overnight (16 hrs). The cells are harvested onto glass
fiber filters with an automatic harvester (Ilacon, Bucks) using 3
cycles of washing: a) distilled H_2O; b) 0.5% Triton X-100 in H_2O;
and c) methanol.

^{3}H-thymidine incorporation into insoluble material in the absence
of a growth stimulus declines progressively with time. If LCM is
added, a lag period of approximately 12 hrs is observed and thereafter H-thymidine incorporation increases progressively over a period of 3-5 days. The duration and magnitude of incorporation is a
function of LCM concentration. This observation is probably indicative of the net consumption and degradation of CSF that accompanies
macrophage growth (6). The routine growth assay decided upon involves
preincubation of the cells for 30 hrs followed by overnight labeling.
We have used this assay to monitor the partial purification of an
active fraction from concentrated serum-containing LCM obtained from
roller bottle L-cell cultures. Roller-bottle LCM is up to 20 x more
active in our assay than LCM from stationary cultures because of the
high surface area/volume ratio but unfortunately L-cells will not
grow in roller bottles under serum-free conditions. The growth-
stimulating activity elutes as a single peak from Sephacryl S-300
(in 0.5M NaCl; 50 mM Tris HCl pH 7.4) with a molecular weight of
about 150,000. This is much higher than the molecular weight of
purified CSF (70,000, Ref. 5). Others have also observed that CSF
runs at a much higher apparent molecular weight in impure samples
and have attributed this fact to association with other proteins
(5). An alternative possibility, considering the apparent homogeneity
of the peak in serum-containing medium compared to its heterogeneity
in serum-free medium (5) is that serum protects the molecule against
proteolytic cleavage.

The partially purified CSF fraction stimulates growth very poorly in the absence of serum. This situation is partly resolved if a
richer medium, RPMl-1640, is used. The residual active component of
serum is not dialysable and, like CSF, is apparently consumed during
growth. BSA-linoleate, insulin, transferrin and selenium have been
tested alone or in combination and none was active in replacing serum
in this assay system.

Thymidine incorporation into 4-7 day old bone marrow derived macrophages in the presence of excess LCM is a linear function of cell number in the range $2 \times 10^3 - 5 \times 10^4$ cells/well. 2×10^4 cells/ well is a convenient starting cell number, giving 20,000 - 70,000 cpm/well of ^{3}H-thymidine incorporation during a 16 hr labeling period starting at 30 hrs. This level is routinely 5-10 fold greater than in controls incubated without LCM. Purification of a responsive macrophage population is a necessary feature of the method. The non-adherent bone marrow cells from the bulk cultures are less active and less responsive to LCM, whilst peritoneal macrophages (resident, thioglycollate or BCG-elicited) incorporate less than 1000 cpm in the same labeling period. Bulk bone marrow cultures can be main-tained for extended periods (up to 35 days) by continuous subculture in fresh LCM every 7 days. With time (eg. after 14 days) the rate of ^{3}H-thymidine incorporation declines and becomes a non-linear function of cell number. Cells from such aged cultures suppress the proliferation of early adherent cells. Degradation of CSF by the unresponsive cells may be a likely mechanism, the aged macrophages are very much larger than the proliferating early adherent cells.

DISCUSSION

A major reason for developing this assay was to study the role of adherence in macrophage proliferation. Lin et al. (8) have re-ported that blood monocytes are incapable of growth in soft agar ("anchorage independent" growth) but can acquire this capacity in vitro or after migration into an inflammatory exudate (since thio-glycollate-elicited peritoneal macrophages will grow in soft agar). The growth stimulatory effect of LCM in our assay is associated with a marked increase in adherence and spreading that is visible even within 2 hrs. The spreading response to LCM is also obvious using peritoneal macrophage populations (thioglycollate-elicited; BCG-activated). To investigate whether morphology was in any way cor-related with macrophage proliferation (as has been reported for fibro-blasts in reference 9), we replated cells from bulk cultures into microtitre wells coated with either gelatin or poly (hydroxyethyl-methacrylate) (9).

Bone marrow macrophages were unable to adhere to either surface in the absence of LCM and instead formed large aggregates of rounded cells as observed when fibroblasts are plated on the same surfaces (9). LCM was able to promote some adherence to gelatin but this ef-fect could be abolished by passing the medium over gelatin-sepharose to remove fibronectin. The stimulation of ^{3}H-thymidine incorporation by LCM was almost unaffected by plating on these non-adherent sur-faces, despite the radical change in cell morphology observed. The residual LCM-independent proliferation was, however, strongly sup-pressed so that the magnitude of the LCM response was greatly increased. This result suggests that adherence either stimulates thymidine incorporation directly or it stimulates endogenous CSF production. The possibility was considered that adherence to other cells (in the aggregates) was substituting for cell-substratum attachment in permitting an LCM response. Aggregation was inhibitable by addition of the Ca^{2+}-chelator EGTA but this agent had very little effect on thymidine incorporation.

The results show that there is no correlation between thymidine incorporation and morphology and adherence in early adherent bone marrow derived macrophages. These cells are apparently facultatively adherent and in this, their lack of an absolute serum requirement, and their insensitivity to Ca^{2+} chelators, they differ from other strongly adherent cells such as fibroblasts but resemble malignantly transformed cells (9-12).

We have also used the thymidine incorporation assay to investigate the influence of glucocorticoids on macrophage proliferation. Dexamethasone was able to inhibit LCM-induced ^{3}H-thymidine incorporation only partially and even at 10^{-5}M only a 60% inhibition was observed. In the absence of LCM dexamethasone almost completely suppressed thymidine incorporation. The most likely explanation for these results is that LCM in some way mitigates against an inhibitory effect of dexamethasone.

In conclusion, the automated thymidine incorporation assay has many advantages over autoradiography, DNA estimation or colony assays for macrophage proliferation and can be used as a bioassay for CSF to monitor purification protocols. Using the assay we have elucidated several features of the regulation of macrophage proliferation.

REFERENCES

1. Van Furth,R., Annals N.Y. Acad. Sci. 278, 161.
2. Metcalf, D., in: "In Vitro Cloning of Normal and Leukemic Cells" Springer-Verlag, Berlin, 1971.
3. Nicola, N. A., Burgess, A. W., and Metcalf, D., J. Biol. Chem. 254:5290, 1979.
4. Byrne, P. V., Guilbert, L. J., and Stanley, E. R., J. Cell Biol. 91:848, 1981.
5. Stanley, E. R., and Guilbert, L. J., J. Immunol. Methods 42:253, 1981.
6. Tushinski, R. J., Oliver, I. T., Guilbert, L. J., Tynan, P. W., Warner, J. R., and Stanley, E. R., Cell 28:71, 1982.
7. Lin, H-S, and Gordon, S., J. Exp. Med. 150:231, 1979.
8. Lin, H-S, Gordon, S., Chen, D. M., and Kurtz, M., J. Exp. Med. 153:488, 1981.
9. Folkman, J., and Moscona, A., Nature (London) 273:345, 1978.

10. Holley, R. W., _Nature_ (London) 258:487, 1975.
11. Boynton, A. L., and Whitfield, J. F., _Cancer Res._ 38:1237, 1978.
12. Swierenga, S. H. H., Whitfield, J. F., and Karasaki, S., _Proc. Natl. Acad. Sci. U.S.A._ 75:6069, 1978.

DISCUSSION

UNKNOWN: Can your assay be adapted to study GM-CSF, where you don't have just macrophages growing but also granulocytes?

HUME: This system is used to monitor the activity in WEHI-3 conditioned medium which is predominantly GM activity. The non-adherent cells respond better to WEHI-3 than the adherent cells, but both will respond to WEHI-3 conditioned medium. It should be possible to develop an assay system whereby you could distinguish the two types of activity from the shape of the dose response curves.

UNKNOWN: If you have a low dose of GM-CSF and low numbers of granulocytes being induced, won't this affect the tritiated thymidine incorporation because at higher concentrations of CSF, you will get a lot of granulocytes responding?

HUME: I haven't looked at that question in detail. With the adherent cells and WEHI-3 conditioned medium, the dose response rises over a very wide range of concentrations, maybe as much as a hundred fold. In contrast, with L-cell conditioned medium, the maximum dose response occurs over a ten fold range with adherent cells. With the non-adherent cells, the reverse is true. Non-adherent cells are very sensitive to WEHI-3 conditioned medium over a ten fold range whereas the L-cell conditioned medium has a very long range effect.

UNKNOWN: So, if you're working with WEHI-3 conditioned medium, is the peak also very early (two to three days of culture) or can you look at them after 7 days?

HUME: It depends. If you take the non-adherent cells after they've already been through 3 or 4 days in the bulk culture to deplete them of adherent cells, then the peak thymidine incorporation is within about 48 hours.

METCALF: The liquid culture systems are quite useful if one knows the determinants of the system. In our system, we purify the progenitor cells, freeze them in liquid nitrogen and then run out micro culture wells, much as you've described. We put just 50 cells in each dish with or without CSF and look 24 or 48 hours later. In the absence of CSF, the cells are dead. In the presence of CSF, you generally get more than 50 colonies. One of the virtues of your system is the low total amount of material required to achieve the needed concentration. We found this to be important in attempts to clone the gene for CSF production. We have miniscule amounts to be assayed and it really requires microcultures of the sort you have described or the ones we're using.

DE VRIES: What kind of serum did you use?

HUME: Fetal calf system.

DE VRIES: The use of fetal calf serum is important because almost all commercial batches of adult serum have CSF activity. So, if you use horse serum, the background will go up.

USE OF GLUCAN TO ENHANCE HEMOPOIETIC RECOVERY AFTER EXPOSURE TO COBALT-60 IRRADIATION

Myra L. Patchen and Thomas J. MacVittie

Experimental Hematology Department
Armed Forces Radiobiology Research Institute
Bethesda, Maryland   20814 USA

INTRODUCTION

Glucan is a B-1,3 polyglucose isolated from the inner cell wall of the yeast Saccharomyces cerevisiae (1). Administration of glucan to rodents significantly enhances reticuloendothelial and immune responses (2-4) and represses tumor growth and experimental infections (5-7). Glucan also alters bone marrow and splenic hemopoietic proliferation and differentiation (8-13). Similar to glucan-induced immunomodulation, glucan-induced hemopoietic regulation depends on the route of glucan administration, the glucan dose administered, and the source of the glucan preparation (8,10,12). In general, following intravenous administration of currently available particulate glucan preparations in the dose range of 0.1 to 2.0 mg per mouse, the proliferation of bone marrow and splenic pluripotent stem cells (CFU-s), splenic macrophage and granulocyte-macrophage colony-forming cells (M-CFC, GM-CFC) and splenic erythroid colony and burst-forming cells (CFU-e, BFU-e) is stimulated in a direct, dose-dependent manner (12). The exact mechanisms through which glucan mediates its stimulatory effects on hemopoiesis are still largely unknown; however, both macrophages and T-lymphocytes have been reported to be involved (9,11,14).

Because of glucan's ability to stimulate hemopoiesis, and in particular, pluripotent hemopoietic stem cell proliferation, we have investigated the ability of this agent to enhance hemopoietic recovery following exposure to cobalt-60 irradiation.

METHODS AND MATERIALS

Ten-to-twelve-week-old female C3H/HeN mice were used in all

experiments.   All mice were quarantined and acclimated to laboratory
conditions for 2 weeks before experimentation, during which time they
were examined and found to be free of murine pneumonia complex and
oropharyngeal Pseudomonas.   Particulate, endotoxin-free glucan was
obtained from Accurate Chemical and Scientific Corporation (Westbury,
NY), and was prepared according to DiLuzio's modification (15) of
Hassid's original procedure (1).   Glucan was diluted in sterile
saline, and either 0.4 or 1.5 mg was injected intravenously into
experimental mice.   Control mice were injected with an equivalent
volume of sterile saline.

     At the indicated times either before or after glucan injection,
650 rads of total-body irradiation from the AFRRI cobalt-60 source
was administered to mice at a dose rate of 150 rads per minute.   The
post-irradiation survival and proliferation of pluripotent hemopoietic
stem cells were measured by the endogenous spleen colony assay (E-CFU)
(16-17).   The spleens of experimental and control mice were removed,
fixed in Bouin's solution, and the number of visible spleen colonies
counted.   Spleens of post-irradiated glucan-treated mice were removed
8 days after irradiation.   Spleens of pre-irradiated glucan-treated
mice were removed 8 days after glucan injection.

RESULTS

     Figure 1 illustrates the effects of pre-irradiation glucan treat-
ment on E-CFU proliferation.   Both 0.4 and 1.5 mg of glucan, injected
into mice 1 hour or 1, 5, 11 or 17 days prior to irradiation, enhanced
E-CFU proliferation.   More dramatic effects were produced by the
higher glucan dose.   At either glucan dose, the greatest enhancement
occurred with glucan administered 1 day and 1 hour prior to irradiation.   As the interval between glucan administration and radiation
exposure increased, less enhancement was observed.

     The effects of post-irradiation glucan treatment on E-CFU pro-
liferation are presented in Figure 2.   Glucan injected 1 hour after
irradiation was as effective in enhancing E-CFU proliferation as
glucan injected 1 hour or 1 day before irradiation.   Glucan administered 1, 5, 11 or 17 days after irradiation had progressively less
effect.   For example, in comparison to the respective radiation controls, E-CFU proliferation was enhanced approximately 1250% after 1.5
mg of glucan injected at 1 hour but only 350% and 250% at 1 and 5
days, respectively.   By the time spleens were harvested from mice
receiving glucan 11 and 17 days after irradiation, these spleens (as
well as the spleens of the respective radiation controls) exhibited
confluent colony growth, which precluded accurate quantitation of
spleen colony numbers.   However, spleens of glucan-treated mice were
larger and weighed more than those of radiation controls.

     In animals that had been irradiated before glucan treatment,

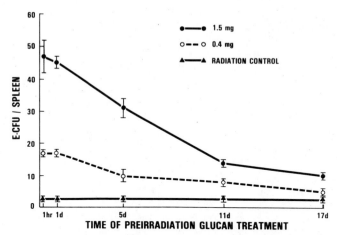

Fig. 1. Effect of pre-irradiation glucan treatment on E-CFU. Data points represent pooled mean values ± 1 standard error from 5 replicate experiments.

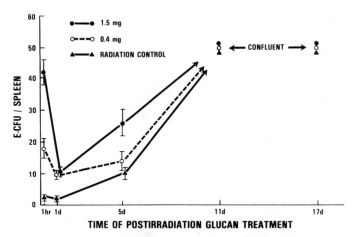

Fig. 2. Effect of post-irradiation glucan treatment on E-CFU. Data points represent pooled mean values ± 1 standard error from 4 replicate experiments.

further enhancement of E-CFU proliferation was attempted by adminis-
tering multiple glucan treatments. Figure 3a illustrates that when
mice received three consecutive 0.4 mg glucan injections at 1 hour,
1 day, and 2 days after irradiation (i.e., total cummulative dose
of 1.2 mg), E-CFU proliferation was significantly enhanced above
that observed with single 0.4 mg glucan injections. If mice also
received 0.4 mg of glucan 1 day prior to irradiation (i.e., total
commulative dose of 1.6 mg), an even greater enhancement of E-CFU
proliferation was observed (Figure 3b).

DISCUSSION

        These studies have shown that the immunomodulating agent glucan
enhances the proliferation of pluripotent hemopoietic stem cells
when administered either before or after exposure to a hemopoietically
damaging dose of cobalt-60 irradiation. Enhancement depends on both
the dose of glucan administered and the time of administration with
respect to the time of irradiation. The timing of glucan administration after irradiation is more critical than before irradiation.

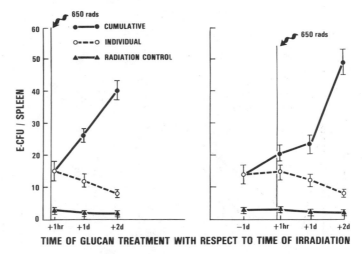

TIME OF GLUCAN TREATMENT WITH RESPECT TO TIME OF IRRADIATION

Fig. 3.   Effect of multiple glucan injections on E-CFU. Data points
          represent pooled mean values ± 1 standard error from 3 rep-
          licate experiments. (3a) Three consecutive 0.4 mg glucan
          injections administered 1 hour, 1 day and 2 days after irra-
          diation. (3b) Four consecutive 0.4 mg glucan injections
          administered 1 day before and 1 hour, 1 day and 2 days after
          irradiation.

Using a single glucan injection, maximum E-CFU proliferation was produced by glucan administered 1 day or 1 hour before or 1 hour after irradiation. Further enhancement could be produced by multiple glucan treatments.

The mechanism(s) by which glucan enhances proliferation of E-CFU in irradiated mice is not known. However, since macrophages are relatively radio-resistant cells (18-19) and since macrophages in glucan-treated animals have been shown to enhance at least some aspects of hemopoiesis (9,11,14), it is possible that these cells may contribute to the regulation of hemopoiesis in irradiated animals.

Acknowledgement

We acknowledge the excellent technical assistance of Glorianne Davis and editorial and typing skills of Junith Van Deusen and Terrie Hunt.

REFERENCES

1. Hassid, W. Z., Joslyn, M. A., and McCready, R. M., J. Am. Chem. Soc. 63:295, 1941.
2. Riggi, S. S., and DiLuzio, N. R., Am. J. Physiol. 20:297, 1961.
3. DiLuzio, N. R., in "The Reticuloendothelial System in Health and Disease: Functions and Characteristics" (S. M. Reichard, M. R. Escobar and H. Friedman, eds.), p. 412, Plenum Publishing Corp., New York, 1976.
4. Wooles, W. R., and DiLuzio, N. R., Science 142:1078, 1963.
5. DiLuzio, N. R., McNamee, R., Jones, E., Cook, J. A., and Hoffmann, E. O., in "The Macrophage in Neoplasia" (M. A. Frank, ed.), p. 181, Academic Press Inc., New York, 1976.
6. DiLuzio, N. R., McNamee, R., Browder, W., and Williams, D., Cancer Treat. Rep. 62:1857, 1978.
7. Reynolds, J. A., Kastello, M. D., Harrington, D. G., Crabbs, C. L., Peters, C. J., Jemski, J. V., Scott, G. H., and DiLuzio, N. R., Infect. Immun. 30:51, 1980.
8. Burgaleta, C., and Golde, D. W., Cancer Res. 37:1739, 1977.
9. Niskanen, E. O., Burgaleta, C., Cline, M. J., and Golde, D. W., Cancer Res. 38:1406, 1978.
10. Patchen, M. L., and Lotzova, E., Exp. Hematol. 8:409, 1980.
11. Deimann, W., and Fahimi, H. D., Lab. Invest. 42:217, 1980.
12. Patchen, M. L., and MacVittie, T. J., Exp. Hematol. 9:118, 1981.
13. Patchen, M. L., and MacVittie, T. J., Exp. Hematol., submitted.
14. Patchen, M. L., and Lotzova, E., Biomedicine 34:71, 1981.
15. DiLuzio, N. R., Williams, D. L., McNamee, R. B., Edwards, B. F., and Kilahama, A., Int. J. Cancer 24:773, 1979.
16. Till, J. E., and McCulloch, E. A., Radiat. Res. 18:96, 1963.
17. Boggs, S. S., Boggs, D. R., Neil, G. L., and Sartiano, G., J. Lab. Clin. Med. 82:727, 1973.

18.   Benacerraf, B., _Bacteriol. Rev._ 24:35, 1960.
19.   Schmidtke, J. R., and Dixon, F. J., _J. Immunol._ 108:1624, 1972.

DISCUSSION

FRIEDLANDER: What happens in the HeJ mouse?

PATCHEN: We have not used glucan in HeJ mice, but that would be an
interesting experiment because the HeJ mouse does not respond to
endotoxin.   Whether or not it would respond to glucan, I'm not sure.
I might add that our glucan has been tested for endotoxin and it's
negative.

GORDON: How does glucan stimulate hemopoiesis?   Is there any circulating substance?

PATCHEN: We used a particulate glucan preparation which is rapidly
phagocytized by macrophages.   Within 24 hours, it's definitely cleared
from the blood.   We've looked at the effect of glucan on GM-CSF and
found that it is related to its macrophage activity.   CSF is present
in the sera of these mice.

THE ROLE OF HELPER MONOCYTES IN MITOGENIC

ACTIVATION OF HUMAN T-LYMPHOCYTE SUBPOPULATIONS

Artur J. Ulmer and Hans-Dieter Flad

Forschungsinstitut Borstel
D-2061 Borstel F.R.G.

INTRODUCTION

It is well established that monocytes/macrophages provide helper
signals necessary for the induction of proliferation of T-lymphocytes
by mitogens (1). This helper function consists in the elaboration
of a growth factor known as Interleukin 1 (IL-1). IL-1 does not
support the proliferative response of T-lymphocytes directly but
acts as a signal which helper T-cells require for the production of
a further growth factor, Interleukin 2 (IL-2). The subsequent interaction of IL-2 with specific receptors on mitogen-activated T-lymphocytes generally initiates cell proliferation (2).

In view of the heterogeneity of human T-lymphocytes, we antic-
ipated that different T-lymphocyte subsets would behave differently
with regard to helper requirements for mitogen-induced proliferation.
To approach this problem, human T-lymphocytes of the peripheral blood
were fractionated by centrifugation on a discontinuous Percoll gradient as described elsewhere (3). It will be shown that (a) the various subsets differ markedly in their capacity to proliferate in
agar culture and to secrete IL-2, and (b) they exhibit distinct helper
requirements for proliferation and IL-2 production.

MATERIALS AND METHODS

Isolation and subfractionation of peripheral blood mononuclear
cells and T-lymphocytes were performed as described elsewhere (4).
In brief, peripheral blood mononuclear cells (PBMC) were isolated from
heparinized human peripheral blood by Ficoll-Paque density gradient
centrifugation. Monocytes were obtained from PBMC by adherence in

273

disposable Falcon tissue culture flasks. Purified T-cells were isolated by rosetting PBMC with sheep erythrocytes and subsequent Ficoll-
Paque gradient centrifugation. The purified T-cells were then further fractionated by discontinuous density gradient centrifugation
on Percoll. This procedure yielded T-cells of the following densities (g/ml): Fraction 1 (F1) 1.080, F2 1.080 to 1.070, F3 1.070 to
1.068, F4 1.068 to 1.066, F5 1.066 to 1.064, F6 1.064 to 1.062, and F7
1.062 to 1.004.

    Growth factors were produced as described previously (5).
IL-2 activity was tested using cultured human T-lymphocytes stimulated
by PHA. A quantitative determination of IL-2 activity was performed
by probit analysis according to Gillis et al. (6). An IL-2 standard
was defined to contain 100% = 1 unit of IL-2. IL-1 activity was
tested by measuring the proliferative response of murine thymus cells
to PHA (5).

    The proliferation of T-lymphocytes was determined by the forma-
tion of T-lymphocyte colonies after stimulation of 18×10^3 T-cells
with PHA in an agar micro culture as described elsewhere (7).

RESULTS

    T-lymphocytes were separated into fractions of different density.
As shown in Table 1, only T-cells from low density fractions could
be stimulated by PHA to form colonies. However, high density T-
lymphocytes responded to PHA when monocytes were added. It should
be noted that various T-lymphocyte subpopulations differed not only
in their thresholds for monocytes but also in the proliferative response obtained at a given concentration of monocytes.

    With respect to the production of IL-2 by the cells it was found
that only unseparated and low density T-lymphocytes produced detectable amounts of IL-2 (Table 2). After addition of monocytes, high
density T-lymphocytes can be stimulated by PHA to produce IL-2.
Dose-response experiments with various numbers of monocytes added to
the system gave a similar relationship as found for the activation of
colony forming T-lymphocytes: the higher the density of the fractionated T-cells, the higher the number of monocytes required for
IL-2 production. Furthermore, this experiment demonstrates that high
density T-lymphocytes (e.g. cells of fraction 2), when supplemented
with high numbers of monocytes (20%), produced less IL-2 than cells
of fraction 6 which contained only 0.6% monocytes.

    In a further experiment we compared the response of human T-
lymphocytes (IL-2 assay) with the response of murine thymus cells
to PHA (IL-1 assay) in the presence of the different culture supernatants. As shown in Table 3, neither supernatants derived from high
density T-lymphocytes nor those derived from PHA-stimulated mono-

Table 1. Dose Effect of Monocytes on Colony Growth of T-lymphocytes of Different Densities

Monocytes added	Unsep. T-cells	Fraction Number of T-cells from Percoll Gradient				
		2	3	4	5	6
0%	0 ± 0	0 ± 0	0 ± 0	0 ± 0	3 ± 1	36 ± 10
5%	0 ± 0	0 ± 0	0 ± 0	19 ± 6	26 ± 6	-
10%	1 ± 1	7 ± 2	16 ± 2	75 ± 12	43 ± 14	-
20%	2 ± 1	18 ± 1	44 ± 2	107 ± 2	86 ± 10	62 ± 8
40%	5 ± 2	64 ± 6	87 ± 1	97 ± 6	61 ± 5	-
Monocytes	5%	0.6%	<0.2%	<0.2%	1.0%	3.0%

Different T-cell populations were stimulated by PHA in agar culture in the presence or absence of different numbers of additional monocytes. Density decreases from fraction 2 to 6. The results are expressed as colonies per culture (mean $\pm$ SEM, n=3). The number of monocytes in the different T-cell populations is given in the bottom line.

Table 2.   Dose Effect of Monocytes on IL-2 Production by T-lymphocytes of Different Densities

Monocytes added	Unsep. T-cells	Fraction Number of T-cells from Percoll Gradient				
		2	3	4	5	6
0%	0.19	0.01	0.01	0.01	0.03	0.66
1%	0.22	0.01	0.03	0.06	0.13	-
3%	0.29	0.03	0.13	0.23	0.15	-
10%	0.41	0.12	0.15	0.50	0.45	-
20%	0.66	0.20	0.20	0.87	0.93	-
Monocytes	0.8%	0.2%	0.2%	0.2%	0.2%	0.6%

Different T-cell populations were stimulated by PHA in the presence of various concentrations of additional monocytes. The results show the IL-2 units produced by these cells. Density decreases from fraction 2 to 6. The number of monocytes in the different cell fractions is given in the bottom line.

Table 3. Response of Cultured Human T-lymphocytes and Murine Thymocytes to Culture Supernatants of Different Sources

Source of Supernatant	CPM Per Culture	
	Murine Thymic Cells (IL-1 Assay)	Human T-lymphocytes (IL-2 Assay)
Monocytes	19226 + 329	171 + 33
Unseparated T-cells	1372 + 489	2113 + 37
T-cell Subsets Fraction 2	2203 + 286	141 + 29
3	2386 + 156	293 + 40
4	1834 + 187	323 + 31
5	2134 + 84	409 + 76
6	5381 + 1697	5070 + 224
7	23960 + 1697	14224 + 870

Different cell populations (non-fractionated or Percoll gradient separated subsets) were stimulated by PHA. After 24h of culture, cell free supernatants were harvested and tested in the IL-1 and IL-2 assays at a concentration of 25%. Each value represents the mean + SEM of 3 cultures.

cytes contained detectable IL-2 activity. Also, murine thymocytes did not proliferate markedly in the presence of culture supernatant derived from PHA-stimulated high density T-lymphocytes, but did so in the presence of the culture supernatant derived from PHA-stimulated monocytes. It should be noted that murine thymocytes responded equally low to culture supernatants derived from PHA-stimulated T-cells of fractions 2 to 5, indicating that these supernatants contained equally low activities of IL-1 or IL-2.

DISCUSSION

These results demonstrate that different T-lymphocyte subsets as defined by density differ in their requirements for helper monocytes to proliferate and produce IL-2 upon activation by PHA. Dose-response experiments with various numbers of monocytes added to the culture show the following patterns of reactivity: the higher the density of T-cells, the higher the number of monocytes required for their activation. The lower the density of T-cells, the higher was their response to PHA in the presence of a given number of additional monocytes. Low density T-lymphocytes need no additional monocytes for activation. Whether these low density T-lymphocytes are actually monocyte-independent, cannot be determined because a small number of monocytes were still present in these T-cell fractions.

Table 3 shows no differences in the thymocyte-stimulating activity in the culture supernatant derived from PHA-stimulated T-cells of fractions 2 to 5. This result indicates that neither the various number nor the various activities of a few residual monocytes in the T-cell fraction could be responsible for the differential effects on IL-2 production and proliferation of these cells. The response of murine thymocytes in the presence of the culture supernatant derived from fractions 6 and 7 cells may be due to the presence of IL-2 as documented by the response of human lymphocytes to these supernatants. As the response of murine thymocytes is influenced by IL-1 as well as by IL-2, the amount of IL-1 produced by these low density T-lymphocytes can only be determined after biochemical separation of the growth factors.

Because residual monocytes in the T-cell subsets cannot explain the differential helper effects observed in Tables 1 and 2, we conclude that the various T-cell fractions differ in number or activity of at least one additional unknown subpopulation presumably of lymphoid origin which regulates the PHA response of T-cells. The target cells of this regulating cell may be the IL-1 producing monocytes or the IL-2 producing T-helper cell. The nature of this regulating cell remains unclear; it could be a suppressor cell population enriched in the high density fractions of the gradient or a helper cell population enriched in the low density fractions.

Acknowledgement

The technical assistance of Mrs. M. Grünefeld, S. Heeg, and R. Guth is gratefully acknowledged. This work was supported by Deutsche Forschungsgemeinschaft SFB 111/C4.

REFERENCES

1. Rosenstreich, D. L., and Mizell, S. B., Immunol. Rev. 40:102, 1978.

2. Smith, K. A., Lachman, L. B., Oppenheim, J. J., and Favata, M. F.,
      J. Exp. Med. 151:1551, 1980.
3. Ulmer, A. J., and Flad, H.-D., J. Immunol. Methods 30:1, 1979.
4. Flad, H.-D., and Ulmer, A. J., Europ. J. Cell Biol. 25:16, 1981.
5. Ulmer, A. J., and Flad, H.-D., Immunobiology, 1982, in press.
6. Gillia, S., Ferm, M. M., Ow., W., and Smith, K. A., J. Immunol.
      120:2027, 1978.
7. Ulmer, A. J., and Flad, H.-D., Immunology 38:393, 1979.

DISCUSSION

STEVENSON: We've done similar studies with somewhat different results. We used elutriator purified, monocyte depleted, lymphocyte
preparations. When stimulated with a variety of lectins including
pokeweed mitogen and PHA, we found that the T cell populations were
not able to respond at all, either in terms of proliferation or B
cell activation. When we added just a few monocytes back to the
system, then the T cells were able to respond. I'm not convinced
that the T cells have to actually encounter the mitogen. For example,
we took monocytes pulsed with mitogen, washed them extensively and
then added them back to our responder T cells. They completely reconstituted the responses of the lymphocytes. Thus, I'm not sure
whether T cells must actually contact the mitogen in addition to
interleukin 1.

ULMER: You can stimulate monocytes by a mitogen other than PHA to
produce interleukin 1. On the other hand, it has been reported that
interleukin 1 production is not necessary for T cell proliferation.
The contention supported by these studies is that the main function
of the monocyte is activating the T helper cell by direct cell contact.

VALENTINE: If I understand your culture system correctly, you're
using an agar culture system which should allow you to observe whether
cell interaction does or does not take place. Presumably it does
not, if it's done at a fairly low cell density. When you add your
monocytes to the lymphocytes, do they form clusters? Are they
separated in the agar from the lymphocytes? If so, it would imply
a soluble factor rather than cell contact.

ULMER: We did not separate the cells. We can induce colony proliferation after addition of interleukin 1 or interleukin 2. We can
stimulate these high density T cells with these factors and obtain
proliferation in the agar culture.

VALENTINE: Presumably, the monocytes and the lymphocytes are
separated in the culture even though you don't purposely separate
them beforehand. After all, they are added as single cell suspensions
and at a fairly low density.

ULMER: It is difficult to decide whether the lymphocytes and the
monocytes are really separated in an agar culture. I believe that
we have some microclusters of monocytes, perhaps helper cells, and
the proliferating cells. These clusters are necessary in order to
form colonies in the agar culture. I don't believe that just one
cell can proliferate in our system.

WUSTROW: How pure were your preparations of interleukin 1 and 2?
Did you use any supernatants from mitogen induced cells or conditioned
media from cell lines?

ULMER: No. We have only the crude supernatant of PHA stimulated
peripheral blood mononuclear cells for interleukin 2 or LPS stimulated
cells for interleukin 1.

KOREN: You're using Percoll gradient separated cells and this same
gradient has been used to enrich for natural killer cells. It is
intriguing that the low density fraction that has the majority of
the NK activity was the same fraction which did not require monocytes
for proliferation. Have you looked at the phenotypes of your T
cells? Could it be that the proliferating cells are indeed natural
killer cells?

ULMER: We have looked for natural killer cells and did not find them
in our highly purified T cell fraction. In our experience, the NK
cells are not in the very low density fraction but in the middle
fraction. And we have looked at the proliferating cells to see that
they really are T cells and not other cells by E rosetting.

KOREN: But NK cells will form E rosetts. Have you used any monoclonal antibodies to identify your cells?

ULMER: No.

SELECTION OF AN IMMUNOGENIC 3LL TUMOR SUBLINE
FOLLOWING SERIAL GROWTH IN VIVO IN THE LOCAL
PRESENCE OF PERITONEAL MACROPHAGES

Patrick De Baetselier, Ahuva Kapon, Shulamit Katsav,
Michael Feldman and Shraga Segal

Instituut voor Moleculaire Biologie
VUB, Paardenstraat 65
Sint-Genesius-Rode, Belgium

Department of Cell Biology
Weizmann Institute of Science
Rehovot, Israel

INTRODUCTION

Macrophages may play a role in host defense against neoplasia
since they inhibit the growth of tumors at primary (1) and metastatic
sites (2). In contrast, some reports indicate that macrophages may
promote tumor development in several ways such as induction of clonal
proliferation of malignant cells (3) or activation of tumor mediated
suppressor mechanisms (4). The diverse and often contradictory effects of macrophages can be explained by the functional heterogeneity
of macrophage populations (5).

To assess whether resistance or adaptation to the local presence
of activated macrophages would modulate the tumorigenic properties
of a polyclonal tumor cell population, we serially transplanted the
metastasizing 3LL Lewis lung carcinoma in the presence of thioglycollate induced peritoneal macrophages. Data are presented indicating
that the resulting tumor cell population, designated 3LL(R) generates
immunogenic signals through interactions with the peritoneal macrophages.

RESULTS

The 3LL tumor cells were transplanted subcutaneously in C57Bl/6
syngeneic recipients in the local presence of peritoneal macrophages

at a tumor to macrophage ratio of 1:10.   After six consecutive passages in vivo, we observed a progressive regression of tumor growth
as well as a lower incidence of tumor take in animals inoculated with
the 3LL(R) cells and macrophages.   The 3LL(R) cells and the original
3LL population were compared as to their local growth properties
(subcutaneous and intra-footpad growth) and their capacity to generate
spontaneous metastases (after amputation of the local tumor) and
experimental metastases (after intravenous inoculation).

INFLUENCE OF PERITONEAL MACROPHAGES ON THE LOCAL GROWTH OF 3LL AND
3LL(R) CELLS

    The results outlined in Table 1 indicate that 3LL(R) cells have
a lower capacity to generate subcutaneous tumors provided peritoneal

Table 1.   Influence of Peritoneal Macrophages on the Local
           Growth of 3LL and 3LL(R) Tumor Cells in Syngeneic
           C57B1/6 Mice

Cells Inoculated	Trials	Tumor Take
(a) Subcutaneous		
3LL	50	45/50 (90%)
3LL + macrophages	20	18/20 (90%)
3LL(R)	80	74/80 (92%)
3LL(R) + macrophages	80	38/80 (48%)
(b) Intra-footpad		
3LL	60	59/60 (98%)
3LL + macrophages	60	55/60 (92%)
3LL(R)	80	80/80 (100%)
3LL(R) + macrophages	80	74/80 (82%)

For subcutaneous inoculation, 10^6 tumor cells were used in
the presence or absence of 10^7 peritoneal macrophages.   For
intra-footpad inoculation, 10^5 tumor cells were used in the
presence or absence of 10^6 peritoneal macrophages.

macrophages were locally present during inoculation. However, when such cells were inoculated intra-footpad, no inhibition of tumor growth and tumor take was observed. The observed inhibition by peritoneal macrophages on the subcutaneous growth of 3LL(R) cells was not due to a tumoricidal effect since in all animals inoculated with 3LL(R) and macrophages a tumor mass appeared faster than in animals inoculated with 3LL(R) cells alone. Thereafter, however, the tumors regressed totally in 60% of the animals inoculated with 3LL(R) plus macrophages. The early acceleration of tumor cell proliferation was also observed in vitro when 3LL(R) cells, but not 3LL cells, were cultured in the presence of peritoneal macrophages. These data suggest that 3LL(R) cells were selected to grow faster in a macrophage enriched microenvironment, and this fast tumor progression leads in many instances to a rejection of the tumor mass. However, this pattern of tumor development was strongly dependent on the anatomical site of inoculation.

INFLUENCE OF PERITONEAL MACROPHAGES ON THE METASTATIC GROWTH OF 3LL AND 3LL(R) CELLS

    Amputation of the different tumors inoculated into footpads provokes an acceleration of metastatic development in the lungs. The data presented in Table 2 indicate that the local presence of peritoneal macrophages during growth leads to an inhibition of metastatic development both for 3LL and 3LL(R) tumors. When 3LL cells were inoculated i.v. simultaneously with macrophages an acceleration of pulmonary metastases was observed. This pronounced effect of macrophages on metastases formation was not observed with 3LL(R) cells when the animals were analyzed two weeks after inoculation. Examination of the inoculated mice at 22 days following i.v. inoculation also revealed an enhancing effect of macrophages on the formation of pulmonary metastases by 3LL(R) cells. Additional experiments have demonstrated that the arrest of 3LL cells in the lungs when inoculated i.v. with macrophages did not differ significantly from 3LL(R) cells. These results indicate no differential trapping between 3LL and 3LL(R) cells in the lung capillaries in the presence of macrophages but rather a difference in subsequent tumor growth and proliferation.

EVIDENCE FOR ANTI-3LL IMMUNITY IN ANIMALS INOCULATED WITH 3LL(R) AND PERITONEAL MACROPHAGES

    Since the local or simultaneous presence of peritoneal macrophages leads to a rejection of 3LL(R) cells grown subcutaneously and to a retardation in the development of experimental pulmonary metastases, it was of interest to study whether immunity was induced in such animals. Therefore, animals which rejected the 3LL(R) cells were tested for (a) the presence of anti-3LL antibodies, (b) the presence of cytotoxic anti-3LL cells and (c) the effect of their

Table 2. Influence of Peritoneal Macrophages on the Meta-
          static Growth of 3LL and 3LL(R) Tumor Cells in
          Syngeneic C57Bl/6 Mice

| | | Weight of lungs (mg) | | |
Cells inoculated	Days	Exp.1	Exp.2	Exp.3
(a) Spontaneous metastases				
3LL	21	920	650	810
3LL + macrophages	21	7.70	340	630
3LL(R)	21	1030	620	850
3LL(R) + macrophages	21	660	470	560
(b) Experimental metastases				
3LL	14	290	220	
3LL + macrophages	14	1030	1010	
3LL(R)	14	220	180	
3LL(R) + macrophages	14	300	290	
3LL	23	410		
3LL + macrophages	17	1140		
3LL(R)	22	520	730	
3LL(R) + macrophages	22	1160	1120	

Spontaneous metastasis were induced by intra-footpad inoculation of 10^5 tumor cells in the presence or absence of 10^7
peritoneal macrophages. Experimental metastasis were induced
by intravenous challenge with 10^6 tumor cells in the presence
or absence of 10^7 peritoneal macrophages. The weights of the
lungs were determined at the days specified after either
tumor amputation (spontaneous metastasis) or tumor inoculation (experimental metastasis).

lymphocytes on the growth of 3LL cells _in vivo_. The data compiled
in Table 3 indicate that humoral as well as cellular immunity can
be induced during inoculation of 3LL(R) cells and peritoneal macro-

Table 3. Evidence for Anti-3LL Immunity in Animals Inoculated with 3LL(R) Tumor Cells and Peritoneal Macrophages

Cells inoculated	Tumor take	Antibody % cells stained		Cytotoxic Cells % target cell cytotoxicity			Tumor Growth % mice with 3LL tumors L:T ratio	
		3LL	3LL(R)	3LL	3LL(R)	EL4	50:1	10:1
3LL	yes	8	0	1.6	2.5	-0.6	100	100
3LL + macrophages	yes	5	4	n.d.	n.d.	n.d.	n.d.	100
3LL(R)	yes	12	13	2.6	3.7	-1.1	100	80
3LL(R) + macrophages	no	72	44	11	12.6	1.4	0	30

Presence of antibody was estimated by fluorescence serology. Tumor growth was measured by inoculating 106 3LL tumor cells subcutaneously with spleen lymphocytes at the indicated lymphocyte:tumor (L:T) ratio.

phages.   Furthermore, spleen cells derived from such animals reduced
dramatically the growth of 3LL cells in an in vitro transfer assay.

DISCUSSION

     In the present report, we analysed the interaction between a
population of activated macrophages which are heterogeneous (5) and
a polyclonal tumor cell population.   The outcome of such interaction
was a tumor cell population which had acquired macrophage mediated
growth properties.   Indeed, the selected population 3LL(R) grew faster
in vivo and in vitro when peritoneal macrophages were locally present.
Despite growth enhancement, most of the established 3LL(R) tumors
were finally rejected.   Thus 3LL(R) cells might acquire immunogenic
properties after cell-cell contact with macrophages.   In fact, we
recently demonstrated that the expression of H-2^b antigens on 3LL(R)
cells, but not on 3LL cells, is enhanced following cell-cell contact
with macrophages.   We assume that other membrane antigens, such as
tumor specific antigens, might be expressed on 3LL(R) cells when
peritoneal macrophages are present locally.   In such a case, the
3LL(R) tumor cells would increase in immunogenicity resulting in a
stimulation of humoral and/or cell-mediated immunity.   Hence activated
macrophages can inhibit the growth of malignant cells not only through
tumoricidal mechanisms but also through selection of cells which
acquire macrophage-mediated immunogenic properties.

Acknowledgement

     P. De Baetselier is a fellow of the N.F.W.O., supported by the
A.S.L.K. Kankerfonds.

REFERENCES

1. Den Otter, W., Dullens Hub, F. S., Van Lovern, H., and Pels, E.,
        in "The Macrophage and Cancer" (K. James, B. McBride, and A.
        Stuart, eds.), p. 119, Econoprint, Edinburgh, 1977.
2. Fidler, I. S., Cancer Res. 34:1074, 1974.
3. Schultz, R. M., Chirigos, M. A., and Olkowski, Z. I., Cellular
        Immunology 54:98, 1980.
4. Ting, C., and Rodrigues, D., Proc. Natl. Acad. Sci. USA 77:4265,
        1980.
5. Tzehoval, E., De Baetselier, P., Feldman, M., and Segal, S., Eur.
        J. Immunology 11:323, 1981.

DISCUSSION

MANTOVANI: Does the 3LL subline exist in the original 3LL population?   In other words, if you clone the original 3LL population, do
you get cells which spontaneously regress?

DE BAETSELIER: We haven't cloned the 3LL cells. We are working on that and indeed some clones isolated from the 3LL are more immunogenic than others. So I presume that they are present in the original population.

MANTOVANI: Is it correct that at low cell density macrophages stimulate the parent 3LL line?

DE BAETSELIER: It happens on Tuesday and it won't happen on Friday. It is not consistent.

STEWART: Tumors contain not only cancer cells but also fibroblasts, endothelial cells and other host cells. Is it possible that the rapid growth of some tumors is not due to the rapid growth of the cancer cells alone but rather to the rapid growth of all the cellular components? If you have macrophages in the tumor, is it not possible that they elaborate a growth factor, similar to platelet derived growth factor, that stimulates fibroblast growth? In other words, is enhancement of tumor growth as measured by increased volume or tritiated thymidine uptake really due to growth of neoplastic cells or to an enhanced growth of all cells, including host-defense elements?

DE BAETSELIER: I haven't thought about that. However, we feel that the tumor proliferation we observed by thymidine uptake was really due to the cancer cell.

STEWART: Did you do radioautography and determine that it was the cancer cells which were growing?

DE BAETSELIER: Yes. We also did it in allogeneic conditions where we checked the genetics of the proliferating cell, and the H-2 phenotype of the proliferating cells was that of the tumor cell.

BUYSSENS: We examined subcutaneous implants of tumor fragments of the B16 melanoma. We used tumor fragments rather than tumor cell suspensions because we tried to imitate as closely as possible the natural tissue state. If you look at such implanted tumor fragments after half an hour, one hour and so on, the volume of the tumor remains constant for up to 4 days. However, the proportion of tumor cells to macrophages changes. Thus, if you start with 2 milligrams of tumor cells, after 24 hours you have only one milligram left and the other milligram is macrophages; after 2 days, only 5 percent of the tumor volume in fact is tumor cells and the rest are macrophages. Then granulation tissue grows into the zone occupied by macrophages and replaces it. The granulation tissue then serves as the site of tumor growth. After the sixth, seventh or the eighth days, the tumor is already at a very advanced stage. Nobody knows exactly what happens in the very first stages and it would be nice if we could see what really happens half an hour or an hour after the injection of the cells.

STEWART: We've tested 9 different commonly used murine tumors and
have found similar data. The initial growth of the tumor is not
cancer cells at all, but consists of blood vessel and normal host
cell infiltration. Paul Kramer has developed a system using gel
foam sponges where you can retrieve the entire arena for examination.
The disadvantage of sponge implants is that it could represent a
foreign body reaction and not just a host defense to the tumor.
However, the model system does allow for quantitative information on
the very early propensity for infiltration of host elements into the
tumor.

MONONUCLEAR PHAGOCYTES IN THE CONTROL

OF PRIMARY AND SECONDARY TUMOR GROWTH

R. Keller

Immunobiology Research Group
Institute of Immunology and Virology
University of Zurich
Schönleinstrasse 22
CH-8032 Zurich, Switzerland

INTRODUCTION

Interactions between tumor and host are subtle, manifold and complex (1,2). The outcome of this interaction is determined not only by the growth properties, adaptive potential and immunogenicity of the tumor cells but also by host tumor defense mechanisms (3). A particularly striking example is individuals who appear healthy after surgical removal of their primary tumor, but who later succumb to resident tumor cells that had been dormant for a long time.

As evidence accumulated, the concept of tumor surveillance underwent profound changes. The monolithic concept of T cell-mediated, immunospecific surveillance as the central defense mechanism was increasingly confronted with and then replaced by the concept of a multifaceted cellular defense system, in some respects analogous to the well-established situation in resistance against microbial infection. The present concept (Table 1) postulates that in addition to T lymphocytes, other cells manifest spontaneous or antibody-dependent cellular cytotoxicity (ADCC). Further, an array of soluble mediators also participate in host defense against tumors.

For optimal effectiveness, host defense against cancer must be capable of adapting to the character and properties of the tumor cell and to the site of the host/tumor encounter. The initial, generally decisive encounter of tumor and host can result either in promotion or in suppression, depending upon the actual local conditions. As both host and tumor are entities each with distinct reactivities and properties, it follows that each tumor/host combination expresses

289

Table 1. Host Mechanisms Considered Critical to Tumor
         Defense

Cellular Mechanisms

Induced cytotoxicity   T lymphocytes
 (immunospecific)

Spontaneous   mononuclear phagocytes,
 cytotoxicity   NK cells, mast cells,
                              PMNL, eosinophils,
                              basophils

Antibody dependent   mononuclear phagocytes,
 cell cytotoxicity   K cells, PMNL, eosinophils,
 (ADCC)   basophils, mast cells,
                              T lymphocytes

Humoral Mechanisms

Antibody: natural or induced

Agents with antibody-like activity (C reactive protein)

Monokines (lymphocyte activating factor, tumor necrosis
 factor)

Lymphokines (macrophage activating factor, lymphotoxin)

Complement components

Oxygen metabolites (hydrogen peroxide, superoxide anion)

Interferons, lysozyme, retinoids, arginase, proteinases

---

its own, unique characteristics. This then is the background against
which a possible role for mononuclear phagocytes in the control of
neoplastic disease has to be considered.

Apart from their traditional role in host defense and tissue

repair, mononuclear phagocytes manifest a large array of other
capabilities, the most persuasive being the governing of differentiation, proliferation and function of other cell types. The operation
of such multifaceted and interlocked regulatory relationships is
particularly meaningful as regards the mononuclear phagocyte and the
lymphoid cell systems. These manifold and occasionally opposing
functions are in part mediated by cell to cell interactions and in
part by secretory products. Among the numerous biologically active
secretory products of macrophages are monokines, lysozyme, growth
factors, α 2-macroglobulin, acid hydrolases, neutral proteinases,
interferon, products of the arachidonic acid lipoxygenase and cyclooxygenase pathways, endogenous pyrogen, complement components, and
platelet activating factor. In view of this extraordinary complexity,
it is not surprising that their contribution to host defense against
tumors is incompletely defined. The present work briefly reviews
in vivo and in vitro evidence for certain host/tumor situations being
under the control of mononuclear phagocytes. Some aspects are exem-
plified by results obtained by us with a rat fibrosarcoma model.

IN VITRO INTERACTION BETWEEN MACROPHAGES AND TUMOR CELLS

    Work from many laboratories attests to the capacity of mono-
nuclear phagocytes either to promote or to impair proliferation and
viability of tumor cells in vitro (4,5). Macrophage cytolytic activity depends on the metabolic and functional state of the cells and
the subset involved (6), the reactivity and repair capability of the
target cell, its ability to subvert host defense, and the actual
ratio of effector to target cells. Some of the basic conditions
now judged critical to the outcome of effector/target interactions
are listed in Table 2. A central problem is that the steps leading
to macrophage activation for tumor killing and its mechanism are
poorly understood. Activated macrophages kill tumor cells with some
selectivity, i.e. they distinguish between the 'normal' and the
'transformed' state by a mechanism independent of immune lymphocytes
(7,8,9). Clearly, the altered surface topography of tumor cells is
somehow registered as 'non-self' or 'altered-self', possibly analogous
to the function of these cells in the removal of damaged or senescent
autochthonous erythrocytes.

MONONUCLEAR PHAGOCYTES IN SURVEILLANCE AGAINST ONCOGENESIS AND PRIMARY
TUMOR GROWTH

    The basic processes underlying the early steps of oncogenesis
are so subtle and complex as to escape detection. In clinical and
experimental work one is forced to deal with discrete, emerging primary tumors or with tumor implants. Nonetheless various findings
both clinical and experimental strongly indicate that surveillance
exists and is particularly efficient in the very early phases of

Table 2. Factors Determining the Outcome of the Inter-
          action between Host and Tumor Cells

Effector: target   low ratio promotes tumor growth
   cell ratio
                           high ratio inhibits tumor growth

Functional activity   low or "resting" state: promotion
   of effectors   of tumor growth

                           "stimulated" state: inhibition of
                           tumor growth

                           "activated" state: elimination
                           of tumor cells

Target cell   adaptive potential
   capabilities
                           repair capability

                           proliferation rate

                           susceptibility to effector products

                           ability to subvert host defense

---

oncogenesis. However, once a malignant cell clone has overcome host
defense and become established (i.e. at a state where clinicians and
experimental investigators begin their observations) it is difficult
to stop its progression.

    In our rat fibrosarcoma model, the efficient operation of natural
rather than immunospecific host antitumor mechanisms is substantiated
by the following:

    (a) Local pretreatment (day -1) with agents such as hydrocorti-
         sone, silica particles, carrageenan, and tumor promoters,
         markedly reduced the survival period following a given tumor
         cell challenge (Figure 1); such reduction in tumor resis-
         tance was achieved only when these agents were administered
         briefly before tumor cell challenge.

(b) Local pretreatment with one of these agents (day -1) de-
     creased the number of day -12 tumor cells required for
     induction of progressive tumor growth (Figure 2).

(c) Reduction in tumor resistance was achieved only when inocu-
     lation of tumor cells and these varied agents was into the
     same compartment.

(d) Local pretreatment with immunostimulants (BCG or C. parvum
     on day -7) or local adoptive transfer of immunostimulant-
     induced adherent cells (day -1) markedly enhanced local
     resistance (Figure 1) without significantly affecting re-
     sistance in other compartments (10).

(e) Immunostimulant-induced, enhanced local resistance was
     readily nullified by local administration of hydrocortisone,
     etc. (Figure 1).

These findings affirm the existence and efficient operation of
spontaneous antitumor mechanisms in this model. The notion that
mononuclear phagocytes are the principal effectors is supported by
various findings: (a) under in vitro-culture conditions, D-12 rat

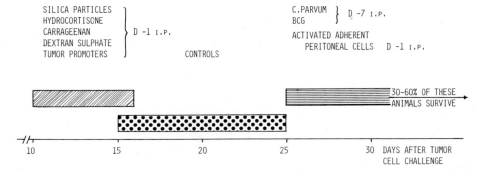

Fig. 1. Modulation of tumor growth by various interventions. DA
         rats, 3 months of age, were challenged i.p. with 10^3 D-12
         fibrosarcoma cells on day 0, and the time to death measured.
         Local administration of BCG or C. parvum (day -7) or of C.
         parvum-induced, adherent peritoneal cells (day -1) resulted
         in marked suppression or complete abolition of local tumor
         growth. On the other hand, local pretreatment (day -1) with
         silica particles, hydrocortisone, carrageenan, or tumor
         promoters decreased local resistance to the tumor in both
         controls and in animals which had been inoculated with
         immunostimulants.

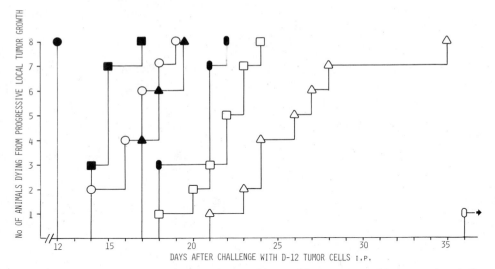

Fig. 2.   Reduction in the number of day-12 tumor cells required for
          induction of progressive tumor growth after local administra-
          tion of tumor-promoting phorbol ester (TPA).   TPA was given
          1 day before tumor implantation at a dose of 10^{-8}g.

O 10 D-12 cells (control);   ● 10 D-12 cells + TPA
△ 10^2 D-12 cells   ▲ 10^2 D-12 cells + TPA
□ 10^3 D-12 cells   ■ 10^3 D-12 cells + TPA
O 10^4 D-12 cells   ● 10^4 D-12 cells + TPA
➤ the remaining cells survived

fibrosarcoma cells are susceptible to killing by macrophages. but are
resistant to NK cells (5,11); (b) macrophage-mediated killing in
vitro is abolished by anti-macrophage agents such as hydrocortisone,
silica particles, carrageenan and tumor promoters; (c) the beneficial
effects achieved by adoptive transfer of C. parvum-induced cells is
abolished by previous interaction of the cells with any of the afore-
mentioned anti-macrophage agents (5,11).

MONONUCLEAR PHAGOCYTES AND SECONDARY TUMOR CELL SPREAD

     Most cancer treatment failures are attributed to the inability
to control metastasis rather than the outgrowth of the primary tumor.
About two thirds of all cancer patients develop disseminated disease.
Varied clinical and experimental evidence indicates that detachment

and dissemination of tumor cells is a continuous ongoing process. Taking this into consideration, the likelihood of metastases is rather small and provides indirect evidence for the presence and efficient functioning of the host defense mechanism. Indeed, to gain entry to the vascular and/or lymphatic tree and to tissues, tumor cells have to cope with a wide variety of hindrances and it is conceivable that the cells which surmount these defenses have been selected for their adaptive capacity. The operation of a single mechanism for parrying the establishment of metastases is therefore unlikely.

The literature suggests that tumor growth is effectively suppressed by a second tumor (primary on secondaries and vice versa), and that removal of either tumor results in enhanced growth of the tumor remaining (12,13,14). The mechanisms underlying these processes are far from understood. It was originally thought that "concomitant immunity", i.e. the ability of the tumor-bearing host to reject a further inoculum of the same tumor at a different site, was strictly a T cell-mediated immune reaction. However, recent data indicate that macrophages are equally essential and may indeed be the effectors in tumor rejection (15). Indirect evidence suggests that macrophages, either alone or in concert with other reactive components, can contribute to resistance against metastasis (Table 3). Particularly informative are studies regarding the degree of macrophage infiltration, immunogenicity and incidence of metastases in a series of rat fibrosarcomas raised in syngeneic hosts. These findings indicated that tumors which metastasize rarely are characterized by extensive macrophage infiltration and immunogenicity whereas tumors which invariably metastasize evoke but little host response (16,17). However, more recent studies by Evans and Eidlen (18) suggest that there was no relationship between the level of macrophages in a tumor and the host immune response.

In the D-12 rat fibrosarcoma, the tumor most extensively studied by us, the percentage of macrophages within the tumor was fairly constant during its progressive subcutaneous growth at a rather low level from 4 to 8% (5). Although in this model the incidence of macroscopic metastases is generally low, it is critically dependent on the site of the primary tumor implant (19,20). However, histologic and biologic findings affirm that dissemination of neoplastic cells is an early and continuing process. Accordingly, the preponderance of tumor cell spread is either eliminated or remains dormant. Removal of the primary tumor implant is frequently associated with the rapid outgrowth of viable tumor cells already present at secondary sites. As the growth characteristics of tumor cells derived from metastatic foci are not measurably different from those inducing primary tumor growth, it is unlikely that metastases involved selection of adaptive tumor cell sublines; rather this is a consequence of diminished tumor resistance of the host (19,20). Thus, although the present rat fibrosarcoma is poorly immunogenic and the percentage of macrophages within the tumor is low, varied evidence, albeit indirect, suggests a role

Table 3. Indications for Involvement of Macrophages in Control of Metastasis

Type of indication	Species and type of tumor
1) Monocytes accumulate with tumors	Rats and mice (21): CBA mice; MCA fibro-sarcoma (22)
2) Content of macrophages within tumor is inversely related to metastatic potential	BALB/c mice: Moloney sarcoma (23) C57B1/6 mice: FS6 sarcoma (24) Rat: various fibrosarcomata (16)
3) Cytolytic activity of macrophages within tumor is inversely related to metastatic potential	
4) Antimacrophage agents (silica, carrageenan, cortisone, etc.) increase rate of metastases	Sprague-Dawley rat: Walker sarcoma (25) DA rat: D-12 fibrosarcoma (Keller, unpublished) C57B1/6 mice: 3LL (26,27); mFS6 sarcoma (27) BALB/c mice: Madison 109 (27)
5) Immunostimulant-induced antimeta-static effects are mediated through macrophages	Hood rat: hepatoma (28) C57B1/6 mice: 3LL (26,29); B16 melanoma (29,30); T241 fibrosarcoma (31) C3H/HeJ mice: Dunn osteosarcoma (29)
6) Inoculation of activated macro-phages decreases incidence of metastases	C57B1/6 mice: B16 melanoma (30); T241 fibro-sarcoma (31) DA rat: D-12 fibrosarcoma (Keller, unpublished)
7) Regression of metastases follow-ing hyperthermia to primary is coun-teracted by silica and cortisone	Rat: MC7 sarcoma (32)

for macrophages in the control of both primary and secondary tumor growth.

SUMMARY

In a variety of defined experimental host/tumor situations there is now considerable evidence that mononuclear phagocytes are critically involved in the control of tumor cell nidation and their establishment as primary tumors. The factors which determine secondary tumor cell spread and the establishment of metastases are clearly complex and make it highly unlikely that tumor cell dissemination is under the control of a single host defense mechanism. Nonetheless, findings in various experimental models suggest a critical role for mononuclear phagocytes in the surveillance of tumor cell dissemination.

Acknowledgement

Supported by the Swiss National Science Foundation (grants No. 3.173.77 and 3.609.80) and the Canton of Zurich.

REFERENCES

1. Castro, J. E., "Immunological Aspects of Cancer", MTP Press, St. Leonard's House, Lancaster, England, 1978.
2. Woodruff, M. F. A., "The Interaction of Cancer and Host", Grune and Stratton, New York, 1980.
3. Nauts, H. C., "The Beneficial Effects of Bacterial Infections on Host Resistance to Cancer and Results in 449 Cases", Monograph No 8, Cancer Res. Institute, Inc., New York, 1980.
4. James, K., McBride, W. H., and Stuart, A., "The Macrophage and Cancer", Econoprint, Edinburgh, 1977.
5. Keller, R., in "Natural Cell-Mediated Immunity Against Tumors" (R. B. Herberman, ed.), pp. 1219-1269, Academic Press, New York, 1980.
6. Förster, O., and Landy, M., "Heterogenicity of Mononuclear Phagocytes", Academic Press, London, 1981.
7. Hibbs, H. B., Lambert, L. H., and Remington, J. S., Proc. Soc. Exp. Biol. Med. 139:1049, 1972.
8. Keller, R., in "Lymphokines 3" (E. Pick, ed.), pp. 283-292, Academic Press, New York, 1981.
9. Cabilly, S., and Gallily, R., Immunology 44:347, 1981.
10. Keller, R. (submitted).
11. Keller, R., in "Mononuclear Phagocytes. Functional Aspects" (R. van Furth, ed.), pp. 1725-1740, M. Nijhoff, The Hague, 1980.
12. Simpson-Herren, L., Sanford, A. H., and Holmquist, J. P., Cancer Treatm. Rep. 60:1749, 1976.

13. Sugarbaker, E. V., Thornthwaite, J., and Ketcham, A. S., in "Cancer Invasion and Metastasis: Biologic Mechanisms and Therapy" (S. B. Day et al., eds.), pp. 227-240, Raven Press, New York, 1977.

14. Gorelik, E., Segal, S., and Feldman, M., Int. J. Cancer 27:847, 1981.

15. Alexander, P., Ann. Rev. Med. 27:207, 1976.

16. Birbeck, M. S. C., and Carter, R. L., Int. J. Cancer 9:249, 1972.

17. Eccles, S. A., in "Immunological Aspects of Cancer" (J. E. Castro, ed.), pp. 123-154, MTP Press, St. Leonard's House, Lancaster, England, 1978.

18. Evans, R., and Eidlen, D. M., J. Reticuloendothelial Soc. 30: 425, 1981.

19. Keller, R., Invasion and Metastasis 1:136, 1981.

20. Keller, R., and Hess, M. (submitted).

21. Evans, R., Transplantation 14:468, 1972.

22. Hopper, K. E., and Nelson, D. S., Cell. Immunol. 47:163, 1979.

23. Russell, S., and McIntosh, A. T., Nature (Lond.) 268:69, 1977.

24. Mantovani, A., Int. J. Cancer 22:741, 1978.

25. Fisher, E. R., and Fisher, B., Acta Cytol. 9:146, 1965.

26. Jones, P. D. S., and Castro, J. E., Br. J. Cancer 35:519, 1977.

27. Mantovani, A., Giavazzi, R., Polentarutti, N., Spreafico, F., and Garattini, S., Int. J. Cancer 25:617, 1980.

28. Proctor, J., Rudenstam, C. M., and Alexander, P., Biomedicine 19:248, 1973.

29. Gatenby, P., and Basten, A., Cancer Immunol. Immunother. 8:103, 1980.

30. Fidler, I. J., Cancer Res. 34:1074, 1974.

31. Liotta, L. A., Gattozzi, C., Kleinerman, J., and Saidel, G., Br. J. Cancer 35:639, 1977.

32. Hanna, M. G., Zbar, B., and Rapp, H. J., J. Natl. Cancer Inst. 48:1441, 1972.

33. Dickson, J. A., Calderwood, S. K., Shah, S. A., and Simpson, A. C., in "Metastasis. Clinical and Experimental Aspects" Developments in Oncology 4 (K. Hellmann, P. Hilgard, and S. Eccles, eds.), pp. 260-265, M. Nijhoff, The Hague, 1980.

DISCUSSION

ROTH: Your observations on metastasis are not confined to animal systems but are seen also in the clinical situation very frequently. Within weeks or months following tumor resection, metastases appear. In our own work, we have noted that tumors produce anti-proliferative factors. In some cases, these anti-proliferative factors are directed specifically against lymphoid cells and result in immunosuppression as measured by mitogen induced proliferation or cell mediated cyto-toxicity. In other cases, we can isolate factors that have a direct anti-proliferative activity on normal cells and even tumor cells.

We could postulate that the primary tumor is producing some anti-proliferative factor which has an inhibitory effect on the metastasis. Have you looked for such factors in your animal studies?

KELLER: We surely have to consider such mechanisms but we haven't done any work so far along these lines.

GOLDMAN: I would like to make a supporting comment for the very early dissemination of tumor cells. We worked with A/J mice and if we inject half a million YAC lymphoma cells into the peritoneum and take the bone marrow two days later, it's very oncogenic. We get 100% tumor takes after inoculation of 2 million bone marrow cells into normal mice.

CUDKOWICZ: In the groups treated with so-called anti-macrophage agents, the possibilities are two-fold at least. One possibility is that the macrophages were impaired, either in number or in function. The second possibility is that the macrophages acted indirectly via a suppressor activity. Have you any data that could distinguish between these two possibilities? To what extent were the early deaths in the treatment groups due to macrophage depletion as opposed to macrophage activation which might render them suppressors for other effector cells?

KELLER: We have no evidence whatever for macrophage depletion. I think that can be ruled out. I cannot comment on the other possibility. It is interesting that perhaps 30% of the animals survived primary tumor growth and, when rechallenged 2 to 3 months later with 10^3 tumor cells, all died just as readily as if they had never experienced the tumor before. That is quite clear. We have no evidence whatever that specific immune resistance develops. The same is true for the animals which survived metastatic tumor growth. These animals have no enhanced resistance to the tumor whatever.

UNKNOWN: What do macrophages recognize on the tumor cells? Are there exceptions in which tumor cells aren't recognized? And are normal cells recognized?

KELLER: This is a difficult question to answer. In 1976, we looked at a variety of normal, mainly epidermal cells for their susceptibility to macrophage mediated killing. Some of these epidermal cells were susceptible to macrophage killing in the first or early passages of in vitro culturing. This seems to indicate that certain normal cells are susceptible, but most normal cells are resistant to macrophage killing. At the present time, I have no idea what determines if a cell will be recognized and killed. There are several different possibilities including susceptibility of the target cell to certain macrophage products. It is certainly not just surface structures that determine whether a cell will be killed or not. But I have no clear cut answer.

GINSBERG: As a microbiologist, I dare to make some analogies between
the microbial world and the tumor world. Certain microbial species
spread very nicely into tissues, while others can not. Duran-Reynals
developed the wonderful idea that these are caused by spreading factors. The spreading factors of bacteria are streptokinase which can
convert plasminogen into plasmin, hyaluronidase, DNase and RNase
that split the thick purulant exudates around the bacteria. Is it
possible that the macrophage modulates the plasminogen/plasmin system
locally, activates hyaluronidase, or releases factors which either
enhance or depress the barrier caused by the fibrin/fibronectin
connective tissue components and in this way manipulates the spreading
of the tumors? The analogy in spreading of these two different
worlds is really striking.

SILVERSTEIN: The accumulated evidence now is that plasminogen
activators aren't involved in the primary take of the tumor but there
may be some association between plasminogen activators and metastases.
However, I think that such evidence is not very strong. What's been
looked at is the capacity of the cell to secrete plasminogen activators. What's not been looked at are all the other things they secrete
or all the other phenotypes they express.

    I wish to ask Dr. Keller for his interpretation of the experiment
in which he removed the retroperitoneal lymph nodes. Although the
lymph nodes contained many tumor cells, the animals did not develop
tumors unless their primary tumor was extrapated. You said that the
tumor itself was not immunogenic, so what do you think is going on?

KELLER: After subcutaneous injection either into the back or into
the leg, the incidence of microscopic metastases was low in all
animals in which no surgery was done. To answer the question as to
whether the spread of tumor cells was an early event, a continuously
occuring event, or whether surgery induced the metastatic spread, we
looked at tumor bearing animals which did not show any sign of macro-
metastases. We made suspensions of these lymph nodes and tested
their capacity to induce tumor growth in normal animals. This biological test showed the presence of viable tumor cells.

SILVERSTEIN: But what is the significance of extrapation of the
primary tumor?

KELLER: That really cannot be answered. You do not induce tumor
cell spread during surgical amputation of the primary tumor on the
leg. Therefore, the metastatic spread must have occured prior to the
surgery.

SILVERSTEIN: I'd suggest that one look for concomittant immunity.

KELLER: A form of concomittant immunity is present, but what type
is difficult to say. We have no evidence for specific immunity whatever.

SILVERSTEIN: Have you examined the lymph nodes of animals with tumors in either Winn type assays or activation type assays?

KELLER: We have and have found no protective activity in either the extrapated or non-extrapated animals.

VAN MAARSSEVEEN: Did you use sham operations? Sham operations may diminish the immune system and contribute to a higher rate of metastasis.

KELLER: We have performed a lot of sham experiments. For example, we removed another leg instead of the tumor bearing leg and there was no increase in incidence of microscopic metastases. The answer was more clouded when large surgical interventions were done such as partial hepatectomy, because here there was a small increase in microscopic metastases.

FELDMAN: Every three years, we are faced with another candidate to be the enzyme which will solve the problem of tumor metastasis. The interesting thing is that everyone who has an idea somehow picks a tumor which fits that particular concept. When you move to another tumor, the general truism seems to fall down. When Bosco Wang implicated plasminogen activator in metastasis formation, he was absolutely right with regard to his particular tumor, but not when 2, 3 or 4 other tumors were examined. Jerry Ross had another idea; that it was due to proteolytic enzymes. Liotta and co-workers indicated beyond any doubt that collagenase type 4 characterizes metastatic tumors and this is absolutely true for those tumors which Liotta studied. Once you move to other tumors, however, then collagenase type 4, a marvelous candidate which splits basement membranes, falls down. What could be a better candidate than that one? In two years time, we will have other candidates. I personally feel that we are not dealing with any universal principles.

   Regarding concomittant immunity, we have found that you can induce concomittant immunity in animals which are incapable of invoking an immune reaction. Nude mice manifest an excellent concomittant immunity. C57B1/6 thymectomized mice depleted of T cells or cyclophosphamide treated mice can do it. Therefore, while we might be dealing with a phenomenon defined as concomittant immunity, we believe that it has nothing to do with the immune reaction. We believe it may be mediated by peptides or glycoproteins which behave as chalone-like substances.

REFERENCES

1. Duran-Reynals, F., Tissue permeability and the spreading factors in infection. A contribution to the host-parasite problem, Bact. Rev. 6:197, 1942.

2.   Wang, B., McLoughlin, G. A., Richie, J. P., and Mannick, J. A.,
     Correlation of the production of plasminogen activator with
     tumor metastasis in B16 mouse melanoma cell lines, Cancer
     Res. 40:288, 1980.

3.   Liotta, L. A., Tryggvason, K., Garbisa, S., Hart, I., Foltz, C.
     M., and Shafie, S., Metastatic potential correlates with
     enzymatic degradation of basement membrane collagen, Nature
     284:67, 1980.

SECTION 5

LEUKOCYTE INFLAMMATORY RESPONSES

SYNTHESIS AND RELEASE OF LYTIC ENZYMES

BY MACROPHAGES IN CHRONIC INFLAMMATION

Marco Baggiolini and Jörg Schnyder

Wander Research Institute, a Sandoz research unit
Wander Ltd., P. O. Box 2747
CH-3001 Berne, Switzerland

Knowledge about the secretory repertoire of macrophages is expanding very rapidly. Many molecules have been identified as products of these cells, and it may sound old-fashioned to keep talking about lytic enzymes. Due to their ability to degrade a great variety of biological molecules, however, the secretory enzymes of macrophages are a major factor in many pathological processes. Two types of effects are particularly important: tissue damage resulting from the breakdown of extracellular connective tissue structures and the conditioning of the pericellular environment. Lytic enzymes are also useful experimental parameters. They are among the best-defined truly secretory molecules of the macrophage and are easy to assay. For this reason, the rate of secretion of certain hydrolases is often adopted as a marker of macrophage function.

There are three classes of lytic enzymes which are secreted by macrophages: lysozyme, lysosomal hydrolases and neutral proteinases. β-Glucuronidase and N-acetyl-β-glucosaminidase are the lysosomal hydrolases which have been studied most frequently. Among the neutral proteinases, plasminogen activator is by far the most active and most commonly used. Enzymes from these three classes are secreted according to different kinetics and respond in different ways to agents or conditions which modify macrophage activity. This indicates that they are confined to separate subcellular compartments and are released by separate mechanisms. Enzyme secretion by macrophages has been studied mainly in cultures of adherent peritoneal cells obtained from normal or treated mice. For all three classes of enzymes as represented by lysozyme, β-glucuronidase and plasminogen activator, secretion is a long-lasting process extending for several days, which depends strictly on de novo synthesis. The latter is shown by the fact that the amounts of enzyme released greatly exceed the intra-

cellular levels and that secretion subsides when cycloheximide is
added to the culture media (1,3,4).

Lysozyme was shown by Gordon et al. (1) to be secreted by various
populations of mononuclear phagocytes. The rate of secretion is similar in macrophages of different sources (1) and is rarely affected
by stimuli of differentiation. In a study of the biochemical properties of mouse macrophages which were elicited by intraperitoneal
treatment with thioglycollate medium, proteose peptone or a suspension
of Streptococcus A cell walls, we found that the lysozyme secretion
rate remained at the level of the non-elicited controls, despite
major changes in other parameters (3). Similarly, macrophage stimulation in culture failed to influence lysozyme secretion (1,4). Owing
to its constancy, the rate of lysozyme secretion is a useful marker
of macrophage viability. Furthermore, since it was not detected in
lymphoid, fibroblast and epitheloid cell lines, lysozyme secretion
appears to qualify as a parameter for the identification of mononuclear phagocytes (1).

β-Glucuronidase and other acid hydrolases are present in macrophages in higher amounts than in most other cells. Cohn and Benson
(5) found many years ago that the activity of these enzymes increases
markedly following macrophage stimulation in culture. This phenomenon, now confirmed by a large casuistic literature (summarized in
Ref. 6), was viewed as an adaptation of the lysosomal apparatus to
an increased endocytic activity (i.e. pinocytosis and phagocytosis)
of the macrophage, since lysosomal enzymes were considered to serve
exclusively the purpose of intracellular digestion. Lysosomal hydrolases, however, are actively secreted both by quiescent and elicited
macrophages which are kept in culture in the absence of added stimuli
(3). As shown in Figure 1a, the rate of secretion appears to reflect
the degree of activation of the cells. Macrophages which are elicited
with thioglycollate medium secrete twice to four times as much β-
glucuronidase as non-elicited cells. There is also some difference
in the time course of secretion which in non-elicited cells begins
after a time lag of up to 2 days. These phenomena are not influenced
by changes in the composition of standard culture media and, in Medium
199, are independent of the presence of serum.

In non-secreting macrophages, i.e. peritoneal cells from normal
mice during the first day of culture, lysosomal enzyme secretion can
be induced by a phagocytic stimulus (Figure 1b). The massive release
of acid glycosidases by phagocytosing macrophages was first described
by Daivs et al. (7). We have studied the mechanism of induction of
lysosomal enzyme secretion and characterized the role of phagocytosis
(4). Two events were found to be essential, phagocytic uptake and
induction of the respiratory burst. If phagocytosis is not accompanied by a respiratory burst, as in the case of latex beads, no
secretion is observed. Phagocytosis, on the other hand, only has
a triggering function; once initiated the release of lysosomal enzymes

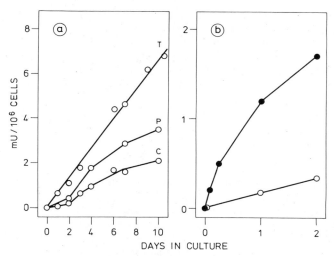

Fig. 1.   Secretion of β-glucuronidase by mouse peritoneal macrophages
          in culture. <u>Panel a</u>: Cumulative secretion curves obtained
          with non-elicited macrophages (C) and with macrophages which
          were elicited by intraperitoneal treatment with proteose
          peptone (P) or thioglycollate medium (T). <u>Panel b</u>:   Effect
          of a phagocytic stimulus. An excess of zymosan particles
          was added to cultures of freshly-harvested, non-elicited
          macrophages at time zero, and non-ingested particles were
          eliminated by washing after one hour. The cumulative secre-
          tion curve of phagocytosis-stimulated cells (●) is compared
          with that of the controls (o) during the first 2 days of the
          experiment, when non-elicited control macrophages secrete
          only minor amounts of lysosomal enzymes.

proceeds at an approximately constant rate, independent of ongoing
particle uptake and of the intracellular fate of the ingested material.

     Some non-phagocytic stimuli, i.e. lymphokines (8) and complement
activation products (9,10), have been reported also to induce lysosomal enzyme secretion. The effects of these agents, however, are
delayed (by 12 to 48 hours) and, particularly in the case of complement products, less pronounced than that of a phagocytic stimulus.
It is therefore conceivable that, in these instances, lysosomal enzyme
secretion is not a direct response, but rather the consequence of
macrophage activation. In fact, lymphokines and complement products
induce spreading of the cells and stimulate cell growth (8). Several

particles, e.g. zymosan, which are known to stimulate macrophages
also activate the alternative pathway of complement (11). It is
clear, however, that the effect of a phagocytic stimulus is not
mediated by complement products since lysosomal enzyme secretion is
induced by particles which do not activate complement and it is
independent of the presence of fresh serum in the culture medium (4).

IgE aggregates or antibodies against IgE were reported to induce
β-glucuronidase secretion from normal and IgE-sensitized macrophages,
respectively (12,13). This secretory response which is mediated by
a stimulus received through the F_c-receptor for IgE is of rapid onset,
similar to that resulting from phagocytosis. Lysosomal enzyme secretion is also induced by weak bases such as ammonium chloride, short-
chain aliphatic monoamines, chloroquine and others. The release
is rapid, extensive and apparently selective. In the presence of
optimal amine concentrations (0.5 mM for chloroquine and 50 mM for
ammonium chloride or methylamine), normal mouse peritoneal macrophages
secrete 60 to 80 percent of their β-glucuronidase and β-galactosidase
in a few hours without appreciable loss of lactate dehydrogenase (14,
15). The mechanism of this process is unknown, but is likely to be
related to the lysosomotropic properties of the eliciting bases (16).
After several hours, the macrophages are filled with large vacuoles
and appear to be damaged since the release of lactate dehydrogenase
increases markedly.

Plasminogen activator was the first neutral proteinase to be
identified as a secretory product of macrophages. Unkeless et al.
(2) have shown that this serine enzyme, which converts plasminogen
into plasmin, is secreted by inflammatory macrophages (elicited in
the mouse by intraperitoneal injection of thioglycollate medium),
but not by quiescent (non-elicited) macrophages (Figure 2a). It was
subsequently found (Figure 2b) that plasminogen activator secretion
could be induced in culture by exposing quiescent macrophages to a
variety of stimuli, e.g. concanavalin A (17), phorbol myristate acetate (17), culture media of proliferating lymphocytes (18-20), phago-
cytosable particles (4) and compounds which activate the hexose monophosphate shunt (21). The induction of plasminogen activator secretion is prevented by actinomycin D (17) and cycloheximide (4), indicating that de novo synthesis of RNA and protein is required.

Thioglycollate-elicited mouse peritoneal macrophages also secrete
an elastolytic enzyme (22) and a specific collagenase (23). Like
plasminogen activator, these two additional proteinases are produced
and secreted continuously and do not appear to be stored intracellularly. They are not produced by quiescent macrophages, but results
obtained with collagenase indicate that secretion can be induced in
these cells by the addition of particles or products of stimulated
lymphocytes (8,24,25).

A very large number of investigations have established beyond

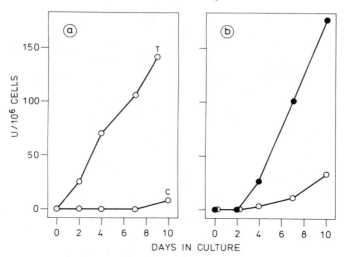

Fig. 2.   Secretion of plasminogen activator by mouse peritoneal macrophages.   Panel a: Cumulative secretion curves with thioglycollate-elicited (T) and non-elicited cells (C).   Panel b:
Effect of a phagocytic stimulus.   An excess of formaldehyde-
treated sheep erythrocytes was added to cultures of freshly-
harvested, non-elicited macrophages at time zero, and non-
ingested particles were eliminated by washing after one hour.
After a lag of at least 2 days, the phagocytic stimulus induces plasminogen activator secretion in non-elicited macrophages which secrete very little if not stimulated.   Phago-
cytosis-stimulated cells (●), non-stimulated controls (o).

doubt that secretion of plasminogen activator and other neutral proteinases is a characteristic property of macrophages which has been
triggered by a stimulus (26), i.e. by cells which have adapted functionally to a challenge in their micro-environment.   Neutral proteinase secretion appears to be a sterotype, unspecific response which
may be required to ensure some basic functions of the stimulated cell,
like mobility in the tissues and ability to contact other cells.
Other properties of the activated macrophages, e. g. the production
of factors which mediate a cytotoxic or an antiparasitic action or
which stimulate bystanding cells, are more selective and appear to
depend on the quality of the triggering event.   An exception to the
rule that macrophages produce and secrete plasminogen activator in
response to stimuli has been found recently.   Endotoxin and muramyl
pentapeptide induce in a dose-dependent fashion a decrease in plasminogen activator secretion and an increase in the production of
interleukin 1 by mouse peritoneal macrophages (27).

ENZYME SECRETION IN INFLAMMATION

Blood monocytes migrate into the tissues in response to chemoattractants, e.g. C5a or C5a-desArg, and become trapped at the origin
of the chemotactic activity, probably as a consequence of overstimula-
tion. Experimental correlates of the trapping of phagocytes in inflamed tissues are the spreading of macrophages induced in culture
by stimulating factors and the well-known inhibition of migration
from capillary tubes induced by lymphocyte activation products (28).

Locally, the young mononuclear phagocytes become activated.
Indeed cells which are collected from old lesions such as granulomata
have many properties of activated macrophages (29). Experiments in
vitro which were discussed in the preceding section have established
that phagocytosis or the exposure to culture media of proliferating
lymphocytes induce the differentiation of quiescent into inflammatory
macrophages. Since phagocytosis is a main function of macrophages
and specific or polyclonal lymphocyte stimulation a frequent consequence of inflammation, it is reasonable to expect that these mechanisms are important for macrophage activation in vivo. These two
triggering events may even be interdependent. Phagocytosis is likely
to be a primary stimulus since mononuclear phagocytes entering inflamed tissues will invariably find and engulf dead cells and denatured tissue fragments. The phagocytic stimuli will induce the
differentiation of the young mononuclear phagocytes into inflammatory
macrophages which, among other factors, will release Interleukin 1
and thus trigger bystanding lymphocytes to produce MAF and other
lymphokines.

Inflammatory macrophages secrete a variety of proteinases capable
of degrading vital connective tissue elements like cartilage proteo-
glycans, collagens and elastin. Extensive tissue destruction as
observed in arthritis, however, could also be an indirect effect of
macrophages. As was shown by Dayer et al. (30), mononuclear phagocytes produce a factor, similar to or even identical with Interleukin
1, which greatly stimulates collagenase secretion by the stellate or
dendritic cells developing in cultures of rheumatoid synovium. A
similar effect is observed with primary cultures of rodent chondrocytes, suggesting that various tissue cells may amplify the pro-
inflammatory activities of mononuclear phagocytes.

REFERENCES

1. Gordon, S., Todd, J., and Cohn, Z. A., J. Exp. Med. 139:1228,
    1974.
2. Unkeless, J. C., Gordon, S., and Reich, E., J. Exp. Med. 139:834,
    1974.
3. Schnyder, J., and Baggiolini, M., J. Exp. Med. 148:435, 1978.

4. Schnyder, J., and Baggiolini, M., J. Exp. Med. 148:1449, 1978.
5. Cohn, Z. A., and Benson, B., J. Exp. Med. 121:153, 1965,
6. Baggiolini, M., in "The Reticuloendothelial System: a Compre-
        hensive Treatise" (S. M. Reichard, and J. P. Filkins, eds.),
        Vol. IV, in press, Plenum Press, New York.
7. Davies, P., Page, R. C., and Allison, A. C., J. Exp. Med. 139:
        1262, 1974.
8. Pantalone, R., and Page, R. C., J. Reticuloendothelial Soc. 21:
        343, 1977.
9. Schorlemmer, H. U., Davies, P., and Allison, A. C., Nature
        261:48, 1976.
10. McCarthy, K., and Henson, P. M., J. Immunol. 123:2511, 1979.
11. Schorlemmer, H. U., Bitter-Suermann, D., and Allison, A. C.,
        Immunology 32:929, 1977.
12. Dessaint, J. P., Capron, A., Joseph, M., and Bazin, H., Cell.
        Immunol. 46:24, 1979.
13. Dessaint, J. P., Waksman, B. H., Metzger, H., and Capron, A.,
        Cell. Immunol. 51:280, 1980.
14. Riches, D. W. H., and Stanworth, D. R., Biochem. J. 188:933,
        1980.
15. Riches, D. W. H., Morris, C. J., and Stanworth, D. R., Biochem.
        Pharmacol. 30:629, 1981.
16. Ohkuma, S., and Poole, B., Proc. Natl. Acad. Sci. USA 75:343,
        1978.
17. Vassalli, J.-D., Hamilton, J., and Reich, E., Cell 11:695, 1977.
18. Vassalli, J.-D., and Reich, E., J. Exp. Med. 145:429. 1977.
19. Newman, W., Gordon, S., Hämmerling, U., Senik, A., and Bloom,
        B. R., J. Immunol. 120:927, 1978.
20. Neumann, C., and Sorg, C., J. Reticuloendothelial Soc. 30:79,
        1981.
21. Schnyder, J., and Baggiolini, M., Proc. Natl. Acad. Sci. USA
        77:414, 1980.
22. Werb, Z., and Gordon, S., J. Exp. Med. 142:361, 1975.
23. Werb, Z., and Gordon, S., J. Exp. Med. 142:346, 1975.
24. Wahl, L. M., Wahl, S. M., Mergenhagen, S. E., and Martin, R. R.,
        Science 187:261, 1975.
25. Horwitz, A. L., and Crystal, R. G., Biochem. Biophy. Res. Commun.
        69, 296, 1976.
26. Cohn, Z. A., J. Immunol. 121:813, 1978.
27. Drapier, J. C., Lemaire, G., and Petit, J. F., Int. J. Immuno-
        pharmacol., 4:21, 1982.
28. Bloom, B. R., and Bennett, B., Science 153:80, 1966.
29. Bonney, R. J., Gery, I., Lin, T.-Y., Meyenhofer, M., Acevedo,
        W., and Davies, P., J. Exp. Med. 148:261, 1978.
30. Dayer, J. M., Goldring, S. R., Robinson, D. R., and Krane, S.,
        in "Collagen in Normal and Pathological Connective Tissues"
        (D. E. Wolley, and J. M. Evanson, eds.), pp. 83-104, John
        Wiley & Sons, New York-Chichester-Brisbane-Toronto, 1980.

DISCUSSION

PICK: Did I understand you to say that the stimulus for neutrophil enzyme release must be capable of inducing an oxidative burst?

BAGGIOLINI: Yes. Phagocytosis is an inducer of lysosomal enzyme secretion and of plasminogen activator secretion. However, the phagocytic stimulus only works to induce secretion if it is accompanied by a respiratory burst. We have studied one example of phagocytosis- the uptake of uncoated, washed latex particles- wherein huge amounts of particles are engulfed without a respiratory burst and there is no enzyme secretion.

PICK: You have a very attractive model of inducing macrophage activation using methylene blue. This is basically a model of stimulation because methylene blue stimulates the respiratory burst. Is the respiratory burst actually an activating burst?

BAGGIOLINI: The driving of the hexose monophosphate shunt with methylene blue or toluidine blue induces plasminogen activator secretion and induces differentiation of the macrophages. It is tempting to say: "Okay, the shunt is the step that you must trigger in order to turn on the macrophages." However, I don't know if that actually occurs.

PICK: Why do you need the burst to get enzyme release?

BAGGIOLINI: Enzyme release such as plasminogen activator secretion is but one aspect of macrophage activation. I don't think that the release itself is triggered by the burst but rather by the products of the burst. Parenthetically, the products of the burst as induced by methylene blue may not be the same products as induced by phagocytosis.

SYNDERMAN: In other words, if you just expose cells to xanthene oxidase and xanthene, you would not see secretion?

BAGGIOLINI: Yes. We have done that.

WEAK-BASE INDUCED LYSOSOMAL SECRETION BY
MACROPHAGES: AN ALTERNATIVE TRIGGER MECHANISM
THAT IS INDEPENDENT OF COMPLEMENT ACTIVATION

David W. H. Riches and Denis R. Stanworth

Rheumatology and Allergy Research Group
Department of Immunology
The Medical School
Vincent Drive, Birmingham B15 2TJ U.K.

INTRODUCTION

The abundance of mononuclear phagocytes at inflammatory sites
has stimulated much interest in their pathological roles. In vitro
investigations have indicated that these cells are amply equipped
to promote an inflammatory response, following suitable stimulation,
by the synthesis and secretion of a variety of biologically active
products such as complement proteins (1), prostaglandins (2), neutral
proteinases (3,4), and lysosomal acid hydrolases (5,6,7). Whereas
the nature of these secretory products has been well characterized,
the biochemical pathways underlying the triggering of their release
remain only poorly understood.

Previous investigations carried out in our own (8) and in other
laboratories (9) have indicated that many of the macromolecular and
particulate stimuli that are capable of inducing the selective secretion of lysosomal enzymes from mouse macrophages also activate the
alternative pathway of complement. This observation led Schorlemmer
et al. (9) to propose that in response to activators of the alternative pathway, the macrophage could assemble, from endogenously produced complement proteins, a self-activating system generating C3b;
which, importantly, has itself been shown to be a potent initiator
of lysosomal secretion (10).

Two years ago (11) we reported that lysosomal secretion could
also be initiated from macrophage cultures that had been exposed to
relatively simple, low molecular weight weak-bases such as ammonia
and methylamine. On the basis of the known capacity of these compounds to interact non-enzymatically with C3, to produce a C3b-like

313

derivative (12,13), we initially suspected that the trigger for lysosomal secretion by them was also endogenously generated C3b. More
recently, however, we have investigated an increasing number of other
weak-bases such as chloroquine (14), and imidazole and benzamidine
(15); and have found that, whilst being capable of initiating potent
selective lysosomal enzyme secretion from macrophages, these compounds
failed to interact with C3. These observations thus led us to reas-
sess our initial supposition about the mechanism of methylamine
induced lysosomal secretion; and to investigate, in more depth, the
relationship between weak-base induced lysosomal secretion and that
induced by exposure of macrophages to inflammatory stimuli such as
zymosan particles. The findings reported here indicate that weak-
base induced lysosomal secretion is functionally distinguishable
from zymosan induced secretion; and, furthermore, occurs independently
of alternative pathway of complement activation.

DIFFERENTIAL INVOLVEMENT OF A Mg^{2+}/Ca^{2+} DEPENDENT SERINE ESTERASE

      Evidence of an involvement of a serine esterase activity in
the initiation of lysosomal β-galactosidase secretion from mouse
macrophages was sought by exposing cells to the irreversible serine
esterase inhibitor phenylmethylsulphonyl fluoride (PMSF) for 30
minutes prior to adding either methylamine (25 mM for 4 hr) or zymosan
particles (200 µg/ml for 6 hr). As illustrated in Figure 1, exposure
of macrophages to PMSF led to a concentration dependent inhibition
of the capacity of the cells to respond to zymosan stimulation, but
failed to influence the secretion of β-galactosidase following stimulation of the macrophages with methylamine. Throughout the experiment, cell viability was maintained at greater than 95% as indicated
by the failure to detect significant levels of the cytoplasmic enzyme
lactate dehydrogenase in culture supernatants. Microscopical examination of those cultures exposed to zymosan particles revealed a graded
inhibition to zymosan binding with increasing concentrations of PMSF.
Thus, the PMSF sensitive process in zymosan stimulated β-galactosidase
secretion is probably the initial interaction between the macrophage
and the activating ligand, rather than being a metabolic event subsequent to ligand binding.

      A similar picture was seen when macrophages were exposed to the
Mg^{2+}/Ca^{2+} chelating agent EDTA (in order to remove calcium (1.8 mM)
and magnesium (0.8 mM) from the culture medium) prior to the addition
of methylamine or zymosan. As can be seen from Figure 2, treatment
of macrophages with EDTA at concentrations between 3-5 mM virtually
abolished the capacity of the cells to secrete β-galactosidase in
response to zymosan stimulation, whilst that initiated by stimulation
with methylamine remained unaffected. As in the previous experiment,
cell viability was maintained at greater than 95% as indicated by
lactate dehydrogenase measurements. By contrast, exposure of the
cells to the calcium chelating agent EGTA prior to challenge with

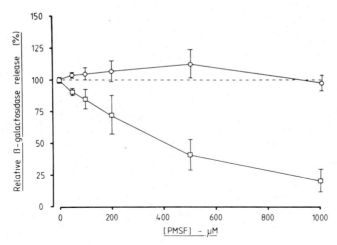

Fig. 1.   Influence of phenylmethylsulphonyl fluoride (PMSF) on
methylamine and zymosan induced β-galactosidase secretion
from macrophages. O = methylamine; □ = zymosan.   Each point
represents the mean ± S.D. for four experiments.

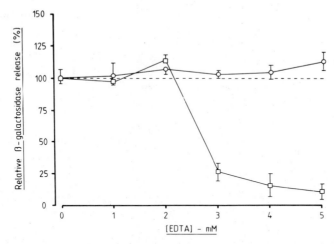

Fig. 2.   Influence of EDTA on methylamine and zymosan induced β-
galactosidase secretion from macrophages. O = methylamine;
□ = zymosan.   Each point represents the mean ± S.D. for
four experiments.

stimulating agents failed to influence either methylamine or zymosan
stimulated β-galactosidase release, thereby indicating that extracellular calcium is not required for the initiation of β-galactosidase
secretion by either type of stimulus. Microscopic examination of
both EDTA and EGTA treated cultures revealed a marked inhibition of
zymosan binding in those cultures exposed to 3-5 mM EDTA; whereas
binding and phagocytosis proceeded normally in cultures exposed to
EGTA.

These findings thus bear a close resemblance to PMSF inhibition
of β-galactosidase secretion in that the EDTA-sensitive (i.e. the
Mg^{2+} dependent) step appears to be the initial interaction of the
zymosan particles with, as yet, an undefined component(s) of the
macrophage plasma membrane. We would like to suggest that in view
of the similarities in the mode of inhibition of β-galactosidase
secretion and zymosan binding by these two agents that the macrophage
plasma membrane component(s) responsible for zymosan particle binding
is dependent upon the activity of a Mg^{2+} dependent ecto-serine esterase activity.

The lack of inhibition of methylamine stimulated β-galactosidase
secretion following exposure of the cells to PMSF or EDTA can be
explained by the capacity of methylamine (and other weak-bases) to
permeate the macrophage plasma membrane in the form of its free base
$R-NH_2$ (16) thus circumventing the requirement for an initial Mg^{2+}
dependent serine esterase dependent receptor-ligand interaction, as
seems to be the case with zymosan particles.

ADDITIVE EFFECTS OF WEAK-BASES AND ZYMOSAN

Further differences in the mode of triggering β-galactosidase
secretion by inflammatory agents and weak-bases were investigated
by examining the additivity of the macrophage secretory response to
each category of stimulus.

The effect of challenging macrophages for 2 hr with increasing
concentrations of methylamine (0-25 mM) in the presence of a fixed
dose of another weak-base, chloroquine (25 μM) is illustrated in
Figure 3. It will be seen that at low concentrations of methylamine,
chloroquine potentiates the secretion of β-galactosidase. However,
as the concentration of methylamine is increased, the additive effect
of chloroquine is progressively diminished until at high concentrations of methylamine, it becomes statistically indistinguishable
($p>0.1$) from that of methylamine alone. This type of non-additive
response is typical of partial agonist full-agonist interactions
and indicates that both weak-bases initiate lysosomal β-galactosidase
secretion via a common pathway.

A somewhat different picture was seen when macrophages were

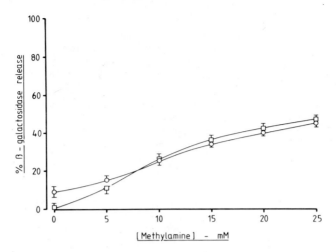

Fig. 3.   Effect of chloroquine 25 μM, (O) on the concentration depen-
          dent release of β-galactosidase induced by exposure of mouse
          macrophages to methylamine (0–25 mM, □).   Each point repre-
          sents the mean ± S.D. for four experiments.   Reproduced by
          courtesy of Biochemical Journal.

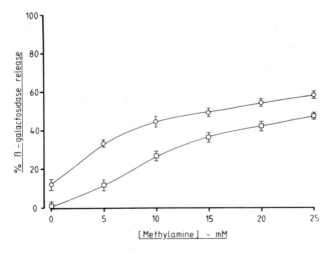

Fig. 4.   Effect of zymosan (100 μg/ml, O) on the dose dependent re-
          lease of β-galactosidase induced by exposure of mouse macro-
          phages to methylamine (0–25 mM, □).   Each point represents
          the mean ± S.D. for four experiments.   Reproduced by courtesy
          of Biochemical Journal.

challenged for 2 hr with methylamine (0–25 mM) in the presence of a
fixed concentration of zymosan particles (100 µg/ml). As will be
seen from Figure 4, zymosan particles were found to additively enhance
the methylamine stimulated secretion of β-galactosidase at all concentrations of methylamine. Likewise, a fixed dose of chloroquine
(20 mM) was also found to additively enhance zymosan stimulated
β-galactosidase secretion at all concentrations of zymosan (0–
100 µm/ml), (data not given), thus indicating that the secretory
response to zymosan is mediated by an activation pathway that is
functionally separable from that initiated by exposure of macrophages
to weak-bases.

COMPLEMENT INDEPENDENCE OF WEAK-BASE INDUCED SECRETION

    As mentioned earlier, we have previously reported that the con-
centration dependence of methylamine induced β-galactosidase secretion
from macrophages is very similar to that required to inactivate C3
(8). This observation initially led us to propose that methylamine
may well promote lysosomal secretion via a trigger mechanism involving
an endogenously generated C3b-like molecule as outlined by Schorlemmer
et al. (9). Recently, however, we have compared the capacity of
several other weak-bases to initiate β-galactosidase secretion from
macrophages with their ability to interact non-enzymatically with C3.
The results of these investigations are summarized in Table 1. They
show that, whilst methylamine exhibited a high degree of association
between the two parameters investigated, chloroquine and benzamidine
failed to interact with C3, but initiated potent β-galactosidase
secretion. Imidazole also initiated a marked secretion of β-galactosidase yet only induced a modest inactivation of C3 and only at higher
concentrations. Thus certain weak-bases such as chloroquine can
trigger lysosomal enzyme release independently of complement activation; and, in view of the findings of the additive experiments reported earlier, it would seem likely that other weak-bases, including
methylamine, promote secretion in this way.

CONCLUSIONS

    In conclusion, our current concepts on the mechanism of weak-
base and zymosan stimulated lysosomal acid hydrolase secretion from
macrophages may be summarized as follows.

    β-galactosidase secretion following stimulation with zymosan
particles is viewed as an obligatory sequel to the initial interaction
of the stimulus with a receptor that is either coupled to, or dependent upon, the activity of a Mg^{2+} serine esterase. Like the so-called
"third receptor" found on human monocytes (18), the macrophage receptor would be expected to display broad specificity towards activators
of the alternative pathway; and may well mediate binding via lectin-

Table 1. Capacity of Various Weak-bases to Initiate β-galactosidase
Secretion from Macrophages and to Activate the Alternative
Pathway of Complement

Weak-base	Concentration	β-galactosidase Secretion %	Complement Activation %
Benzamidine	0	1.5	3.0
	5 mM	3.8	0.2
	10 mM	7.4	1.8
	15 mM	39.0	2.2
	20 mM	70.9	2.8
	25 mM	82.3	4.3
Imidazole	0	0.7	0.1
	10 mM	13.4	1.0
	20 mM	31.0	5.1
	30 mM	41.5	9.9
	40 mM	47.9	17.1
	50 mM	54.2	20.2
Chloroquine	0	0.8	0.4
	25 μM	7.1	4.1
	50 μM	15.9	2.3
	100 μM	28.8	3.9
	300 μM	61.1	3.7
	500 μM	72.0	3.1
Methylamine	0	1.4	3.3
	10 mM	28.2	29.5
	20 mM	51.9	62.2
	30 mM	65.2	84.1
	40 mM	74.0	91.2
	50 mM	79.2	92.0

The capacity to activate the alternative complement pathway was
determined using the method of Riches and Stanworth (17).

like interactions with carbohydrate moieties on the activating stimulus (19). Moreover, on the basis of our recent finding (20) of a
failure of anti-mouse C3b F(ab')$_2$ fragments to influence either
zymosan or methylamine stimulated β-galactosidase release we would
suggest that endogenous C3b generation is not involved in the initiation of this response.

The secretion of β-galactosidase from macrophages in response to
methylamine stimulation, in contrast to that by zymosan, relies on a
receptor independent mechanism of uptake of stimulus by the cells.
de Duve et al. (16), and Ohkuma and Poole (21) view the cellular uptake and concentration of weak-bases as being a consequence of the
low pH inside the lysosomes, and the much greater permeability of the
plasma and lysosomal membranes to the free bases as contrasted to
their protonated forms. Thus, as illustrated in Figure 5, the free
base form of methylamine R-NH$_2$ will traverse the plasma and lysosomal
membranes of the macrophage; and, due to the acidity of the lysosomes
(approximately pH 4.8 in resting macrophages), will become protonated,
thereby trapping it within the lysosomal compartment.

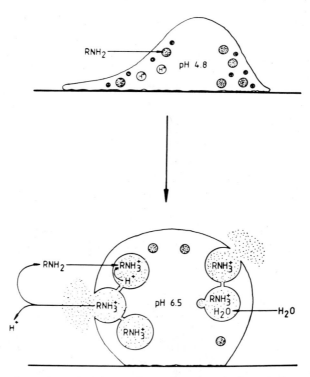

Fig. 5.   Postulated mechanism of weak-base induced lysosomal secretion
           from macrophages.

The second phase of the secretory process is speculated as being shared by both types of stimulus; and is only initiated when the intralysosomal concentration of the stimulating agent becomes sufficiently high. Thus, in the case of methylamine, as the intralysosomal concentration of the weak-base increases, the lysosomes undergo a marked osmotic expansion ultimately resulting in their fusion with the plasma membrane. By analogy, zymosan is also viewed as promoting expansion of the lysosomal compartment; although, in this case, it is postulated as being due to the physical size of the particles, rather than resulting from purely an osmotic effect.

Clearly much of what has been proposed above is of a speculative nature. In work now in progress in our laboratory, the substrate specificity of the serine esterase is being examined more closely, in order to obtain more objective data as to the involvement of this enzyme in macrophage acid hydrolase secretion.

REFERENCES

1. Bentley, C., Fries, W., and Brade, V., Immunology 35:971, 1978.
2. Bonney, R. J., Wightman, P. D., Davies, P., Sadowski, S. J.,
        Kuehl, F. A., and Humes, J. L., Biochem. J. 176:433, 1978.
3. Werb, Z., and Gordon, S., J. Exp. Med. 142:346, 1974.
4. Unkeless, J. C., Gordon, S., and Reich, E., J. Exp. Med. 139:834,
        1974.
5. Cardella, C. J., Davies, P., and Allison, A. C., Nature 247:46,
        1974.
6. Schorlemmer, H. U., Edwards, J. H., Davies, P., and Allison, A.
        C., Clin. Exp. Immunol. 27:198, 1977.
7. Schnyder, J., and Baggiolini, M., J. Exp. Med. 148:435, 1978.
8. Riches, D. W. H., and Stanworth, D. R., Immunology 44:29, 1981.
9. Schorlemmer, H. U., Bitter-Suermann, D., and Allison, A. C.,
        Immunology 32:929, 1977.
10. Schorlemmer, H. U., Davies, P., and Allison, A. C., Nature 261:
        48, 1976.
11. Riches, D. W. H., and Stanworth, D. R., Biochem. J. 188:933,
        1980.
12. von Zabern, I., Nolte, R., and Vogt, W., J. Immunol. 124:1543,
        1980.
13. Pangburn, M. K., and Muller-Eberhard, H. J., J. Exp. Med. 152:
        1102, 1980.
14. Riches, D. W. H., Morris, C. J., and Stanworth, D. R., Biochem.
        Pharmacol. 30:629, 1981.
15. Riches, D. W. H., and Stanworth, D. R., Biochem. J. 202:639,
        1982.
16. de Duve, C., de Barsy, T., Poole, B., Trouet, A., Tulkens, P.,
        and van Hoof, F., Biochem. Pharmacol. 23:2495, 1974.
17. Riches, D. W. H., and Stanworth, D. R., Immunol. Letts. 1:363,
        1980.

18. Czop, J. K., Fearon, D. T., and Austen, K. F., J. Immunol. 120:
        1132, 1978.
19. Weir, D. M., and Ogmundsdottir, H. M., Clin. Exp. Immunol. 30:
        323, 1977.
20. Riches, D. W. H., Morris, C. J., and Stanworth, D. R., Immunology
        45:473, 1982.
21. Ohkuma, S., and Poole, B., J. Cell Biol. 90:656, 1981.

DISCUSSION

GINSBERG: What evidence do you have that the amines are getting into
the lyzosomes? Is it possible that they just trigger surface phenomenon linked to an oxidative burst? Have you measured chemiluminescence with these amines?

RICHES: We haven't actually measured chemiluminescence. As I understand it, however, hexose monophosphate shunt activity which has been
measured is not stimulated by these amines.

UNKNOWN: When we administered methylamine to bone marrow derived
macrophages, we found that the hexose monophosphate shunt was not
stimulated. Furthermore, superoxide and hydrogen peroxide secretion
were not enhanced.

GINSBERG: What is the smallest amine that is needed to trigger
chemiluminescence versus penetration into a macrophage?

RICHES: Spermine and spermidine presumably have surface effects,
for they are not freely diffusible into the cells in the same way
that the other compounds are. We've done experiments with spermine
and spermidine and find that they don't induce secretion of lyzosomal
enzymes as is the case with the weak base methionine. Furthermore,
if we expose the cells to diamines, such as putrescein or cadaverine,
we don't see a secretion of lyzosomal enzymes and they too are very
poorly taken up by the cells.

UNKNOWN: Presumably, you're getting phagocytosis of the zymosan by
the macrophages. I'd like to suggest that part of the reason they
cause secretion is due to mechanical or osmotic swelling following
the breakdown of the zymosan particles within the phagolyzosomes.

RICHES: In our electron micrographs, we really don't see much evidence of osmotic swelling of the lyzosomes using zymosan.

BAGGIOLINI: Zymosan is not broken down. If you do the same with
sheep erythrocytes, then you would expect the sort of swelling induced
by osmosis which you are postulating. After a week or so, the swelling subsides because after digestion the small molecules have leaked
out. My question is why do macrophages secrete while fibroblasts
which vacuolate in exactly the same way with methionine don't?

RICHES: Yes, correct. We find that as well, but we don't know why.
An important question as well is whether the secretion of lyzosomal
enzymes actually occurs through the phagocytic vacuole or whether
it's a true secretory process. I suspect it's a secretory process,
but I'm not entirely sure.

BAGGIOLINI: In my opinion, this points out again the uniqueness of
the macrophage in terms of secretion of lyzosomal enzymes. Any other
cell when exposed to these reagents does not secrete as abundantly
as the macrophage.

INFLAMMATORY RESPONSE OF LPS-HYPORESPONSIVE AND LPS-RESPONSIVE MICE TO CHALLENGE WITH GRAM-NEGATIVE BACTERIA SALMONELLA TYPHIMURIUM AND KLEBSIELLA PNEUMONIAE

T. J. MacVittie,[1] A. D. O'Brien,[2] R. I. Walker,[3] and S. R. Weinberg[1]

[1]Experimental Hematology Department, Armed Forces Radiobiology Research Institute, Bethesda, Maryland 02814 USA
[2]Department of Microbiology, Uniformed Services University of Health Sciences, Bethesda, Maryland 21814 USA
[3]Medical Microbiology Branch, Naval Medical Research Institute, Bethesda, Maryland 20814 USA

INTRODUCTION

The murine response to lipopolysaccharide (LPS) or endotoxin is determined by the allelic form of the Lps gene carried by the host (1,2). Mice that are homozygous for the defective Lps^d allele, such as C3H/HeJ (HeJ) animals, respond to only high doses of endotoxin, whereas mice that are homozygous or heterozygous for the normal Lps^n allele, e.g., C3H/HeN (HeN) mice, react to low-dose challenge. Thus, HeJ mice are insensitive to quantities of LPS that elicit mitogenic, inflammatory, hemopoietic, or lethal effects in HeN mice (3–10). Furthermore, the nature of the cellular influx into the peritoneum of LPS-inoculated HeN and HeJ mice differs. Low doses of LPS (1–10 µg) induce an early polymorphonuclear (PMN) increase, followed by a rapid rise in macrophages and macrophage colony-forming cells in the HeJ peritoneal cavity (7,11,12). By contrast, a relatively small PMN infiltrate is evident in the HeN peritoneal inflammatory response, and the onset of the macrophage influx is delayed compared to the HeJ response.

The purpose of this investigation was to determine whether the inflammatory responses of HeJ and HeN mice to killed or viable gram-negative bacteria mimicked those seen after administration of LPS. This question was of particular interest because endotoxin-hypore-

sponsive HeJ mice are very susceptible to infection with the gram-
negative bacteria Salmonella typhimurium (13-15) and Klebsiella
pneumoniae (16), whereas endotoxin-sensitive HeN mice are relatively
resistant to these LPS-containing microbes. In addition, O'Brien
et al. (15) have found that the gene that controls the response of
the HeJ mice to S. typhimurium is closely linked to or the same as
the single gene defect that controls their response to LPS.

METHODS AND MATERIALS

     Peritoneal cell suspensions were obtained by lavage from 8-to-
12-week-old male or female mice of the strains C3H/HeN (Charles River,
Wilmington, MA) and C3H/HeJ (Jackson Lab, Bar Harbor, ME). The
animals were maintained on a 12-hour light-dark cycle. Wayne Lab-
Blox and acidified (pH 2.5) water were available ad libitum. All
mice were acclimated to laboratory conditions for 2 weeks before
experimental treatment. During this time, the mice were examined
and found to be free of lesions of murine pneumonia complex and of
oropharyngeal Pseudomonas sp. Lavage was accomplished by i.p injection and subsequent withdrawal of 4 ml of Hanks' Balanced Salt Solution (calcium- and magnesium-free). Total and differential cell
counts were performed on samples of exudate cells. Smears of cells
were prepared with the use of a cytospin centrifuge (Shandon Southern
Institute, Ltd., Sewickley, PA), air-dried, and stained with a Wright-
Giemsa solution.

     Peritoneal exudates were induced by injection of (a) 10 μg of
E. coli 055.B5 lipopolysaccharide-W (LPS-W) (List Biological Labs,
Campbell, CA), (b) 1.5 mg (wet weight) of formalin-killed Klebsiella
pneumoniae, or (c) viable S. typhimurium or K. pneumoniae at doses
of 10 organisms for HeJ mice or 1000 organisms for HeN animals.
The S. typhimurium (strain TML) (17) was grown overnight at $37^{\circ}C$
with shaking in Penassay broth (Difco Labs), whereas the encapsulated
K. pneumoniae strain was cultured in brain heart infusion (BHI) broth
(Difco). Bacterial challenge doses were prepared in sterile pyrogen-
free saline and the actual inoculum size verified by plate count on
BHI (K. pneumoniae) or tryptic soy agar (S. typhimurium). For killed
preparations of K. pneumoniae, 18-hour broth cultures were treated
with 0.75% formalin for 2 hours at room temperature with constant
stirring. An aliquot of cells was incubated in BHI broth to determine
sterility of the killed-cell suspension.

     The 50% lethal dose (LD_{50}) of an agent for mice was determined
by the method of Reed and Muench (18). Groups of five age-matched
mice were inoculated i.p. with graded doses of each substance, and
deaths were recorded daily for 28 days.

RESULTS

Lethal Dose 50 Values of LPS or Bacteria for Mice

HeJ strain mice were significantly more resistant to LPS and killed K. pneumoniae than were mice of the HeN strain. The LD_{50} of LPS or killed K. pneumoniae for HeJ mice was 2000 µg and 65 mg, respectively, whereas only 150 µg of LPS or 7.5 mg of killed bacteria was sufficient to kill 50% of HeN mice (Table 1). Conversely, all HeJ mice succumbed to S. typhimurium or K. pneumoniae challenge with <10 organisms, but the LD_{50} of these microbes for endotoxin-sensitive HeN animals was 2,000.

Total Peritoneal Cell Influx after Challenge with Bacteria or LPS

As shown in Table 2, the total number of resident peritoneal nucleated cells was equivalent in the HeN and HeJ mice (range = 3.8 to 4.7×10^{6}). When mice were given LPS-W or viable K. pneumoniae or S. typhimurium, the magnitude of the total cellular influx was also similar for HeN and HeJ mice. However, administration of killed K. pneumoniae elicited a significantly greater (p <0.01) response in HeJ than in HeN animals.

Differential Inflammatory Response of C3H/HeN and C3H/HeJ Strain Mice

The inflammatory response of the HeJ strain mouse to LPS was characterized by an early PMN response followed by a subsequent and predominant monocyte-macrophage influx that peaked at day 4 after

Table 1. Comparative Lethality of S. typhimurium, K. pneumoniae, LPS-W, and killed K. pneumoniae for C3H/HeN and C3H/HeJ Strain Mice

Mouse Strain	LPS/W	Killed K. pneumoniae	S. typhimurium	K. pneumoniae
C3H/HeN	150 µg	7.5 mg	1,000–10,000	1,000–10,000
C3H/HeJ	2000 µg	65.0 mg	10	10

Comparative lethality is reported as the LD_{50} dose by weight for LPS-W (obtained from S. typhimurium) and formalin killed K. pneumoniae (wet weight) and by numbers of organisms for viable K. pneumoniae and S. typhimurium injected intraperitoneally.

Table 2.   Peak Peritoneal Nucleated Cell Influx in C3H/HeN
           and C3H/HeJ Strain Mice at Day 4 after Challenge

Challenge	Resident Cells	Day-4 Cells
LPS-W:		
HeN mice	4.2±0.5	10.5±1.5
HeJ mice	4.4±0.7	13.0±2.0
Killed K. pneumoniae:		
HeN mice	3.8±0.6	10.3±1.2
HeJ mice	4.2±0.8	15.3±1.8
K. pneumoniae:		
HeN mice	4.7±0.6	15.0±2.1
HeJ mice	4.3±0.5	13.0±1.7
S. typhimurium:		
HeN mice	4.1±0.7	20.2±2.2
HeJ mice	4.2±0.7	17.1±2.1

Mice were challenged with 10 µg of LPS-W; 1.5 mg of formalin
killed K. pneumoniae; 10 organisms of K. pneumoniae or S.
typhimurium into HeJ or 1000 organism into HeN mice. The data
are reported as mean total nucleated cells ± standard error
x 10^6.

injection (Figure 1). Lymphocytes also increased twofold over resident values. Influxes of all three cell types in the HeN mice were
delayed relative to the HeJ mice, but HeJ and HeN mice attained
equivalent numbers of mononuclear cells by 72-96 hours after injection. These results confirmed the findings of Sultzer and Goodman
(11) and Moeller et al. (7).

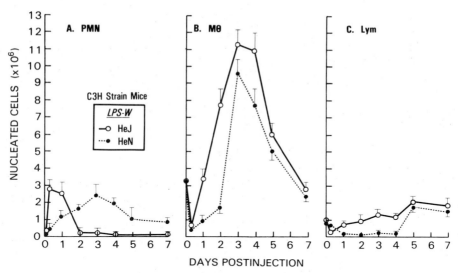

Fig. 1. Peritoneal cellular influx of (A) polymorphonuclear leukocytes (PMN), (B) macrophages (MΘ), and (C) lymphocytes (Lym)
in C3H/HeN (●) and C3H/HeJ (○) mice injected with 10 µg of
LPS-W. Mean values (±SEM) of four replicate experiments.

     Similar responses were observed after injection of killed K̲.
pneumoniae. Both murine strains showed a predominantly mononuclear
influx, although the HeJ mice responded to a significantly greater
degree than did HeN mice (Figure 2).

     Markedly different responses were observed following infection
of HeJ or HeN mice with either live K. pneumoniae (Figure 3) or S̲.
typhimurium (Figure 4). The HeN mouse showed an early influx of
PMNs followed by a rise in mononuclear cells. In contrast to the
cellular response of HeN mice to LPS-W and killed K. pneumoniae,
the mononuclear fraction elicited in response to live bacteria contained a greater share of lymphocytes (Figures 3 and 4). The HeJ
mouse also responded with a significant PMN influx, but unlike the
HeN mouse, the influx continued to increase until death of the HeJ
animal within 5 days after infection with either K. pneumoniae or
S. typhimurium (Figures 3 and 4). The mononuclear response, although
initiated in the HeJ mouse, failed to maintain itself as the infection progressed.

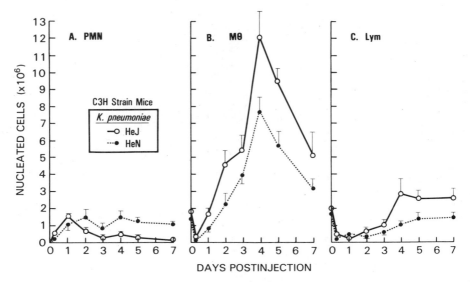

Fig. 2.   Peritoneal cellular influx of (A) polymorphonuclear leukocytes (PMN), (B) macrophages (MΘ), and (C) lymphocytes (Lym)
in C3H/HeN (●) and C3H/HeJ (o) mice injected with 1.5 mg
(wet weight) killed K. pneumoniae. Mean values (± SEM) of
four replicate experiments.

DISCUSSION

    In this investigation, the inflammatory responses of endotoxin-
hyporesponsive C3H/HeJ mice and endotoxin-responsive C3H/HeN mice
were compared after the animals had been given an LPS preparation,
killed K. pneumoniae as a source of endotoxin, or viable LPS-
containing S. typhimurium or K. pneumoniae. Three parameters of the
response were examined: the quantitative cellular response, the
types of cells involved, and the kinetics of the cellular influx.

    The data indicated that in response to LPS or killed K. pneu-
moniae as a source of endotoxin, the HeJ mouse responded with a predominantly mononuclear influx that was equivalent to or greater than
that of the endotoxin-responsive HeN strain. Thus, the HeJ mouse
was sensitive to the presence of endotoxin as a soluble molecule
or as formalin-killed gram-negative K. pneumoniae. However, infection
of the HeJ mice with S. typhimurium or K. pneumoniae produced an
inflammatory response composed predominantly of PMNs rather than
mononuclear cells as was seen in resistant HeN mice. Since the total
cellular influx was equivalent between the two C3H strains, the PMN

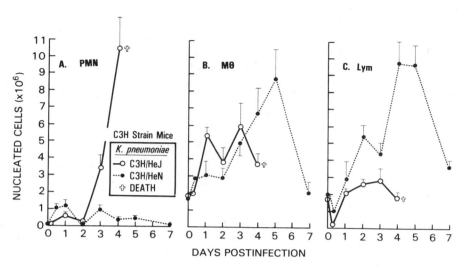

Fig. 3.   Peritoneal cellular influx of (A) polymorphonuclear leukocytes (PMN), (B) macrophages (MΘ), and (C) lymphocytes (Lym)
in C3H/HeN (●) and C3H/HeJ (o) mice infected with 1,000 or
10 K. pneumoniae, respectively. Mean values (± SEM) of
five replicate experiments.

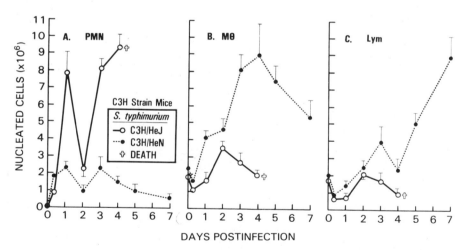

Fig. 4.   Peritoneal cellular influx of (A) polymorphonuclear leukocytes (PMN), (B) macrophages (MΘ), and (C) lymphocytes (Lym)
in C3H/HeN (●) and C3H/HeJ (o) mice infected with 1,000 or
10 S. typhimurium, respectively. Mean values (± SEM) of
five replicate experiments.

influx represented a marked shift away from the normal immigration
of mononuclear cells. Jerrels and Osterman (19) have recently described a similar PMN response in a strain of C3H mice susceptible
to infection with Rickettsia tsutsugamushi. Thus, the phenotypic
expression of the Lps^d allele, as measured by the nature of the
peritoneal cell infiltrate, is markedly different when endotoxin is
presented to the hyporesponsive host on viable, replicating gram-
negative bacteria rather than on killed microbes or as an extract.

Two lines of evidence suggest that the failure of endotoxin-
hyporesponsive HeJ mice to mount a normal macrophage reaction to S.
typhimurium infection may be directly related to their innate susceptibility to this microbe. First, macrophages, not PMNs, are the
effector cells in murine resistance to salmonellosis (20). Second,
recent studies by O'Brien et al. (21) indicate that for S. typhimurium, susceptibility of Lps^d animals is a consequence of a macrophage abnormality. That such a macrophage dysfunction may be quantitative as well as qualitative is suggested by the data presented
herein.

Acknowledgement

Supported by Armed Forces Radiobiology Research Institute,
Defense Nuclear Agency, under Research Work Unit MJ 00029. The views
presented in this paper are those of the authors; no endorsement
by the Defense Nuclear Agency has been given or should be inferred.
The authors gratefully acknowledge the excellent technical assistance
of James L. Atkinson, Brenda Watkins, and Daniel P. Dodgen. We also
thank Junith A. Van Deusen and Doris M. Pateros for editing and
preparing this manuscript.

REFERENCES

1. Watson, J., Riblet, R., and Taylor, B. A., J. Immunol. 118:2088,
       1977.
2. Watson, J., Kelly, K., Largen, M., and Taylor, B. A., J. Immunol.
       120:422, 1978.
3. Sultzer, B. M., J. Immunol. 103:32, 1969.
4. Watson, J., and Riblet, R., J. Exp. Med. 140:1147, 1974.
5. Glode, M. L., Jacques, A., Mergerhagen, S. E., and Rosenstreich,
       D. L., J. Immunol. 119:162, 1977.
6. Doe, W. F., and Hensen, P. M., J. Immunol. 123:2304, 1979.
7. Moeller, G. R., Terry, L., and Snyderman, R., J. Immunol. 120:
       116, 1978.
8. Apte, R. M., and Pluznik, D. H., J. Cell. Physiol. 89:313, 1976.
9. Boggs, S. S., Boggs, D. R., and Joyce, R. A., Blood 55:444, 1980.
10. MacVittie, T. J., and Weinberg, S. R., in "Experimental Hematology
       Today" (S. J. Baum and G. D. Ledney, eds.), pp. 19-28,
       Springer-Verlag, New York, 1980.

11. Sultzer, B.M., and Goodman, G. W., in "Microbiology" (D.
        Schlesinger, ed.), p. 304, American Society of Microbiology,,
        Washington, D. C., 1977.
12. MacVittie, T. J., and Weinberg, S. R., in "Genetic Control of
        Natural Resistance to Infection and Malignancy" (E. Skamene,
        P. A. L. Kongshavn and M. Landy, eds.), pp. 511–518, Academic
        Press, New York, 1980.
13. Robson, H. G., and Vas, S. I., J. Infect. Dis. 126:378, 1972.
14. Von Jeney, N. E., Gunter, E., and Jann, K., Infect. Immun. 15:
        26, 1977.
15. O'Brien, A. D., Rosenstreich, D. L., Scher, I., Campbell, G. H.,
        MacDermott, R. P., and Formal, S. B., J. Immunol. 124:20,
        1980.
16. Chedid, L., Parent, M., Damais, C., Juy, D., and Galelli, A.,
        Infect. Immun. 13:722, 1976.
17. Gianella, R. A., Broitman, S. A., and Zamcheck, N., Am. J. Dig.
        Dis. 16:1007, 1971.
18. Reed, L. J., and Muench, H., Am. J. Hyg. 27:493, 1938.
19. Jerrells, T. R., and Osterman, J. V., Infect. Immun. 31:1014,
        1981.
20. O'Brien, A. D., Scher, I., and Formal, S. B., Infect. Immun. 25:
        513, 1979.
21. O'Brien, A. D., Metcalf, E. S., and Rosenstreich, D. L., Cell.
        Immunol., 67:325, 1982.

DISCUSSION

GINSBERG: When you use LPS, it is usually either commercial material
or that produced yourself by phenol extraction or some similar method.
Is this the real LPS which is released in vivo? What do we know about
the mechanisms by which LPS is released from gram negative organisms
in vivo? Is it by autolysis of the gram negative organism? Is LPS
attacked by lysozomal enzymes or by antibody and complement on the
surface of the cell?

    The dichotomy you observed between resistance or susceptibility
to commercial LPS versus the whole, killed organism raises another
question. Is it possible that some humoral substance or some enzymes
are missing in vitro which cleave the LPS off the bacterial cells
in vivo? The reason for raising this question comes from work done
in our laboratory recently showing that LPS can be obtained from gram
negative organisms by lysozomal factors from granulocytes. If you
kill the microbe by antibiotics, heat or ultraviolet radiation, you
no longer release the natural LPS by treatment with lysozomal factors
from granulocytes. I suggest that you look at the LPS released in
vivo by natural mechanisms. Such probably would shed more light on
the various genetic groups of mice that you are dealing with.

MacVITTIE: Thank you for that comment. I agree with you 100%. We did perform several studies using LPS-W (Westphal) and LPS-B (Boivin) as well as a purified LPS but we wondered if the information had relevance to the whole viable organism? In other words, how does the animal see this organism and what are the interacting factors?

GINSBERG: There are certain strains of gram negative mutants, which are either very susceptible or very resistant to the effect of antibody and complement mediated bacteriolysis. Since bacteriolysis may be the origin of the genuine LPS, it would be prudent to examine highly resistant, serum resistant and serum susceptible strains of infectious organisms.

CHARACTERIZATION OF THE LEUCOCYTIC INFILTRATE OF RHEUMATOID SYNOVIUM FROM TISSUE SECTIONS AND SYNOVIAL ELUATES

J. G. Saal,[1] P. Fritz,[1] J. Müller,[2] and M. Hadam[3]

[1] Robert Bosch Hospital, Stuttgart F.R.G.
[2] Institute of Biochemistry, University of Stuttgart F.R.G.
[3] Immunology Laboratories, University of Tübingen F.R.G.

INTRODUCTION

Connective tissue diseases and rheumatoid arthritis (RA) are characterized by polyclonal B-cell hyperreactivity associated with an oligoclonal, disease-related pattern of autoantibody production. The underlying defects in immune regulations may be found either in the peripheral blood or locally at sites of lymphocyte/plasma cell infiltration. Our interest has focused on the local micro-environment of the inflamed joint as the main site for RA. The dynamic changes in histopathology-cellular infiltration and fibro-blast proliferation- characteristic of RA-synovitis require separate analysis of the different states of the disease. In contrast to the morphological approach, a functional analysis of joint inflammation requires isolating the infiltrating cells. In this study, we report first on the distribution of mononuclear and polymorphonuclear (PMN) phagocytes in paraffin sections of different stages of RA-synovitis. Second, infiltrating lymphocytes were isolated from RA-synovium and the cellular eluates analysed for T-cell subpopulations.

MATERIAL AND METHODS

Synovial tissues from 20 patients with osteoarthritis and 60 patients with definite RA were investigated. Before synovectomy, medication with anti-rheumatics was discontinued. The histological "typing" of RA-synovitis was done as described in Reference 1. In brief, three patterns of synovitis could be distinguished: namely, synovitis with (a) lymphocyte and plasma cell infiltration (b) cel-

lular infilatration plus fibroblast proliferation or (c) cellular
infiltration with or without fibroblast proliferation plus deep
ulceration. The demonstration of lysozyme in paraffin sections
was done as described (1) using the peroxidase-anti-peroxidase (PAP)
method. For isolating T-cells, dissected synovial membranes were
digested with collagenase and T-lymphocytes isolated by standard procedures. Purity was determined by morphological evaluation. The few
contaminating synovial cells could be identified easily by their
non-lymphoid morphology and were excluded from analysis. Cell surface markers were evaluated by rosetting (Fc-receptors) and by indirect immunofluorescence according to published procedures (2).

RESULTS

       In this study, lysozyme production was used as a marker for in-
filtrating PMN and mononuclear cells. Analysis of the synovial lining
cells revealed that some of these cells were also lysozyme positive.
As shown in Table 1, the number of such cells in RA exceeded by far
that seen in osteoarthritis. Lysozyme positive synovial lining cells
were not distributed evenly throughout the lining layer but concentrated in small areas. It was not possible to identify the positive
cells as A, B, or C type lining cells. Only a few positive cells were
identified as PMN or mononuclear cells. Fibroblasts were lysozyme
negative. The proliferation of fibroblasts in the subsynovial tissue
was associated with a significant decrease in the proportion of
positive synovial lining cells. For PMN, this relationship did not
reach statistical significance.

       In the subsynovial tissue, lysozyme containing cells were by
morphology PMN, mononuclear cells or giant cells. Fibroblasts and
cartilage were lysozyme negative. Lysozyme positive mononuclear cells
and PMN were both found to be more abundant in RA- than in osteoarthritis synovitis (Table 2). PMN infiltration was not dependent
upon the state of RA-synovitis, whereas infiltration by mononuclear
cells was significantly lower in synovitis with fibroblastic proliferation compared to the other types.

       We also analysed the lymphocytic infiltration of RA-synovitis.
Here we report data from 14 patients with the lymphocyte-plasma cell
pattern of RA-synovitis. In these patients the lymphocytic infiltrate consisted predominantly of T-lymphocytes ($87 \pm 17\%$) which were
further purified by E-rosetting. As shown in Table 3, the proportion
of T-lymphocytes bearing Fc-gamma receptors for IgG (FcγR) (Tγ)
exceeded by far that of the patient's peripheral blood used as reference. Previous studies have shown that the latter was not significantly different from healthy control values (3). Conversely, only
a few Fc mu-receptor FcμR) bearing T-cells (Tμ) were detected in
synovial tissue as compared to the autologous peripheral blood.

Table 1. Lysozyme Positive Synovial Lining Cells in
          Osteoarthritis and Different States of
          Rheumatoid Arthritis

Disease	Number of Patients	% Lysozyme Positive Cells
Osteoarthritis	20	3 + 7
Rheumatoid arthritis		
All patients	60	18 + 19
Lymphoid-Plasma cell infiltrate	20	21 + 17
Infiltrate + fibroblast proliferation	20	9 + 12
Synovitis with ulceration	20	25 + 24

Data reported as mean + standard deviation.

When isolated T-lymphocytes were incubated at 37°C, cells eluted
from RA-synovitis could not be induced to express Fc mu receptors
upon incubation as was easily possible with either the patient or
normal peripheral T-lymphocytes. The high numbers of T -cells remained more or less unchanged upon incubation. When synovial T-cells
were characterized by the use of monoclonal antibodies, OKT 8-positive
cells were found in a proportion similar to that of Tγ-cells.

DISCUSSION

     The use of lysozyme as a cell marker allows specific and quanti-
tative evaluation of cells which by morphological criteria alone may
escape identification. The data reported here demonstrate that all
types of lysozyme positive cells are more abundant in RA- than in
osteoarthritis irrespective of the type of RA-synovitis analysed.
It is tempting to speculate that the few lysozyme positive synovial
lining cells were A-type and activated in RA.

Table 2. Lysozyme Positive Cells in the Subsynovial
          Tissue of Osteoarthritis and Different
          States of Rheumatoid Arthritis

Disease	Number of Patients	Number of Lysozyme Positive Cells/mm^2	
		Mononuclear Cells	PMN
Osteoarthritis	20	67 $\pm$ 111	25 $\pm$ 37
Rheumatoid arthritis			
All patients	60	267 $\pm$ 187	115 $\pm$ 140
Lymphoid-plasma cell infiltrate	20	343 $\pm$ 281	118 $\pm$ 83
Infiltrate + fibroblast proliferation	20	56 $\pm$ 108	95 $\pm$ 18
Synovitis with ulceration	20	417 $\pm$ 337	135 $\pm$ 129

Data are reported as mean $\pm$ standard deviation.

     With regard to the lymphocytic infiltrate of RA-synovium, it
has been shown that FcγR, OKT8-positive T-lymphocytes accumulate
in rheumatoid arthritis. This T-cell phenotype contains a T-cell
subpopulation that functions as a cytotoxic or suppressor cell. In
view of extensive immunoglobulin production in RA joints, the accumulation of such suppressor-phenotye cells is unexpected. However,
surface marker phenotyping should not be overemphasized in terms of
functional correlation. The phenotypic profile described here is
partly supported by Burmester et al. (4). Using immunohistology in
cryostat sections of synovial tissue, Janossy et al. (5) found only
a few OKT8-positive "suppressor" cells but large areas of OKT4-
positive "helper" T-cells. Differences in methodology and patient
selection may account for such differences.

     Synovial T-lymphocytes in RA differed from autologous peripheral
T-cells in that they could not be induced to express the Fc mu re-

Table 3.   Fc Receptor and OKT-8 Positive T-lymphocytes
in Rheumatoid Arthritis Patients with Lymphocyte-plasma Cell Type Synovitis

Determination	Peripheral T-lymphocytes	Synovial T-lymphocytes
Fc-gamma Receptor Positive		
0 hour	14 ± 4	53 ± 18
16 hours	12 ± 4	37 ± 10
Fc-mu Receptor Positive		
0 hour	0 ± 0	0 ± 0
16 hours	56 ± 11	7 ± 5
OKT-8 Positive Cells	ND	59 ± 16

Data are reported as mean $\pm$ standard deviation (14 patients).

ceptor. The lack of this physiological T-cell response may point
to a functional defect of the "suppressor"- phenotype bearing cells
possibly resulting in loss of suppression.   Indeed, a local suppressor T-cell dysfunction in RA has been described by Chattopadhyay et al. (6).   Alternatively, in view of our own previous studies
on Fc gamma receptor heterogeneity (2), the synovial Tγ-cells may
belong to a separate subset of T-lymphocytes, characterized by a
receptor with high affinity for monomeric IgG.

Acknowledgement

This study has been supported by Robert-Bosch Foundation.   We
thank F. Frank and U. Braun for excellent technical assistance.

REFERENCES

1.   Fritz, P., Müller, J., Braun, U., et al., Rheum. Int., in press.
2.   Saal, J. G., Hadam, M. R., Feucht, H. E., et al., Scand. J.
     Immunol., in press.

3. Saal, J. G., Rautenstrauch, H., and Seybold, G., Verh. Dtsch.
        Ges. Rheumatol. 7:503, 1981.
4. Burmester, G. R., Yu, D. T. Y., Irani, A. M., Kunkel, H. G., and
        Winchester, R. J., Arthr. Rheum. 24:1370, 1981.
5. Janossy, G., Duke, O., Poulter, L. W., Pavayi, G., Bofill, M., and
        Goldstein, G., Lancet, 839, 1981.
6. Chattopadhyay, C., Chattopadhyay, H., Natvig, J. B., Michaelsen,
        T. E., Mellbye, O. J., Scan.J. Immuno. 10:309, 1979.

DISCUSSION

PRUZANSKI: I am pleased that you have confirmed our results, which
we published about 12 years ago in the book called Lysozyme (edited
by E. F. Osserman, R. E. Canfield and S. Beyclok, Academic Press,
N.Y., p. 419, 1974). At that time, we reported that the synovial
lining cells had plenty of lysozyme. However, we found that lysozyme
in the synovial fluid derives from 6 or 7 different sources including
blood, PMNs as well as synovial cells. It has also been reported
that osteoclasts and chondrocytes may have some lysozyme. The traffic
of lysozyme is in both directions. As you know, lysozyme is a somewhat sticky substance and is also pinocytosed by PMNs and other cells.
Since you reported that different types of synovitis have different
quantities of lysozyme, what is the significance of this observation?
Can you apply this observation to any distinction or differentiation
between types of rheumatoid synovitis?

SAAL: I must confess that up to now we have used lysozyme mainly
as a marker for polymorphonuclear and mononuclear cell infiltration.
As more cells accumulate, more lysozyme is detected.

UNKNOWN: Two colleagues working with me, Dr. Muller and Dr. Sylinski,
are working with synovial cells and have found a very large suppressive activity on T cell plaque formation.

SAAL: But there are other data showing a lack of suppressor cells
in the infiltrate. This is a matter of confusion.

UNKNOWN: How do you know that the cells which you isolated from your
synovial membrane are really representative of cells in the membrane?
Is it possible that you're just dealing with a small sample of cells
simply because you could get them out and that the major functional
population is still in the synovial membrane? Secondly, we know
now that quite a number of cells in the inflamed synovial membrane
consist of monocytic cells. How do you measure the contribution of
the monocytes together with your T cells?

SAAL: We are aware of the problems of selective elution or recovery,
but so far we only lose plasma cells and not T cells. We did controls
showing that the T cells obtained had the same surface characteristics

independent of the elution process. If you perform elution in different ways, you get always the same proportion of cells with the
same proportion of receptors. But this is not true for plasma cells.
Therefore, we isolated only the T cells. Human T cells are characterized by forming E rosettes. All the T cells which were analyzed
were E rosette forming cells and all had the OKT3 marker.

INHIBITORS OF MONOCYTE RESPONSES TO

CHEMOTAXINS ARE ASSOCIATED WITH HUMAN NEOPLASMS

Ralph Snyderman and George Cianciolo

Laboratory of Immune Effector Function
Howard Hughes Medical Institute
Division of Rheumatic and Genetic Diseases
Department of Medicine and
Department of Microbiology and Immunology
Duke University Medical Center
Durham, North Carolina   27710 USA

Macrophages can destroy cancer cells in vivo and in vitro (1-3) but low numbers of macrophages are generally found within progressively growing tumors. Moreover, a tumor's metastatic potential is often inversely related to the number of macrophages which it contains (4,5). As a group, cancer patients have abnormally functioning monocytes which regain normal function after successful cancer therapy or surgical tumor removal (6-10). Therefore, we have hypothesized that cancer cells may produce factors that alter monocyte-macrophage functions, thereby subverting immune surveillance (11). Indeed, in rodents neoplasms produce agents which inhibit macrophage accumulation in vivo and macrophage chemotaxis in vitro (12-15).

An obstacle to characterizing the effects of human neoplasms on monocyte function has been the lack of a suitable assay. In vitro chemotaxis assays can give spurious results when materials tested for their effects on leukocyte function are not highly purified and contain both chemotactic and inhibitory activities. We therefore developed a murine model to assess the effects of tumor cells or their products on macrophage function in vivo. We showed that tumor cells, low molecular weight extracts of tumor cells, or the plasma and urine of tumor-bearing mice were all capable of depressing macrophage accumulation at sites of inflammation (16-18). Other laboratories have reported similar defects in macrophage function in tumor-bearing animals (19-23). In addition, we found that low molecular weight extracts of certain oncogenic murine viruses also inhibited macrophage accumulation in vivo (24) and we reported that the inhib-

343

itory activity of the murine viruses was associated with the structural component of the virus envelope termed $P_{15}(E)$.

We recently developed an assay which measures the initial morphological responses of monocytes to chemotaxins (25), i.e., their change
in shape from round to an elongated, triangular, "polarized" config-
uration (Figure 1). This assay has allowed us to study the effects
of effusions from humans with cancer on monocyte function. We have
found that human cancerous effusions contain novel and potent inhibitors of monocyte responses to chemotaxins. Interestingly, these
inhibitors are recognized by monoclonal antibody reactive to the
type C retrovirus protein, $P_{15}(E)$.

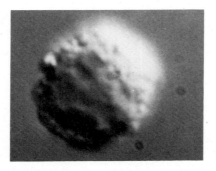

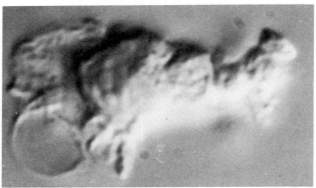

Fig. 1.   Change in shape of human monocytes exposed to f-Met-Leu-Phe.
           Cells were suspended in GBSS at 1×10^6 peroxidase-positive
           cells/ml, incubated at 37°C for 20 minutes with GBSS (top)
           or 10 nM f-Met-Leu-Phe (bottom), fixed, then examined.
           Nomarski interference contrast optics. X 1,400. (Reproduced
           with permission from Journal of Clinical Investigation 67,
           January 1981.)

The monocyte polarization assay was used as previously described (25). Mononuclear cells were isolated from venous blood of healthy volunteers by sedimentation with dextran followed by Ficoll-Hypaque density gradient centrifugation of the leukocyte-rich supernatant. The cells were then washed twice in modified Gey's balanced salt solution (GBSS, pH 7.0) and resuspended to 1.6×10^6 peroxidase-positive cells per ml. The cell suspension was then incubated for 10 minutes at $37^{\circ}C$ with either media alone or media containing the material being tested. A chemotactic stimulant, either the synthetic peptide N-formyl-methionyl-leucyl-phenylalanine (f-Met-Leu-Phe), zymosan-activated human serum, or lymphocyte-derived chemotactic factor (LDCF), was then added at a concentration sufficient to induce maximal polarization. The cells were incubated for an additional 7.5 minutes and fixed by the addition of buffered ice-cold 10% form-aldehyde. The percentage of monocytes polarized was calculated as:

$$\% \text{ monocytes polarized} = \frac{\% \text{ total cells polarized}}{\% \text{ peroxidase positive cells in cell suspension}} \times 100$$

Percent inhibition of monocyte polarization was calculated as:

$$\% \text{ Inhibition} = 1 - \frac{(\% \text{ Polarized in Chemotaxin})_{TX} - (\% \text{ Polarized in GBSS})}{(\% \text{ Polarized in Chemotaxin}) - (\% \text{ Polarized in GBSS})} \times 100$$

where TX represents cells treated with effusion or some fraction thereof. Polymorphonuclear leukocyte (PMN) polarization was performed in a similar fashion but at pH 7.2.

Effusions were obtained from patients with a variety of neoplasms or non-cancerous diseases. These diagnoses and other information regarding these patients are described elsewhere (26) but this study included patients with: adenocarcinoma of the pancreas, ovary, lung, breast, or colon; hepatoma; melanoma; lymphoma; squamous cell car-cinoma of the lung; melanosarcoma; liposarcoma; renal cell carcinoma; undifferentiated carcinoma of the lung; and acute myelocytic leukemia. Non-cancerous diseases included: cirrhosis; endometriosis, uremia; congestive heart failure; pulmonary embolus; lupus serositis; bac-terial empyema; bacterial peritonitis; and fibrocystic breast disease. Effusions were standardized using absorbance at 280 nm and the pH adjusted to 7.0 for compatibility with the assay conditions.

Incubation of cancerous effusions with human monocytes caused inhibition of their subsequent polarization to f-Met-Leu-Phe (Figure 2). At the highest concentration of effusions tested the average inhibition (n=16) was 55.9 ± 12.7% while the non-cancerous fluids (n=17) inhibited only 6.2 ± 4.2%. The inhibitory activity in the cancerous effusions affected the monocyte rather than the chemotactic factor since preincubation of the cells with effusion for 10 minutes followed by exposure to f-Met-Leu-Phe resulted in inhibition of

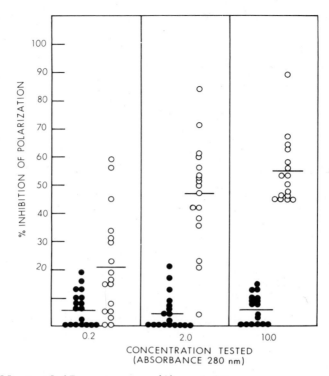

Fig. 2.   Effect of 17 cancerous (0) and 17 noncancerous (●) effusions
          on monocyte polarization. Monocytes (10^6/ml) were preincu-
          bated with various effusion concentrations for 10 minutes
          at 37°C, stimulated with 10 nM f-Met-Leu-Phe for 7.5 minutes,
          fixed, and the percentage of polarized monocytes determined.
          Percentage of inhibition of polarization was calculated as
          described in the text. At 10.0 absorbance 280 nm, all
          cancerous effusions, but no noncancerous effusions, caused
          statistically significant inhibition of f-Met-Leu-Phe in-
          duced polarization (P<0.01 by analysis of variance, Q-test,
          and chi-square). (Reproduced with permission from Journal
          of Clinical Investigation 68, October 1981.)

polarization, while preincubation for 10 minutes of f-Met-Leu-Phe
with the effusion followed by the addition of cells did not inhibit
polarization (26). Moreover, preincubation of monocytes with cancerous effusion, followed by extensive washing of the cells, still
resulted in inhibition of polarization.

The inhibitory activity for monocyte polarization in cancerous effusions did not affect PMNs (26). This specificity is identical to that observed for the defect in the response of leukocytes from cancer patients in that their monocyte but not PMN chemotactic responses are abnormal (27). In addition to inhibiting monocyte polarization to the synthetic chemotactic peptide f-Met-Leu-Phe, cancerous effusions also inhibited polarization to the naturally-derived chemoattractants zymosan-activated human serum or LDCF (26).

Fractionation of cancerous effusions using gel-filtration by high pressure liquid chromatography suggested that there are three peaks of inhibitory activity with approximate molecular weights of $\geq 200,000$ daltons, ca. 46,000 daltons, and ca. 21,000 daltons (26). Fractionation of effusion from patients with non-cancerous diseases revealed only a single peak of inhibitory activity with molecular weight of $\geq 200,000$ daltons. The low molecular weight inhibitory activity of cancerous effusion was stable at $56^\circ C$ for 30 minutes and trypsin-sensitive (26).

Since studies in mice had suggested that the retroviral structural component $P_{15}(E)$ inhibited macrophage accumulation _in vivo_ we tested the effect of monoclonal antibody reactive to $P_{15}(E)$ on the inhibitory activity in human cancerous effusions. Monoclonal antibody (7.5 µg) was incubated for 15 minutes at $22^\circ C$ with 500 µl of effusion or ultrafiltrate of effusion. The antibody was then removed by incubation of the reaction mixture with formalin-fixed _Staphlococcus aureus_ followed by centrifugation. The inhibitory activity in the ultrafiltrate of all eight different cancerous effusions tested was removed by absorption with monoclonal anti-$P_{15}(E)$ while 7 different monoclonal antibodies had no effect (26). Table 1 is a representative example of the effects of absorption with monoclonal anti-$P_{15}(E)$. A low molecular weight extract of Rauscher leukemia virus (RLV), a virus known to contain $P_{15}(E)$, was used as a positive control. Monoclonal antibody to $P_{15}(E)$ but not monoclonal anti-GP $_{70}$ or murine IgG_{2a} was capable of removing almost all of the inhibitory activity of the virus preparation.

To further purify the inhibitory activity in cancerous effusions, we prepared an affinity column by coupling monoclonal anti-$P_{15}(E)$ to cyanogen-bromide activated Sepharose 4B. This column was used to absorb cancerous and non-cancerous effusions. Four different cancerous effusions and three benign effusions were individually passed through the affinity column and the column was then washed extensively. The bound material was eluted by lowering the pH to 2.8. Inhibitory activity for monocyte polarization could be detected at dilutions of up to 1:8000 in the eluted material of the four cancerous effusions while no activity was detected in the eluted material of the benign effusions (28). Since the protein concentration required to attain 50% inhibition of polarization was about 5 mg/ml with the unfractionated cancerous effusions and <50 ng/ml with the affinity

Table 1. Effect of Absorption by Monoclonal Anti-P_{15}E on Inhibitory
          Activity of Ultrafiltrates of Cancerous Effusions

Source of Inhibitor	% Inhibition (±SE) of Monocyte Polarization after Preincubation with			
	Buffer	Anti P_{15}(E)	Anti GP_{70}	IgG_{2a}
Pleural fluid, squamous carcinoma of lung	58.4±4.3	4.3±1.2	50.0±0	56.7±12.0
Ascites fluid, melanoma	67.6±1.1	9.4±0	63.0±12.3	63.0±0
Pleural fluid, metastatic carcinoma unknown origin	46.5±8.6	7.9±1.5	52.4±9.0	57.9±9.0
Sonicated Rauscher leukemia virus (2 mg/ml)	60.6±2.1	0.8±1.3	55.5±7.2	55.5±3.1

Five-tenths ml of ultrafiltrate was incubated for 15 minutes at 22°C
with 10 µl of GBSS or GBSS containing 7.5 µg of the indicated antibody. Twenty-five µl of formalin-fixed Staph. aureus was added, the
mixture incubated an additional 12 minutes at 22°C, and the Staph.
aureus removed by centrifugation. One-tenth ml of absorbed ultrafiltrate then was incubated for 10 minutes at 37°C with 0.3 ml of cell
suspension containing 4.8×10^5 peroxidase-positive mononuclear cells.
One-tenth ml of GBSS or 50 nM f-Met-Leu-Phe in GBSS was added to each
of duplicate tubes, the tubes incubated an additional 7.5 minutes at
37°C, the cells fixed and the percent of polarized monocytes determined. The percent inhibition of polarization was calculated as
described in the text. (Reproduced with permission from Progress in
Cancer Research and Therapy, Volume 20, "Lymphokines and Thymic Hormones", Raven Press, pp. 205-213, 1981.)

column eluates, this procedure resulted in a >100,000-fold increase
in specific activity of the inhibitory factors.

Our recent development of a rapid, quantitative assay to measure
monocyte responsiveness to chemotactic stimuli has permitted us to
begin investigating the mechanisms of the monocyte chemotactic defect
in cancer patients. Of the seventeen cancerous effusions thus far

tested, all contained significant inhibitory activity for monocyte polarization while none of the 17 benign effusions tested had such activity (26). The inhibitory activity for monocytes in cancerous effusions did not affect the polarization of PMNs. This is identical to the cell specificity previously observed for the chemotactic defects in cancer patients and in tumor-bearing mice (12,27). Initial characterization of the inhibitory activity from cancerous effusions indicates that a significant portion of it is of low molecular weight, is relatively heat stable and is, at least in part, proteinaceous. Furthermore, monoclonal antibody reactive to the $P_{15}(E)$ component of type C retroviruses is capable of specifically absorbing the inhibitory activity from cancerous effusions. Use of anti-$P_{15}(E)$ affinity columns on a large scale basis should allow purification of the inhibitory factors in quantities sufficient to complete their characterization.

The significance of an antibody reactive to a retroviral protein being capable of absorbing the inhibitory activity from human cancerous effusions is not yet clear. Our studies in mice had indeed suggested that $P_{15}(E)$ played a role in the ability of retrovirus extracts to inhibit macrophage function (24). Furthermore, data suggested that both viral and murine tumor-derived inhibitors of macrophage accumulation shared several physicochemical and antigenic characteristics (29). Studies by Thiel et al. (30) have recently shown that $P_{15}(E)$ of many species is broadly cross-reactive. Therefore, the inhibitory factors in human cancerous effusions could be $P_{15}(E)$. However, the possibility exists that malignant transformation induces the production of a protein which is not $P_{15}(E)$ but merely is antigenically similar to it.

It is too early to determine whether isolation and characterization of the monocyte polarization inhibitory factor will have practical implication for the immunotherapy of various neoplasms but this could be the case. Evidence thus far accumulated suggests, however, that screening for the inhibitor might provide a basis for a useful diagnostic tool. In any case, these studies show that human cancerous effusions contain potent inhibitors of monocyte function and that these inhibitors are recognized by antibody to a structural component of oncogenic viruses.

REFERENCES

1. Cerottini, J. C., and Brunner, K. T., Adv. Immunol. 18:67, 1974.
2. Levy, M. H., and Wheelock, E. F., Adv. Cancer Res. 20:131, 1974.
3. Shin, H. S., Hayden, M., Langley, S., Kaliss, N., and Smith,
        M. R., J. Immunol. 114:1255, 1975.
4. Eccles, S. A., and Alexander, P., Nature 250:667, 1974.
5. Russell, S. W., and McIntosh, A. T., Nature 268:69, 1977.

6. Boetcher, D. A., and Leonard, E. J., J. Natl. Cancer Inst. 52: 1091, 1974.
7. Hausman, M. S., Brosman, S., Snyderman, R., Mickey, M. R., and Fahey, J., J. Natl. Cancer Inst. 55:1047, 1975.
8. Rubin, R. H., Cosimi, A. B., and Goetzl, E. J., Clin. Immunol. Immunopathol. 6:376, 1976.
9. Snyderman, R., Seigler, H. F., and Meadows, L., J. Natl. Cancer Inst. 58:37, 1977.
10. Snyderman, R., Meadows, L., Holder, W., and Wells, S., J. Natl. Cancer Inst. 60:737, 1978.
11. Adams, D. O., and Snyderman, R., J. Natl. Cancer Inst. 62:1341, 1979.
12. Snyderman, R., and Pike, M. C., Science 192:370, 1976.
13. Normann, S. J., and Sorkin, E., J. Natl. Cancer Inst. 57:135, 1976.
14. Stevenson, M. M., and Meltzer, M. S., J. Natl. Cancer Inst. 57: 847, 1976.
15. Nelson, M., and Nelson, D. S., Immunology 34:277, 1978.
16. Snyderman, R., Pike, M. C., Blaylock, B. L., and Weinstein, P., J. Immunol. 116:585, 1976.
17. Snyderman, R., and Cianciolo, G. J., J. Reticuloendothelial Soc. 26:453, 1979.
18. Cianciolo, G. J., Herberman, R. B., and Snyderman, R., J. Natl. Cancer Inst. 65:829, 1980.
19. Fauve, R. M., Hevin, B., Jacob, H., Gaillard, J. A., and Jacob, F., Proc. Natl. Acad. Sci USA 71:4052, 1974.
20. North, R. J., Kirstein, D. P., and Tuttle, R. L., J. Exp. Med. 143:559, 1976,
21. Snyderman, R., Siegler, H., and Meadows, L., J. Natl. Cancer Inst. 58:37, 1977.
22. Brozna, J. P., and Ward, P. A., J. Clin. Invest. 64:302, 1979.
23. Normann, S. J., Schardt, M., and Sorkin, E., J. Natl. Cancer Inst. 63:825, 1979.
24. Cianciolo, G. J., Matthews, T. J., Bolognesi, D. P., and Snyderman, R., J. Immunol. 124:2900, 1980.
25. Cianciolo, G. J., and Snyderman, R., J. Clin. Invest. 67:60, 1981.
26. Cianciolo, G. J., Hunter, J., Silva, J., Haskill, J. S., and Snyderman, R., J. Clin. Invest. 68:831, 1981.
27. Wilson, J., and Snyderman, R., Clin. Res. 26:378A, 1978.
28. Cianciolo, G. J., and Snyderman, R., Fed. Proc., in press, 1982.
29. Cianciolo, G. J., Bolognesi, D. P., and Snyderman, R., Fed. Proc. 39(3):478A, 1980.
30. Thiel, H. J., Broughton, E. M., Matthews, T. J., Schafer, W., and Bolognesi, D. P., Virology 111:270, 1981.

DISCUSSION

WILKINSON: It worries me that you're not using an assay of locomotion.

SNYDERMAN: In our initial studies, we did use Boyden chambers to measure chemotaxis and found that most effusions had inhibitory activity. However, with the Boyden chamber assay we had perhaps 25% false positives and 25% false negatives. I'm not sure why. We did match standard chemotaxis assays against our polarization assays and found that in both systems we could differentiate the group of patients with cancer from those with benign disease. In terms of individuals, however, we were not able to use chemotaxis tests as a screening procedure. I think it probably relates to adherence or to the fact that chemotaxis assays are more stringent.

WILKINSON: I accept that a non-polarized cell probably isn't moving. What I'm worried about is that a polarized cell may be either moving or not moving.

SNYDERMAN: I don't think one can infer from this assay whether macrophages will migrate into a tumor. I think it's an interesting phenomenon which may or may not relate to cell movement.

UNKNOWN: I have been working with MSV-induced tumors in the mouse and those induced by sarcoma and leukemia viruses. One striking feature is that the regressing tumor has about 50% or 60%, even 70% macrophages. How does this relate to the fact that the $P_{15}(E)$ could be inhibitory for the accumulation of macrophages?

SNYDERMAN: I really don't know. However, one possibility is a phasic release of $P_{15}(E)$ and perhaps at some point the systemic release decreases enough to allow macrophage influx. Alternatively, macrophages may become resistant to $P_{15}(E)$ or a new population of macrophages develop that are resistant. It would be fascinating to study the serum at various times or perhaps the tumor itself for this activity.

UNKNOWN: Is the blockade specific for chemotaxis? And do the cells eventually recover?

SYNDERMAN: Let me answer first that it is specific for monocytes. It does not affect polymorphonuclear leukocytes. It affects superoxide production, but the dose seems to be shifted perhaps about a log so that it takes a lot more material to depress superoxide production than polarization. In terms of recovery, we have not cultured the cells long enough to see if they will recover. Over the short period of time, an hour or so, they do not recover.

EVANS: I'm not sure it's a general phenomenon that murine tumors induce this monocyte defect in inflammation. All the tumors we work with are C type virus negative and, as you might expect, they do not induce a monocyte defect. Under these conditions, moreover, the influx of monocytes into the tumor mass is a continuous process.

MANTOVANI: I've been working with ovarian effusions and find that the concentration of macrophages in the effusion is very variable. It ranges anywhere between 5% and 95% of the recovered cells. Is there any relationship between the levels of your inhibitor and the macrophage content of the effusion?

SNYDERMAN: That's a very important question. Our studies with ovarian effusions were done in collaboration with Steve Haskill. We are hoping to do the studies you suggest but have not done so yet.

PRUZANSKI: When monocytes come in contact with a malignant effusion, do they release a lot of lysozyme? Lysozyme is a highly cationic protein and might inhibit chemotactic activity.

SNYDERMAN: That's a very interesting suggestion. I haven't studied lysozyme content within a malignant effusion.

CZARNETSKI: Have you tried to identify the high molecular weight chemotactic inhibitor? Is it possible that this factor is just the smaller inhibitor linked to an ascites protein like alpha-2 macroglobulin?

SNYDERMAN: When we treat the effusions with anti $P_{15}(E)$ antibody and pull out the activity, then refractionate it, we don't see the high molecular weight material. So, I think your contention is plausible.

DEPRESSION OF MONOCYTE INFLAMMATORY RESPONSE WITH PREGNANCY, CANCER, REGENERATION AND SURGICAL WOUND REPAIR

S. Normann,[1] J. Cornelius,[1] M. Schardt,[2] and E. Sorkin[2]

[1]Department of Pathology, University of Florida, Gainesville, Florida 32610 USA
[2]Schweizerisches Forschungsinstitut Medizinische Abteilung, Davos, Switzerland

INTRODUCTION

Macrophage accumulation at inflammatory sites is depressed by tumor bearing (1,2) and this phenomenon may constitute a mechanism by which tumors abrogate host surveillance (3,4). However, is this anti-inflammatory effect unique to cancer? Could the tumor utilize an already established mechanism for regulating monocyte traffic and function? We speculate that the diversity of macrophage functions necessitates some form of biological control mechanism to prevent untoward reactions. The purpose of this communication is to explore this issue by examining anti-inflammation consequent to several non-cancerous conditions as well as during growth of both spontaneous and transplanted tumors.

MONOCYTE TRAFFIC DURING PREGNANCY

Intrauterine growth of a fetus is a model of rapid but highly regulated new growth occuring in an immunocompetent mother despite expression of paternal membrane antigens on the fetus. While lymphocyte responses occur in the mother (5), pregnancy proceeds undisturbed. Further, implantation occurs without involving any inflammatory response despite trophoblastic cell invasion of the uterine wall. These factors invite consideration of whether or not the products of conception inhibit the inflammatory response.

Pregnancy was induced in Wistar rats and macrophage accumulation measured in the peritoneal cavity 48 hours after injecting 200 ugm of phytohemagglutinin (PHA). Gestation in rats is normally 21 days and by day 18, pregnancy had decreased macrophage yields by 60%. Immediately following delivery, macrophage responses returned to normal and actually exceeded normal values 5 days postpartum. The yields of PMN's induced by sodium caseinate also decreased during pregnancy but the change was not significant. An anti-inflammatory effect during pregnancy may explain the clinical observation that certain inflammatory diseases such as rheumatoid arthritis subside during late pregnancy (6).

MCNOCYTE TRAFFIC DURING LIVER REGENERATION AND SURGICAL WOUND REPAIR

Certain tissues are capable of regeneration after injury as illustrated by partial hepatectomy which induces rapid new growth that ceases when liver size is restored. In contrast, hind limb amputation results only in a local and limited fibroblastic and endo-thelial proliferation associated with wound repair. We selected these two conditions in order to study the effect on inflammation of different degrees of non-cancerous growth.

A 70% hepatic resection in rats dramatically decreased the yield of peritoneal macrophages induced by proteose peptone. Although hepatic regeneration had ceased by 5 days, the inhibition in macro-phages persisted, and normal responses were not observed until 30 days after surgery. The fact that partial hepatectomy did not alter PMN responses suggests that the inhibition was not due to altered adherence to serosal surfaces.

Surgical resection of the rat hind limb also depressed the macro-phage inflammatory response. Three days after resection, macrophage accummulation in the peritoneal cavity to PHA was reduced by nearly 60% although PMN responses were essentially normal. Thus major sur-gery outside the abdomen had a selective effect on the inflammatory process within the peritoneal cavity.

Whereas macrophages possess certain growth regulating activities, it might be desirable to limit their accumulation during liver regen-eration. The biological significance of limiting macrophage accumula-tion during surgical wound repair is more difficult to understand unless it relates to the means by which the body regulates the inflam-matory process. Inhibition of inflammation following surgery has been described by ourselves (7) and others (6,8) and may be mediated by a serum factor.

MONOCYTE TRAFFIC DURING CANCEROUS GROWTH: SPONTANEOUS TUMORS

Lymphoreticular Neoplasms

We have examined inflammation during development of two lympho-

reticular neoplasms: histiocytic lymphoma in SJL/J mice (9) and acute
T cell leukemia in AKR mice (10).   SJL/J mice with tumors accumulated
fewer macrophages than did age matched control mice without tumors
either on sc inserted nitrocellulose filters or in peritoneal exudates
induced with PHA.   By implanting filters at different times before,
during, and after cancer emergence, we determined that in some animals
the anti-inflammatory effect developed abruptly whereas in others
it occured slowly and progressively.

Young AKR mice had macrophage responses equivalent to that
observed in C57B1/6 or BALB/C mice which do not develop leukemia,
suggesting that the presence of AKv virus in animals destined to
develop leukemia did not cause defects in monocyte responses.   However, the onset of the leukemia was associated with an average depression in macrophage responses of 66%.   When mice were followed sequentially, it was observed that abnormal macrophage responses usually
occured either concurrently with or immediately following the onset
of each leukemic episode.

From these studies, we concluded that abnormal macrophage responses occur in association with spontaneously developing lymphoreticular neoplasms.   The critical question of whether or not these
abnormalities were associated with tumor emergence or arose as a
consequence of tumor bearing was not resolved.   The defect in SJL/J
mice arose early when the lymph nodes were only slightly enlarged,
but such nodes contained upwards of 10^8 cells.   Onset of leukemia
in AKR mice correlated with abnormal macrophage responses but this
was not always seen, particularly during short cycles of leukemia
and remission.

Chemically Induced Sarcomas

We used 3-methylcholanthrene (3-MCA) to induce tumors in DA
strain rats by implanting the carcinogen sc in paraffin pellets.
Tumors developed in 40% of these inbred rats with a latency period
of $2\frac{1}{2}$ to $9\frac{1}{2}$ months.   During the latency period, animals with carcinogen had a higher frequency of depressed macrophage responses than
did controls implanted with paraffin but no carcinogen.   However,
there was no apparent correlation between these abnormal responses
and subsequent tumor emergence.   Inflammatory responses were examined
also during early and late tumor growth.   Emerging tumors had a
slightly lower accumulation of macrophages on sc placed filters compared to controls but this difference was not statistically significant.   Large tumors induced no apparent abnormalities.   In animals
bearing either a small or a large tumor, the frequency of abnormal
macrophage responses was equivalent to controls.   Therefore, tumor
presence per se did not produce inflammatory defects.

MONOCYTE TRAFFIC DURING GROWTH OF TRANSPLANTED TUMORS

Recently Transplanted Tumors

The primary tumors induced by 3-MCA and described above were
transplanted to syngeneic recipients through at least 15 generations.
The macrophage inflammatory response was quantified by macrophage
accumulation on sc placed filters implanted at a site distant to the
tumor (11). Primary tumors and first generation transplants did not
produce a depression in macrophage responses in contrast to subsequent
tumor transplants. After the second generation, defects in macrophage inflammation were observed that persisted through all subsequent
transplants and ranged from 24 to 45% depression.

Multiply Transplanted Tumors

We and others have described defects in multiple transplanted
tumors (1-3) and the phenomenon is generally reproducible. Currently,
there are at least 15 different tumors which have been reported to
produce or be associated with defects in macrophage inflammation.
Recent studies have established that surgical excision of the tumor
rapidly corrects the defect but it reappears with metastasis (12).
Further, the phenomenon is biphasic (13). A transient early defect
develops within 2 days of tumor inoculation and requires a threshold
number of tumor cells. After a variable interval of normal responses,
a persistent but late defect develops, approximately midway in the
clinical course, which is more severe than the early defect.

STUDIES ON THE MECHANISM OF ANTI-INFLAMMATION

Chemotaxis

Seventy-five percent of patients with various types of cancer
have either elevated (15%) or depressed (60%) monocyte chemotactic
responses (14,15). It is logical to assume that the impaired inflammation observed in cancer bearing animals arises from defective monocyte chemotaxis. We examined this assumption by correlating macrophage inflammatory responses in two rat transplanted tumor models with
blood monocyte chemotactic responses. In addition, we examined the
chemotactic response of monocytes obtained from pregnant rats to
determine if altered chemotaxis explained their inflammatory defect.

All chemotaxis tests were performed with a constant number of
monocytes ($2x10^6$), chemotactic attractants of either pooled rat serum
or lymphocyte dependent chemotactic factor (LDCF), and chemotaxis
filters of either polycarbonate or nitrocellulose incubated for 90
or 300 minutes respectively.

Progressive growth of a DMBA induced fibrosarcoma depressed inflammation but increased monocyte chemotactic responses. At 14-16

days after tumor transplantation, monocyte inflammatory responses
were depressed by 42% whereas chemotactic responses were increased
300 to 800 percent depending upon the type of filter used and the
choice of attractant. In the second experiment, a methylcholanthrene
induced fibrosarcoma of recent origin was used which inhibited inflammation by 39% but produced only a modest and not statistically
significant reduction in chemotaxis of 9%. In the third experiment,
pregnancy decreased macrophage accumulation in the peritoneal cavity
by 66% but diminished monocyte chemotaxis only 13%. These studies
indicate that defective monocyte chemotaxis may not be the only or
even significant mechanism involved in anti-inflammation consequent
to either transplanted tumors or pregnancy.

Soluble Mediators

     Snyderman and Pike (3) described a dialysable, anti-inflammatory
factor obtained from homogenates of sonicated cancerous tissues. We
found that P-815 mastocytoma cells grown ip contained anti-inflammatory activity in their tumorous ascites fluid (16). In vitro, the
same tumor cells elaborated soluble anti-inflammatory factors into
the culture medium. P-815 mastocytoma cells elaborated at least
2 soluble anti-inflammatory factors. One factor was a low molecular
weight (MW) peptide, as judged by ultrafiltration, failure of extraction by lipid solvents, partial inactivation by trypsin and complete
inactivation by carboxypeptidase B. The second anti-inflammatory
factor had a MW between 30,000 and 100,000 daltons. Anti-inflammatory
factors were not found in culture supernatants from splenocytes or
peritoneal exudate cells. Tumorous ascites fluid contained the low
but not the high MW anti-inflammatory factor. Homogenates prepared
from the products of conception also contained anti-inflammatory
activity consisting of only a low MW factor, which was inactivated
by trypsin and carboxypeptidase A but not B.

SYNTHESIS

     Our experiments provide evidence that anti-inflammation directed
against macrophages exists during cancerous growth, normal embryonic
growth, hepatic regeneration and surgical amputation of the lower
extremity. The anti-inflammatory phenomenon for each condition is
similar in that it affects macrophages more than PMNs, is not related
to levels of circulating monocytes or their mobility, and is associated
with a soluble anti-inflammatory factor. Thus, the anti-inflammatory
phenomenon originally described for tumors is not unique to cancer.

     Because anti-inflammation is rapidly reversed by tumor excision
(12), and requires a threshold number of tumor cells (16), we speculated that cell proliferation was a determinant of the anti-inflammatory effect. However, the relationship between anti-inflammation
and cell growth is not clear. First, the defect associated with
tumors is biphasic (13) with a variable interval during which the

tumor is proliferating rapidly but no abnormalities in inflammation are observed. Second, the defect in macrophages initiated by hepatic resection persisted long after the cessation of DNA synthesis. Third, the defect associated with pregnancy develops late although rapid embryonic growth has existed for some time.

Similarly, it is unlikely that altered monocyte numbers or chemotactic mobility are involved. Cancer, surgical resections, and pregnancy are all associated with normal or elevated levels of circulating monocytes. Although defective monocyte chemotaxis might predispose to altered accumulation of macrophages, it is incisive that we observed elevated monocyte chemotaxis in one tumor system concurrent with defective macrophage inflammation. Chemotaxis of monocytes obtained in late pregnancy was normal, despite failure to emigrate into the peritoneal cavity in response to PHA. In these experiments, a dissociation appears to exist between monocyte chemotaxis measured in vitro and the appearance of the cells in inflammatory foci in vivo.

On the other hand, it appears probable that some form of soluble mediator is involved. Anti-inflammatory factors have been obtained from ascites fluid (16), homogenates of tumorous tissues (3) supernatants of cultured tumor cells, homogenates of pregnant uteri, and in blood following major surgery (6-8). It now appears likely that more than one mediator is involved. Cultured P-815 mastocytoma cells produced at least two mediators that differed in molecular weight. Pregnancy was associated with a mediator of very low molecular weight similar to that found in the supernatant of cancerous cells. If the two mediators are peptides, they probably differ at least at their terminal amino acids as judged by susceptibility to carboxypeptidase B. Additional investigation is needed to clarify the relationship between the various factors and to determine if the different effects on macrophages are due to different mediators.

It has now emerged that the presence of cancer is not always sufficient to produce anti-inflammation. We were unable to demonstrate a defect in recently induced MCA tumors prior to transplantation. Then too, the biphasic nature of the defect clearly demonstrated that cancer can exist in the absence of an apparent abnormality in macrophages. In the MCA induced tumor model, it is curious that the macrophage defect did develop upon multiple transplantation. Did transplantation select for a more virulent tumor cell population? Are defects in macrophages only seen in spontaneously arising lymphoreticular neoplasms? Is so, what then is the relevance of the defect to cancer surveillance?

Acknowledgement

This work was supported by Swiss National Science Foundation grant no. 3.603.80, the Swiss Cancer League and by grant 22517-05 from the National Cancer Institute (USA).

REFERENCES

1. Normann, S. J., and Sorkin, E., J. Natl. Cancer Institute 57:
    135, 1976.
2. Snyderman, R. M., Pike, M. C., Blaylock, B. L, and Weinstein,
    P., J. Immunology 116:585, 1976.
3. Pike, M. C., and Snyderman, R., J. Immunology 117:1243, 1976.
4. Adams, D. O., and Snyderman, R., J. Natl. Cancer Institute 62:
    1341, 1979.
5. Gill, R. J., and Repetti, C. F., Am. J. Pathology 95:465, 1979.
6. Persellin, R. H., Vance, S. E., and Peery, A., Br. J. Exp. Path.
    55:25, 1974.
7. Normann, S. J., Schardt, M., Cornelius, J., and Sorkin, E., J.
    Reticuloendothelial Soc. 30:89, 1981.
8. Boers, W., van Gool, J., and Zwart, N. A., Br. J. Exp. Path. 60:
    239, 1979.
9. Normann, S. J., Schardt, M., and Sorkin E., J. Natl. Cancer Inst.
    63:825, 1979.
10. Normann, S. J., Schardt, M., and Sorkin E., J. Natl. Cancer Inst.
    66:157, 1981.
11. Normann, S. J., and Schardt, M., J. Reticuloendothelial Soc.
    23:153, 1978.
12. Normann, S. J., Schardt, M., and Sorkin, Int. J. Cancer 23:110,
    1979.
13. Normann, S. J., Schardt, M., and Sorkin, E., Int. J. Cancer 28:
    185, 1981.
14. Snyderman, R., Seigler, H. F., and Meadows, L., J. Natl. Cancer
    Inst. 58:37, 1977.
15. Snyderman, R., Meadows, L., Holder, W., and Wells, S., J. Natl.
    Cancer Inst. 60:737, 1978.
16. Normann, S. J., J. Natl. Cancer Inst. 60:1091, 1978.

PULMONARY ALVEOLAR MACROPHAGE CHEMO-

TAXIS IN MALIGNANT TUMORS OF THE LUNG

G. Renoux, E. Lemarie, M. F. Legrand, M. Renoux
and M. Lavandier

Laboratoire D'Immunologie et Service de Pneumologie
Centre Hospitalier Bretonneau
37044 Tours Cedex France

INTRODUCTION

It has been shown that the presence of a tumor suppresses chemotactic responses of macrophages (1-3). Pulmonary alveolar macrophages function as primary effector cells against microorganisms that reach the lungs, yet their function in the control of primary and/or metastatic tumors in man is poorly understood.

In the present report, we show that in patients with primary lung carcinoma the response of alveolar macrophages to chemotactic stimuli is severely altered, whereas the response of macrophages obtained from patients with metastasis is equivalent to that of normal control donors.

MATERIALS AND METHODS

Broncho-alveolar lavage fluids were recovered after injection of three 50 ml aliquots of 0.9 per cent sterile saline into selected pulmonary lobar segments, through the aid of an Olympus bronchoscope, and under topical anesthesia with 2 percent xylocaine. Total cell counts were performed and differential cell counts were made after staining with Wright Giemsa. The cells were collected by centrifugation, resuspended in phosphate buffered saline (PBS), centrifuged again for 10 min at 500 g, and finally resuspended at a concentration of 1.5×10^6 viable cells (trypan blue exclusion) per ml of RPMI-1640 medium. Random migration and chemotaxis were evaluated in modified Boyden chambers using a 8 μ pore-sized Millipore filter (4,5). The chemotactic attractant was zymosan-activated autologous serum (1 mg

361

zymosan/ml serum, 1 hr at 37°C, then inactivated by 30 min at 56°C
and diluted 1:20 in sterile PBS). Autologous serum diluted in PBS
was used as control. In addition, alveolar macrophages from 12 lung
cancers were tested in the presence of 3×10^{-8}M formyl-methionine-
leucyl-phenylalanine (F-Met-Leu-Phe) per ml of PBS. Only those cells
that had migrated across the entire thickness of the filter in an
incubation time of 90 min at 37°C were counted after staining with
hematoxylin. Average number ($\pm$ SD) were evaluated in 10 microscopic
fields (x100) in duplicate assays in two different chambers. Significance of the data was assessed by the paired Student's $\underline{t}$ test.

Alveolar macrophages were obtained from four groups of individuals. Group I consisted of 12 normal volunteers (mean age of 30
years), of which 7 were smokers and 5 non-smokers. Group II was made
up of 20 normal patients between the age of 34 and 72 years (mean 47)
of which 13 were smokers and 7 non-smokers, who underwent broncho-
alveolar lavage for a long-standing bronchopulmonary infection; no
bronchial suppuration was present at the time of sampling, performed
in the lingula. Group III consisted of 38 patients, aged 40 to 71
years (mean 59) with primary bronchogenic carcinoma, of which 24 were
smokers and 14 were non-smokers. Alveolar lavage was performed in
the vicinity of the tumor and prior to specific therapy. Tumor types
by histology were 22 squamous cell carcinomas, 5 small cell cancers,
4 adenocarcinomas and 7 undifferentiated tumors. In 8 cases of
squamous cell carcinoma, a lavage was also performed in the lung
opposite the tumor. Group IV was made up of 8 cases (mean age 55
years) with hematogenous pulmonary metastases from different primary
origins: 2 carcinomas of the oropharynx, 1 testicular embryonal
carcinoma, 1 fibrosarcoma, 1 prostate cancer, 1 breast cancer and 2
metastases of unknown origin.

RESULTS

As shown in Table 1, the presence of a primary lung cancer or
metastases did not significantly modify the volume of recovered fluid,
the percentage of alveolar macrophages or cell viability. The absolute numbers of macrophages which were recovered per cubic milliter
were 304 ± 190 in healthy controls, 273 ± 250 in patient controls,
427 ± 330 in primary cancers and 216 ± 180 in metastases.

It is known that normal serum contains substances that increase
macrophage migration above the value of random migration measured in
balanced salt solutions. We used autologous serum diluted 1:20 in
PBS to perform control tests which are here reported as random migration. Under these conditions, random migration was not affected by
cancer (2.3 ± 1.6) or metastases (3.8 ± 1.9) in comparison with
healthy controls (3.4 ± 1.6). Similar results have been reported
previously (6).

Table 1.   General Characteristics of Recovered Broncho-alveolar Fluids

	Volume ml	Mean Total Cell Yield	Percent		
			Lympho-cytes	Polymorphonuclear Leukocytes	Macro-phages
Healthy controls	83±19	25,232	7±6	1±1	91±9
Patient controls	69±24	18,837	10±7	6±5	89±17
Bronchogenic cancers	66±23	28,182	6±5	2±2	89±7
Pulmonary metastases	68±24	14,688	5±2	1±1	93±2

Percent viability of macrophages was 70±14.

    In contrast, the health of the patient influenced chemotactic
migration. Figure 1 summarizes data from 38 primary lung cancers,
8 lung metastases, 20 patient controls with non-neoplastic pulmonary
infections and 12 healthy controls. The presence of a primary tumor
impaired chemotactic activity (5.7 ± 3.6) of macrophages obtained
as near as possible to the tumor site, to either zymosan-activated
serum or F-Met-Leu-Phe as chemoattractants. These values are significantly (p<0.01) below that of healthy controls (13 ± 2.5), and even
more different from the responses of patient controls (16.1 ± 9.1).
The dysfunction of macrophages from patients with primary lung cancer
was independent of the histological type of tumor or the smoking
habit of the patient. In contrast, macrophages from normal controls
and from patients bearing metastases (11.4 ± 3.8) responded similarly
to chemotactic attractants. The data are suggestive of an intrinsic
defect in macrophages from untreated lung cancer patients, as random
migration in the presence of autologous serum was not affected by
malignancy.

    That alveolar macrophages from lung cancer patients undergo a
marked change in their activity is further supported by an assay

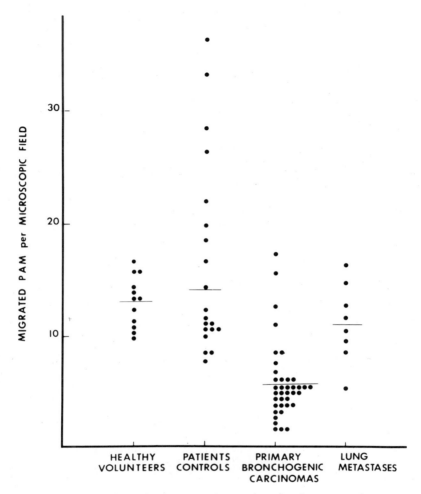

Fig. 1.   Pulmonary alveolar macrophage (PAM) chemotactic responsiveness of normal volunteers, hospitalized patients with non-
neoplastic diseases, primary lung cancer patients and hematogenous lung metastases.

reported in this congress (7).   Eight squamous cell carcinoma patients
have had alveolar lavage in the tumor-affected lung and in the opposite tumor-free lung.   The chemotactic activity of macrophages from
the tumor territory (4.5 ± 2.5) was significantly (p<0.05) below that
of macrophages from the opposite lung (6.8 ± 2.3), although these
latter values were still significantly (p<0.01) below those of healthy

controls. This latter finding suggests that the defective pulmonary
macrophage activity in lung cancer patients could be caused by components of the tumor cells or by suppressive factors from nearby
lymphocytes.

DISCUSSION

Our study shows that the ability of alveolar macrophages to
respond to chemoattractant stimuli is diminished in the presence of
a primary lung neoplasm, but not in the presence of a hematogenous
metastasis.

The findings support the hypothesis that primary lung cancer is
associated with an intrinsic dysfunction of alveolar macrophages.
The tumor did not influence spontaneous random migrations as such
activity in the presence of autologous serum was very similar in all
groups of patients. Our data confirm the report of Stone and Fidler
(8) that the number or function of alveolar macrophages is not suppressed by the presence of metastatic tumor cells.

It has already been shown that approximately 60% of patients
with cancer have depressed peripheral blood monocyte chemotactic
responsiveness, which can be corrected by successful treatment (9).
The present work provides information on the activity of the macrophages sampled from the vicinity of a tumor site.

As the accumulation of macrophages at local sites is important
for host protection against tumors, the functional activity of alveolar macrophages may be a major factor in the destruction of tumor
cells arising in the lung. It is tempting to speculate that an
intrinsic functional defect of alveolar macrophages could favor the
development of an *in situ*-born lung malignancy.

REFERENCES

1. Normann, S. J., and Sorkin, E., J. Natl. Cancer Inst. 57:135,
       1976.
2. Snyderman, R., Pike, M. C., Blaylock, B. L., and Weinstein, P.,
       J. Immunol. 116:585, 1976.
3. Rhodes, J., Cancer Immunol. Immunother. 7:211, 1980.
4. Zigmond, S. H., and Hirsch, J. G., J. Exp. Med. 137:387, 1973.
5. Snyderman, R., and Pike, M. C., in "Leukocyte Chemotaxis" (J. I.
       Galin, P. G. Quie, eds.), pp. 357-375, Raven Press, New York,
       1978.
6. Lemarié, E., Lavandier, M., Legrand, M. F., Anthonioz, P., Renoux,
       M. and Renoux, G., Am. Rev. Resp. Dis. 123:39, 1981.
7. Lemarié, E., Legrand, M. F., Renoux, M., Lavandier, M., and
       Renoux, G., Abst. 9th Internat. RES Congress, 1982.

8. Sone, S.,and Fidler, I. J., Cancer Res. 41:2401, 1981.
9. Snyderman, R., and Mergenhagen, S. A., in "Immunobiology of the
        Macrophage" (D. S. Nelson, ed.), pp. 323-348, Academic Press,
        New York, 1976.

DISCUSSION

GINSBERG: Did you look at the macrophages to see whether they contained intracellular material?

RENOUX: Yes, we looked of course; there were no infections in these
patients. The only thing we have seen is the constant existence of
a defect in the tumor affected lung if it was a primary lung cancer.

PUSSELL: Did you discriminate between the smokers and non-smokers,
since smokers have sometimes five to sixfold more alveolar macrophages?

RENOUX: There was no difference in the reduction of chemotactic
activity by cancer in smokers versus non-smokers.

PUSSELL: Yes, but the number of alveolar macrophages is much higher
in smokers.

RENOUX: No, unfortunately. In our series, the number of macrophages
was almost identical in all patients.

SILVERSTEIN: Have you or Dr. Snyderman looked to see whether these
macrophages bind chemotactic factors?

RENOUX: No.

SNYDERMAN: We have looked at peripheral blood monocytes with tritiated f-Met-Leu-Phe. They do bind the peptide but probably not
normally. The binding assay for human monocytes is not nearly as
good as with polymorphonuclear leukocytes. Nonetheless, it appears
that as with cancer there is a shift in monocyte binding, either
in the number of sites or in the affinity of binding.

SILVERSTEIN: You're saying they don't bind normally after they've
been treated with the effusion?

SNYDERMAN: Yes. After they've been treated with the effusion in
vitro and washed compared to those that have been treated with control effusions.

VAN MAARSSEVEEN: What do you lavage first, the site with carcinoma
or the control site?

RENOUX: In some patients the tumor site first and in others the control site, depending on the ease of bronchoscopy. The amount of recovered fluid was identical from both sites. It was always 2/3 of the administered fluid. General characteristics were identical except in the chemotactic activity of the cells.

SERUM BLOCKING FACTORS IN TUMOR BEARING HOSTS:
THE VALUE OF EXTENDED PLASMA EXCHANGE ON IN VITRO
IMMUNE PARAMETERS OF PATIENTS WITH ADVANCED CANCER

Peter Schuff-Werner, Jörg-Herbert Beyer and
Gerd Arno Nagel

Division of Oncology, Department of Medicine
University Clinics
Robert-Koch-Strasse 40
D-3400 Goettingen F.R.G.

INTRODUCTION

   Experimental and clinical studies (reviewed in Ref. 1) indicate
that serum blocking factors may alter immunological and inflammatory
responses promoting tumor growth. In recent years, many plasma components including as yet undefined soluble factors especially glycoproteins have been described as inhibitory in in vitro and in vivo
bio-assays. It is still controversial if these factors are tumor
products or if they are consequent to a tumor induced host response.
In the present study, we investigated the possibility of reducing
or eliminating such factors by large volume plasma-exchange with the
aim of modulating the host tumor response.

PATIENTS AND METHODS

   Encouraged by studies of Israel and co-workers (2) we performed
plasma-exchange in patients with advanced tumors resistant to conventional chemotherapy. All patients had an expected survival of
more than two months. Directly after plasma-exchange, the same chemotherapy was reinstituted. Clinical response was estimated by x-ray
or sonographic reexamination, computer-tomography, and clinical reexamination four weeks later. This procedure was repeated at least
two times and was stopped when progression was evident. The first
clinical results of 32 patients have been published recently (3).

   Before and one day after plasma-exchange, plasma samples were
drawn and aliquots frozen at -80°C. Cellular reactivity was evaluated

369

by the conventional in vitro assays of E-rosette formation; mitogenic
and allogeneic lymphocyte stimulation (MLC); and monocyte and granulocyte activation as measured by chemiluminescence. The same bio-assays
were performed in the presence of patient's serum to detect serum
blocking factors in the samples as described in Ref. 4. Blocking
activity was expressed as inhibitory index or percent inhibition as
compared to pooled normal control plasma.

RESULTS

      The dose dependent response of lymphocytes to phytohemagglutin in
(PHA) (Figure 1) showed no significant difference before and after
plasma-exchange when they were incubated in the presence of AB-plasma
from healthy donors. Further, the ratio of ^{3}H-thymidine uptake did
not exceed the normal range of healthy controls. Only in our patient
with Hodgkin's disease were the peripheral mononuclear cells significantly less reactive. E-rosette formation of lymphocytes showed a
slight increase within the normal range of controls after the plasma-
exchange procedure. There was no measurable change in monocyte and
granulocyte activation by zymosan as measured by chemiluminescence.

      Tritiated thymidine uptake was markedly decreased when autologous
cells were stimulated by mitogens (PHA) or alloantigens in the presence of plasma taken before plasma-exchange, indicating serum blocking
activity. This effect could be demonstrated also in cultures containing the patient's plasma and homologous cells from healthy donors
(Figure 2). However, the effect was not observed with Con A as the
mitogen. Blocking activity could be reduced by the exchange of 200%
of the calculated plasma volume with 5% human albumin salt solution
("effective" plasma-exchange).

      Forty seven patients, who underwent the described therapy regi-
men, have been checked for serum blocking factor (Table 1). Thirty-
two out of 47 patients were positive in the MLC-assay, but only in
13 patients was it possible to reduce the blocking activity by "effective" plasma-exchange. Interestingly the clinical response to
the above mentioned regimen was better when compared to patients with
"ineffective" plasma-exchange (Table 2). The response rate of patients with "effective" plasma-exchange is comparable to those without
any blocking activity in the MLC assay. As shown in Figure 3, "effective" plasma-exchange results in a clinical response, but eight
weeks later serum blocking factors could not be removed efficiently
by large volume plasma exchange leading again to clinical progression.

      In one patient with no serum blocking factors but who had a
mucous producing adenocarcinoma of the colorectal region, the combined regimen of large volume plasma exchange and chemotherapy did
not prevent the progressive growth of the tumor. In contrast, a
non-mucous producing, serum blocking factor negative adenocarcinoma

Table 1. Serum Blocking Factors in Patients with Advanced
          Chemotherapy Resistant Cancers of Different Origins
          and the Clinical Response to Large Volume Plasma
          Exchange plus Chemotherapy

| | | Number of Patients | | |
Diagnosis	Total	Serum Blocking Factors	Clinical Responder	Clinical Non-responder
Colorectal Cancer	16	23	11	5
Breast Cancer	3	2	3	0
Lung Cancer	5	1	4	1
Teratocarcinoma	3	3	0	3
Hypernephroma	3	3	0	3
Melanoma	5	3	2	3
Hodgkin's Disease	2	1	0	2
Leiomyofibrosarcoma	2	2	1	1
Plasmocytoma	2	2	2	0
Miscellaneous	6	3	2	4
	47 (100%)	32 (70%)	25 (54%)	22 (46%)

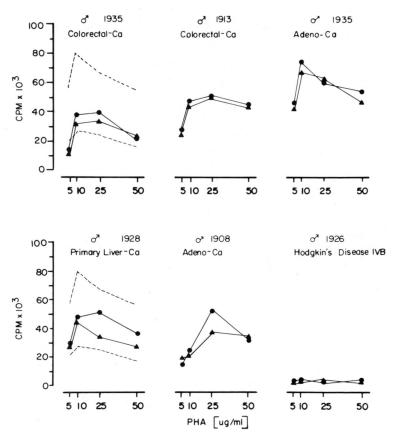

Fig. 1. PHA-induced lymphocyte proliferation before and after plasma-
exchange. ▲ before plasma exchange; ● after plasma exchange; ---- normal range of lymphocyte proliferation.

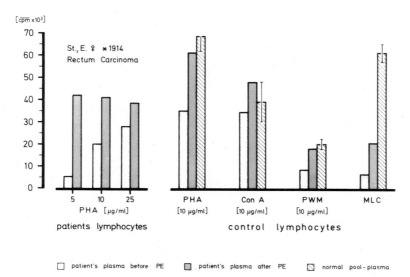

Fig. 2.   Serum blocking activity in different lymphocyte proliferation
          assays.

responded very well (Figure 4).   Such data suggest the possibility
that factors not detectable by our assay were eliminated by plasma
exchange.   Neither response rates nor serum blocking activity were
correlated with immune complex levels as assessed by a C_1q binding
assay or Raji-cell assay respectively.   The immune absorption of
elevated carcinoembryonic antigen (CEA) in our samples by an anti-
CEA coated Sephadex column resulted in persistent serum blocking
activity, indicating that in colorectal cancer CEA seems to have no
important suppressive value.   In some cases, acute phase reactants
seem to correlate with serum blocking activity.   For example, α1-
antitrypsin in Hodgkin's disease and α2-macroglobulin in breast
carcinoma showed a statistical correlation.

Table 2.   Correlation between Serum Blocking Factors in Advanced
           Cancer and the Clinical Response to Large Volume Plasma-
           exchange

	Number of Patients		
	Serum Blocking Factor		
Clinical Status	Positive	Positive	Negative
Plasma Exchange	Effective	Ineffective	
Clinical Responder			
Partial remission	0	0	1
Minor remission	4	5	4
No change	5	3	6
Total	9 (69%)	8 (42%)	11 (73%)
Clinical non-responder			
Progressive cancer	4 (31%)	11 (58%)	4 (27%)

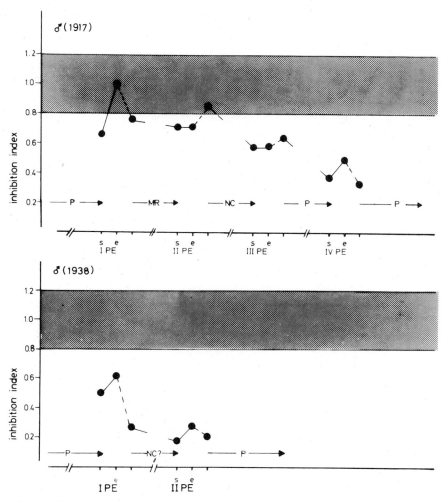

Fig. 3. Behavior of serum blocking factors and clinical response
to repeated large volume plasma exchange and chemotherapy
in two patients with chemotherapy resistant progressive
colorectal cancer.

inhibition index = $\dfrac{\text{cpm test assay}}{\text{cpm control assay}}$

P = progression
MR = minor response
NC = no charge

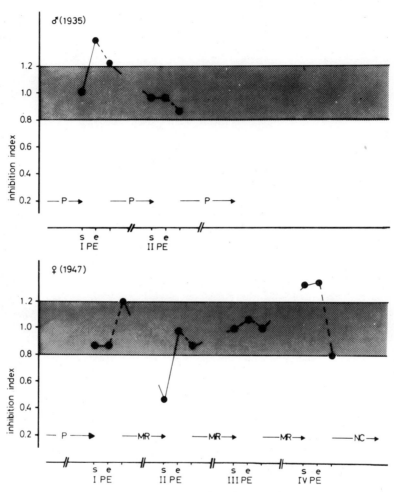

Fig. 4.   Different clinical response to large volume plasma exchange
          and chemotherapy in two serum blocking factor negative pa-
          tients with mucous producing (top) and non mucous producing
          colorectal adenocarcinoma.

CONCLUSION

Large volume plasma exchange is still an experimental procedure in cancer patients but our group showed recently that it is possible to overcome for a short time the progression of chemotherapy resistant tumors by the combination of plasma exchange and chemotherapy. The mechanisms involved in this phenomenon are unknown. We have shown that plasma exchange does not influence significantly autologous cellular immune reactions in vitro. So far in 32 out of 47 patients, we have demonstrated serum blocking factors in the patient's plasma using the mitogenic and allogeneic stimulation assay. Serum blocking factors could be eliminated or reduced by large volume plasma exchange only in 13 out of 32 patients (41%).

The clinical response to large volume plasma exchange plus chemo-therapy is not directly correlated with serum blocking factor activ-ity. However, the latter may have prognostic value since the highest responder rate was in serum blocking factor positive patients. It will be necessary to use more subtle and specific bio-assays before we will be able to analyze the real value of immune modulation by large volume plasma exchange in tumor bearing hosts.

Acknowledgement

We thank U. Neumeyer and Dr. J. Borghardt who performed the CEA absorption, and Dr. G. A. Krieger who evaluated the IC-levels.

REFERENCES

1. Kamo, I., Friedmann, H., Adv. Cancer Res. 25:271, 1977.
2. Israel, L., Edelstein, R., Mannoni, P., Radot, E., Greenspan, E. M., Cancer 40:3146, 1977.
3. Beyer, J.-H., Schuff-Werner, P., Kaboth, U., Klee, M., Köstering, H., Nagel, G. A., Schweiz. med. Wschr. 111:1522, 1981.
4. Schuff-Werner, P., Brattig, N., Beyer, J.-H., Bartel, J., Köstering, H., Berg, P. A., Nagel, G. A., Plasmapheresesym-posium Göttingen, 5.-6.12.1980, S. Karger Verlag, Basel, 1982.

DISCUSSION

ROTH: How much plasma did you remove from these patients and what was the total volume of plasma exchanged? Were you able to look at any patients that had resectable tumors to see if your blocking activity disappeared after tumor resection and reappeared with tumor progression?

SCHUFF-WERNER: We made a plasma exchange of 200% of the calculated plasma volume. In 8 patients we stopped at 150% exchange volume

because of coagulation problems.   We cannot answer your last question
because the patients we studied had advanced malignancy which was no
longer possible to treat either by surgery or by radiation or by
chemotherapy.

THE ROLE OF THE INFLAMMATORY

RESPONSE DURING TUMOR GROWTH

Robert Evans and Lawrence G. Eidlen

Jackson Laboratory
Bar Habor, Maine   04609 USA

When tumor cells are injected into a syngeneic recipient, an inflammatory response occurs at the site of injection (1). The sequence of events and the types of cells entering the lesion are fairly typical of that seen during wound healing (2). Both tumor growth and wound healing involve changes in blood vessel permeability, as well as proliferation of vascular endothelial cells, influx of leukocytes and plasma constituents, and proliferation of cells such as fibroblasts and epithelial cells. In both situations these reactions probably involve the release of vasoactive amines (by platelets and mast cells), arachidonic acid or prostaglandins (by macrophages, mast cells and perhaps tumor cells), activation of the complement, coagulation, fibrinolytic and kinin systems, and the release of mediators controlling or stimulating cellular proliferation. It has been reported in tumor model systems (3) and during wound healing (2) that prevention of macrophage infiltration at the site may result in impaired growth or repair, implying an important role for macrophages in these reactions.

The experiments described below were designed to influence the accumulation of inflammatory cells, with special reference to the macrophage, at the tumor site. Experiments using C57BL/6J (B6), BALB/cByJ and C3H/HeJ mice have been carried out involving such agents as hydrocortisone, azathioprine, indomethacin, silica, carrageenan and whole body irradiation. Selected experiments in B6 mice will be described.

INFLAMMATION DURING EARLY TUMOR GROWTH

An analysis of the cell types infiltrating the tumor site during

the first few days was performed after B6 mice were injected im with
B6 sarcoma cells (MCA/76-9). At intervals, the muscle containing the
tumor was excised, weighed and disaggregated with a mixture of papain,
collagenase and DNase. Figure 1 shows the relative proportion of
granulocytes (PMNs) and macrophages in relation to the rest of the
cell population. Neoplastic and other cell types declined to about
25% of the total by 24 hrs but thereafter increased progressively
to reach a stable level of about 60%. Evaluation of cytocentrifuge
preparations indicated an increase in the proportion of non-neoplastic, non-lymphoid, Fc-receptor and non-specific esterase negative
cells, which on morphological criteria probably were fibroblasts.

THE EFFECT OF DEPRESSION OF THE INFLAMMATORY RESPONSE ON TUMOR GROWTH

Hydrocortisone

        Ten mg of hydrocortisone suspended in incomplete Freund's adju-
vant were injected subcutaneously prior to an intramuscular injection
of thioglycollate medium to elicit an inflammatory response. Preliminary experiments indicated that this dose of hydrocortisone totally inhibited the inflammatory response for at least 4-7 days. To
demonstrate the influence of hydrocortisone on tumor growth, B6 mice
were injected subcutaneously with hydrocortisone 24 hours before an

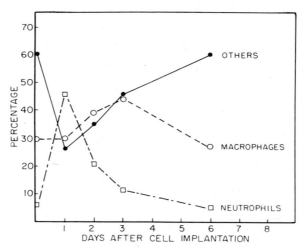

Fig. 1.   The inflammatory response during the early phase of tumor
          growth. MCA/76-9 sarcomas were disaggregated and analyzed
          for cellular composition. "Others" refers to neoplastic
          cells, fibroblasts and any other cell not identified as
          a neutrophil or macrophage.

intramuscular injection of 10 MCA/76-9 tumor cells. At intervals
tumors were excised, weighed, disaggregated with enzymes and analyzed
for cellular composition. Table 1 summarizes a typical experiment,
and shows that the hydrocortisone induced effects were strongest
during the initial 6 days when tumors were smaller, and yielded
fewer cells, including macrophages, than control tumors. After
this time tumors grew at the same rate in hydrocortisone treated
mice as in control mice. Histological evaluation showed that injection of tumor cells into control mice resulted in a pronounced localized inflammatory response. Initially, infiltrating leukocytes (PMNs
and macrophages) were mixed with the neoplastic cells but as the
number of tumor cells increased PMNs became less readily discernible,
while macrophages appeared in large numbers peripherally. By days
3-5 a capsule was clearly developing. In hydrocortisone treated mice,
inflammation was much reduced at 24 hours with PMNs comprising the
majority of the infiltrating cells. Few macrophages were discernible

Table 1.   The Effect of Hydrocortisone on Growth of the MCA/76-9
            Sarcoma and on its Macrophage Content

Treatment	Tumor Weight (g)	Cell Yield per Tumor x 10^6±SEM	Macrophages per Tumor x 10^6±SEM
		Day 6	
Hydrocortisone[a]	0.13	3.5±1.7	0.8±0.4
Controls[b]	0.29	11.4±2.3[c]	2.3±0.6
		Day 13	
Hydrocortisone	0.71	151±35	24±11
Controls	1.01	303±23[c]	65±11

[a] 10 mg of hydrocortisone were injected subcutaneously in 0.1 ml
incomplete Freund's adjuvant 24 hours before mice (5 per group)
were injected with 10^6 tumor cells.
[b] Pooled data from untreated mice or those receiving incomplete
Freund's adjuvant alone.
[c] Values significantly different from hydrocortisone group
(P <0.001) (Student's t test).

until about 7-10 days. Capsule formation was not evident until 7-10
days of tumor growth. Compared with control tumors, the number of
mitotic figures in sections of tumors from hydrocortisone treated
mice was reduced. For example, on day 5 there were 3 times as many
figures per field in controls.

Indomethacin, Silica, Carrageenan

B6 mice were treated with indomethacin (daily or contained in
the drinking water), silica, or iota-carrageenan (single intraperitoneal injections) and injected intramuscularly with 10^6 MCA/76-9
cells. The data indicated that tumor growth was impaired in all
cases during the first week, but, thereafter, growth paralleled that
of controls and in the case of silica appeared to be faster. The
total cell yields were significantly reduced during the early phase
with a concomitant reduction in macrophage numbers. The percentage
of macrophages, however, was much the same as in control tumors at
all stages of tumor growth.

Whole Body Irradiation

When the B6 sarcomas, MCA/76-9 and 77-23, were injected into
irradiated mice, tumor growth rates and macrophage contents were
related to the dose of irradiation. This is illustrated in Table 2.
It was apparent that as long as the bone marrow was releasing monocytes at a critical level, as was presumably the case after exposure
of mice to 400 R whole body irradiation, the tumor-macrophage content
could reach control tumor levels, even though the peripheral blood
monocyte counts were still depressed (4). However, it was evident
that higher doses of radiation resulted in a chronic monocytopenia,
which influenced the macrophage content of tumors. These events
were associated with impaired tumor growth. Admixing the tumor cell
inoculum with macrophages overcame the radiation-induced impairment.
Moreover, enhanced tumor growth could be demonstrated in normal mice
by injecting similar mixtures of tumor cells and macrophages (Table 3)
or their culture supernates. These experiments indicated the potential of macrophages or their products to promote tumor growth.

CONCLUSION

This report summarizes some preliminary data on the potential
role of macrophages during the inflammatory response resulting from
the intramuscular injection of syngeneic tumor cells. The evidence
suggests several different pathways that influence the inflammatory
response, tumor growth and, in particular, the immunobiology of the
monocyte-macrophage lineage.

Table 2. The Effect of Whole Body Irradiation on Subsequent
Growth of Two C57BL/6J Sarcomas and on their Macrophage
Content

Treatment	Tumor Weight (g)	Cell Yield per Tumor x 10^6±SEM	Macrophages per Tumor x 10^6±SEM
MCA/76-9 Sarcoma			
None	2.7	490±21	79±5
400R	2.6	390±20	53±6
600R	1.6	200±40	15±5
800R	0.9	59±10	0.5±0.1
MCA/77-23 Sarcoma			
None	1.5	184±19	40±4
400R	1.2	134±12	30±3
600R	1.1	76±6	11±2
800R	0.3	9±1	1±0.2

Mice injected intramuscularly with 10^6 sarcoma cells. Tumors
excised and disaggregated on day 14 of growth.

Table 3. Macrophage Enhancement of Syngeneic Tumor Growth in
          C57BL/6J Mice

Tumor	Admixed Macrophages[a]	Tumor Weight g±SEM	P Value[b]
MCA/76-24 Sarcoma	+	1.70±0.10	0.006
	-	1.27±0.11	
MCA/76-45 Sarcoma	+	0.77±0.04	0.001
	-	0.51±0.03	
MCA/77-23 Sarcoma	+	1.49±0.17	0.004
	-	0.86±0.09	
MCA/76-9 Sarcoma	+	0.91±0.06	0.005
	-	0.58±0.05	

[a]Mice were injected intramuscularly with 2.5×10^4 sarcoma
cells either with (+) or without (-) 2.5×10^6 thioglycollate-
induced peritoneal macrophages.
[b]P value was determined by the Mann-Whitney U test in which a
P value of <0.05 is considered significantly different. Tumor
weights are the mean of 9-12 tumors per group.

Acknowledgement

    This research was supported by PHS Grant number CA 27523 awarded
by the National Cancer Institute, DHHS. The Jackson Laboratory is
fully accredited by the American Association for Accreditation of
Laboratory Animal Care.

REFERENCES

1. Evans, R., J. Reticuloendothelial Soc. 26:429, 1979.
2. Leibovitch, J. T., and Ross, R., Amer. J. Pathol. 78:71, 1975.
3. Evans, R., Brit. J. Cancer 37:1086, 1978.
4. Evans, R., and Eidlen, D. M., J. Reticuloendothelial Soc. 301:
      425, 1981.

DISCUSSION

NORMANN: The agents you used (hydrocortisone, silica and radiation) decrease the inflammatory response. If you're assessing tumor growth by weight or by total cell number, how do you know that the changes you're observing are not due to the diminished inflammatory response while the tumor itself grows unimpaired?

EVANS: I can only answer that question by saying that we've actually taken the trouble to count the neoplastic cells and in fact it's not due to a lack of inflammatory cells. It's an actual reduction in the number of neoplastic cells.

NORMANN: To me, this would present a major technical problem. How many tumor cells do you inject? If you're looking at day 1, day 2, or day 3 after injection, how do you find and count them?

EVANS: It is a tough technical problem, but we spent a lot of time trying to overcome these problems. We have worked extensively on the erythidium bromide technique for assessing DNA in tissues and our cell recovery, even with very small tumors, is now about 55%. So we're losing 45% of our cells. But this percentage loss tends to be the same from very early to late stages of tumor growth. We have no problem in recovering cells if we inject a million tumor cells intramuscularly. The next day we will recover, it's variable, but certainly a million cells.

MANTOVANI: Are these general phenomenon such that you get the same answer with all the tumors that you have? When we did our experiments with carrageenan and silica, we had some tumors in which we didn't get that type of answer.

EVANS: Were you looking during the first few days?

MANTOVANI: No.

EVANS: You probably wouldn't see it. It's not uncommon for these tumors to start growing after the first week at the same rate as controls. And it's conceivable that in some cases they may even catch up to the controls because most tumors go through a logarithmic phase and then plateau. Now, if you take a carrageenan or silica treated mouse, whose tumor growth has been delayed, that tumor will not reach the same size until later. So, if you examine the tumors too late, you'll get no difference. If you do it at the right time, then it's possible you will see a difference.

MANTOVANI: You gave cortisone prior to tumor inoculation and monocyte counts should recover in about 10 days. I'm wondering whether the catching up of the cortisone treated tumor with the control is related to the fact that you now have an adequate supply of monocytes.

EVANS: Of course this invokes the argument that monocytes or macrophages are important in stimulating tumor growth. It is possible that that is the reason the tumors do catch up.

KAMBER: Did you inject your tumor cells with the host cells or did you separate out the different populations?

EVANS: In fact, it makes no difference. If you use tissue culture cells free of any contaminating leukocytes you get an identical inflammatory response.

KAMBER: So, you always get the inflammatory response?

EVANS: You always get the inflammatory response. Whether that's the tumor or a wounding effect, I don't really know. But you don't need the host cells, at least not in the inoculum.

KAMBER: In our experience, if you inject tissue culture cells subcutaneously which are all viable and don't show any sign of dying, such as a diffused granulation of pigment, we don't get an inflammatory response at all. These experiments were done with the B-16 melanoma into C57B1/6 mice. We only find an inflammatory response when some of the early tumor cells die.

EVANS: You've looked at the histology presumably and you do not get inflammation.

UNKNOWN: Yes, we've looked at the histology very carefully.

EVANS: Well, our site is generally the muscle and maybe that's a preferential site for inducing inflammation.

KAMBER: You're probably damaging the muscle quite badly by injection.

EVANS: I think that's a generalization. If you just insert a needle, you do not get an intense inflammatory response. And if you just inject saline, you don't. In terms of tumor inoculation, it's very dose dependent. If we cut down our inoculum ten-fold, we would not detect an inflammatory response. I'm not saying it wouldn't be there, but it would be very difficult to measure.

KAMBER: And if you inject only host cells from the tumors?

EVANS: We haven't injected only host cells from the tumors, but we've injected peritoneal macrophages, thymus cells and dead cells. Here, we don't really see an acute inflammation and this is different from tumor cells.

KAMBER: Is it possible that the host cells in the tumor are activated and inducing an inflammatory response?

EVANS: No. You don't need the host cells. You can use tissue culture lines and produce the same effects.

UNKNOWN: Do these anti-inflammatory agents also affect tumor cells?

EVANS: Yes. Hydrocortisone affects all manner of things, such as
prostaglandin synthesis, which in turn would affect vascular permeability, the egress of plasma constituents, including leukocytes.
The effects of silica or carrageenan are well documented. Silica
is supposed to be macrophage poison, but it is also an immunosup-
pressant, and the list goes on. So, I think the interpretation of
my data should not be confined to the macrophage. I'm just using
the macrophage as a means of detecting an inflammatory response.

GINSBERG: Have you tried to inject the tumor cells together with
macrophages which have been activated and transformed into secretory
cells? By injection, irradiation or putting in steroids, silica
or carrageenan, you actually cause a lot of macrophage rupture. Have
you tried to take products of macrophages and coat the tumor cells
to see if this enhances tumor growth?

EVANS: If you activate thioglycollate macrophages with LPS in suspension, wash them thoroughly, and then mix them with the tumor cells,
you get impaired tumor growth. We know that activated macrophages
can suppress tumor growth. We have, in fact, taken supernatants from
macrophage cultures to coat tumor cells. This is the work of my
colleague, Dr. Larry Eidlen. We've taken the culture supernates,
incubated the tumor cells with the supernates, and then put these
into mice and we still get stimulation of growth.

SECTION 6

DEFINING MACROPHAGE HETEROGENEITY

DIFFERENTIATION ANTIGENS

AND MACROPHAGE HETEROGENEITY

Siamon Gordon and Stan Hirsch

Sir William Dunn School of Pathology
University of Oxford
South Parks Road
Oxford OX1 3RE. U.K.

INTRODUCTION

The macrophage provides a favorable system to study differentiation in eukaryotic cells. Multipotential hemopoietic precursors proliferate and differentiate in response to ill-defined signals and are then widely distributed via the blood stream as mature members of the "Mononuclear Phagocyte System". Macrophages in the periphery display heterogeneous properties depending on their site and adaptation to local inflammatory and microbial agents. Although different lineages derive from the same stem cell, mature macrophages can be readily distinguished from polymorphonuclear leukocytes (PMN), other granulocytes, lymphocytes and dendritic cells. The myelomonocytic cells derive from a common bipotential precursor (CFU-C) which gives rise to colonies of macrophages and/or PMN in culture in the presence of specific colony stimulating factors (CSF). However, commitment to the macrophage or PMN lineage is poorly understood.

Surface antigens are ideal markers to define cell lineages and the mechanism and control of differentiation. Monoclonal antibodies provide homogenous reagents to study macrophage development and heterogeneity (Figure 1). The central question is whether macrophage heterogeneity reflects stages in maturation or modulation by extrinsic stimuli or whether independent subsets of macrophages separate as stable sublineages. We have chosen to study this problem in the mouse since relatively simple model systems are available to analyze macrophage differentiation and activation in genetically defined animals. Progenitors are cultivated in conditioned media which

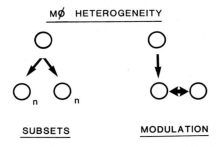

Figure 1.   The nature of macrophage (Mφ) heterogeneity.

contain the growth factors required for growth and differentiation
or they can be maintained as cycling precursors under special conditions (1).   Cell lines derived from tumors or by in vitro transformation express macrophage and/or PMN markers constitutively or
after exposure to inducing agents such as phorbol myristate acetate
or LPS.   Infection with BCG or treatment with lymphokines activate
macrophage cytotoxicity in vivo or in vitro.

Most investigators attempting to isolate monoclonal antibodies
against differentiation antigens of macrophages and PMN have employed
a "shotgun" approach, using heterogeneous cell mixtures or cell lines
as immunogen and screening for selective binding to mature stages
or their precursors.   We adopted a systematic strategy in which we
first isolated an antibody (F4/80) specific for mature mouse macrophages and then used successive hybridizations to isolate antibodies
specific for immature states (Figure 2).   Bone marrow cultivated
for 3 days in L cell fibroblast conditioned medium, a source of
M-CSF, contains mature proliferating adherent macrophages and non-
adherent PMN (>85%) and precursors committed to the macrophage
lineage.   For convenience we shall call the latter "monoblasts" since
they give rise with high frequency to colonies of adherent macrophages
after further cultivation in L cell conditioned medium.   After removal
of adherent macrophages, an antibody was raised against mouse PMN
(7/4) and used to purify monoblasts.   Another hybridoma (MB1) was
produced subsequently against the monoblasts.   Sensitive microassays
were required to isolate this antibody since monoblasts are present
in small numbers.   Once specificity had been established on primary
cells, a cell line which expresses this antigen, ADIII, a factor-
dependent myelomonocytic line developed by Dr. M. Dexter, has been
used to study the antigen.   These and other antibodies which show
restricted expression to macrophages and PMN have been used to study
macrophage differentiation and heterogeneity.

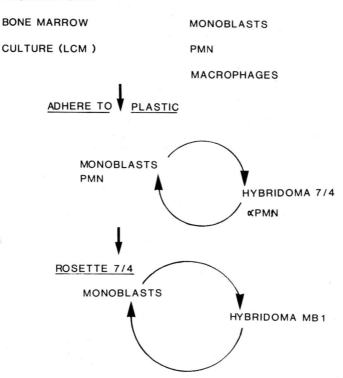

BONE MARROW   MONOBLASTS

CULTURE (LCM)   PMN

                               MACROPHAGES

ADHERE TO ▼ PLASTIC

MONOBLASTS
PMN

HYBRIDOMA 7/4

αPMN

ROSETTE 7/4

MONOBLASTS

HYBRIDOMA MB1

Fig. 2.   Hybridoma strategy for macrophage precursors.

MACROPHAGE DIFFERENTIATION ANTIGENS

Antigen F4/80

    The hybridoma F4/80 was isolated after hyperimmunizing a rat
with purified, adherent mouse peritoneal macrophages obtained after
injection of thioglycollate broth and cultivation for 2 days.   The
antibody binds to a novel macrophage antigen of 160,000 via the Fab
site and not via the Fc portion of the molecule.   The antigen F4/80
is found on the external surface of the plasma membrane and on the
luminal surface after endocytosis (3).   The antigen is resistant to
proteolytic digestion, although the molecule is cleaved into larger
fragments.   All mature mouse macrophages and macrophage cell lines
express antigen F4/80, unlike PMN, lymphocytes, dendritic cells (4)
and fibroblasts.   Expression of antigen F4/80 varies with macrophage
maturation (5) and activation of microbial agents such as BCG (6)
and is also subject to modulation during adherence to tissue culture
plastic surfaces.   Its function is unknown.

Antigen Mac-1

The antibody M1/70 directed against Mac-1 was isolated by Springer and co-workers (7) after immunization with non-adherent mouse spleen cells, but similar antibodies have been isolated by several groups after immunization with macrophage cell lines or primary macrophages. After immunoprecipitation and analysis by SDS-PAGE, a characteristic doublet (190,000 and 105,000) is observed. The Mac-1 antigen is expressed by mouse macrophages, PMN, human monocytes and ?NK cells (8). Recent studies by Trowbridge and Omary (9) have shown that the glycoprotein bearing the Mac-1 antigenic determinant belongs to a family of antigenically and structurally related glyco-proteins analogous to the T200 glycoprotein family. Antibody M1/70 immunoprecipitated a lower Mr polypeptide of 95,000 from a dexametha-sone treated macrophage cell line (M1) which was indistinguishable from that precipitated by different antibodies (C71/16 or I21/7) from M1 cells or BW5147 cells, a T lymphoid line. However, the higher Mr polypeptide of 170,000 precipitated by these antibodies from M1 and BW5147 cells was different.

Antigen 2.4G2

The rat anti-mouse FcRII monoclonal antibody 2.4G2 isolated by Unkeless blocks rosetting of sheep erythrocytes coated with monoclonal anti-SRBC antibody of a defined subclass (IgG2b) (10). The antigen is expressed by mouse macrophages, PMN and B lymphocytes and the molecule immunoprecipitated from these sources displays significant microheterogeneity of 47,000 to 60,000 (11). The molecule is gly-cosylated, binds to concanavalin A and has been purified from deter-gent solubilized cell membranes. The FcR antigen modulates within the plane of the membrane when macrophages adhere to surface-bound immune complexes and can be internalized by immune phagocytosis, which enhances its rate of degradation.

Antigen 7/4

This antigen is expressed on mouse PMN and 50% of BCG-activated macrophages but is not found on resident peritoneal macrophages and on less than 10% of thioglycollate elicited peritoneal macrophages or bone-marrow macrophages obtained after cultivation in L cell conditioned medium (12). PMN obtained from bone marrow, blood or from the peritoneal cavity during acute inflammation express antigen 7/4. FACS analysis of mouse bone marrow has revealed bright and dull labeled populations indicating that the PMN population is hetero-geneous. Spleen and thymus lymphocytes and fibroblasts are unlabeled. The antigen has not been characterized.

Antigen MB1

This antigen is expressed on a large proportion of non-adherent macrophage precursors found in bone marrow cultures in L cell conditioned medium, but is not expressed on adherent macrophages in these cultures, on peritoneal macrophages or on PMN (13). Studies on this antigen are in progress.

ANTIGEN EXPRESSION DURING MACROPHAGE DIFFERENTIATION

Only antigen F4/80 has been studied in sufficient detail (5). FACS sorting experiments combined with cloning in L cell conditioned medium have shown that CFU-C and CFC progenitors lack antigen F4/80 (Figure 3). The antigen first appears on non-adherent macrophage precursors in bone marrow cultured for 3 days in L cell conditioned medium, then increases progressively on adherent, proliferating macrophages. Subsequent exposure to lymphokines down-regulates antigen F4/80, as well as a specific lectin-like receptor for mannose/fucose terminated glycoproteins (14). Preliminary observations indicate that Mac-1 is expressed at an earlier stage of differentiation than F4/80. MB1 expression, in contrast, decreases with maturity (Figure 3). Studies are in progress to determine expression of this antigen on progenitors.

HETEROGENEITY AMONG MACROPHAGES

This has been studied by single cell analysis of various primary macrophage populations obtained from different sites or after different treatments, by analysis of independent colonies derived from bone marrow cultivated in the presence of L cell conditioned medium, and by assays of macrophages and related cell lines. Our results are summarized in Tables 1 and 2. With the possible exception of

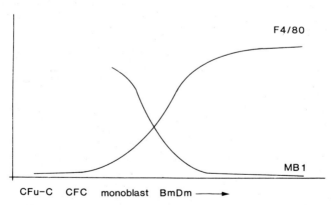

Fig. 3.   Antigen expression during macrophage differentiation.

antigen 7/4 which has not yet been studied in sufficient detail, the
heterogeneity observed can be ascribed to differences in maturation
or modulation rather than to independent subsets. Colonies derived
from bone marrow in vitro express antigen F4/80, Mac-1 and 2.4G2
uniformly (5) and similar results have been obtained with Ia antigen
(22). The ectoenzyme 5'nucleotidase may show heterogeneity under
similar conditions (23). Other workers have reported heterogeneous
expression of antigens found on various macrophage populations from
mouse (24-26), rat (27) and man (28-32), but the effects of maturation
and modulation have not been examined systematically.

THE BRANCH POINT BETWEEN MACROPHAGE AND PMN

        The evidence is incomplete, but a provisional assignment of
antigens is illustrated in Figure 4. It is not surprising that macro-

Table 1. Expression of Antigens on Primary Macrophages

Source of Macrophages	Antigen			
	F4/80	Mac-1	2.4G2	7/4
Bone marrow	100	100	100	0
Monocytes	50-100[a]	100	100	0
Peritoneum				
Resident	100	100	100	0
Thioglycollate elicited	100	100	100	10
BCG-activated	100[b]	100	100[b]	50

Data are expressed in percentage of cells labeled. Assessment
of bone marrow macrophages was performed on independent adherent colonies cultivated in L cell conditioned medium as
described in Ref. 5. Antigens F4/80, Mac-1 and 2.4G2 are also
detectable on lung, spleen, and liver macrophages.
a. Reported in Ref. 2,4.
b. Amount of antigen per cell decreased (6).

Table 2. Expression of Antigens on Macrophage and Related Cell Lines

Cell Line	Ref	Antigen				Comment
		F4/80	Mac-1	2.42G	7/4	
M1	15	Trace	+	+	-	Undifferentiated
416B	16	Trace	+	ND	Trace	"stem-cell" like
ADIII	17	+	+	+	+/Trace	Factor-dependent
427E	18	+	+	+	Trace	Myelo-monocytic
WEHI-3	19	+	+	+	-	"
J774.2	20	+	+	+	-	Mature macrophage
P388.D1	21	+	+	+	-	"

Antigen 2.4G2 also binds to B cell lines. Data for antigen 7/4 are in part unpublished observations (S. Hirsch).
ND, not determined.

phage and PMN should share antigens (Mac-1, 2.4G2, 7/4). It would be useful to obtain a marker specific for PMN, comparable to F4/80. Further experiments are needed to define the relationship between macrophage differentiation and activation and to confirm precursor-product relationships between MB1 positive cells and their negative progeny. Antigens provide markers to study commitment to each lineage and its regulation by specific colony-stimulating factors (33). Antigen 7/4 can be induced in some lines (e.g. ADIII) by WEHI-3 conditioned medium or post-endotoxin serum and studies are in progress to determine the nature of the inducing factor and its effect on primary PMN and macrophages. Expression of myelomonocytic differen-tiation antigens on progenitors requires further study.

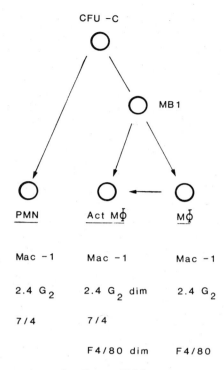

Fig. 4.   Antigens expressed after differentiation of macrophages and
           PMN.

REFERENCES

1.   Dexter, T. M., and Lajtha, L. G., Brit. J. Haematol. 28:525, 1974.
2.   Austyn, J. M., and Gordon, S., Eur. J. Immunol. 11:805, 1981.
3.   Mellman, I. S., Steinman, R. M., Unkeless, J. C., and Cohn, Z. A.,
       J. Cell Biol. 86:712, 1980.
4.   Nussenzweig, M. C., Steinman, R. M., Unkeless, J. C., Witmer,
       M. D., Gutchinov, B., and Cohn, Z. A., J. Exp. Med. 154:168,
       1981.
5.   Hirsch, S., Austyn, J. M., and Gordon, S., J. Exp. Med. 154:713,
       1981.
6.   Ezekowitz, R. A. B., Austyn, J. M., Stahl, P. D., and Gordon, S.,
       J. Exp. Med. 154:60, 1981.
7.   Springer, T., Galfre, G., Secher, D. S., and Milstein, C., Eur.
       J. Immunol. 9:301, 1979.
8.   Ault, K. A., and Springer, T. A., J. Immunol. 126:359, 1981.
9.   Trowbridge, I. S., and Omary, M. B., J. Exp. Med. 154:1517, 1981.
10. Unkeless, J. C., J. Exp. Med. 150:580, 1979.

11. Mellman, I. S., and Unkeless, J. C., J. Exp. Med. 152:1048, 1980.
12. Hirsch, S., and Gordon, S., submitted.
13. Hirsch, S., and Gordon, S., unpublished.
14. Ezekowitz, R. A. B., and Gordon, S., J. Exp. Med., in press.
15. Ichikawa, Y., J. Cell Physiol. 74:223, 1969.
16. Dexter, T. M., Allen, T. D., Scott, D., and Teich, N. M., Nature
       277:471, 1979.
17. Dexter, T. M., Garland, J., Scott, D., Scolnick, E., and Metcalf,
       D., J. Exp. Med. 152:1036, 1980.
18. Testa, N. G., Dexter, T. M., Scott, D., and Teich, N. M., Br. J.
       Cancer 41:33, 1980.
19. Ralph, P., Moore, M. A. S., and Nilsson, K., J. Exp. Med. 143:
       1528, 1976.
20. Ralph, P., Prichard, J., and Cohn, M., J. Immunol. 114:898, 1975.
21. Koren, H. S., Handwerger, B. S., and Wunderlich, J. R., J.
       Immunol. 114:894, 1975.
22. Unanue, E. R., personal communication.
23. Goldman, R., personal communication.
24. Deming, S., and Lohmann-Matthes, M-L., submitted.
25. Akagawa, K. S., Maruyama, Y., Takano, M., Kasai, M., and
       Tokunaga, T., Microbiol. Immunol. 25(11):1215, 1981.
26. Katz, H. R., LeBlanc, P. A., and Russell, S. W., J. Reticulo-
       endothelial Soc. 30(5):439, 1981.
27. Rumpold, H., Sewetly, P., Boltz, G., and Förster, O., in
       "Heterogeneity of Mononuclear Phagocytes" (O. Förster and
       M. Landy, eds.), Academic Press, New York, p. 47, 1981.
28. Breard, J., Reinherz, E. L., Kung, P. C., Goldstein, G., and
       Schlossman, S. F., J. Immunol. 124:1943, 1980.
29. Raff, H. V., Picker, L. J., and Stobo, J. D., J. Exp. Med. 152:
       581, 1980.
30. Haynes, B. F., Hemler, M. E., Mann, D. L., Eisenbarth, G. S.,
       Schelhamer, J., Mostowski, J. S., Thomas, C. A., Strominger,
       J. L., and Fauci, A. S., J. Immunol. 126:1409, 1981.
31. Todd, R. F. III, Nadler, L. M., and Schlossman, S. F., J.
       Immunol. 126:1435, 1981.
32. Ugolini, V., Nunez, G., Smith, R. G., Stastny, P., and Capra,
       J. D., Proc. Natl. Acad. Sci., USA 77:6764, 1980.

DISCUSSION

LIEBOLD: Is F-480 antigen expressed by human monocytes or macrophages?

GORDON: Not as far as we can determine.

STEWART: Does F-480 or 24G2 resemble Mac 2 or 3?

GORDON: As far as I can tell from the published information, they're
different.

GOLDMAN: Using an enzyme marker, we have evidence for at least 2
subsets of monocyte precursor cells. Using 5' nucleotidase, we can
show that there are at least 2 kinds of colonies in bone marrow cells
growing in vitro. One subset consists of small colonies in which
100% of the cells express high 5' nucleotidase activity and a second
subset of bigger colonies, around 1,000 cells per colony, having a
different shape which is 5' nucleotidase negative. So, we feel that
there are 2 subsets of precursor cells: one gives rise to macrophages
that express high 5' nucleotidase activity and the other probably
gives rise to inflammatory macrophages which lack this enzyme marker.

GORDON: You know of course that some macrophages can be induced to
down regulate their nucleotidase by treating them with lymphokines.

VAN DER MEER: I'm not sure what you mean by a monoblast and adherence. In our system we have good evidence that a monoblast is
slightly adherent. You can wash it off the surface but it's still
an adherent cell.

UNKNOWN: Is F4/80 antigen a glycoprotein?

GORDON: We think it is, but we are not absolutely certain.

UNKNOWN: We have very strong evidence, at least in erythroid chicken
systems, that many monoclonal antibodies will precipitate the same
molecule from different cells at different stages of differentiation.
The only difference is glycosylation.

SURFACE PROPERTIES OF ACTIVATED MACROPHAGES: SENSITIZED LYMPHO-
CYTES, SPECIFIC ANTIGEN AND LYMPHOKINES REDUCE EXPRESSION OF
ANTIGEN F4/80 AND FC AND MANNOSE/FUCOSYL RECEPTORS, BUT INDUCE Ia

R. A. B. Ezekowitz and S. Gordon

Sir William Dunn School of Pathology
University of Oxford
South Parks Road, Oxford   U.K.

The term macrophage activation was introduced by Mackaness in
the 1960's to describe the enhanced microbicidal activity of macrophages from animals with acquired immunity to infection with facultative intracellular pathogens (1).   Attempts have been made since
that time to define the properties of macrophage activation in terms
of morphological changes, biochemical or membrane events and correlate
these with microbicidal and tumoricidal activity (2,3,4).   After
infection with bacillus Calmette-Guérin (BCG), the host may acquire
immunity to specific secondary challenge, protection against unrelated
virulent organisms such as Listeria monocytogenes and increased resistance to transplantable tumors (5-7).   Macrophages from such animals
spread rapidly in cultures, secrete high levels of hydrogen peroxide
(8) and plasminogen activator (9) and display enhanced tumoricidal
activity (10).   We have shown that infection of the mouse peritoneal
cavity by BCG markedly alters the surface properties of the macrophages induced, compared with uninfected controls, or after injection
of thioglycollate broth (11).   Quantitative binding assays with radiolabeled ligands or monoclonal antibodies showed that BCG-activated
peritoneal macrophages (BCG-PM) express reduced antigen F4/80, (a
macrophage specific antigen of MW 160,000), Fc receptors and mannose
specific receptor activity, but have enhanced Ia antigen and increased
secretion of hydrogen peroxide and plasminogen activator.   We have
also shown that (a) these alterations in the plasma membrane make it
possible to distinguish between activated and non-activated macrophages, and (b) all changes of macrophage activation by BCG are dependent upon 'T' lymphocytes and specific antigen.   However, studies
with nude mice indicate that the activation phenotype may also arise
by an independent pathway.   The altered surface properties are stable,
occur in a co-ordinate manner, independent of a particular agent and
can be induced in vivo and in vitro.

401

DOWN-REGULATION OF ENDOCYTIC RECEPTORS IN ACTIVATED MACROPHAGES
MANNOSE/FUCOSYL RECEPTOR

Rat alveolar macrophages (12), mouse peritoneal and bone marrow
macrophages and some macrophage hybrids bind and internalize a variety
of glycoproteins and glycoconjugates terminating in mannose or fucose
via a specific membrane receptor (13). Binding and uptake of mannose-
BSA or β glucuronidase by macrophages can be specifically prevented
by the mannose-rich yeast mannan. Figure 1 shows mannose-specific
binding and uptake of ^{125}I-mannose-BSA by BCG and thioglycollate
elicited peritoneal macrophages under saturation conditions. Both
binding and uptake of ^{125}I-mannose-BSA by BCG macrophages were markedly depressed: 6% and 25% of thioglycollate elicited macrophage
activity respectively. Radioautographic studies showed that all
thioglycollate macrophages were heavily labeled by ^{125}I-mannose-BSA
(>10 grains/cell) and that BCG-macrophages were less heavily but
uniformly labeled. Experiments carried out in the presence of a
cocktail of inhibitors of oxygen reactive products, (catalase and
superoxide dismutase) as well as indomethacin which inhibits formation of some prostaglandins failed to reverse down-regulation of the
mannose/fucosyl receptor in BCG macrophages. The role of macrophage
proteinase secretion on receptor expression was examined by cultivat-
ing thioglycollate macrophages in serum free media and co-cultivating
them with BCG macrophages. Proteolysis at the surface of cells could
not account for the stable loss in mannose-specific receptor activity. Peritoneal macrophages from uninfected animals, harvested 48
h after injection of BCG sensitized lymphocytes and purified protein
derivative (PPD), or co-cultivated with sensitized lymphocytes and
PPD display a 40% reduction in mannose/fucosyl receptor activity.   on
The addition of a lymphokine-rich supernatant prepared from immune
BCG spleen cells and PPD induces a 70% reduction in mannose/fucosyl
receptor activity at 24 h. The effect can be noted as early as 4 h
and half-time of the decay is 16 h. The residual population remains
stable and is not due to a lymphokine-resistant subpopulation.

FC RECEPTOR

We examined the expression of Fc receptors which bind and internalize immune complexes using the rat monoclonal antibody 2.4G2 (14)
in a quantitative indirect binding assay. BCG-macrophages show reduced expression of Fc receptors ($3 \pm 2 \times 10^4$ molecules/cell) compared with thioglycollate macrophages ($1 \pm 0.6 \times 10^5$ molecules/cell)
or resident macrophages ($5 \pm 1 \times 10^4$ molecules/cell). Single cell
analysis using both autoradiography and uptake of opsonized sheep
red blood cells confirmed that Fc receptor expression was reduced
in 80% of BCG-macrophages.

Although activated macrophages express reduced levels of mannose/
fucosyl and Fc receptors compared with elicited or resident macro-

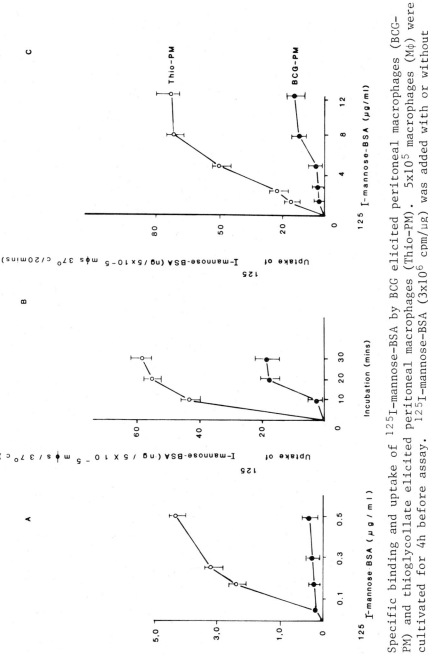

Fig. 1. Specific binding and uptake of ^{125}I-mannose-BSA by BCG elicited peritoneal macrophages (BCG-PM) and thioglycollate elicited peritoneal macrophages (Thio-PM). 5×10^5 macrophages (Mϕ) were cultivated for 4h before assay. ^{125}I-mannose-BSA (3×10^6 cpm/μg) was added with or without 1.25 mg mannan/well. Results show average $\pm$ SD of pooled results of five independent experiments. O = Thio-PM; ● = BCG-PM

phages, and endocytosis via these receptors is reduced, the receptors remain functional at the cell surface. The Fc receptor mediates extracellular lysis of antibody-coated target cells by stimulating hydrogen peroxide release, and the mannose/fucosyl receptor effectively triggers the respiratory burst upon contact with zymosan particles which have not been opsonized (G. Berton, unpublished).

SURFACE EXPRESSION OF OTHER MEMBRANE ANTIGENS AND SECRETION PRODUCTS

Quantitative indirect binding assays were performed with monoclonal antibodies F4/80 (16); Mac I/70 (18), which binds to polymorphonuclear leukocytes and macrophages; MRC OX6 (17), an anti Ia K, S antibody; and an affinity purified ^{125}I labeled rabbit anti rat F (ab')$_2$ second antibody. BCG-macrophages expressed reduced levels of F4/80, little change in Mac I, but enhanced Ia antigen when compared with thioglycollate elicited or resident macrophages. The secretion of plasminogen activator by thioglycollate and BCG-macrophages is comparable, but BCG-macrophages secrete fifteen fold more hydrogen peroxide after further stimulation with phorbol myristate acetate.

OTHER STIMULI

Macrophages obtained after injection of C. parvum resemble BCG-macrophages in their ability to secrete hydrogen peroxide, and plasminogen activator and with respect to tumoricidal and antimicrobial properties. Four hour adherent peritoneal macrophages from C. parvum injected animals have similar properties to BCG-macrophages (Table 1). An in vitro pulse of C. parvum to a 4 h adherent monolayer of resident peritoneal macrophages failed to induce the surface changes seen after in vivo injection. Phase contrast microscopy of coverslip preparations confirmed that the macrophages had phagocytosed organisms. The effects of intraperitoneal injection of endotoxin, proteose peptone and muramyl dipeptide (MDP) were also studied. Table 1 shows that MDP elicited macrophages displayed properties which were intermediate between resident and BCG-macrophages in that there was induction of Ia expression, some decrease in mannose/fucosyl receptor activity (50% of resident macrophages) and, as reported by others (19), enhanced secretion of superoxide, whereas fibrinolytic activity was not increased. Endotoxin and proteose peptone elicited macrophages, however, were similar to resident macrophages. Addition of MDP or LPS in vitro to 4 h adherent resident macrophages and further cultivation for 24 or 48 h failed to induce any change in macrophage surface properties.

Although not exhaustive, these studies confirmed that the activation phenotype described was not unique to BCG infection, but that it was confined to agents such as BCG and C. parvum with known ability to induce enhanced antimicrobial and cytocidal activity.

Table 1. Effects of Corynebacterium Parvum and Other Agents on Macrophage Properties[a]

Treatment[b]	Mannose/Fucosyl Receptor[c]	Antigen Expression Molecules/Macrophage x10⁴			Superoxide nM/mg/60 min	Plasminogen Activator % Fibrinolysis[d]
		F4/80	0x6	MacI		
C. Parvum	24±3	4±2	18±3	36±10	126±12	62±8
LPS	62±8	42±5	6±3	81±5	49±5	18±5
MDP	40±15	62±10	15±3	80±12	80±1	16±4
Proteose peptone	76±12	45±12	9±3	70±20	50±24	22±6

(a) Data are representative of three independent experiments and are reported as mean ± standard deviation.

(b) Cells that had adhered for 4 hours were washed and assayed. C. parvum (0.2 ml) was injected intraperitoneally 14 days prior to harvest. Lipopolysaccharide (LPS, 30 μg), muramyl dipeptide (MDP, 200 μg), or proteose peptone (10 mg/ml) was injected intraperitoneally 4 days prior to harvest.

(c) Mannose/fucosyl receptor is reported in ng uptake per 5x10⁵ macrophages in 30 minutes at 37°C.

(d) Fibrinolysis by 2x10⁵ macrophages was measured over 180 minutes.

CONCLUSIONS

These and earlier studies establish that activated mouse macrophages display complex changes in surface and endocytic function which accompany enhanced secretion of reactive intermediates of oxygen and plasminogen activator. The altered surface properties are stable, occur in a co-ordinate manner, independent of a particular agent, and can be induced in vivo and in vitro. Down-regulation of the mannose/fucosyl receptor provides an attractive new marker to study heterogeneity of lymphokines and the mechanisms of macrophage activation.

REFERENCES

1. Mackaness, G. B., J. Exp. Med. 116:381, 1962.
2. North, R. J., J. Immunol. 121:806, 1978.
3. Karnovsky, M. L., and Lazdins, J. K., J. Immunol. 121:809, 1978.
4. Cohn, Z. A., J. Immunol. 121:813, 1978.
5. Blanden, R. N., Lefford, M. J., and Mackaness, G., J. Exp. Med. 129:1079, 1969.
6. Nathan, C. F., in "Mononuclear Phagocytes" (R. van Furth, ed.), p. 1165, Martinus, Nijhoff Publishers, 1980.
7. Nelson, D. S., "Immunobiology of the Macrophage", Academic Press, Inc., London, 1976,
8. Nathan, C. F., and Root, R. K., J. Exp. Med. 146:1648, 1977.
9. Gordon, S., and Cohn, Z. A., J. Exp. Med. 147:1175, 1978.
10. Old, L. J., Benacerraf, B., Clarke, D. A., Carswell, C. E., and Stockert, E., Cancer Res. 21:1281, 1961.
11. Ezekowitz, R. A. B., Austyn, J., Stahl, P. D., and Gordon, S., J. Exp. Med. 154:60, 1981.
12. Stahl, P. D., Schlesinger, P. H., Sigardson, E., Rodman, J. S., and Lee, Y. S., Cell 19:207, 1980.
13. Stahl, P., and Gordon, S., J. Cell Biol. 93:49, 1982.
14. Unkeless, J. C., J. Exp. Med. 150:580, 1979,
15. Nathan, C., and Cohn, Z. A., J. Exp. Med. 152:198, 1980.
16. Austyn, J., and Gordon, S., Eur. J. Immunol. 11:805, 1981.
17. McMaster, W. R., and Williams, A. F., Immunol. Rev. 47:117, 1979.
18. Springer, T., Galfre, G., Secher, D. S., and Milstein, C., Eur. J. Immunol. 4:91, 1979.
19. Cummings, N. P., Pabst, M. J., and Johnston, R. B. Jr., J. Exp. Med. 152:1659, 1980.

DISCUSSION

METCALF: Since the type of lymphocyte conditioned medium you described contains GM-CSF, what happens when you put pure GM-CSF onto thioglycollate elicited macrophages? Do they express less antigen F4/80?

EZEKOWITZ: No, not with CSF.

RUCO: What is the percentage of activated macrophages which shows the phenotype for activation?

EZEKOWITZ: If you look at the data from a BCG induced peritoneal exudate, about 60% of the cells express the phenotype. With lymphokine treatment, antigen F4/80 and 2.4G2 appear to generally decrease on all the cells. With respect to the induction of Ia with lymphokine, you can induce between 40 and 50% of the cells in culture to express Ia.

FRIEDLANDER: Have you looked at antibody to interferon to see whether any of the lymphokine induced events are mediated by interferon?

EZEKOWITZ: Our preparation of lymphokine was a crude supernate prepared from immune BCG spleen cells. Also, we've done this with lectin induced supernatants. But we haven't looked at interferon which is a likely candidate to produce these changes.

ANTIGENIC MARKERS ON HUMAN

MONOCYTES AND MACROPHAGES

Robert F. Todd, III and Stuart F. Schlossman

Division of Tumor Immunology
Sidney Farber Cancer Institute and
Department of Medicine
Harvard Medical School
Boston, Massachusetts   02115 USA

INTRODUCTION

In an effort to examine heterogeneity within the human reticulo-endothelial system, we have developed a panel of monoclonal antibodies that define the expression of five distinct, nonpolymorphic antigens displayed on the surface of monocytes, macrophages and platelets (1,2). Antigen Mo1 is shared by peripheral blood monocytes, granulo-cytes, and a subset of Null cells. The expression of Mo2 and Mo3 is restricted to the monocyte-macrophage lineage. The density of Mo3, as distinguished from Mo2, is low on freshly isolated monocytes, but increases as a function of time in culture. Mo4 is borne by both monocytes and platelets, while Plt-1 is a platelet-specific antigen whose detection on monocytes reflects adsorption of platelets to monocyte surfaces. Mo1-4 are expressed by cultured macrophages derived from peripheral blood monocytes while only Mo1, 2, and 4 are borne by human peritoneal macrophages. Evidence for these conclusions will be presented and the significance of the Mo antigen series as differentiation markers within the monocyte-macrophage lineage will be discussed.

MATERIALS AND METHODS

Production of Monoclonal Antibodies: A six week old female BALB/c mouse was immunized intraperitoneally with $5x10^6$ adherent human monocytes. Six months later, the mouse was boosted intra-venously with $8x10^6$ monocytes from the same individual, and somatic cell hybridization was carried out as described previously (1).

After fusion and subsequent growth, individual hybridoma cultures were tested for the synthesis of antibodies reactive with peripheral blood monocytes but negative for T and B lymphocytes, using an indirect immunofluorescence assay (1). Selected cultures were cloned twice by the limiting dilution method and subsequently maintained as ascites tumors. Antibody-containing ascites were used in all subsequent experiments.

Isolation of Human Peripheral Blood Cells: Peripheral blood mononuclear cells were obtained by venapuncture of volunteer donors or from leukocyte-rich residues of blood bank platelet bags (2). Venapuncture specimens were either heparinized or defibrinated over glass beads (2). Ficoll-Hypaque density gradient centrifugation was used to separate granulocytes and erythrocytes from monocytes and lymphoid cells. Monocytes were obtained either by adherence to plastic dishes or by single step Percoll density gradient centrifugation (2,3). In selected experiments, isolated monocytes were cultured for varying lengths of time at 4° or 37^oC. Granulocytes were extracted from Ficoll-Hypaque pellets by 1 x G sedimentation in medium containing 0.4% dextran (4). Platelets were obtained by differential centrifugation of plateletpheresis residues and fixed in paraformaldehyde 1% in phosphate buffered saline. T lymphocytes were extracted from nonadherent peripheral blood mononuclear cells by sheep erythrocyte rosetting. Rosette-negative nonadherent cells were separated into B and Null cell fractions by affinity chromatography over Sephadex G-200 anti-human $F(ab')_2$ (5). Purity of fractionated blood cells was ascertained by morphological examination and/or the expression of surface markers specific for human T and B lymphocytes.

Human Leukemia Cells: Tumor cells were isolated from the peripheral blood or bone marrow of 43 patients with acute leukemia. Subclassification of these leukemic cells into acute lymphocytic (ALL), acute myeloblastic (AML), acute monocytic (AMoL), and acute myelomonocytic (AMML) variants was performed by morphological, histochemical, and surface immunological examination.

Cell Lines: Human cell lines of B cell (Laz 156 and 388), T cell (MOLT 4, CCRF-CEM, and HPB-ALL), histiocytic (U937), promyelocytic (HL-60), and erythrocytic (K562) origin were provided by Dr. Herbert Lazarus, Sidney Farber Cancer Institute.

Peritoneal Macrophages: Peritoneal macrophages were obtained from effluent dialysis fluid from patients undergoing chronic peritoneal dialysis.

Indirect Immunofluorescence: Analysis of monoclonal antibody binding to human cells was performed by indirect immunofluorescence using fluorescein-conjugated goat anti-mouse immunoglobulins (1). Antigen expression is represented by the percent of 10,000 cells demonstrating fluorescence above that of cells exposed to a negative control antibody, as analyzed by a fluorescence activated cell sorter.

Enzymatic Treatment of Monocytes: Monocytes ($60-100 \times 10^6$) were suspended in serum-free medium containing varying concentrations of trypsin or papain and incubated for 30 minutes at 37°C. All suspensions contained deoxyribonuclease (100 µg/ml) to prevent clumping by released DNA from dead cells. Enzymatic action was stopped by washing cells in medium containing serum and trypsin inhibitor. In selected experiments, enzyme treated cells were cultured overnight at 37°C in enzyme-free medium.

HL-60 Differentiation: HL-60 cells were cultured at 37°C for 96 hours in the presence of 10% PHA-leukocyte-conditioned medium (PHA-LCM) or $1-2 \times 10^{-9}$ M 12-0-tetradecanoyl phorbol-13-acetate (TPA) (6). Macrophage differentiation was ascertained by morphological and histochemical criteria.

RESULTS

Reactivity of Monoclonal Antibodies for Human Peripheral Blood Cells

Figure 1 summarizes the reactivity for circulating cells (by indirect immunofluorescence) of the 5 monoclonal reagents generated by immunization with human peripheral blood monocytes. Anti-Mo1, Mo2, and Mo4 bind to the majority of monocytes of all individuals tested (>14 determinations), whereas anti-Mo3 and Plt-1 demonstrate less reactivity. In the case of anti-Mo3, there is considerable variability between determinations. Anti-Mo1, in addition to monocytes, binds to granulocytes and a substantial fraction of Null cells, while both anti-Mo4 and anti-Plt-1 bind to platelets. None of the reagents react with a significant number of T and B lymphoid cells, erythrocytes or activated T lymphoblasts (data not shown). Established human cell lines of various hematopoietic lineages are also negative with the exception of moderate binding of anti-MO3 to U937.

Increase in Mo3 Antigen Density on Cultured Monocytes

The wide variability of Mo3 antigen expression on monocytes seen in Figure 1 (<10 to >80%) is a function of the time of culture prior to antibody binding assay and not, as was originally presumed, a result of polymorphism for this marker within the blood donor population. As shown in Figure 2, if monocytes are isolated after 60 minutes adherence and immediately screened for antigen expression ($M\phi_{t0}$), anti-Mo3 binding is low. Monocytes cultured overnight at 37°C ($M\phi_{t1}37^\circ$C) prior to assay demonstrate high Mo3 antigen density comparable to Mo1 and Mo2. Cells cultured overnight at 4°C ($M\phi_{t1}4^\circ$C) are essentially negative for Mo3 suggesting that the culture-dependent augmentation of antigen density requires metabolic

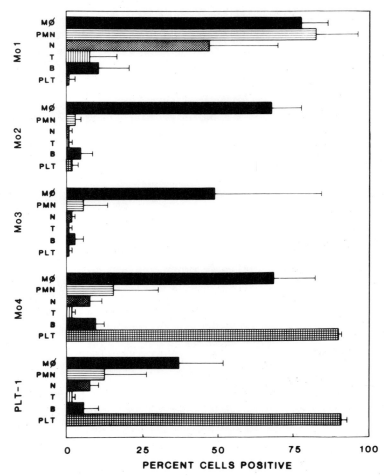

Fig. 1.   Reactivity of monoclonal antibodies for human peripheral
           blood cells.   Length of bars represents mean percentage of
           cells positive by indirect immunofluorescence ± SD of multi-
           ple determinations.   Mφ, monocytes; PMN, granulocytes; N,
           sheep erythrocyte rosette −, surface Ig− Null cells; T,
           sheep erythrocyte + cells; B, sheep erythrocyte rosette −,
           surface Ig+ cells; PLT, platelets.

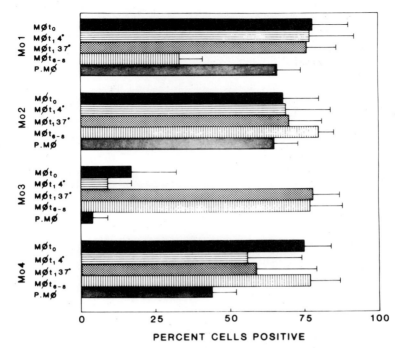

Fig. 2.   Antigen expression on fresh or cultured monocytes and peri-
          toneal macrophages. Length of bars represents mean percent-
          age of cells positive by indirect immunofluorescence $\pm$ SD
          of multiple determinations. $M\phi_{t0}$, monocytes assayed after
          60 minutes adherence; $M\phi_{t1}37^{\circ}C$, monocytes assayed after
          overnight culture at $37^{\circ}C$; $M\phi_{t1}4^{\circ}C$, monocytes assayed after
          overnight culture at $4^{\circ}C$; $M\phi_{t6-8}$, monocytes assayed after
          culture for 6–8 days at $37^{\circ}C$; P. $M\phi$, peritoneal macrophages.

activity (also supported by the fact that culture in the presence of
puromycin partially inhibits the density increase (2)). Adherence
per se appears to play no role in this phenomenon since monocytes
isolated by Percoll density sedimentation and cultured in "nonad-
hereable" polypropylene tubes either at 4° or $37^{\circ}C$ showed the same
difference in antigen density. As seen in Figure 2, Mo1, Mo2 and Mo4
demonstrate no significant change in antigen expression as a function
in culture.

Macrophages derived from monocytes cultured for 6–8 days
$(M\phi_{t6-8})$ are positive for Mo1–4 although the density of Mo1 is reduced. Binding of anti-Plt-1 approaches background levels. Peri-

toneal macrophages express Mol, Mo2, and Mo4, but are negative for
Mo3.

Platelet Binding of Anti-Mo4 and Anti-Plt-1

        As seen in Figure 1, both anti-Mo4 and anti-Plt-1 bind to human
platelets. The fact that platelets may adsorb to monocytes (7)
prompted the question of whether the reactivity of these antibodies
for monocytes merely reflects the adsorption of platelet antigens
to monocyte membranes. As shown in Figure 3, monocytes derived from
peripheral blood depleted of platelets by defibrination no longer
express Plt-1, whereas binding of anti-Mo4 is unchanged. These data
suggest that Mo4 is shared by both monocytes and platelets, while
Plt-1 is a platelet-specific antigen.

Regeneration of Mo2, Mo3, and Mo4 after Trypsin Digestion

        Figure 4 demonstrates that the expression of Mo2, Mo3, and Mo4
is abrogated following trypsin treatment of peripheral blood monocytes; Mol is resistant to degradation by either trypsin or papain.
Culture of trypsin-treated cells overnight at $37^\circ C$ resulted in a
restoration of Mo2, Mo3 and Mo4 expression. These results suggest
that Mo2-4 are synthesized by monocytes and are not, as in the case
with Plt-1, adsorbed "contaminating" antigens.

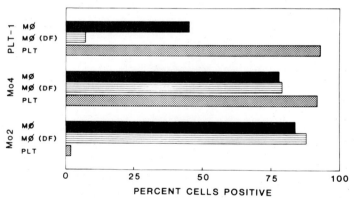

Fig. 3.   Expression of Mo4 and Plt-1 on monocytes and platelets.
          Length of bars represents mean percentage of cells positive
          by indirect immunofluorescence. Mϕ, monocytes isolated from
          heparin anticoagulated blood; Mϕ (DF), monocytes isolated
          from platelet-depleted defibrinated blood. Reactivity of
          both antibodies for platelets (PLT) is also shown.

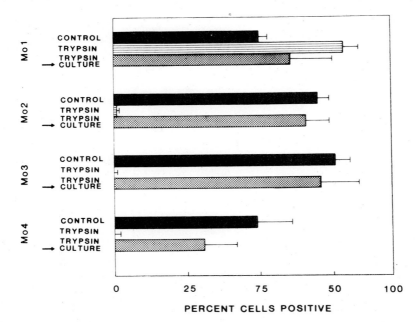

PERCENT CELLS POSITIVE

Fig. 4. Regeneration of Mo2, Mo3, and Mo4 after trypsin treatment
of monocytes. Length of bars represents mean percentage of
cells positive by indirect immunofluorescence ± SD of multiple determinations. Control, sham-treated control monocytes;
trypsin, monocytes treated with trypsin 1 mg/ml for 30 minutes at 37°C; trypsin→culture, trypsinized monocytes incubated for 18 hours at 37°C.

Expression of Mo1-4 on Tumor Cells from Patients with Acute Leukemia

Figure 5 demonstrates the expression of Mo1-4 on tumor cells
from patients with various forms of acute leukemia. It can be seen
that acute lymphocytic leukemia (ALL) cells are negative for these
antigens whereas myeloid leukemia cells show considerable reactivity.
Myeloid leukemia cells demonstrating monocytoid characteristics (acute
monocytic leukemia, AMoL; acute myelomonocytic leukemia, AMML) tend
to express Mo1, Mo2 and Mo4 with a higher frequency than the more
primitive acute myeloblastic leukemia cells. Using >20% of cells
positive as an arbitrary cut-off point, Mo1 is found on 95% of AMoL
and AMML cells. Mo3 is not expressed on AML cells and only a small
fraction of AMoL/AMML cells. Out of 39 patients with myeloid leukemia, Plt-1 has been uniformly negative (data not shown).

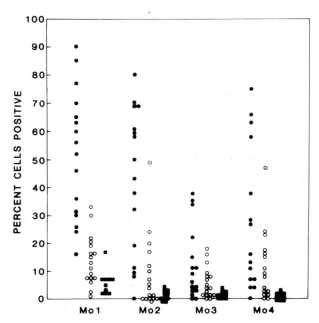

Fig. 5. Expression of Mol, Mo2, Mo3, and Mo4 on acute leukemic cells.
Each symbol represents the percentage of leukemic cells
positive (as indicated on the Y axis) by indirect immunofluorescence from an individual patient. (●), AMML or
AMoL cells; (○), AML cells; (■), ALL cells.

Reactivity of Monoclonal Antibodies for Differentiated HL-60

The promyelocytic leukemia cell line HL-60 when exposed to phorbol diester or certain conditioned media differentiates into cells
with macrophage characteristics (8,9). Figure 6 shows that conditioned media-differentiated cells (PHA-LCM) express Mol-4 to a variable extent but are negative for Plt-1; in the case of TPA-induced
differentiation, only Mol and Mo3 demonstrate significant antigen
density.

DISCUSSION

The monocyte-macrophage lineage plays an integral role in immune
function including phagocytosis, processing, and presentation of
antigenic material in such a way that it is stimulatory toward T
lymphocytes (10). In addition, these cells secrete factors that

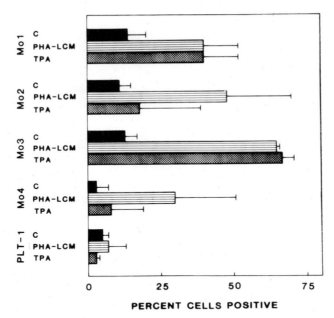

PERCENT CELLS POSITIVE

Fig. 6.   Expression of Mo1, Mo2, Mo3, Mo4, and Plt-1 on HL-60 cells
          induced toward macrophage differentiation. Length of bars
          represents mean percent of cells positive by indirect immuno-
          fluorescence ± SD of 2 determinations. C, control HL-60
          cells; PHA-LCM, HL-60 cells cultured for 4 days in 10% PHA,
          leukocyte-conditioned medium; TPA, HL-60 cells cultured for
          4 days in $1-2 \times 10^{-9}$M TPA.

modulate the immune response and they may serve as effector cells
for the in vivo correlates of in vitro ADCC and NK reactions (11-13).
To better understand the differentiation and function of this important class of cells, we and others have sought to characterize specific surface markers which distinguish members within the human
monocyte-macrophage lineage (1,2). In this paper, we have summarized
our experience with 5 monoclonal reagents that define surface antigens
expressed by various myeloid elements.

     Mo1 is an antigen borne by mature peripheral blood monocytes
and granulocytes as well as a fraction of Null cells. Among immature
myeloid forms, it is found on a significant proportion of normal
nucleated bone marrow cells and myeloid leukemia cells with partial
monocytic differentiation. It is absent on the CFU-GM stem cell
(Griffin and Todd, unpublished). The antigen itself is found on a

glycoprotein consisting of two subunits of 95,000 and 155,000 MW
(Todd and van Agthoven, unpublished); antigenic activity is not degraded by trypsin or papain.

Mo2 is restricted to the monocyte-macrophage series being expressed by adherent bone marrow cells, peripheral blood monocytes,
and macrophages derived from cultured monocytes, and peritoneal macrophages. Among leukemic forms, it is found with increased frequency
on those cells expressing monocytic differentiation. As a marker
for human peripheral blood monocytes, anti-Mo2 has been used successfully to deplete or purify monocytes from peripheral blood for functional studies (14; Todd, unpublished). The antigen, which is
protease-sensitive, is present on a glycoprotein structure with an
apparent MW of 55,000 (Todd and van Agthoven, unpublished).

The expression of Mo3 within the monocyte-macrophage series is
more restricted than Mo2 in that Mo3 is virtually absent on freshly
harvested monocytes; with culture under conditions promoting metabolic
activity, Mo3 density increases dramatically (2). This change in
antigen density is blocked by cold temperatures and is partially
inhibited by the protein synthesis inhibitor puromycin (2). Mo3
expression is retained by monocytes that have differentiated into
macrophages during 6-8 days in culture, but it is absent on peritoneal
macrophages. Mo3 is not expressed by bone marrow cells nor is it
found on myeloid leukemia cells except for a small subset of AMoL/
AMML cells. It is intriguing to speculate that the expression of Mo3
is among the morphological/biochemical events that occur during
differentiation from monocyte to macrophage (15). Except for its
sensitivity to proteases, the structural characteristics of Mo3 are
unresolved.

Mo4 is an antigen that is shared between a significant fraction
of peripheral blood monocytes and platelets. Among bone marrow cells,
preliminary studies indicate that megakaryocytes are also positive
(Griffin and Todd, unpublished). A significant fraction of monocytoid leukemic forms (50%) express Mo4. Mo4 is protease sensitive
and resides on a polypeptide structure of MW 100,000 (Todd, unpublished).

Plt-1 is a platelet-specific antigen whose expression on monocytes is an in vitro artifact resulting from the absorption of platelets to monocyte membranes. Among normal bone marrow cells, it too
is expressed by megakaryocytes (Griffin and Todd, unpublished). It
is not found on myeloid leukemia cells.

Several other investigators have recently reported the characterization of other monocyte-macrophage markers (16-25). The development
of these macrophage-specific monoclonal reagents and their relationship to the Mo series has been the subject of our recent review (26).

In summary, we have developed a panel of monoclonal antibodies detecting distinct nonpolymorphic antigens on cells of the human monocyte-macrophage series. These antigens appear to delineate stages of macrophage differentiation and are presently being used to determine the functional characteristics of the cells that bear them.

Acknowledgement

The authors thank Mr. Todd Abrams and Ms. Eve Leeman for their assistance and Ms. Felice Coral for her aid in phlebotomizing normal volunteers.

REFERENCES

1. Todd, R. F., Nadler, L. M., and Schlossman, S. F., J. Immunol. 126:1435, 1981.
2. Todd, R. F., and Schlossman, S. F., Blood 59:775, 1982.
3. Pertoft, H., Johnsson, A., Warmegard, B., and Seljeld, R., J. Immunol. Meth. 33:221, 1980.
4. Levy, P. C., Shaw, G. M., and Lobuglio, A. G., J. Immunol. 123: 594, 1979.
5. Chess, L., MacDermott, R. P., and Schlossman, S. F., J. Immunol. 113:1113, 1974.
6. Todd, R. F., Griffin, J. D., Ritz, J., Nadler, L. M., Abrams, T., and Schlossman, S. F., Leukemia Res. 5:491, 1981.
7. Perussia, B., Jankiewicz, J., and Trinchieri, G., Blood, in press.
8. Elias, L., Wogenrich, F. J., Wallace, J. M., and Longmire, J., Leukemia Res. 4:301, 1980.
9. Rovera, G., Santoli, D., and Damsky, C., Proc. Natl. Acad. Sci. USA 76:2779, 1979.
10. Unanue, E. R., Immunol. Rev. 40:227, 1978.
11. Nathan, D. F.,Murray, H. W., and Cohn, Z. A., New Engl. J. Med. 303:622, 1980.
12. Poplack, D. G., Bonnard, G. D., Holiman, B. J., and Blaese, R. M., Blood 48:309, 1976.
13. Mantovani, A., Jerrells, T. R., Dean, J. H., and Herberman, R. B., Int. J. Cancer 23:18, 1979.
14. Morimoto, C., Todd, R. F., Distaso, J. A., and Schlossman, S. F., J. Immunol. 127:1137, 1981.
15. Zuckerman, S. H., Ackerman, S. K., and Douglas, S. D., Immunol. 38:401, 1979.
16. Breard, J., Reinherz, E. L., Kung, P. C., Goldstein, G., and Schlossman, S. F., J. Immunol. 124:1943, 1980.
17. Raff, H. V., Picker, L. J., and Stobo, S. D., J. Exp. Med. 152: 581, 1980.
18. Hogg, N., Slusarenko, M., Cohen, J., and Reiser, J., Cell 24: 875, 1981.
19. Ugolini, V., Nunez, G., Smith, R. G., Stastny, P., and Capra, J. D., Proc. Natl. Acad. Sci. USA 77:6764, 1980.

20. Dimitriu-Bona, A., Burmester, G. R., Waters, S. J., and Winchester, R. J., Fed. Proc. 40:988, 1981.
21. Linker-Israeli, M., Billing, R. J., Foon, K. A., Fitchen, J. H., and Terasaki, P. I., J. Immunol. 127:2473, 1981.
22. Springer, T., Galfre, G., Secher, D. S., and Milstein, C., Eur. J. Immunol. 9:301, 1979.
23. Griffin, J. D., Ritz, J., Nadler, L. M., and Schlossman, S. F., J. Clin. Invest. 68:932, 1981.
24. Perussia, B., Trinchieri, G., Lebman, D., Jankiewicz, J., Lange, B., and Rovera, G., Blood, in press.
25. Austyn, J. M., and Gordon, S., Eur. J. Immunol. 11:805, 1981.
26. Todd, R. F., and Schlossman, S. F., in "Immunology of the Reticuloendothelial System. A Comprehensive Treatise. Vol. VI" (J. A. Bellante and H. B. Herscowitz, eds.), Plenum Press, New York, in press.

DISCUSSION

UNKNOWN: Have you looked at the expression of Mo antigens on bone marrow cells of leukemic patients?

TODD: Many of our samples were derived from the bone marrow. We selected either peripheral blood or bone marrow from patients that had at least 70-80% malignant forms in the blood.

KOREN: Do you think that Mol, which reacts against null cells, detects the same molecule as OKM-1 or 4F2?

TODD: We have evidence that anti-Mol and OKM-1 probably bind to the same membrane structure. Both antibodies precipitate a glycoprotein with 2 subunits of approximately 95,000 and 155,000 molecular weight. However, OKM-1 does not block the binding of fluorescinated Mol suggesting that the antibodies may be reacting to different antigenic sites on the same molecule.

STEVENSON: Have you looked at the binding of any of your Mo reagents to activated T cells?

TODD: Yes, and all are negative. We examined T cells activated by either PHA or the mixed lymphocyte reaction.

MANTOVANI: Is Mo3 expressed by more mature macrophages from different anatomical sites, such as bronchoalveolar macrophages or liver macrophages? And how were the peritoneal macrophages obtained?

TODD: The peritoneal macrophages were obtained from two sources. One source was effluent fluid from patients undergoing peritoneal dialysis. The other source was peritoneal washouts from women undergoing intraperitoneal chemotherapy before they had received various

activating substances.   Concerning the expression of Mo3 on macrophages from different sites, I have information only about peritoneal
cells. We have tried to look at pulmonary macrophages but technically
it's a problem because they seem to have a higher background.

WEINER: Have you tested Mo3 reactivity against any non-adherent
monocytes? Have you tested it against Ficoll-Hypaque isolated monocytes in suspension?

TODD: The question is whether this antigen is brought about or
stimulated by adherence? Our approach was to use a Ficoll-Hypaque
preparation followed by a partial purification step on Percoll.
At that point the cells were negative. We then cultured them overnight in polypropylene tubes to which they don't adhere very well.
Yet we still observed an increase in Mo3 antigen expression. So,
adherence per se doesn't appear to play a major role.

LEONARD: Have you taken human peritoneal cells that don't express
Mo3, placed them in culture and then examined for antigen expression?

TODD: We have indeed. We cultured them overnight, but even then
there was no antigen expression. So peritoneal macrophages appear
to be truly negative.

FÖRSTER: You looked at your HL-60 promyelocytic cells after 4 days
in TPA or lymphokine. But have you looked at earlier time periods?

TODD: We actually did a kinetic study examining the cells daily from
day 1 up through day 7. The expression of these antigens is gradual
and appears to correspond with the gradual decrease in specific
esterase and the gradual increase in non-specific esterase as these
cells transform into more macrophage-like cells. You start to see
the antigen expression around day 2 and it reaches a peak around day
4.

HUMAN MONONUCLEAR PHAGOCYTES

DIFFERENTIATION ANTIGENS

Alexandra Dimitriu-Bona, Robert J. Winchester
and Gerd R. Burmester

Hospital for Joint Diseases
Mount Sinai School of Medicine
New York, New York   10003   USA

Through a series of irreversible events, the progenitor cell committed to the mononuclear phagocyte lineage differentiates into blood monocytes and fluid or tissue macrophages. These successive maturational events, as well as reversible modulations induced by certain local stimuli, reflect the expression of specific gene products and the repression of others. Taken together they explain the morphological, functional and biochemical heterogeneity that characterizes the diverse members of this cell lineage. The analysis of these gene products found on the mononuclear phagocyte membrane has recently been approached through the use of monoclonal antibody technology (1-3). This report is a summary of studies on the mononuclear phagocyte cell surface phenotype utilizing a series of monoclonal antibodies developed in this laboratory (4,5).

Ten monoclonal antibodies specifically reacting with mononuclear phagocytes were selected from four different hybridomas produced by polyethylene glycol induced fusion of Sp2/0 myeloma cells with BALB/c mice splenocytes immunized against one of the following cell populations: (a) blood monocytes from a normal individual (R), (b) blood monocytes from a leukopheresed rheumatoid arthritis patient (P), (c) pleural mononuclear phagocytes (S) and (d) U937 cells, a histiocytic lymphoma cell line with the phenotypic features of an immature monocytoid cell line (U). The selected reagents were designated MoR-17, MoP-7, MoP-9, MoP-15, MoS-1, MoS-39 and MoU-26, MoU-28, MoU-48, and MoU-50. Antibody binding to the cell surface was detected by a F(ab')2 rabbit anti-mouse Ig reagent in indirect immuno-fluorescence as described (6) and evaluated in conventional microscopy as well as by cytofluorometric analysis. (FACS IV, Becton Dickinson, Mountain View, CA).

DETECTION OF MULTIPLE DIFFERENTIATION ANTIGENS ON THE SURFACE OF
U937 MONOCYTOID CELLS, BLOOD MONOCYTES AND FLUID MACROPHAGES

Antigens defined by the above antibodies were assayed on the
surface of U937 monocytoid cell line, blood monocytes and fluid macrophages (Table 1). The ten reagents defined distinct antigenic determinants. Despite their diversity, however, the following general
patterns were evident. (a) Primitive monocytoid surface antigens
were detected by the reagents MoU-26, MoU-28, MoU-48 and MoU-50.
In this group of surface determinants, the antigen MoU-50 was restricted to the U937 cell line whereas the antigens MoU-26, MoU-28
and MoU-48, mainly expressed on U937 cells, were also identified
in small quantities on blood monocytes, macrophages and a minor
fraction of granulocytes. (b) A monocyte-macrophage restricted
antigen was defined by the MoP-15 antibody. This antigen was present
on 38 to 80% of blood monocytes, with marked variations from individual to individual, and was found on nearly all fluid mcarophages.
The MoP-15 antigen appears to be a highly modulated gene product.
(c) Primary monocyte-macrophage antigens were defined by the reagents
MoP-9, MoS-1 and MoS-39 but they were identified also in minute
amounts on a fraction of blood granulocytes. These antigens are
similar to that recognized by 63d3, a previously reported reagent (1).
As demonstrated by simultaneous or sequential additive experiments,
each of the antigens in this group detects distinct, antigenic determinants (4). These four antigens are well represented on virtually
all monocytes but increase in expression on fluid macrophages. (d)
Antigens associated with late maturational states are defined by the
reagents MoP-7 and MoR-17. These antigens have minimal expression
on blood monocytes but are dominant surface moieties on mature, fluid
or tissue macrophages. These latter antigens can also be detected
on a fraction of activated T cell blasts but not on resting T or B
lymphocytes.

MODULATION OF DIFFERENTIATION ANTIGENS FOLLOWING IN VITRO CULTURE

In vitro culture of blood monocytes is associated with significant modulation of antigens on the cell surface. Marked differences
were encountered for the antigens MoP-15, MoP-9, MoS-1 and MoS-39
with significant enhancement for each demonstrable by flow cytometry.
Primitive monocytoid antigens of the MoU series and maturation antigens MoP-7 and MoR-17 do not undergo consistent modulation following
in vitro culture of monocytes, even for periods of time as long as
28 days.

EXPRESSION OF MONONUCLEAR PHAGOCYTE RELATED ANTIGENS ON VARIOUS ACUTE
MYELOID LEUKEMIAS (AML)

The expression of different primitive or maturational mononuclear

Table 1. Human Mononuclear Phagocyte Differentiation Antigens as Defined by Monoclonal Antibodies

Cell Populations	Monoclonal Antibodies								
	MoU-50	MoU-26	MoU-28	MoP-15	MoS-1	MoS-39	MoP-9	MoP-7	MoR-17
U-937	+	+	2+	−	−	−	−	−	−
Monocytes	−	±	+	+	2+	2+	2+	±	±
Macrophages	−	−	±	2+	3+	3+	3+	+	+
T and B Cells	−	−	−	−	−	−	−	−	−
T cell blasts	−	−	−	−	−	−	−	±	±
Granulocytes	−	−	±	−	±	±	±	−	−
Cultured Monocytes	−	±	+	2+	3+	3+	3+	±	±

Data were collected using cytofluorometric analysis and graded as follows: ± as less than 30 channels above control, + as 30 to 60 channels above control, 2+ as 60 to 100 channels above control, and 3+ as more than 100 channels above control.

phagocyte differentiation antigens on AML cells was related to the degree of leukemic cell maturity. Undifferentiated leukemic cells (FAB M1) expressed primarily MoU and little or no MoP or MoS defined antigens. The absence of MoU markers and the presence of MoP or MoS defined antigens characterized more mature acute monocytic leukemias (FAB M5). Upon in vitro culture, the membrane phenotype of the FAB M1 leukemic cells progressively acquired MoP and MoS maturational markers while at the same time the expression of MoU antigens gradually declined until, in most instances, the antigens became undetectable.

SUMMARY

     Cell surface antigenic analysis using 10 different monoclonal antibodies indicates the following. (a) The membrane antigenic phenotype of various members of the mononuclear phagocyte lineage is characterized by the co-expression of multiple but distinct antigenic determinants. (b) The maturational evolution of the cells is associated with qualitative and/or quantitative modifications of cell surface antigens resulting in the definition of different phenotypic entities. (c) In vitro culture of blood monocytes is accompanied by significant modulation of these antigens. In particular, the expression of MoP-15 appears to be modulated following in vitro activation. (d) The occurrence of these antigens on various normal and leukemic cell populations provides a basis for the establishment of defined patterns of antigenic expression corresponding to successive maturational states of the human mononuclear phagocyte cell lineage. (e) The reactivity of AML cells with various antibodies defines patterns of antigen expression related to the degree of leukemic cell maturity. These patterns change during culture.

Acknowledgement

     This study was supported by USPHS Grant CA20107, an Arthritis Foundation Clinical Research award, and the Milton Petrie Endowment Fund. We thank R. Batorsky and K. Kelley for technical assistance, and N. L. Rodriguez and R. Berger for secretarial assistance.

REFERENCES

1. Ugolini, V., Nunez, G., Smith, R., G., Stastny, P., Capra, J. D.,
      Proc. Natl. Acad. Sci 77:6764, 1980.
2. Breard, J., Reinherz, E. L, Kung, P. C., Goldstein, G., Schloss-
      mann, S. F., J. Immunol. 124:1943, 1980.
3. Todd, R. F., Nadler, L. M., Schlossman, S. F., J. Immunol. 126:
      1435, 1981.
4. Dimitriu-Bona, A., Burmester, G. R., Waters, S. J., Winchester,
      R. J., Fed. Proc. 40:988, 1981.

5. Dimitriu-Bona, A., Kelley, K., Winchester, R. J., Fed. Proc.
        (Abstr), 1982.
6. Winchester, R. J., Wang, C. Y., Halper, G., Hoffman, T., Scand.
        J. Immunol. 5:747, 1976.

DISCUSSION

RUMPOLD: Monoclonal antibodies MoP-7 and MoR-17 obviously stained
only a low percentage of cells. Did they also react weakly in
experiments with complement mediated lysis which is probably more
sensitive?

DIMITRIU-BONA: Yes. They reacted very weakly with blood monocytes
and much more intensely with macrophages.

RICHTER: Your MoP-7, MoP-9, and MoP-15 were derived from a patient
with rheumatoid arthritis. Was there any difference between the
monocytes taken from that patient and monocytes from a normal patient
used to produce MoR-17?

DIMITRIU-BONA: Not as far as I know.

RICHTER: Rheumatoid monocytes are supposed to express more Ia antigens.

KOREN: Do you have any evidence for the specificity of your monoclonal antibodies, checking them against other cell types or cell
lines?

DIMITRIU-BONA: Yes. MoP-15 is specifically expressed on monocytes
and macrophages. Several reacted very weakly to a sub-population
of granulocytes whose antigen is comparable to 63D3. MoP-7 and MoR-17
are expressed by activated T cells but very, very weakly.

A CELL SURFACE MARKER EXPRESSED ON CYTOTOXIC

PERITONEAL MACROPHAGES AND NORMAL LUNG MACROPHAGES

Tohru Tokunaga[1], Kiyoko S. Akagawa[1]
and Takashi Momoi[2]

[1]National Institute of Health, Shinagawa-ku
Tokyo 141, Japan
[2]Institute of Medical Science, University of Tokyo
Minato-ku, Tokyo 108, Japan

The Ml cell line (Clone M1/436.7) established from a myeloid leukemia of an SL strain mouse by Y. Ichikawa (1) is known to be inducible to differentiate into mature macrophages by exposure to various inducers, such as lipopolysaccharide (LPS) (2). Employing these clonal cells, we tested various macrophage activating agents for their possible differentiation inducing ability. As summarized in Figure 1, crude lymphokines, obtained by stimulating normal mouse spleen cells with concanavalin A-agarose, induced Ml cell differentiation, while neither purified interferon (IFN; $10^{7.3}$ units/mg; given by Haruo Kono, NIH, Tokyo) nor synthetic muramyl dipeptide (MDP) did so (3,4). Instead, IFN enhanced Ml cell differentiation triggered by LPS or lymphokines, and MDP acted on matured Ml cells to produce factor(s) which could trigger Ml cell differentiation (4).

Comparative biochemical analysis on membrane glycolipids of original and differentiated Ml cells were carried out (5). ^{3}H-Glycosphingolipids extracted from nontreated Ml cells and Ml cells preincubated with lymphokines for 48 hours at 37°C were applied to a high-performance silica gel thin-layer plate, and developed with chloroform-methanol-water (60:30:5). A band comparable to asialo GM1 was clearly visible in the glycosphingolipid from differentiated Ml cells, while it was only faintly detected in control Ml cells. Bands corresponding to GbOse$_3$-Cer and to monohexosylceramide were more strongly visible also in glycolipids from Ml cells incubated with lymphokines than from nontreated Ml cells (5).

F(ab')$_2$ fragment of anti-asialo GM1 antiserum (anti-asialo GM1) was prepared as an ammonium sulfate precipitate of whole antiserum

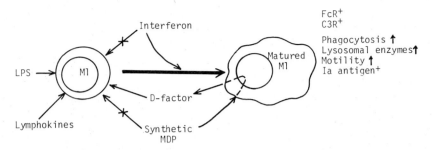

Fig. 1.   Induction of differentiation of M1 cells into mature macro-
          phages.

of rabbits immunized with purified asialo GM1. This fraction reacted
with asialo GM1, but not with other glycolipids, such as GM1, GM2,
GD1b and asialo GM2 in immunoflocculation tests (6). The immunofluorescence microscopic assay showed that more than 50% of M1 cells
incubated with lymphokine were stained, while none of the original
M1 cells reacted (5).

     These findings led us to test whether or not mouse peritoneal
macrophages change their reactivity to anti-asialo GM1 during activation (7). Resident peritoneal cells and peritoneal exudate cells
induced by either proteose peptone, viable BCG (Japanese strain) or
heat-killed Corynebacterium parvum were adhered onto coverslips, and
then stained indirectly with anti-asialo GM1. These cells were also
tested for viability after treatment with antiserum plus complement.
Simultaneously, cytotoxic activity of these cells against EL4 leukemic cells was tested by an in vitro 18 hour ^{51}Cr release assay.
It has been reported that peritoneal exudate cells induced by BCG
or C. parvum contain not only activated macrophages but also activated
natural killer cells (NK cells) (8,9) and that NK cells are reactive
to anti-asialo GM1 (10). To distinguish NK cell activity from cytotoxic macrophage activity, we used both a 4 hour ^{51}Cr release assay
from labeled YAC cells and an 18 hour ^{51}Cr release assay from labeled
EL4 cells; cytotoxicity of activated macrophages was detected only
in the latter assay and NK activity was detected only in the former.
Our results showed that BCG-induced peritoneal exudate cells showed
cytotoxicity in both assay systems, and pretreatment with anti-asialo
GM1 plus complement destroyed cytotoxicity in both assays (7).

     Reactivity to anti-asialo GM1 of macrophages from various sources
and their cytotoxic activity as measured by the 18 hour ^{51}Cr release
assay are shown in Table 1. While resident and peptone-induced macrophages showed no or weak cytotoxicity, macrophages induced by BCG
or C. parvum showed strong cytotoxic activity. Pretreatment of the

macrophages with anti-asialo GM1 plus complement destroyed their cytotoxicity. An intravenous injection of a small amount of anti-asialo GM1 was given one day before harvesting peritoneal cells from mice injected intraperitoneally with BCG or C. parvum 5 days before. The antisera destroyed cytotoxicity of the harvested cells. (Table 1.)

Peritoneal exudate cells activated in vitro by incubating with lymphokines or LPS for 4 hours at 37°C showed a strong cytotoxic activity against EL4 and the number of asialo GM1 positive cells also increased. Treatment of these cells with anti-asialo GM1 plus complement diminished the activity. Thus, there was a correlation between asialo GM1-positive cells and cytotoxic activity of mouse peritoneal macrophages (7).

Table 1.   Correlation Between Asialo GM1 Positive Cells and Tumor Cytotoxicity

Type of Cell	Asialo GM1	Cyto- toxicity
Resident macrophages	−	−
Peptone induced macrophages	±	±
BCG or C. parvum induced macrophages	++	++
pretreated with anti-asialo GM1 + complement in vitro	−	−
pretreated with anti-asialo GM1 in vivo	−	−
Peritoneal exudate cells plus lymphokine or LPS in vitro	++	++
treated with anti-asialo GM1 + complement	−	−
pretreated with anti-asialo GM1 + complement, then incubated with lymphokine or LPS	++	++

Time course analysis of asialo GM1 positive cells contained in peritoneal exudates during culture with lymphokine or LPS showed that they gradually increased for the first 2 days and then decreased.

The tissue distribution of asialo GM1 positive macrophages was examined. To avoid possible nonspecific adsorption of the anti-serum onto macrophages, we used the $F(ab')_2$ fragment of anti-asialo GM1. More than 95% of the lung macrophages strongly reacted to the $F(ab')_2$ fragment, while about 1% of the peritoneal resident macrophages reacted. Only low percentages of asialo GM1 positive cells were found in peritoneal exudate cells and splenic macrophages. Treatment with anti-asialo GM1 plus complement killed only 2% of the peritoneal resident macrophages, but 98% of the lung macrophages. These results were true for macrophages from different mouse strains, including nu/nu mice.

These facts suggest that lung macrophages are always activated, possibly from continuous infection with exogenous microorganisms, and that this results in expression of asialo GM1 on their cell surface. To test this hypothesis, we examined lung macrophages from germ-free mice and fetal mice (18 days' gestation). Contrary to the hypothesis, the results showed that more than 90% of lung macrophages of these mice reacted to the $F(ab')_2$ fragment of anti-asialo GM1.

Lung macrophages from a strain 2 guinea pig or from a Cynomologus monkey did not react to this antiserum.

We next tested cytotoxic activity of lung macrophages. Lung and peritoneal exudate macrophages were incubated with lymphokine or LPS for 5 hours at 37°C, washed and assayed for cytotoxicity against EL4 cells. Cytotoxicity of lung macrophages was less than 1% while that of peritoneal macrophages was more than 25%.

We wondered if the antigen of lung macrophages that reacted to anti-asialo GM1 was not asialo GM1 itself, because it had been reported that glycophorin, a major erythrocyte membrane glycoprotein having the sugar residue of $Gal\beta(1\rightarrow3)GalNAc\beta$, reacted to this antiserum if it was pretreated with neuraminidase (11). However, autoradiofluorography of glycolipids extracted from lung macrophages clearly showed that these cells contained a relatively large amount of asialo GM1. It may be interesting to study the role of the glycolipids on lung macrophages in relation to "surfactant".

In summary, macrophages are heterogeneous in terms of surface glycolipids, including the marker lipid asialo GM1. Asialo GM1 appears on mouse peritoneal macrophages during activation. Treatment of the macrophages activated with BCG, C. parvum, lymphokine or LPS either with anti-asialo GM1 plus complement in vivo or in vitro

destroys their cytotoxic activity. In contrast, mouse alveolar macrophages, including those from germ-free and fetal mice, always express
asialo GM1 but are not cytotoxic for tumor cells.

REFERENCES

1. Ichikawa, Y., J. Cell Physiol. 74:223, 1969.
2. Hozumi, M., Homma, Y., Tomida, M., Okabe, J., Kasukabe, T.,
        Sugiyama, K., Hayashi, M., Takenaga, K., and Yamamoto, Y.,
        Acta Haematol. Japan 42:941, 1979.
3. Akagawa, K. S., and Tokunaga, T., Microbiol. Immunol. 24:1005,
        1980.
4. Tokunaga, T., Akagawa, K. S., Kasai, M., Momoi, T., and Nagai, Y.,
        in "Monograph: Naito Foundation International Symposium on
        Self-Defense Mechanisms - Role of Macrophages",in press.
5. Akagawa, K. S., Momoi, T., Nagai, Y., and Tokunaga, T., FEBS
        Letters 130:80, 1981.
6. Kasai, M., Iwamori, M., Nagai, Y., Okumura, K., and Tada, T.,
        Eur. J. Immunol. 10:175, 1980.
7. Akagawa, K. S., and Tokunaga, T., in press.
8. Wolfe, S. A., Tracey, D. E., and Henney, C. S., Nature 262:584,
        1976.
9. Ojo, E., and Haller, O., Int. J. Cancer 21:444, 1978.
10. Kasai, M., Yoneda, T., Habu, S., Maruyama, Y., Okumura, K., and
        Tokunaga, T., Nature 291:334, 1981.
11. Momoi, T., Tokunaga, T., and Nagai, Y., in press.

DISCUSSION

VAN FURTH: Have you checked whether your asialo GM1 antiserum
reacts with surfactant? Since all your cells are positive, it might
well be that you just have a reaction with this protein.

TOKUNAGA: I've not done so yet, but studies with surfactant would
be very interesting.

SILVERSTEIN: Did you use the asialo GM1 antiserum also against
glycoproteins?

TOKUNAGA: Yes. Glycoproteins treated with neuraminidase reacted
with this antiserum.

SILVERSTEIN: And if the antibody is used directly against cells
without neuraminidase treatment, are there glycoproteins that react
with this antiserum?

TOKUNAGA: Yes.

UNKNOWN: Is the asialo GM1 marker analogous to the material that binds peanut agglutinin?

TOKUNAGA: In essence yes. But asialo GM1 antiserum seems to have a more narrow range than peanut agglutinin.

RUCO: Is the expression of asialo GM1 antigen reversible on cytotoxic macrophages? If you culture the macrophages, do they lose the antigen?

TOKUNAGA: After 3 days, they lose the antigen.

STEVENSON: Is there any unifying theory relating the ability of cells to kill and the expression of this marker?

TOKUNAGA: Asialo GM1 is expressed on natural killer cells, immature thymocytes, fetal liver cells and some other cells. For peritoneal macrophages, asialo GM1 is a differentiation marker.

MACROPHAGE FUNCTIONAL HETEROGENEITY

William S. Walker

Division of Immunology
St. Jude Children's Research Hospital
Memphis, Tennessee   38101 USA

INTRODUCTION

Macrophages are ubiquitous in mammalian tissues where they participate in such vital processes as homeostasis (1,2), the secretion
of chemical mediators (3), the development of immune competence (4,5),
the regulation of immune induction (6), and as effector cells in
certain expressions of efferent immunity (7). It is now evident that
macrophages are not uniformly capable of all these functions; indeed,
evidence accumulated over the past several years shows that populations of macrophages, like populations of lymphocytes, are functionally heterogeneous.

In this article, I present examples of macrophage inter- and
intra- population heterogeneity and discuss some of the possible
reasons for this diversity. The reader interested in a more detailed
accounting of macrophage heterogeneity is referred to recent reviews
(8-10).

EVIDENCE FOR MACROPHAGE HETEROGENEITY

Interpopulation

There are essentially two types of macrophage heterogeneity.
The first is termed interpopulation heterogeneity and refers to differences among populations of macrophages located at different tissue
sites. Such differences, as shown by so-called "free" macrophages
of the lung and peritoneal cavity, have been studied and compared
extensively (Table 1). Peritoneal and alveolar macrophages differ,
for example, in the manner by which they derive energy, the former

435

Table 1. Some Comparisons of Macrophages from Different Tissues

Tissue Source	Activity	References
Peritoneum and lung	Energy metabolism	11, 35
	In vivo activation	36
	Accessory cell activity	8, 12
	Complement receptors	37, 38, 39
	Microbicidal	8
Spleen and liver	Removal of senescent RBC	1
	Accessory cell activity	8, 40, 41, 42

relying on anaerobic and the latter on aerobic metabolism (11). Further examples include the finding that rat alveolar macrophages inhibit T-cell proliferation in vitro, whereas their peritoneal counterparts provide a necessary accessory cell function in the same T-cell response (12). Another example of interpopulation diversity may be found in the spleens of normal animals, whose macrophages are responsible for the recognition and removal of senescent red blood cells - an activity not solely due to the anatomy of this organ since the liver Kupffer cell is the primary site for removal of aged red blood cells in splenectomized animals (1).

Local tissue adaptation accounts for some of these interpopulation differences. For instance, the activities of oxygen metabolizing enzymes of either peritoneal or alveolar macrophages can vary depending on whether the cells are maintained in an aerobic or anaerobic environment (13), and similar results have been reported for culture-derived macrophages from bone marrow progenitors (14). Whether all interpopulation differences are due to a similar adaptation process is unknown.

Intrapopulation

Direct evidence for intrapopulation heterogeneity has come from demonstrations that populations of elicited peritoneal macrophages can be separated into subpopulations of cells with different phago-cytic activities (15), Fc-receptors (16,17), and immunogenic RNA molecules that induce specific antibody formation in vitro (15,18). Subsequent work by numerous investigators has further documented intrapopulation heterogeneity within groups of peritoneal, alveolar and splenic macrophages and within the pool of circulating monocytes (Table 2). For instance, there are a number of morphologically

distinguishable types of splenic macrophages located at distinct
anatomical sites (19). Humphrey and Grennan (20) recently noted that
those of the red pulp differ from those in the white pulp marginal
zones. About one-half of the former bore Ia-antigens, whereas few,
if any, of the latter bore or could be induced to express these determinants. In addition, marginal zone macrophages selectively ingested
and retained polysaccharides that were not recognized by red pulp
cells. The functional significance of these differences are unknown,
and the lineal relationship of the two types of macrophages has yet
to be explored.

Populations of macrophages from animals previously injected
with such activating agents as C. parvum show marked heterogeneity
in tumoricidal activity. Miller and associates (21) examined subpopulations of activated rat peritoneal macrophages, distinguished
by size and density, and observed that tumoricidal activity was associated almost exclusively with high-density macrophages, whereas
those in the light-density subpopulation promoted rather than inhibited the growth of the tumor cells both in vitro and in vivo (21).

Table 2. Examples of Intrapopulation Heterogeneity

Property	Reference
General	
Morphology	19
Size-density	15, 43, 44
Enzyme content	45, 46
Progenitors	47
Surface Features	
Fc-receptors	16, 17, 22
Complement receptors	39
Ia-antigen	48, 49, 50, 51
Other antigens	52
Functions	
Accessory cell	15, 50, 51, 53
Effector cell	54, 55, 56, 57
Secretion	58, 59, 60

MECHANISMS OF INTRAPOPULATION DIVERSITY

Traditionally, macrophage diversity has been attributed to the stage of differentiation-maturation within a single lineage of cells (Figure 1). In support of this view, the heterogeneity of Fc-receptor activity seen among populations of alveolar and elicited-peritoneal cells is largely eliminated by in vitro culture (22). In recent work, Gordon and co-workers (23) examined individual colonies of culture-derived bone marrow macrophages for evidence of the clonal expression of the macrophage-restricted antigen F4/80. They found that while considerable heterogeneity of expression characterized young cultures, all the cells in all colonies bore the antigen after several additional days of incubation, an observation most consistent with a single lineage scheme.

Alternatively, the existence of branch points leading to multiple lineages of macrophages would afford the opportunity for true functional specialization (Figure 1). Despite lack of conclusive evidence for the existence of differentiation branch points or multiple macrophage lineages, there are data to support the existence of self-renewing autonomous populations of macrophages (24-26) which, of course, could evolve activities and functions distinct from macrophages derived directly from the bone marrow. Evidence for a lineal independence of the resident tissue macrophage apart from the circulating pool of monocytes is cogently discussed by Volkman (24) and has been reviewed (10).

The cell cycle may also play a role in generating intrapopulation heterogeneity. Under normal steady-state conditions, a relatively minor proportion of macrophages are actively synthesizing DNA (27,28). Populations of inflammatory macrophages, on the other hand, may con-

I. SINGLE LINEAGE WITH DIFFERENTIATION – MATURATION

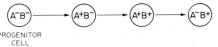

2. MULTIPLE LINEAGE WITH TRUE FUNCTIONAL SPECIALIZATION

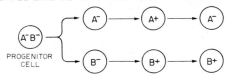

3. CELL CYCLE FOR SELF-RENEWING POPULATIONS

Figure 1.   Possible explanations for functional heterogeneity of macrophages.

tain more than 30% of their cells in cycle, and Berlin and co-workers (29) found that macrophages in mitosis fail to ingest particles. Recent studies in the author's laboratory (30,31) have confirmed and extended this finding by showing that macrophages in the G2 phase of the cell cycle exhibit an enhanced antibody-mediated phagocytic activity which is linked to an increase of Fc-receptors.

CONCLUSIONS AND FUTURE PROSPECTS

Incomplete understanding of macrophage immunobiology poses a major obstacle to identifying the origins of interpopulation and intrapopulation heterogeneity. Any of three possibilities could account for macrophage functional diversity - (i) local tissue adaptation and heterogeneity of differentiated states within a single lineage of cells, (ii) the presence of separate cell lineages, and (iii) heterogeneity related to cell-cycle phase. The second alternative, multiple cell lineages, appears to be the most formidable to assess.

Although providing baseline information (18,32), studies with subpopulations of primary macrophages separated on the basis of either size or density, or with transformed cell lines, have yielded few insights into the origins of macrophage diversity. One approach that should prove fruitful is the use of long-term cultures of cloned, nontransformed macrophages. Methods to obtain such cultures using progenitors isolated from a variety of tissues (33) are at hand; the only shortcoming is that present techniques do not yield a sufficient number of progeny for multiple functional analyses. Nevertheless, recent studies (33) indicate that this limitation in growth is due to insufficient concentrations of the necessary growth factors and not to the proliferative capacity of the progenitor. Thus, it seems reasonable to predict that in the foreseeable future it will be possible, as it is now with T-cells (34), to obtain large numbers of clonogenically derived macrophages to resolve important issues about macrophages: their origins, developmental biology and roles in host defense.

Acknowledgement

The author's work has been supported by the National Institutes of Health, Grant CA 16652; the American Cancer Society Institutional Grant IN-99, and by ALSAC. Mr. John Gilbert provided editorial assistance and Ms. Chris Winston typed the manuscript.

REFERENCES

1. Berlin, K., and Berk, P. D., in "The Red Blood Cell" (D. Mac-
      Surgenor, ed), pp. 957-1019, Academic Press, New York, 1975.

2. Kay, M. M. B., Nature 289:491, 1981.
3. Rocklin, R. E., Bendtzen, K., and Greineder, D., Adv. Immunol.
    29:56, 1980.
4. Van den Tweel, J. G., and Walker, W. S., Immunology 33:817, 1977.
5. Beller, D. I., and Unanue, E. R., J. Immunol. 118:1780, 1972.
6. Unanue, E. R., Adv. Immunol. 31:1, 1981.
7. Adams, D. O., and Snyderman, R., J. Natl. Cancer Inst. 62:1341,
    1979.
8. Walker, W. S., in "Immunobiology of the Macrophage" (D. S. Nelson,
    ed.), pp. 91-110, Academic Press, New York, 1976.
9. Hopper, K. E.,Wood, P. R., and Nelson, D. S., Vox Sang. 36:257,
    1979.
10. Walker, W. S., and Hester, R. B., in "Immunology of the Reticulo-
    endothelial System: A Comprehensive Treatise" (J. A.
    Bellanti and H. B. Herscowitz, eds.), Plenum Press, New York,
    in press.
11. Oren, R. A., Farnham, A. E., Saito, K., Milofsky, E., and Kar-
    novsky, M. L., J. Cell Biol. 17:487, 1963.
12. Holt, P. G., and Batty, J. E., Immunology 36:257, 1980.
13. Simon, L. M., Robin, E. D., Phillips, J. R., Acevedo, J.,Axline,
    S. G., and Theodore, T., J. Clin. Invest. 59:443, 1977.
14. Bar-Eli, M., Territo, M. C., and Cline, M. S., Blood 57:95, 1980.
15. Walker, W. S., Nature New Biol. 229:211, 1971.
16. Walker, W. S., Immunology, 26:1025, 1974.
17. Kavai, M., Laczko, J., and Csaba, B., Immunology 36:729, 1979.
18. Rice, S. G. and Fishman, M., Cell. Immunol. 11:130, 1974.
19. Carr, I., "The Macrophage", Academic Press, New York, 1973.
20. Humphrey, J. H., and Grennan, D., Eur. J. Immunol. 121:221, 1981.
21. Morahan, P. S. and Miller, G. A., in "Heterogeneity of Mono-
    nuclear Phagocytes" (O. Förster and M. Landy, eds.), pp. 161-
    164, Academic Press, New York, 1981.
22. Rhodes, J., J. Immunol. 114:976, 1975.
23. Hirsch, S., Austyn, J. M., and Gordon, S., J. Exp. Med. 154:713,
    1981.
24. Volkman, A., J. Reticuloendothelial Soc. 19:249, 1976.
25. Daems, W. Th. and van der Rheem, H. J., in "Mononuclear Phago-
    cytes- Functional Aspects, Part I" (R. van Furth, ed.),
    pp. 43-60, Martinus Nijhoff, The Hague, 1980.
26. Widmann, J. J., and Fahimi, H. D., Am. J. Path. 80:349, 1975.
27. Shands, J. W., and Axelrod, B. J., J. Reticuloendothelial Soc.
    21:69, 1977.
28. van Furth, R., Diesselhoff-den Dulk, M. M. C., Roeburn, J. A.,
    van Zwet, T. L., Crofton, R., van Oud Altlas, A. B., in
    "Mononuclear Phagocytes- Functional Aspects, Part I" (R. van
    Furth, ed.), pp. 279-298, Martinus Nijhoff, The Hague, 1980.
29. Berlin, R. D., Oliver, J. M., and Walter, R. J., Cell 15:327,
    1978.
30. Gandour, D. M., Grant, W. D., and Walker, W. S., in "The Cell
    Cycle and Macrophage-Antibody-Dependent Effector Cell
    Activities of Macrophage-like Cell Lines" (H. S. Koren, ed.),
    Marcel Dekker, New York, in press.

31. Gandour, D. M., and Wlaker, W. S., Proc. 18th Natl. Mtg., Retic-
     uloendothelial Soc. Abst. 96, 1981.
32. Morahan, P. S., J. Reticuloendothelial Soc. 27:223, 1980.
33. Stewart, C. C., in "Macrophage Regulation of Immunity (E. R.
     Unanue and A. S. Rosenthal, eds.), pp. 455-476, Academic
     Press, New York, 1980.
34. Paul, W. E., Sredni, B., and Schwartz, R. H., Nature 294:697,
     1981.
35. Cohn, Z., Adv. Immunol. 9:163, 1968.
36. Ryning, F. W., Krahenbuhl, J. L., and Remington, J. S., Immunol-
     ogy 42:513, 1981.
37. Ross, G., J. Immunol. Methods 37:197, 1980.
38. Hearst, J. E., Warr, G. A., and Jakab, G. J., J. Reticuloendo-
     thelial Soc. 27:443, 1980.
39. Walker, W. S., and Yen, S.-E., J. Cell. Physiol. 1982, in press.
40. Inchley, C. J., and Howard, J. G., Clin. Exp. Immunol. 5:189,
     1969.
41. Rogoff, T. M., and Lipsky, P. E., J. Immunol. 124:1740, 1980.
42. Nadler, P. I., Klingenstein, R. J., Richman, L. K., and Ahmann,
     G. B., J. Immunol. 125:2521, 1980.
43. Gorczynski, R. M., Scand. J. Immunol. 5:1031, 1976.
44. Normann, S. J., and Weiner, R., in "Heterogeneity of Mononuclear
     Phagocytes" (O. Förster and M. Landy, eds.), pp. 496-500,
     Academic Press, London, 1981.
45. Fishman, M., and Weinberg, D. S., Cell. Immunol. 45:437, 1979.
46. Schroff, G., Neuman, C., Sorg, C., Eur. J. Immunol. 11:637, 1981.
47. McCarthy, K. F., and MacVittie, T. J., J. Reticuloendothelial
     Soc. 24:263, 1978.
48. Schwartz, R. J., Dickler, H. B., Sachs, D. H., and Schwartz, B.
     D., Scand. J. Immunol. 5:731, 1976.
49. Yamashita, V., and Shevach, E. M., J. Immunol. 119:1584, 1977.
50. Lee, K-C. and Wong, M., J. Immunol. 125:86, 1979.
51. Tzehoval, E., De Baetselier, P., Feldman, M., and Segal, S.,
     Eur. J. Immunol. 11:323, 1981.
52. Raff, H. V., Picker, L. J., and Stobo, J. D., J. Exp. Med. 152:
     581, 1980.
53. Lee, K-C., Wong, M., and McIntyre, D., J. Immunol. 126:2474,
     1981.
54. Walker, W. S., J. Reticuloendothelial Soc. 20:57, 1976.
55. Weinberg, D. S., Fishman, M., and Veit, B. C., Cell. Immunol.
     38:94, 1978.
56. Lee, K-C., and Berry, D., J. Immunol. 118:1530, 1977.
57. Campbell, M. W., Sholley, M. M., and Miller, G. A., Cell. Im-
     munol. 50:153, 1980.
58, Pelus, L. M., Borxmeyer, H. E., De Sousa, M., and Moore, M. A.
     S., J. Immunol. 126:1016, 1981.
59. Tice, D. G., Golberg, J., and Nelson, D. A., J. Reticuloendo-
     thelial Soc. 29:459, 1981.
60. Yasaka, T., Mantich, N. M., Boxer, L.A., and Baehner, R. L.,
     J. Immunol. 127:1515, 1981.

DISCUSSION

LIEBOLD: Was your P388D1 cell line free of mycoplasma?

WALKER: Yes.

LIEBOLD: How did you compensate for the size of your cells in cal-
culating phagocytic activity? Big cells will phagocytize more than
small ones.

WALKER: We don't. In fact, we said that cell size can play a
definite role and that the number of receptors on the cells increased
in G2. Obviously, there is an enhanced capacity for phagocytosis,
you would agree with that?

LIEBOLD: Not entirely. You showed that the Fc receptor activity is
more or less identical throughout the cell cycle.

WALKER: The affinity of binding is identical.

LIEBOLD: Did you try to separate the large cells from the smaller
ones in order to see if you could get a sub-clone with different
properties?

WALKER: No, we have not done that.

OLIVER: One of the critical questions concerns why mitotic cells
don't phagocytize. Does the mitotic cell express the Fc receptor?
Have you been able to subfractionate the G2+M cells, which are
obviously mostly G2, and obtain only the mitotic cells?

WALKER: Not yet.

KAPLAN: If you calculate the number of Fc receptors per unit area,
it seems to me that you probably have fewer receptors on the G2+M
cells, not more.

WALKER: If you actually calculate it, it does appear to go up, but
there are a lot of ruffles and foldings in the cell membrane. Therefore, we really can't calculate the value per surface area precisely.

RUMPOLD: Have you looked at the cytotoxicity of the P388D1 cells?

WALKER: This cell line does not carry out ADCC. However, it has an
IgG2b-Fc receptor.

LEONARD: Have you distinguished binding from ingestion? That would
be very interesting as related to Dr. Oliver's question.

WALKER: The only binding data we have is with the myeloma.

FÖRSTER: How long does your phagocytosis assay take?   Could the
cells proceed in cycle during that time?

WALKER: We do two types of phagocytosis assays.   One lasts 1 hour
and the other lasts about 20 minutes.   One is done by measuring the
ingestion of chromium labeled cells, which takes about 60 minutes,
and the other assay is done by counting the number of red cells
per phagocytic cell.   The data from both assays correlate.

MACROPHAGE HYBRIDOMAS: AN APPROACH TO THE

ANALYSIS OF THE FUNCTIONAL HETEROGENEITY OF MACROPHAGES

E. Tzehoval, S. Segal, N. Zinberg, B. Tartakovsky
and M. Feldman

Department of Cell Biology
The Weizmann Institute of Science
Rehovot, 76100 Israel

Macrophages are wandering phagocytic cells known to participate in a wide variety of immunological processes, including phagocytosis, antigen presentation and cytostasis or killing of tumor cells. Despite their seemingly common histological origin, they do not consist of a homogeneous population. Macrophages seem to differ in morphology, expression of membrane markers (H-2I, Fc and C3 receptors), enzymatic contents and biological activities. Our previous studies concerning control and immunogenic properties of distinct subpopulations of macrophages revealed the existence of two major subpopulations, only one of which was highly efficient in presenting antigen to specific T lymphocytes (1). The other subpopulation, while highly phagocytic, was devoid of antigen-presenting capacity (1). The heterogeneity of cell types, the inability of macrophages to grow in culture, particularly as cloned populations, and the limitations of the methods for separation of distinct macrophage subpopulations do not enable a precise and detailed analysis of the molecular mechanisms and controlling signals which regulate the biological activity (i.e., phagocytosis, chemotaxis, secretion of interleukin 1, antigen presentation) in correlation to membrane molecules which characterize defined subsets of macrophages.

We recently attempted to approach the study of the immunogenic function of macrophages, using the hybridization technology. We believe that "immortalization" of the various types of macrophages offers an experimental system which is most suitable for the analysis of macrophage properties and functions which were hitherto unapproachable.

GENERATION AND IDENTIFICATION OF MACROPHAGE-LIKE HYBRIDOMAS

Aiming at the immortalization of different functional macrophage-like cells, we used a C3H splenic population enriched for two main cell types: dendritic cells and macrophages. Such cells, prepared according to the procedure of Steinman et al. (2), were fused with the B myeloma cell line 4T00.1.L1 (3), following the PEG fusion protocol of Galfre et al. (4). The hybrids were selected by growth in HAT medium (5). Four to five weeks after fusion, small clones of cells appeared, differing in morphology. Hybrids which morphologically resembled the parental myeloma were discarded and only hybridomas exhibiting macrophage morphology (large ameboidal adherent cells with a ruffled plasma membrane and a cytoplasm rich in vacuoles and granules (Figure 1)) were further analyzed. Since morphological criteria alone do not ensure an accurate identification of cells with macrophage functions, it was essential to screen the emerging macrophage hybridomas by other rapid yet sensitive methods. Therefore, we analyzed the selected hybrids for the presence of two macrophage-specific enzymes, esterase (6) and lysozyme (7).

From the potential macrophage-like hybridomas which were found to stain for the presence of a macrophage intracytoplasmic nonspecific esterase and also to secrete lysozyme, we selected the most positive line, E2-7, for further investigations. To minimize overgrowth by contaminating hybrids and nonfunctional variants, we immediately

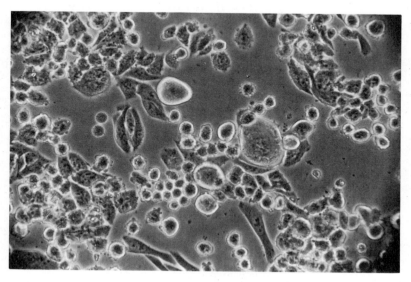

Fig. 1.   Phase microscopy of E2-7 macrophage-hybridoma cells.

cloned this cell line in soft agar. Various clones derived from the
E2-7 were further characterized.

CHARACTERIZATION OF THE CLONES OF THE E2-7 MACROPHAGE-HYBRIDOMA

Five clones derived from the original E2-7 cell line were subjected to an extensive analysis aimed at the identification of various
macrophage-like characteristics.

Molecular Surface Markers

Two sets of cell surface markers were used in differentiating
E2-7 clones from each other and in confirming their macrophage origin.
These were the Fc and C3 receptors and the lymphocyte H-2I associated
alloantigens. The presence of C3 and Fc receptors was detected by
rosette formation with opsonized erythrocytes, E(IgM) (8) and E(IgG)
(9), while the expression of $H-2I^k$ antigens was analyzed using indirect fluorescence serology. Table 1 indicates that heterogeneity
existed at the clonal level, as detected by differential expression
of membrane markers. Indeed, although all E2-7 clones expressed the
Fc receptor (at different levels), the C3 receptor was found to be
completely absent from the E2-7.8 and E2-7.33 clones. Furthermore,
the level of expression of $H-2I^k$-encoded antigens differed quantitatively. Hence, some of the E2-7 clones were found to express low
levels of $H-2I^k$, while others expressed high levels.

Functional Activities

Two major physiological roles of macrophages are phagocytosis
and generation of immunogenic signals (i.e., antigen presentation).
Hence, it was of obvious importance to test whether the different
macrophage-like hybridomas would manifest these functions. First,
we analyzed the different E2-7 clones for opsonin-dependent phagocytosis. Two series of experiments were conducted, using two particulate antigens of different sizes, fluorescein-labeled Micrococci
(10) and IgG-coated sheep erythrocytes (9). The results of two
typical experiments, shown in Table 2, indicated that different
clones may represent selected variants, some of them defective in
phagocytosis (E2-7.8 and E2-7.33).

Then we tested the immunogenic, namely the antigen-presenting,
capacity of the E2-7 macrophage-hybridoma clones. Cells of the different clones were fed KLH as antigen, treated with mitomycin C and
washed extensively. These cells served then as restimulators of
primed lymph node cells. The proliferative response to KLH was
determined by ^{3}H-thymidine incorporation. The results (Figure 2)
indicated that cells of high $H-2I^k$ clones (E2-7.7 and E2-7.21) were
also the most potent in antigen presentation. Cells of the E2-7.8,
E2-7.15 and E2-7.33, which were found to lack $H-2I^k$ alloantigens,
were also devoid of immunogenic capacity.

Table 1. Properties of Different Clones of the E2-7 Macrophage
          Hybridoma

| Clone • | Percent of Cells Possessing | | |
	Fc Receptor[a]	C3 Receptor[a]	H-2I^k Alloantigens[b]
E2-7 (parent)	59	8	53
E2-7.7	59	17	42
E2-7.8	65	0	12
E2-7.15	19	23	16
E2-7.21	47	24	51
E2-7.33	11	0	2

[a]Cells adherent to coverslips (48 h after mitomycin C treatment) were incubated with either (a) E(IgG) or (b) E(IgM) in
the presence of C_5-deficient fresh mouse serum for 30 min at
37°C, then examined for the number of rosette-forming cells.
[b]3×10^6 cells in 0.1 ml phosphate buffered saline were incubated for 30 min at 4°C with 20 μl of anti H-2I^k (ATH anti-
ATL). After washings, the cells were further incubated with
20 μl of FITC-conjugated rabbit anti-mouse immunoglobulin
and analyzed on the FACS II.

     To conclude, it seems that the active clones of our macrophage-
hybridomas were phagocytic cells. Only those which expressed H-2I
could present antigen to memory T cells.

SUMMARY

     We approached problems concerning the immunogenic function of
distinct macrophage subpopulations, using the hybridization technology. We succeeded in obtaining several different clones of macro-
phage-B myeloma hybridomas. These clones resembled macrophages
morphologically, adhered to plastic and possessed some macrophage
characteristics, such as esterase activity and high intracellular

Table 2.   Fc-mediated Phagocytosis by Different Clones of
           Macrophage-hybridomas

Clone Tested	Cells Phagocytizing Opsonized FITC-Micrococci[a] %	Cells Phagocytizing IgG Coated Erythrocytes[b] %
E2-7 (parent)	14	7
E2-7.7	39	14
E2-7.8	Not tested	0
E2-7.15	14	21
E2-7.21	20	24
E2-7.33	15	3

[a]2×10^6 cells suspended in 15% FCS-PBS were incubated for 30
   min at 37°C with 2.5×10^8 FITC-bacteria (previously opsonized
   with Micrococcus in the presence of fresh mouse serum) and
   analyzed on the FACS II for their fluorescence.
[b]Cells were treated with mitomycin C and adhered to coverslips.
   48 h later the cells were incubated in the presence of E(IgG)
   for 1 h at 37°C and evaluated for ingestion of at least 3
   erythrocytes per cell.

and secretory levels of lysozyme. Cloned macrophage hybrids representing different subsets were further characterized according to
presence of Fc and C3 receptors, H-2I encoded antigens, and Fc-mediated
phagocytosis of either IgG-coated SRBC or opsonized Micrococcus.
Furthermore, some of them manifested antigen-presenting capacity in
a secondary antigen-specific proliferation assay. The immunogenic
capacity seemed to be dependent on the presence of H-2I antigens.

Acknowledgement

      We thank Ms. Varda Segal for her excellent technical assistance.
This work was supported by a grant from the United States-Israel
Binational Science Foundation, Jerusalem (Grant #2645).

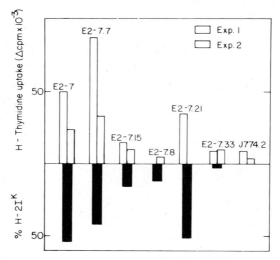

Fig. 2. Antigen presented on different macrophage-hybridomas induces
         specific proliferation of primed lymph node cells. C3H–Hen
         mice were immunized intra-footpad with 100 ug of KLH in
         complete Freund's adjuvant. Ten days later, 5×10^5 draining
         popliteal lymph node cells were stimulated with KLH-pulsed,
         mitomycin C-treated macrophage-hybridoma cells. The pro-
         liferative response to KLH was assessed by incorporation of
         ^{3}H-thymidine as expressed in cpm.

REFERENCES

1. Tzehoval, E., De Baetselier, P., Feldman, M., and Segal, S., Eur.
      J. Immunol. 11:323, 1981.
2. Steinman, R. M., and Cohn, Z. A., J. Exp. Med. 139:380, 1974.
3. Margulies, D. H., Kuehl, W. M., and Scharff, M. D., Cell 8:405,
      1976.
4. Galfre, G., Howe, S. C., and Milstein, C., Nature 266:550, 1977.
5. Littlefield, J. W., Science 145:709, 1964.
6. Koski, I. R., Poplack, D. G., and Blaese, R. M., in "In Vitro
      Methods in Cell-Mediated and Tumor Immunity" (B. R. Bloom and
      J. R. David, eds.), p. 359, Academic Press, New York, 1976.
7. Parry, R. M., Chandan, R. C., and Shanhani, K. M., Proc. Soc.
      Exp. Med. 119:384, 1965.
8. Bar-Shavit, Z., Raz, A., and Goldman, R., Eur. J. Immunol. 9:385,
      1979.
9. Diamond, B., Bloom, B. R., and Scharff, M. D., J. Immunol. 121:
      1329, 1978.
10. Vray, B., Hoebeke, J., Saint-Guillain, M., Leloup, R., and
      Strosberg, A. D., Scand. J. Immunol. 11:147, 1980.

DISCUSSION

GORDON: Can I ask whether lysozyme secretion segregated with the other markers or whether this was quite autonomous?

TZEHOVAL: Only cells which were esterase positive secreted lysozyme.

DISTINCTION OF MACROPHAGE SUBPOP-
ULATIONS: MEASUREMENT OF FUNCTIONAL
CELL PARAMETERS BY FLOW CYTOMETRY

Alexander Raffael and Günter Valet

Arbeitsgruppe Krebszellforschung
Max-Planck-Institut fuer Biochemie
D-8033 Martinsried F.R.G.

INTRODUCTION

Macrophages represent a functionally heterogeneous group of cells
which belong to the mononuclear phagocyte system (1). Heterogeneity
may exist between macrophages from different organs as well as among
macrophages within one organ (2,3). Heterogeneity has been defined
by differences of Ia-antigen expression (4,3); monoclonal antibodies
(5,6) against cell surface determinants; receptors for the C3 complement component or the Fc part of IgG molecules; cell size (7,3);
enzyme activities e.g. phosphatase, nucleotidase (8) peroxidase (9)
and transglutaminase (10); wheat germ lectin binding (11); tumor
cytoxicity (12); or phagocytosis (13) and adherence. Classification
according to several parameters is necessary to identify small subpopulations of macrophages (1). Flow cytometry is a particularly
useful method for this purpose, especially because functional parameters of living cells can be measured simultaneously at the single
cell level in a fast and accurate way. Such parameters include cytoplasmic (14,15) or lysosomal (16) enzyme activities, transmembrane
potential (17,18),intracellular pH (19) and phagocytosis. The use
of vital stains also permits cell sorting. Sorted cells can be re-
cultivated and further analyzed. Macrophages are often characterized
as cells with high esterase activity (20,21,22), although there are
some reports on low esterase activity in macrophages (23,24,25).
It was the purpose of this study to characterize the low activity
macrophages in more detail.

MATERIAL AND METHODS

Preparation and Incubation of Peritoneal Macrophages: Resident
peritoneal macrophages or paraffin oil induced macrophages were re-

moved from the peritoneal cavity under sterile conditions. Paraffin
oil induced macrophages were collected between 3 and 7 days after
injection of 20 ml sterile paraffin oil into the peritoneal cavity
of 600 to 900 g female Pirbright guinea pigs. The cells were eluted
with RPMI 1640 medium buffered with 20mM HEPES to pH 7.35, washed
once by centrifugation for 5 minutes at 200 x g and maintained in
the same medium. Two ml containing 1.5 to $1.8x10^6$ cells/ml were
incubated in 35x10 mm uncoated plastic petri dishes (Greiner, Solin-
gen, Germany) at 37°C in humified air. Non-adherent cells were removed
at various times by gentle shaking and aspiration of the supernatant.
Adherent cells were either mechanically scraped off or enzymatically
detached using 1% Dispase (Boehringer, Mannheim, Germany) for 8
minutes at 37°C. Cell viability was assessed by exclusion of trypan
blue (4.5 mg/ml), or exclusion of DNA stain mithramycin (100 µg/ml,
15 mM $MgCl_2$ Serva, Heidelberg, Germany). May-Grünwald-Giemsa stained
cytocentrifuge slides were prepared to determine the portion of lymphocytes and granulocytes.

        Cell Staining: The esterase activity and the intracellular
pH were both determined with the recently developed dye 1,4-diacetoxy-
2,3-dicyanobenzene (ADB), final concentration 25 µg/ml (19).   The
intracellular pH was determined from the ratio of fluorescent light
of intracellularly hydrolyzed ADB collected between 420 to 440 and
500 to 580 nm. In some experiments fluorescein-diacetate (FDA),
final concentration 4 µg/ml (15), was used for the esterase activity
measurement. Esterase inhibition studies were performed by incubating
the cells for 50 minutes at 4°C with 70 mM NaF (26). The phagocytic
activity of macrophages was quantified from the red fluorescence of
ingested monosized 1.2 µm latex particles.

        Flow Cytometry: Flow cytometric measurements were performed
at 25°C with a Fluvo-Metricell flow cytometer developed earlier in
our laboratory (27).   The electronic cell volume and two fluorescence
signals of each cell were measured simultaneously, amplified by 2.5
decade logarithmic amplifiers, and collected in list-mode on magnetic
tape. Data acquisition and display were accomplished by FORTRAN IV
computer programs (28).   A hydrodynamically focused orifice of 95 µm
diameter and 100 µm length with a current of 0.229 mA was used for
the electronic cell volume determination.

RESULTS

        The volume versus esterase activity (ADB as substrate) display of
six day paraffin stimulated peritoneal exudate cells immediately after
cell preparation (Figure 1A) showed four cell populations: Population
I to III consisted of small, middle and large macrophages (10%, 42%,
39%) with high esterase activity, while Population IV comprised 9%
of all cells and contained macrophages with esterase activity approximately 3% that of high activity cells. Their volume was between the

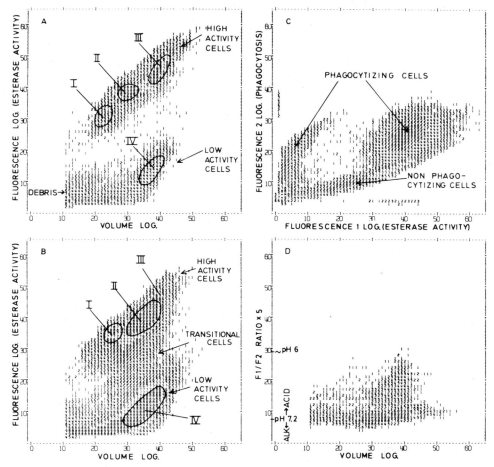

Fig. 1.   A- Esterase activity versus cell volume of g. pig peritoneal
           macrophages 6 days after paraffin oil injection.   Each scale
           comprises 2.5 log decades divided into ten equal steps and
           numbers between 1 and 9 assigned according to the number of
           cells in the channel.   Volume class 23 corresponds to 160
           μm^3.   Approximately 6 classes correspond to doubling of the
           cell volume or fluorescence signals.   Each histogram is
           standardized to the channel 'M'.   The histograms contain
           between 40 to 80,000 cells.   Four populations of macrophages
           (I-IV) are distinguishable.   B- Esterase activity of macro-
           phages after 24 h culture in uncoated plastic petri dishes.
           Transitional cells are apparent between the high and low ac-
           tivity cells.   C- Esterase activity versus phagocytosis of
           rhodamine stained latex particles after 24 h of culture.
           D- Intracellular pH versus cell volume of the cells of
           Fig. 1A.

values of population II and III. Similar esterase activity patterns
were obtained with FDA as substrate. The esterase activity was not
significantly reduced by 50 minutes incubation of the cells with
3 mg/ml NaF. The macrophage preparation contained 17% lymphocytes,
4% monocytes and 1% granulocytes. Comparative measurements of the
leukocytes from the peripheral blood of the guinea pig showed that
lymphocytes and granulocytes were superimposed upon macrophage population I and II.

The following experiments were performed with resident and
paraffin oil induced macrophages to determine whether macrophages
with low esterase activity were living or dying cells.

a. Adherence: The portion of cells with low esterase activity
during the first 25 hours of culture increased both in the non-
adherent and adherent cell fraction from 5 to 27% of all cells
(Figure 2A,B) although the viability (trypan blue exclusion) only
slightly declined from 95 to about 80% (Figure 2C). Mechanically
detached, adherent cells behaved similarly. A corresponding decrease
in the proportion of intermediate and small cells occurred at the
same time (Figure 2A-C). The transition of small and intermediate
cells from a higher to lower esterase activity is also seen in Figure
1B.

b. Viability: Adherent and non-adherent cells with low esterase
activity excluded trypan blue. Low esterase activity macrophages
also excluded the fluorescent DNA stain mithramycin which brightly
stained the nuclei of dead cells.

c. Phagocytosis: FDA esterase activity when plotted against
phagocytosis showed that approximately 50% of both the low and high
esterase activity cells were phagocytic (Figure 1C).

d. Intracellular pH: Figure 1A shows that a portion of intermediate and large cells had an acid intracellular pH (Figure 1D).
By thresholding the esterase data of Figure 1A, it became   apparent
that the high esterase activity cells had a pH around 7.2, whereas
the majority of the low esterase activity cells had an acidic pH
between 6 and 6.9.

DISCUSSION

A significant number (5 to 15%) of macrophages with low esterase
activity exist in guinea pig resident and paraffin induced peritoneal
macrophage populations. Similar results were obtained in the rat
(unpublished data). The low esterase activity cells are living cells
as assessed by adherence, phagocytosis, and trypan blue and mithramycin dye exclusion. They have a low intracellular pH, occur in the
non-adherent and adherent cell fractions and increase in number during

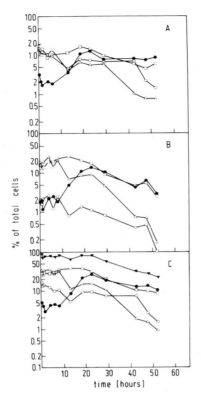

Fig. 2. Viability and esterase activity of non-adherent (A) and
         adherent (B) cells as a % of total cells in culture. Data
         are for paraffin oil induced g. pig peritoneal macrophages.
         Total cells (▼); high esterase activity macrophages: small
         population I (O), intermediate sized population II (△),
         and large sized population III (☐); low esterase activity
         population IV macrophages (●).

the first 24 hours of culture by recruitment mostly from interme-
diately sized cells. The esterases which cleave FDA and ADB are
located mainly in the cytoplasm of the cell (15). The localization
of these esterases is similar to the non specific esterases which
cleave alpha-naphthyl substrates.

   It is of interest that flow cytometry distinguished several
populations of high and low activity cells. Almost no cells with
intermediate activity were observed immediately after removing resi-

dent or paraffin oil induced macrophages from the peritoneal cavity (Figure 1A). The low activity cells were often slightly enriched in the adherent cell fraction and their number increased by 24 hours cultivation.

REFERENCES

1. van Furth, R., in "Mononuclear Phagocytes. Functional Aspects" (R. van Furth, ed.), pp. 1-30, Martinus Nijhoff Publishers, The Hague, 1980.
2. Hopper, K. E., Wood, P. R., and Nelson, D. S., Vox Sang. 36: 257, 1979.
3. Lee, K.-C., Molec. Cell. Biochem. 30:39, 1980.
4. Cowing, C., Schwartz, B. D., and Dickler, H. B., J. Immunol. 120: 378, 1978.
5. Springer, T. A., in "Monoclonal Antibodies. Hybridomas: A New Dimension in Biological Analyses" (R. H. Kennett, Th. J. McKearn and K. B. Bechtol, ed.), pp. 185-217, Plenum Press, New York, 1980.
6. Sun, D., and Lohmann-Matthes, M.-L., Eur. J. Immunol, 1982, in press.
7. Kwan, D., Epstein, M. B., and Norman, A., J. Histochem. Cytochem. 24:355, 1976.
8. Suga, M., Dannenberg, A. M. Jr., and Higuchi, S., Am. J. Pathol. 99:305, 1980.
9. Daems, W. Th., and van der Rhee, H. J., in "Mononuclear Phago- cytes. Functional Aspects" (R. van Furth, ed.), pp. 43-60, Martinus Nijhoff Publishers, The Hague, 1980.
10. Schroff, G., Neumann, Ch., and Sorg, C., Eur. J. Immunol. 11: 637, 1981.
11. Water, R. de, Noordende, J. M. van't, Ginsel, L. A., and Daems, W. Th., Histochem. 72:333, 1981.
12. Hopper, K. E., Harrison, J., and Nelson, D. S., J. Reticulo- endothelial Soc. 26:259, 1979.
13. Roubin, R., Kennard, J., Foley, D., and Zolla-Pazner, S., J. Reticuloendothelial Soc. 29:423, 1981.
14. Malin-Berdel, J., and Valet, G., Cytometry 1:222, 1980.
15. Rotman, B., and Papermaster, B. W., Proc. Natl. Acad. Sci. USA 55:134, 1966.
16. Tsou, K. C., Yip, K. F., and Miller, E. E., J. Histochem. Cytochem. 28:1032, 1980.
17. Shapiro, H. M., Natale, P. J., and Kamentsky, L. A., Proc. Natl. Acad. Sci. USA 76:5728, 1979.
18. Valet, G., Jenssen, H.-L., Krefft, M., and Ruhenstroth-Bauer, G., Blut. 42:379, 1981.
19. Valet, G., Raffael, A., Moroder, L., Wünsch, E., and Ruhenstroth- Bauer, G., Naturwiss. 68:265, 1981.
20. Yam, L. T., Li, C. Y., and Crosby, W. H., Am. J. Clin. Pathol. 55:283, 1971.

21.   van Furth, R., Raeburn, J. A., and van Zwet, Th. L., Blood 54:
        485, 1979.
22.   Bozdech, M. J., and Bainton, D. F., J. Exp. Med. 153:182, 1981.
23.   Kaplow, L. S., and Lerner, E., J. Histochem. Cytochem. 25:590,
        1977.
24.   van Furth, R., Diesselhoff-den Dulk, M. M. C., Raeburn, J. A.,
        van Zwet, Th. L., Crofton, R., and Blussé van Oud Alblas,
        A., in "Mononuclear Phagocytes. Functional Aspects" (R.
        van Furth, ed.), pp. 280-298, Martinus Nijhoff Publishers,
        The Hague, 1980.
25.   Raffael, A., and Valet, G., Immunobiol. 160:88, 1981.
26.   Fischer, R., and Schmalzl, F., Klin. Wochenschr. 42:751, 1964.
27.   Kachel, V., Glassner, E., Kordwig, E., and Ruhenstroth-Bauer,
        G., J. Histochem. Cytochem. 25:804, 1977.
28.   Benker, G., Kachel, V., and Velet, G., in "Flow Cytometry IV"
        (O. D. Laerum, T. Lindmo, E. Thorud, eds.), pp. 116-119,
        Universitetsforlaget, Oslo, 1980.

DISCUSSION

OLIVER: I'm concerned about a technical point. The fluorescein
diacetate substrate which you used to measure esterase produces
fluorescein whose emission is strictly pH dependent. Is it possible
that you have the same esterase activity in both populations but
different pH's which could account for the apparent difference in
activity?

RAFFAEL: You might be right in this case, but not with the substrate
ADP.

OLIVER: But that only measures pH, not esterase.

RAFFAEL: We measured simultaneously the esterase activity and the
intracellular pH.

VAN FURTH: I'm surprised that in culture your esterase activity goes
down. With guinea pig cells using a conventional assay method after
24 hours incubation, most of the cells are heavily stained for the
non-specific esterase. Do you have any explanation for this observation?

RAFFAEL: Perhaps it is dependent on the substrate you use to measure
the esterase activity. The non-specific esterase activity of macrophages is localized only on the membrane, but we are measuring intracellular enzyme activity.

FLOW CYTOMETRIC CHARACTERIZATION OF MACROPHAGES: FLUORESCENT SUBSTRATES AS MARKERS OF ACTIVATION AND DIFFERENTIATION

Stephen Haskill, Susanne Becker and Jouko Halme

Department of Obstetrics/Gynecology
and Cancer Research Center
University of North Carolina
Chapel Hill, North Carolina   27514 USA

INTRODUCTION

Biochemical criteria have been used to characterize blood monocytes as well as resident and elicited macrophages. While biochemical
assays determined on cell lysates or cell free supernatants provide
important information, similar studies carried out at the single cell
level could assess both activity and heterogeneity. If several assays
could be carried out simultaneously, the relationship between different markers on the same cell could be studied.

We report the successful development of a flow cytometric method
which uses various fluorescent substrates to quantify a number of
important macrophage features. By selecting the appropriate excitation wave length (365 nm) and a variety of reagents with distinct
emission spectra, we have been able to assess up to three distinct
markers at the same time.

MATERIALS AND METHODS

Instrumentation: We utilized a modified Ortho Instruments ICP-
22 flow cytometer equipped with a third photo multiplier tube and a
new electronic cell volume flow cell. Mercury arc illumination
through a UG-1 filter provided at least 10 mw of UV excitation.
Data from the four different sensors were processed through a combination of preselection (gating circuits) and computer storage.

Fluorescent Probes of Macrophage Differentiation and Activation:
4-methoxy-β-naphthylamide substrates are frequently utilized for bio-

461

chemical and histologic assessments of a variety of enzymes (1-3). Dolbeare and Smith have made an extensive list of these and other classes of substrates useful in the detection of proteinases, ester- ases, acid hydrolases and peroxidases (4). These reagents together with fluorescent coupling agents could be useful in flow cytometry.

Simultaneous determination of different markers requires that each probe can be excited by the same wave length but emit distinct fluorescent spectra. It is feasible with many of these markers to use UV excitation (<400 nm) and measure blue, green and red emissions. A summary of some commonly employed probes is given in Table 1.

Cells reacted with appropriate combinations of reagents emit characteristic spectra which can be selectively monitored with the

Table 1. Summary of Fluorescent Probes Useful in Phagocyte Analysis

Determination	Fluorescent Probe	Spectral Emission
Phagocytosis	Zymosan stained with isatoic anhydride	Blue
Acid phosphatase	Flavone diphosphate substrate	Green
Leucine aminopeptidase	4-methoxy-B-naphthylamide (MNA) coupled to NSA	Green
	Methyl coumarin derivative	Blue
Plasminogen activator	MNA-NSA coupling	Green
	Methyl coumarin derivative	Blue
Cathepsin B	MNA-NSA coupling	Green
	Methyl coumarin derivative	Blue
Myeloperoxidase	Homovanillic acid rhodamine red	Red
Ribonucleic acid Deoxyribonucleic acid	Propidium iodide	Red

aid of interference or high pass filters. The emission spectra derived from an example of one such combination, zymosan (blue), acid
phosphatase AP (green) and RNA/DNA (red) is given in Figure 1.

Heterogeneity of Human Peritoneal Macrophages

Cytochemical analysis suggests a marked difference in macrophages
between various normal human donors in contrast to normal mice.
Many human peritoneal fluid specimens contain mostly peroxidase positive macrophages; others had a spectrum of cells with weak to intense
acid phosphatase activity (5). Simultaneous phagocytic and leucine
aminopeptidase (LAP) determinations were carried out on adherent cell
preparations from three different mouse strains and several human
peritoneal fluid preparations.

The three mouse samples gave identical cytograms, (Figure 2),
indicating moderate levels of both markers. In contrast, (Figure 3),
one normal human (C) showed a wide spread of activity with a clear
relation between the two markers. Another donor (D), although having
strong LAP activity had little phagocytic activity. Normal blood
monocytes (A), while having moderate phagocytic activity were markedly
low in LAP. The peroxidase positive inflammatory macrophages (B)
derived from an ovarian cancer patient given intraperitoneal C. parvum
three days previously had markedly increased phagocytic activity
compared to LAP.

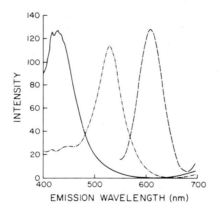

Fig. 1.   Fluorescence emission spectra for isatoic anhydride stained
           zymosan (-); flavone (-.-) and propidium iodide (--).   Day
           4 human blood monocyte cultures were stained individually
           for each marker.   Excitation 365nm.   Cells stained with all
           3 had the expected composite spectra.

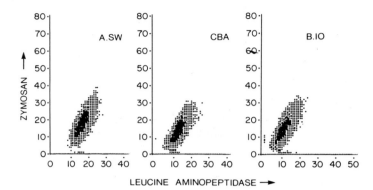

Fig. 2.   Leucine aminopeptidase and phagocytosis cytograms deter-
          mined on adherent resident peritoneal macrophages from 3
          mouse strains. The small and large dots represent thresh-
          olds of 1 and 10 cells; 5000 cells analyzed.

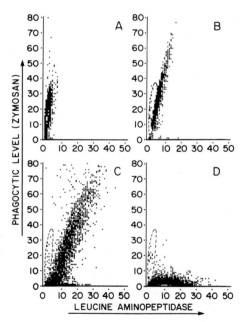

Fig. 3.   Leucine aminopeptidase and phagocytosis cytograms determined
          on adherent human cells: (A) monocytes; (B) peritoneal macro-
          phages, day 3 post ip C. parvum; (C and D) normal female per-
          itoneal macrophages. Instrument settings as in Figure 2.

Table 2. Effect of Various Doses of Fibroblast Interferon and
          Time of Administration on the Development of Macro-
          phage Markers in Human Monocyte Cultures

Interferon Added		Mean Fluorescence Intensity		
Time	Amount	Acid Phosphatase	Leucine Amino- peptidase	Zymosan
None	--	36.4	20.3	41.7
Day 0	100μ	29.8	14.4	29.0
	200μ	16.4	10.6	25.4
	300μ	10.3	7.2	22.7
None	--	50.1	27.3	31.0
Day 0	300μ	18.6	16.8	16.3
Day 0,2,3	3x100μ	20.4	15.9	15.2
Day 3,5	2x150μ	32.7	26.2	26.4
None	--	44.9	30.8	26.4
24h	100μ	30.4	21.7	22.2
	200μ	23.8	26.6	32.3
	400μ	20.4	22.5	33.3
Monocytes		1.5	4.3	10.1

Acid phosphatase was assayed at day 6.

     LAP and cell volume analysis of a heterogeneous peritoneal cell
preparation clearly indicated that increasing cell volume and LAP
activity were closely associated (Figure 4 top).   Comparison with
purified human blood lymphocytes, monocytes and granulocytes (overlay
of the neutrophil data) indicated that lymphocytes and a few granulocytes were present in the peritoneal sample (Figure 4 bottom) and
this conclusion was confirmed by cytologic analysis.

Interferon Mediated Inhibition of Monocyte Maturation

     Normal blood monocytes undergo a series of morphologic, bio-
chemical and functional changes during culture in human serum (6,7).
We quantified several of these changes by flow cytometry. Highly
purified monocytes (>95%) rapidly lost peroxidase activity and gained

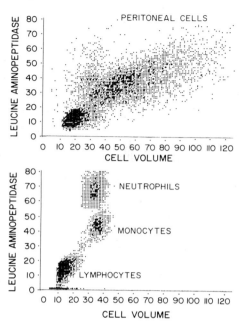

Fig. 4.   Leucine aminopeptidase (LAP) and cell volume cytograms. Top: normal human peritoneal cells (not purified); bottom: composite cytogram for neutrophils, monocytes, and lymphocytes. Instrument setting for LAP on blood cells was twice that of peritoneal cells.

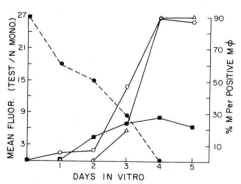

Fig. 5.   Development of leucine aminopeptidase, (O), acid phosphatase (Δ), and phagocytic activity (■) of human blood monocytes cultured in 10% AB serum. Weighted means of each computer derived histogram provided marker quantitation. Myeloperoxidase levels were assessed cytochemically.

LAP and AP activity. A 15-30 fold increase in these enzymes occurred around day 4 of culture. Phagocytic activity increased somewhat earlier than that of the enzymes. (Figure 5).

Interferon has been reported to inhibit differentiation of a number of cells, including monocytes (8). We were interested to know if interferon inhibition was specific for a particular macrophage function or if it was a general suppressive effect. We found that interferon dramatically inhibited development of AP and LAP activity and only modestly suppressed the level of phagocytic activity (Table 2). This inhibitory effect was dependent upon the dose of interferon added. The effect was maximal if monocytes were exposed to interferon at the time of culture establishment (day 0). Exposing the monocytes to interferon for only the first 24 hours still decreased AP and LAP activity at day 6.

CONCLUSIONS

The data presented indicate the use of fluorescent substrates in the flow cytometric analysis of macrophage heterogeneity and differentiation.

Acknowledgement

We acknowledge the assistances of Dr. Wolfgang Gohde in incorporating the third photo multiplier tube and Dr. Dominic Marro of Ortho Instruments for the electronic cell volume flow cell. This work was supported by an NIH Program Project Grant #CA 09156 and ACS Grant #IM 84 E.

REFERENCES

1. Gossrau, G., J. Histochem. Cytochem. 29:464, 1981.
2. Smith, R. E., and Dean, P. N., J. Histochem. Cytochem. 27:1499, 1979.
3. Dolbeare, R. A., and Vanderlann, M., J. Histochem. Cytochem. 27: 1493, 1979.
4. Dolbeare, F. A., and Smith, R. E., in "Flow Cytometry and Sorting" (M. R. Melamed, P. F. Mullaney, and M. L. Mendelsohn, eds.), Wiley Press, 1979.
5. Becker, S. E., Halme, J., and Haskill, J. S., Manuscript submitted for publication.
6. Johnson, W. D., Meiand, B., and Cohn, A. Z., J. Exp. Med. 146: 1613, 1977.
7. Musson, R. A., Shafran, H., and Henson, P. M., J. Reticuloendothelial Soc. 28:249, 1979.
8. Lee, S. H., and Epstein, L. B., Cellular Immunol. 50:177, 1980.

DISCUSSION

VAN FURTH: You considered your human peritoneal cells as resident
cells. The nature of such cells from females varies with the
menstrual cycle and further the cells have enzyme reactivities by
conventional methods similar to exudate induced macrophages. So, I
wondered if you could have correlated your data on phagocytosis and
enzyme content with the stage of the menstrual cycle?

HASKILL: In the first few days after the beginning of menstruation,
there is certainly a higher number of cells and they are more peroxidase positive. But by no means are they absent by day 20. There
is an increase during the first few days, but throughout the entire
cycle you see peroxidase positive cells. Unfortunately, we have no
male donors.

SCHNYDER: You claim to have a reaction for plasminogen activator,
but plasminogen activator is secreted by macrophages. How can you
determine released enzymes in your system?

HASKILL: Plasminogen activator is secreted, but in some circumstances, the macrophages very clearly reacted with the specific
substrate for plasminogen activator. Your question is that the enzyme
is secreted and therefore it should not be detected. But in fact,
with some macrophages, we seem to be able to do this. We will have
to do parallel studies to see if a correlation exists between surface
and supernatant activities. We do know that some macrophages contain
enough plasminogen activator in the cytoplasm for us to detect it.
It's quite possible that during secretion there's trapping at the
surface membrane whose efficiency may be very low, but sufficient
for detection.

GORDON: The real question is how specific your substrates are for
various proteolytic enzymes that are closely related, for example
thrombin or other trypsin-like enzymes.

HASKILL: I have not proven the specificity of our substrates but
other groups have who have been using them for a number of years now.

SECTION 7

MACROPHAGE REGULATION

EXTRINSIC REGULATION OF MACROPHAGE FUNCTION
BY LYMPHOKINES - EFFECT OF LYMPHOKINES ON THE
STIMULATED OXIDATIVE METABOLISM OF MACROPHAGES

Edgar Pick, Yael Bromberg and Maya Freund

Section of Immunology
Department of Human Microbiology
Sackler School of Medicine
Tel-Aviv University
Tel-Aviv 69978, Israel

INTRODUCTION

     Macrophage function is subject to multifactorial regulation,
the details of which are only partially understood. The main determinants are: the stage of cell differentiation (maturation), the
tissue localization, and the acute or chronic exposure to soluble
or cellular elements interacting with the macrophage membrane. This
last category is composed of a large number of unrelated agents and
includes immunoglobulins, complement components, clotting factors,
hormones and neurotransmitters, serum-derived lipoproteins and antiproteases, bacterial products and, of special relevance to this presentation, lymphokines. Macrophages are also subject to autoregulatory influences by such products as oxygen radicals, prostaglandins
(and other arachidonate metabolites), complement components and interferon.

     The subject of lymphokine-dependent regulation of macrophage
function is intimately linked to the concept of activation. This
term was introduced to define the complex of changes in macrophage
physiology resulting in an enhanced capacity to destroy intracellular
pathogens (reviewed in Ref. 1-4). Mostly through work by Mackaness and colleagues, it soon became apparent that, although the
bactericidal capacity of activated macrophages lacks immunological
specificity towards the target organism, the induction of the activated state is linked to the existence of a state of cell-mediated
immunity and is strictly dependent on the temporal proximity of specific antigenic challenge (5). Further work showed that T cell-
derived lymphokines represent the link between specific cell-mediated

immunity and the enhanced effector function of activated macrophages. A considerable number of in vitro models were established in which preformed lymphokines were found that activated resting macrophages, as demonstrated by an enhanced capacity to kill Listeria monocytogenes, Trypanosoma cruzi or Toxoplasma gondii. A parallel development was the extension of the concept of activation to cover the acquired capacity of such cells to damage tumor targets and other rapidly proliferating cells, most notably, T and B lymphocytes in the process of clonal expansion. There is also evidence for the ability of activated macrophages to damage larger multicellular parasites.

The rather schematic nature of the concept of macrophage activation has obscured a number of essential questions. Thus, we do not know whether enhanced killing of intracellular parasites is the property of the same cells which also exhibit tumor cytocidal or cytostatic properties and whether it is effected by the same mechanism. Secondly, we tend to overlook the existence of both lymphokine-dependent and lymphokine-independent mechanisms of activation. Thirdly, the relationship (and possible identity?) between several lymphokines is still controversial. This refers to the relative importance of migration inhibitory factor (MIF), macrophage activating factor (MAF), macrophage cytotoxicity factor (MCF), macrophage fusion factor (MFF), colony stimulating factor (CSF) and the interferons. The issue is further complicated by the difficulty of distinguishing in tissue culture short-term and long-term (differentiation-inducing) effects of lymphokines, due to interference by cell death and selective survival of cellular minorities and to the absence of cellular dynamics (chemotaxis, cell traffic) and hormonal influences in in vitro models. Additional interpretational difficulties are generated by the existence of a delicate interplay of non-lymphokine activating signals, such as lipopolysaccharide (LPS) with lymphokines in the intact animal and by the virtual elimination of macrophage-lymphocyte feedback circuits in vitro.

In spite of these impediments, experimental evidence has accumulated in support of a unifying theory of macrophage activation. The essence of this theory is that, whatever the nature of the inducing agent, activated macrophages share certain characteristic biochemical (enzymatic) markers which are causally related to the "hypercidal" qualities of these cells. In the past five years, support has accumulated for the proposal that the essential feature of activation is priming for an enhanced cyanide-resistant oxidative metabolism in response to certain forms of membrane stimulation (6-10). Such priming can be achieved by both lymphokines and non-lymphokine activating agents (10). The important conceptual advance contained in this proposal is that activation represents a cellular potential and not an ongoing active process. This latent potential becomes expressed only upon challenge, normally represented by bacteria, fungi, protozoa or other target cells. Although the cytocidal action of activated macrophages is essentially non-specific, its induction by T cell-derived lymphokines generates a topical and temporal link to the immunologically specific interaction between pathogen or its antigens and sensitized T lymphocytes. Thus, while MAF produced by lymphocytes encountering M. tuberculosis is capable of activating macrophages situated at a distance, it is obvious that activation is more likely to occur in the proximity of the mycobacteria, where specific T cells are being activated and lymphokine concentrations are maximal. We would like to suggest that the ability to develop a certain degree of activation by non-lymphokine agents such as bacterial products (LPS) may have had survival value in the course of evolution, especially in the presence of a less developed specific immune system. A quasi-symbiotic relationship between the immune system and the animals' own enteric bacteria, involving LPS as a "normal" differentiation signal for macrophages, was recently proposed (11). Later in evolution, macrophage activation probably developed into a more specific and restricted mechanism, heavily dependent on T cell function and lymphokine production. However, the less focussed forms of activation, such as those accompanying chronic infections with intracellular pathogens, may confer protection against another infectious or parasitic disease and, perhaps, against malignancy. An interesting example for such a situation is the enhanced killing of schistosomula by monocytes of tuberculous patients (12).

In this presentation we shall review briefly recent work from our laboratory supporting the contention that lymphokine action is associated with modulation of the cyanide-resistant oxidative metabolism of macrophages. Our basic thesis is that both early (migration inhibitory) and late (activating) effects of lymphokines are the result of the enhanced production by the macrophage of reactive oxygen radicals in response to membrane (receptor) stimulation, such as that provided by bacteria, parasites or tumor cells.

THE OXIDATIVE BURST OF MACROPHAGES: A MECHANISM OF ACTIVATION

For the past three years, we have studied the oxidative metabolism of guinea pig peritoneal macrophages elicited by the injection of sterile mineral oil. This particular type of macrophage was chosen for study because it is notoriously responsive to MIF and has been used extensively as a target cell for MAF in both the bactericidal and the tumor cell cytotoxicity models (reviewed in Ref. 13). We found that such macrophages produce abundant amounts of superoxide (O_2^-) and hydrogen peroxide upon stimulation by particulate agents, such as opsonized zymosan, and by a variety of soluble activators, such as phorbol myristate acetate (PMA), the lectins concanavalin A (ConA) and wheat germ agglutinin (WGA), several N-formyl-methionyl peptides, the Ca^{++} ionophore A23187, NaF, phospholipase C from Cl. welchii and microtubule disrupting drugs, such as colchicine (14). O_2^- and H_2O_2 production was elicited also by a number of unsaturated fatty acids (arachidonic, linolenic, linoleic and oleic)

(15). All these observations were made by using methods of assay
capable of detecting only O_2^- and H_2O_2 released into the culture medium
surrounding the cells. With the aid of a novel densitometric assay
for the reduction of nitroblue tetrazolium (NBT), it became apparent
that oil-elicited guinea pig macrophages exhibit intense intracellular NBT reduction in response to PMA, A23187, N-formyl-methionyl-
leucyl-phenylalanine (FMLP) and ConA (16). The relative intensities
of NBT reduction stimulated by various agents did not correspond to
those of extracellular O_2^- release, suggesting that the two tests
reflect distinct aspects of the oxidative burst. We also found that
oil-elicited macrophages stimulated by agents inducing an oxidative
burst became intensely cytolytic towards syngeneic erythrocytes (17).
The principal lytic agent in this system appeared to be H_2O_2.

    It is well established that the primary macrophage enzyme acti-
vated in the course of an oxidative burst is an O_2^- generating NADPH
oxidase, localized in the plasma membrane (18). A prerequisite for
elucidating how lymphokines facilitate the oxidative response is
understanding the mechanism by which the NADPH oxidase is activated.

    We were impressed by the multiple correlations between phospho-
lipid metabolism and the oxidative burst. (a) The majority of agents
eliciting an oxidative burst in macrophages also induce arachiodonic
acid release and prostaglandin or thromboxane production. (b) Several
unsaturated fatty acids activate the oxidative burst. (c) Exogenous
bacterial phospholipase C and interference with membrane architecture
by detergents (deoxycholate, digitonin) elicit an oxidative burst.
In the light of this, we blocked phospholipase A_2 action by drugs
and examined the ability of macrophages to produce O_2^- following membrane stimulation. We found that mepacrine (50 µM) reduced the oxidative response to all stimulants by 50% (15). These results were
difficult to interpret because mepacrine also affects phospholipase
C and its mechanism of action might involve the Ca^{++} -calmodulin
complex. Therefore, we examined the effect of p-bromophenacyl bromide, an inhibitor interacting directly with the phopholipase A_2
molecule. The drug (5 µM) totally eliminated O_2^- production in response to the lectins ConA and WGA and reduced by 50% the response
to all other stimulants, except PMA, which was not inhibited (15).
These results indicate the existence of multiple pathways for the
induction of an oxidative burst in macrophages and should warn us
against looking for a simplistic explanation for the enhancement
of the oxidative burst by lymphokines.

    Support for the existence of several ways of activating the O_2^-
producing enzyme in macrophages was also provided by the variable
effect of removing extracellular Ca^{++} (an ion required for the activity of both phospholipases A_2 and C) during O_2^- production. Only
responses to the ionophore A23187 and to exogenous phospholipase C
were significantly depressed by the absence of Ca^{++} in the medium;
the oxidative burst elicited by other stimulants was only partially

depressed and the PMA-elicited O_2^- production was totally independent of extracellular Ca^{++} (15). This suggested that induction of an oxidative burst by most stimulants makes use of intracellular Ca^{++} stores. Further proof for this proposal came from experiments utilizing the intracellular Ca^{++} antagonist, TMB-8 (15).

In many cell types the most common substrate for phospholipase A_2 is arachidonate-containing phosphatidylcholine, which can be synthesized by the S-adenosyl-L-methionine dependent methylation pathway (19). We hypothesized that phospholipid methylation which leads to the formation of phosphatidylcholine is a prerequisite for the elicitation of an oxidative burst. We found that blocking phospholipid methylation by preincubating macrophages with deazaadenosine combined with homocysteine thiolactone resulted in the complete inhibition of O_2^- production in response to stimulation with ConA and WGA and a partial inhibition of the response to the other stimulants (20). The PMA-elicited O_2^- production was particularly resistant to inhibition by deazaadenosine + homocysteine thiolactone. These data, together with those derived from experiments with the phospholipase A_2 inhibitor, p-bromophenacyl bromide, indicate that the lectin-stimulated oxidative burst is representative of one particular mechanism of activation which appears to be essentially different from that of PMA. We have recently proposed that phospholipids synthesized by the methylation pathway are localized in the cell membrane in close association with receptors for some stimulants (ConA, WGA), phospholipase A_2 and the O_2^- forming NADPH oxidase (20). We suggest that lymphokine-activated macrophages are characterized by a change in membrane phospholipid composition and/or topography. Thus, an increase in the overall amount and/or perireceptoral clustering of methylation derived phospholipids would result in the facilitated induction of an oxidative burst. We also suggest that non-lymphokine activating agents (LPS, muramyl depeptide (MDP)) share with lymphokines that capacity to bring about this alteration in membrane phospholipid composition or architecture. This proposal is in agreement with the findings that activated macrophages possess low ecto-5'-nucleotidase and high adenosine deaminase activities (21). These enzymatic characteristics would be likely to cause a decrease in the intracellular levels of adenosine and, consequently, of 5-adenosyl-L-homocysteine (which acts as an endogenous inhibitor of transmethylation) resulting in ehhanced phospholipid methylation.

In summary, the hypothesis presented here proposes that an essential component of lymphokine (and non-lymphokine) mediated macrophage activation is the enhancement of phospholipid methylation resulting in quantitative and topographical changes in membrane phospholipids. As a consequence, there is increased availability of substrate for phospholipase A_2 action and, indirectly, an enhanced responsiveness to stimulants of an oxidative burst acting via phospholipase activation. It is possible that the level of phospholipid methylation is modulated by the intracellular level of adenosine.

It has been recently reported that the action of a chemotactic lymphokine on guinea pig macrophages is prevented by inhibitors of methylation (22).

We next addressed the question of the nature of the product of
phospholipase activity which is responsible for activating the NADPH
oxidase. Phospholipase A_2 liberates arachidonic acid from phospholipids resulting in the formation of lysophosphoglyceride. Addition
of lysophosphatidylcholine to macrophages did not induce an oxidative
burst. It therefore seemed likely that arachidonic acid or one of
its metabolites acted as the intracellular messenger. Macrophages
were exposed to several oxidative burst stimulants in the presence
of the cyclooxygenase inhibitor indomethacin (up to 200 µM) or of
the cyclooxygenase and lypoxygenase blocker, 5,8,11,14 - eicosatetray-
noic acid (ETYA) (up to 100 µM). We found that O_2^- production in
response to all stimulants, with the exception of NaF, was totally
resistant to both indomethacin and ETYA (15). ETYA severely depressed
the NaF-elicited oxidative burst. An unexpected result of these
studies was that 200 µM indomethacin actually enhanced O_2^- production
to most stimulants, most notably to A23187, colchicine, ConA and WGA
(15). At this concentration, indomethacin prevents reacylation of
lysophosphoglycerides in macrophages which results in an elevation
of free fatty acid concentration(23). It therefore appears that
activation of the O_2^- producing enzyme is not mediated by an arachidonic acid oxidation product and that, on the contrary, the level
of free, unoxidized fatty acid might be the determining factor (15).
If this should be the case, the existence of multiple pathways leading to the elicitation of an oxidative burst becomes comprehensible.
Thus, fatty acid liberation from phospholipids can be the result
of both phospholipase A_2 and C action (in the case of phospholipase C,
the subsequent intervention of a diacylglycerol lipase is required).
Thus, it is possible that some stimulants, such as PMA, might activate
phospholipase C. Specific loss of phosphatidylinositol (the usual
substrate for phospholipase C) from the membrane following incubation of the cells with FMLP has been reported (22). Sequential
activation of phospholipases C and A_2 is also feasible, as has been
proposed to occur in platelets. The precise delineation of the molecular mechanisms underlying the activation of the O_2^- forming NADPH
oxidase in the macrophage and its modulation by preexposure to lymphokines is under active investigation.

While we favor the hypothesis that lymphokine treatment affects
the transmission of the activating signal (represented by a phospholipid breakdown product) from the membrane receptor to the NADPH
oxidase, a number of alternatives should be considered. These are:
(a) an increase in the amount of NADPH oxidase; (b) a change in the
characteristics of the enzyme (increase in Vmax or decrease in Km);
(c) an increase in the number, degree of exposure or mobility of
membrane receptors to oxidative burst stimulants; (d) an effect on
"detoxifying" enzymes (superoxide dismutase, catalase or glutathione

peroxidase and reductase); and (e) an effect on transport mechanisms responsible for the extracellular release of O_2^-.

EVIDENCE FOR MODULATION OF THE OXIDATIVE METABOLISM OF MACROPHAGES BY LYMPHOKINES

We recently proposed that lymphokine action on macrophages involves short term effects which are essentially autotoxic and later effects which are manifested by an augmented cytocidal potential on a background of enhanced detoxifying mechanisms that assure maintenance of cellular function in the presence of toxic oxygen-derived radicals (24,25). Short term effects are due either to a direct stimulation of an oxidative burst by the lymphokine or to the facilitation of such by trivial stimuli, such as the surfaces to which macrophages are adhering. Toxic products of the oxidative burst which have escaped the scavenging mechanism inflict reversible damage to yet unidentified cellular elements, principally those involved in motility. We have suggested that lymphokine-induced migration inhibition is a manifestation of reversible autotoxic damage by self-generated oxygen radicals and have examined the effect of antioxidants on the action of MIF. We found that lymphokine-induced inhibition of migration could be prevented by the hydroxyl radical and singlet oxygen scavengers L-methionine and L-histidine and by propyl gallate (25). Exogenous superoxide dismutase and catalase did not prevent MIF action. Migration inhibition was also unaffected by mepacrine, indomethacin and ETYA. However, MIF action was prevented by the thromboxane synthetase inhibitors, imidazole, benzyl-imidazole and U-51605 and by the thromboxane agonist U-44069 (25). We interpreted these findings as indicating that an oxygen-derived radical, the identity of which remains to be determined, is involved in the cessation of movement elicited by MIF. It also appears that thromboxane synthesis by macrophages is a component of MIF action, possibly by promoting cell-surface and cell-cell adhesiveness. The fact that indomethacin, which inhibits the formation of all cyclooxygenase-derived metabolites including thromboxane, did not prevent MIF action suggests that macrophage motility is regulated by the thromboxane/E or I type prostaglandin ratio rather than by the absolute level of thromboxane. Preliminary results indicate that both MIF-containing and control lymphocyte culture supernatants stimulate thromboxane synthesis by guinea pig peritoneal macrophages 3 to 4-fold but supernatants with MIF activity contain more of the stimulatory material (Y. Bromberg, unpublished). Therefore, it is opportune to investigate the effect of MIF on the ratio of thromboxane to prostaglandins E_2 and I_2.

Lymphokine-induced migration inhibition is a reversible process; the cells inevitably escape MIF action and, on prolonged incubation with lymphokine, exhibit a different set of functional and biochemical characteristics. These include increased adherence, enhanced glu-

cose-1 C oxidation, low-5' – nucleotidase activity, augmented bacteriostatic and cytocidal potential (reviewed in Ref. 13) and increased
O_2^- and H_2O_2 production and NBT reduction in response to membrane
stimulation (8,9). The capacity to stimulate oxidative metabolism
is not unique to lymphokines since similar effects are obtained by
incubating of macrophages with LPS (26), MDP (27) and some proteolytic enzymes (28). A similar situation has been described in polymorphonuclear leukocytes; preincubation of the cells with formylated
peptides or complement-related chemotactic factors resulted in enhanced O_2^- production and chemiluminescence in response to such membrane stimulants as PMA, ConA, zymosan or NaF (29-31). This was
accompanied by enhanced bactericidal capacity (30,32).

We have proposed that the passage of lymphokine-treated cells
from the early to the later stage of activation is accompanied by
enhanced antioxidant defense (24,25). Indeed, lymphokine-treated
murine macrophages exhibit elevated levels of superoxide dismutase,
glutathione peroxidase and catalase (33). We have undertaken recently
a study of the effect of lymphokines on the glutathione based antioxidant protection in oil-elicited guinea pig macrophages (34, M.
Freund and E. Pick, in preparation). We first looked for evidence
of autotoxic damage in lymphokine-treated macrophages by assessing
the degree of cellular lipid peroxidation as assayed by the production
of malonyl dialdehyde (MDA). Incubation of macrophages with lymphokine-containing culture supernatants did not increase MDA production;
control preparations consisting of cells exposed to high concentrations of PMA or A23187 for up to 24 h demonstrated only moderate
(1.5-2.5-fold) increases in the level of MDA (34). If oxygen radicals
are produced in the course of lymphokine action, such data indicates
that antioxidant buffering reactions are sufficient to prevent oxidative damage, as manifested by lipid peroxidation. We therefore looked
for changes in the ratio of oxidized/reduced glutathione in lymphokine-treated macrophages in macrophages incubated with stimulants of
the oxidative burst and in cells pretreated with lymphokine and subsequently exposed to oxidative burst stimulants. As apparent in Table
1, brief incubation of macrophages with opsonized zymosan and PMA
resulted in a massive increase in the relative amount of oxidized
glutathione with no significant change in the total glutathione content. Incubation with lymphokine-containing culture supernatants
for 2 or 24 h had no effect on the level of oxidized glutathione
(Table 2). Preincubation with lymphokine for 2 h caused a moderate
enhancement in the accumulation of oxidized glutathione which followed the induction of an oxidative burst by exposure to opsonized
zymosan for 10 min. However, preincubation with lymphokine for 24 h
resulted in a markedly diminished accumulation of oxidized glutathione
in zymosan-stimulated macrophages (Table 2). Control culture supernatants also exhibited lymphokine-like activity, albeit less pronounced. We concluded that prolonged (but not brief) exposure to
lymphokine boosted the antioxidant defense of macrophages, as evident
from a "sparing" effect on the level of reduced glutathione. The

Table 1. Effect of Stimulants of the Oxidative Burst
on Glutathione Metabolism in Macrophages

Stimulant	Concentration	No. of expts	Total glutathione (GSS + GSSG)nmol/mg macrophage protein ± SEM	Oxidized glutathione (GSSG) % of total ± SEM
None			55.4 + 6.7	2.1 + 0.5
		5		
Zymosan	1 mg/ml		43.7 + 3.4	16.9 + 3.8
None			43.6 + 9.7	1.0 + 0.2
		2		
PMA	0.2 µM		59.4 + 2.9	23.9 + 5.5
None			47.1 + 3.9	1.6 + 0.3
		6		
PMA	2 µM		49.1 + 2.5	16.0 + 3.3
None			56.4 + 9.2	1.3 + 0.0
		2		
Con A	50 µg/ml		39.9 + 8.6	6.5 + 1.4

Duration of incubation with zymosan and PMA was 10 minutes
whereas for Con A it was 60 minutes.

mechanism of this effect was investigated by assaying the activities
of glutathione peroxidase and reductase in lymphokine-treated cells.
No differences in the activities of the two enzymes were seen between
macrophages incubated for 24 h to 72 h with lymphokine or with control
culture supernatants (Table 3). Therefore, we examined the enzyme
activities in macrophages pretreated with lymphokine and challenged
with the potent oxidative burst eliciter, PMA. It became apparent
that in the course of an oxidative burst, such as that elicited by
PMA, the activities of both glutathione peroxidase and glutathione
reductase were severely depressed (Table 4). A similar depression of
glutathione peroxidase and catalase was described in guinea pig alveolar macrophages during hyperoxia (35). Attempts to demonstrate
a protective effect of lymphokine pretreatment on glutathione cycle

Table 2.   Effect of Lymphokine on Glutathione Metabolism of Macrophages in the
Presence and Absence of Oxidative Stress

Time	Stimulant	Total glutathione (GSH + GSSG) nmol/mg macrophage protein ± SEM			Oxidized glutathione (GSSG) % of total ± SEM		
		Medium	Control Sup.	Lymphokine Sup.	Medium	Control Sup.	Lymphokine Sup.
2 h	none	48.6+7.3	58.1+3.2	59.9+2.8	2.2+0.7	1.6+0.7	2.2+1.1
	zymosan	39.0+1.6	51.8+7.6	48.7+5.5	17.8+6.0	19.4+3.9	22.1+1.9
24 h	none	48.6+7.3	n.d.	44.7+3.6	2.2+0.7	n.d.	1.2+0.4
	zymosan	39.0+1.6	38.0+4.4	40.2+4.8	17.8+6.0	9.8+1.9	7.2+0.9

Macrophages were incubated with lymphokines for 2 or 24 h. The cells were then exposed to
zymosan (1 mg/ml) for 10 min and glutathione levels determined. Data are for 3 experiments.
n.d.= not done

Table 3. Effect of Lymphokine on the Activities of Glutathione Peroxidase and Reductase in Guinea Pig Macrophages

Time	Glutathione Peroxidase		Glutathione Reductase	
	Control Sup.	Lymphokine Sup.	Control Sup.	Lymphokine Sup.
24 h	78.3 ± 10.7	73.8 ± 7.9	50.6 ± 3.9	51.1 ± 6.5
48 h	58.6 ± 2.7	57.6 ± 4.5	73.2 ± 11.3	63.8 ± 10.1
72 h	50.3 ± 5.8	65.2 ± 3.3	59.7 ± 7.1	69.3 ± 1.8

Data are given in nanomoles NADPH oxidized per ng macrophage protein per minute ± SEM. Number of experiments was at 24 h for glutathione peroxidase 4 and glutathione reductase 2; at 48 and 72 hours 3 in each category. Macrophages were incubated with lymphokine for 24 h, 48 h or 72 h.

Table 4. Effect of the Oxidative Burst on the Activities of Glutathione
Peroxidase and Reductase in Macrophages

Stimulus	Glutathione Peroxidase		Glutathione Reductase	
	Activity	Control	Activity	Control
none	95.2 + 35.8	100	79.8 + 11.1	100
PMA (200 nM)	35.0 + 16.2	37.1 + 12.5	41.9 + 5.5	52.5 + 0.4

Activity is reported in nmoles NADPH oxidized per mg macrophage protein per minute + SEM. Number of experiments was 3 for glutathione peroxidase and 2 for glutathione reductase. PMA = phorbol myristate acetate.

enzymes subject to oxidative stress failed. The effect of lymphokine treatment on the activities of catalase and superoxide dismutase is under investigation.

We conclude with a brief discussion of the methodology utilized by many laboratories to elucidate the mechanism of the lymphokine mediated oxidative burst. Basically, two approaches have been used. First, macrophages (resident or elicited) are pretreated with lymphokines (prepared by activation of lymphocytes by specific antigens or mitogens), and the oxidative burst is elicited by "trivial" stimulants, such as PMA or zymosan, or more specific stimulants, such as Toxoplasma (8,9). The intensity of the oxidative burst is measured by assaying the extracellular release of O_2^- and/or H_2O_2. A variation of this approach is to use a macrophage cell line that has a low potential for oxidative burst induction but which is expected to be augmented by lymphokine treatment (36). In the second approach, macrophages preincubated with lymphokines are examined for their capacity to reduce intracellular NBT spontaneously or following exposure to oxidative burst inducer (trivial or specific) (9, 37-39). In the light of the accumulating evidence for the existance of multiple pathways of oxidative burst induction and the limitations of all assays for quantitating the oxidative burst, we suggest the following strategy for future work. (a) Both extracellular release of O_2^- and H_2O_2 and NBT reduction by lymphokine-treated macrophages should be assayed. This is required because NBT reduction frequently does not correlate with O_2^- release in response to the same stimulant (16); (b) The responsiveness of lymphokine-treated cells to as many oxidative burst eliciting agents as possible should be examined. This should include both soluble and particulate stimulants and agents interacting with different types of membrane receptors (Fc of immunoglobulins, C3b, lectins, formylated peptides, etc.). Agents exhibiting different Ca^{++} requirements and inhibition by phospholipase or methylation blockers should be regarded as representing distinct biochemical activation pathways; (c) The activity of antioxidant enzymes should be assessed. This should involve assaying superoxide dismutase, catalase, glutathione oxidase and reductase; (d) Finally, in the light of the clear link between the oxidative burst and phospholipid metabolism, the effect of lymphokines on phospholipid turnover and the activity of associated enzymes deserves close scrutiny.

Acknowledgement

This research was supported by a grant from the National Council for Research and Development, Israel and the Deutsches Krebsforschungszentrum, Heidelberg, Germany. We thank Mrs. Patricia Bar-On for excellent secretarial assistance.

REFERENCES

1. North, R. J., J. Immunol. 121:806, 1978.
2. Cohn, Z. A., J. Immunol. 121:813, 1978.
3. Ögmunsdottir, H. M., and Weir, D. M., Clin. Exp. Immunol. 40: 223, 1980.
4. North, R. J., Lymphokines 3:1, 1981.
5. Mackaness,G. B., J. Exp. Med. 120:105, 1964.
6. Nathan, C. F., and Root, R. K., J. Exp. Med. 146:1648, 1977.
7. Johnston, R. B., Godzik, C. A., and Cohn, Z. A., J. Exp Med. 148:115, 1978.
8. Nathan, C. F., Nogueira, N., Juangbhanich, C., Ellis, J., and Cohn, Z. A., J. Exp. Med. 149:1056, 1979.
9. Murray, H. W., and Cohn, Z. A., J. Exp. Med. 152:1596, 1980.
10. Johnston, R. B., Lymphokines 3:33, 1981.
11. Vogel, S. N. and Rosenstreich, D. L., Lymphokines 3:149, 1981.
12. Olds, R. G., Ellner, J. J., El Kholy, A., and Mahmoud, A. A. F., J. Immunol. 127:1538, 1981.
13. David, J. R., and Remold, H. G., in "The Immunobiology of the Macrophage" (D. S. Nelson, ed.), pp. 401-426, Academic Press, New York, 1976.
14. Pick, E., and Keisari, Y., Cell. Immunol. 59:301, 1981.
15. Bromberg, Y., and Pick, E., submitted for publication.
16. Pick, E., Charon, J., Mizel, D., J. Reticuloendothel. Soc. 30: 581, 1981.
17. Keisari, Y., and Pick, E., Cell. Immunol. 62:172, 1981.
18. Bellavite, P.,Berton, G., Dri, P., and Soranzo, M. R., J.Reticulo-endothel. Soc. 29:47, 1981.
19. Hirata, F., and Axelrod, J., Science 209:1082, 1980.
20. Pick, E., and Mizel, D., submitted for publication.
21. Soberman, R. J., and Karnovsky, M. L., J. Exp. Med. 152:241, 1980.
22. Pike, M. C., and Snyderman, R., J. Immunol. 127:1444, 1981.
23. Kröner, E. E., Peskar, B. A., Fischer, H., and Ferber, E., J. Biol. Chem. 256:3690, 1981.
24. Pick, E., Keisari, Y., and Bromberg, Y., in "Advances in Aller-gology and Applied Immunology" (A. Oehling, ed.), pp. 399-407, Pergamon Press, London, 1980.
25. Pick, E., Keisari, Y., Jakubowski, A., Bromberg, Y., and Freund, M., in "Lymphokines and Thymic Hormones" (A. L. Goldstein and M. A. Chirigos, eds.), pp. 177-185, Raven Press, New York, 1981.
26. Pabst, M. J., and Johnston, R. B., J. Exp. Med. 151:101, 1980.
27. Cummings, N. P., Pabst, M. J., and Johnston, R. B., J. Exp. Med. 152:1659, 1980.
28. Johnston, R. B., Chadwick, D. A., and Cohn, Z. A., J. Exp. Med. 153:1678, 1981.
29. Allred, C. D., and Hill, H. R., Infect. Immun. 19:833, 1978.
30. Van Epps, D. E., and Garcia, M. L., J. Clin. Invest. 66:167, 1980.
31. English, D., Roloff, J. S., and Lukens, J. N., Blood 58:129, 1981.
32. Issekutz, A. C., Lee, K. Y., and Biggar, W. D., Infect. Immun. 24:295, 1979.

33. Murray, H. W., Nathan, C. F., and Cohn, Z. A., J. Exp. Med. 152: 1610, 1980.
34. Pick, E., Keisari, Y., Bromberg, Y., Freund, M., and Jakubowski, A., in "Heterogeneity of Mononuclear Phagocytes" (O. Förster and M. Landy, eds.), pp. 331-336, Academic Press, New York, 1981.
35. Rister, M., and Baehner, R. L., J. Clin. Invest. 58:1174, 1976.
36. Murray, H. W., J. Exp. Med. 153:1690, 1981.
37. Ando, M., Suga, M., Shima, K., Takenaka, S., and Tokuomi, H., Kumamoto Med. J. 29:88, 1976.
38. Krueger, G. G., Ogden, B. E., and Weston, W. L., Clin. Exp. Immunol. 23:517, 1976.
39. Alföldy, P., and Lemmel, E. M., Clin. Immunol. Immunopathol. 12: 263, 1979.

DISCUSSION

ADAMS: I have a mild demurrer with your conclusion that hydrogen peroxide production is a good marker of activation for tumoricidal activity. The reasons are as follows. First, cells from certain inbred strains of mice that do not kill tumor cells make copious amounts of peroxide when triggered with phorbol myristate acetate. Second, if you take BCG elicited macrophages from normal strains and manipulate them in culture in ways that either promote or inhibit tumoricidal activity, there is no correlation with peroxide production. Depending upon how the system is manipulated, one can find loss of tumoricidal ability and maintenance of peroxide production or the reciprocal effect. Third, when we have examined the in vitro pathway by which these cells are activated for tumoricidal killing, we don't see a correlation with peroxide production. Specifically, lymphokine treatment which effectively leads to a tumoricidal potential does not do so via peroxide production. Now, I think the question of ADCC-type mechanisms is different. I'm merely saying that I don't believe that peroxide production always marks the tumoricidal macrophage.

PICK: I agree entirely. There is no good evidence that lymphokine induced activation is associated with tumor cell killing that is mediated by peroxide. We tend to speak of macrophage activation as one process, whereas it is likely to be several processes. The mechanism which leads to killing of Lysteria monocytogenese probably won't kill a tumor cell. Even if that were the case, and you accept the phospholipase Stage I versus Stage II hypothesis, there are a lot of things which can happen during phospholipase activation which are not necessarily reflected in hydrogen peroxide production. These steps could be involved in tumor cell killing. In my opinion, it would be worth looking at phospholipase even in those cases where you have no correlation.

ACTIVATION OF TUMORICIDAL AND/OR SUPPRESSOR MACROPHAGES: DIFFERENT STIMULATORY SIGNALS TRIGGER EITHER FUNCTION BOTH IN VIVO AND IN VITRO

D. Taramelli, M. B. Bagley, H. T. Holden and
L. Varesio

Laboratory of Immunodiagnosis
National Cancer Institute
Bethesda, Maryland USA

INTRODUCTION

Macrophages upon in vivo and in vitro stimulation become activated and manifest functional and morphological characteristics either not present or differently expressed in non-stimulated macrophage populations (1-4). Antithetical functions can be exerted by a given macrophage population. For instance, cytotoxic macrophages could contribute to the elimination of infectious agents or to the control of neoplastic growth. On the other hand, by inhibiting certain immune responses, suppressor macrophages could impair host defense mechanisms.

Various signals may trigger cytotoxic and/or suppressor macrophages. In vitro studies have indicated that lymphokine as well as other substances such as interferon (IFN), lipopolysaccharide (LPS) or double-stranded RNA can trigger macrophages to become tumoricidal (5-6). It is not clear, however, whether the same array of signals also generates suppressor macrophages. Therefore, we decided to investigate the role of lymphokine, IFN, LPS and polyinosinic-polycytidilic acid (poly I:C) on the in vitro generation of tumoricidal and suppressor macrophages.

RESULTS AND DISCUSSION

Peritoneal exudate cells were elicited in C57BL/6 (B6) mice by injection of proteose-peptone 48 hours earlier. Macrophages purified by adherence to plastic were treated for 18 hours with different stimulatory agents. At the end of the incubation, their cytotoxic activity was tested in an 18 hour ^{51}Cr release assay as previously described (7). The results of a representative experiment are shown in Table 1.

Strong cytotoxic activity was obtained by treating macrophage monolayers with lymphokine(produced by Concanavalin A (Con A stimula-tion of normal spleen cells) LPS, IFN (partially purified fibroblast IFN) or poly I:C. Macrophages treated with medium or control super-natants were not tumoricidal. Moreover, the response of macrophages to lymphokine was dependent upon the presence of trace amounts of endotoxin during the activation. When serum and reagents screened to be endotoxin-free were used (negative in the Limulus Amoebocyte Lysate Assay), no cytotoxic activity by lymphokine treated macro-phages was observed (Table 1). The addition of 10 ng/ml LPS during stimulation completely restored the response. In contrast, the acti-vation of tumoricidal macrophages by IFN or poly I:C did not require the presence of LPS (Table 1). Thus, it appears that different sig-

Table 1. Tumor Cytotoxicity of Proteose Peptone Elicited Macrophages from B6 Mice Activated In Vitro with Different Agents

Activating Agent	Dose	%^{51}Cr Release ± SE	
		With LPS	Without LPS
Medium		2.1±0.4	3.2±0.9
Lymphokine	1:3	34.1±2.4	2.6±0.9
LPS	100 µg/ml		31.5±1.3
Interferon	10^4 U/ml	40.5±3.1	37.4±2.4
Poly I:C	100 µg/ml	36.2±2.4	32.6±1.8

Cytotoxicity was measured using RL♂1 tumor targets at an attacker to target cell ratio of 36:1. Lymphokine was produced by Con A stimulation of B6 normal spleen cells. The cytotoxicity tests were performed in the presence or absence of lipopolysaccharide (LPS) addition to the reagents at a final concentration of 10 ng/ml.

nals can trigger the expression of high levels of macrophage tumoricidal activity in vitro.

We addressed next the question of whether or not macrophages
activated in vitro to a cytotoxic stage could suppress the production
of lymphokine by mitogen-stimulated lymphocytes. Different numbers
of macrophages were stimulated in vitro and added to Con A-pulsed
normal spleen cells at the time of the mitogenic stimulation. After
24 hours, the supernatants were recovered and tested for migration
inhibitory factor (MIF) and macrophage activating factor (MAF).
MIF activity was assayed using a microdroplet assay technique as
previously described (8) and MAF activity was tested in the 18 hour
^{51}Cr release assay. When lymphokine treated macrophages were added
to the lymphokine producing system, strong suppression of both MIF
and MAF production was observed (Figure 1). The addition of the same
number of medium-treated macrophages or macrophages incubated with
control supernatants did not have any effect. The suppression was
mediated by adherent, Thy 1.2 negative, peroxidase positive cells,
with macrophage-like morphology. Furthermore, suppressor macrophages
generated by different lymphokine preparations could inhibit various

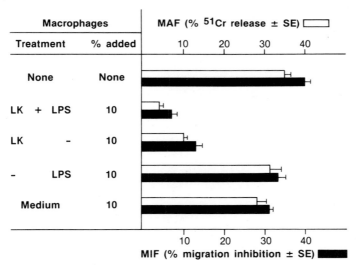

Fig. 1. Suppression of MIF and MAF production by proteose peptone
elicited adherent macrophages from B6 mice treated with lymphokine (1:3 dilution) with or without lipopolysaccharide
(LPS) 10 ng/ml. The lymphokine system consisted of B6 normal spleen cells activated with Con A (2.5 μg/ml) in the
presence of suppressor macrophages. Medium consisted of RPMI
1640 with 10% endotoxin-free, fetal bovine serum.

lymphokine-producing systems (9). Therefore, the in vitro exposure of macrophages to lymphokine caused them to suppress the production of lymphokine by activated lymphocytes.

We observed that macrophages incubated with lymphokine in the absence of detectable amounts of endotoxins (Figure 1) were suppressive although the same treatment was not sufficient to induce cytotoxic macrophages. Therefore, it is possible to produce suppressive macrophages in vitro which lack cytotoxic activity.

To test whether stimuli other than lymphokines were able to induce suppressor macrophages, peptone elicited adherent macrophages were treated for 18 hours with doses of IFN (5×10^3U/ml), poly I:C (100 µg/ml) or LPS (100 µg/ml) optimal for the induction of cytotoxic cells. The addition of IFN, LPS or poly I:C-treated macrophages did not reduce the amount of MAF produced by Con A-stimulated spleen cells, while lymphokine activated macrophages were strongly inhibitory. We concluded that macrophages activated by IFN, poly I:C or high doses of LPS did not acquire suppressor activity for lymphokine production despite their cytotoxic activity. This indicates that different activities can be generated in vitro within the same macrophage population depending on the stimuli used.

Further support for this conclusion came from in vivo experiments in which LPS (20 µg/mouse) or poly I:C (100 µg/mouse) were inoculated ip in B6 mice. After 24 hours, peritoneal macrophages were tested for tumoricidal and suppressive functions. As shown in Figure 2A, in vivo activated macrophages were highly cytotoxic at all attacker to target ratios tested. However, they were unable to inhibit the production of MAF by Con A-stimulated spleen cells (Figure 2B).

Thus, it appears that under the conditions used here poly I:C and LPS can trigger only cytotoxic and not suppressor macrophages both in vivo and in vitro.

In conclusion, we showed that suppressive and cytotoxic functions can be manipulated in vitro by changing the types of stimulating agents. We could preferentially induce suppressor macrophages by treatment with lymphokine in an LPS-free environment. We speculate that the induction of immunosuppressive functions is a secondary event, dependent on the production of lymphokine by activated lymphocytes. The triggering of suppressor macrophages could represent a mechanism through which the expansion of the immune response is internally regulated. It remains to be determined whether the same macrophages could be at the same time cytotoxic and suppressive, or whether different subpopulations of macrophages exist, each with separate functions. In addition, the signals (poly I:C, IFN and LPS) appeared to directly stimulate macrophages to develop tumoricidal activity.

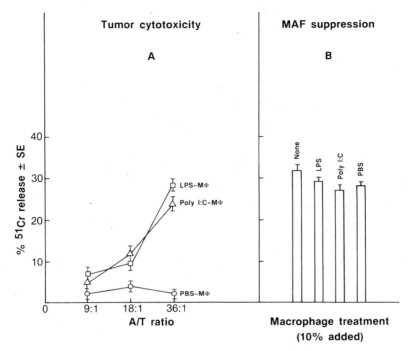

Fig. 2.   Tumor cytotoxicity and suppression of MAF production by
          macrophages from B6 mice injected 24 hours before with lipo-
          polysaccharide (LPS 20 µg/mouse), poly I:C (100 µg/mouse)
          or phosphate buffered saline (PBS).   (A) Direct cytotoxicity
          against RL♂1 target cells by macrophages activated in vivo.
          (B) MAF activity in the supernatants from normal spleen cells
          stimulated by Con A in the presence of macrophages activated
          in vivo by poly I:C or LPS.   The concentration of macrophages
          was 10% relative to that of spleen cells.

Acknowledgement

     The authors gratefully thank Dr. R. B. Herberman for his many
suggestions during this work and for his thorough review of the manuscript.

REFERENCES

1. Mononuclear Phagocytes (R. van Furth, ed.), Martinus Nijhoff
        Publishers, The Hague, Vol. I-II, 1980.

2. Oehler, J. R., Herberman, R. B., and Holden, H. T., Pharmac.
        Ther. A. 2:551, 1978.
3. Varesio, L., and Holden, H. T., J. Immunol. 125:1694, 1980.
4. Varesio, L., Holden, H. T., and Taramelli, D., Cell. Immunol.
        63:279, 1981.
5. David, J. R., Remold, H. G., in "Biology of Lymphokines" (S.
        ·Cohen, E. Pick, J. J. Oppenheim, eds.), pp. 121-140, Academic
        Press, New York, 1979.
6. Taramelli, D., and Varesio, L., J. Immunol. 127:58, 1981.
7. Taramelli, D., Holden, H. T., and Varesio, L., J. Immunol.
        Methods 37:225, 1980.
8. Varesio, L., and Holden, H. T., Cell. Immunol. 56:16, 1980.
9. Taramelli, D., Holden, H. T., and Varesio, L., J. Immunol. 126:
        2123, 1981.

DISCUSSION

PICK: What does LPS really do in the whole animal? You get one type
of response if you have just lymphokine and another response if you
have both lymphokine and LPS.

TARAMELLI: In vivo it is very difficult to think that LPS plays a
role in macrophage activation for tumor killing. Nonetheless, I
believe that two signals are necessary to induce cytotoxicity, at
least in culture.

PICK: In the guinea pig, prostaglandins can depress MIF production.
Can the suppression of lymphokine production be removed by using
cyclo-oxygenase inhibitors?

TARAMELLI: Adding prostaglandin directly to the system doesn't seem
to suppress lymphokine production. However, we didn't use cyclooxygenase inhibitors.

LENZINI: You talked about peptone induced macrophages. What about
resident peritoneal macrophages?

TARAMELLI: We didn't use resident macrophage because it's difficult
to activate them for cytotoxicity with lymphokine. However, they
respond very well to poly IC and interferon.

PRODUCTION OF MACROPHAGE ACTIVATING FACTOR BY CYTOLYTIC AND NONCYTOLYTIC T LYMPHOCYTE CLONES

Anne Kelso, Andrew L. Glasebrook, K. Theodor Brunner, Osami Kanagawa, Howard D. Engers, H. Robson MacDonald and Jean-Charles Cerottini

Department of Immunology, Swiss Institute for Experimental Cancer Research, and The Ludwig Institute for Cancer Research, Epalinges, Switzerland

INTRODUCTION

Antigen-stimulated lymphoid populations release a soluble mediator(s), macrophage activating factor (MAF), which acts synergistically with nonactivating doses of lipopolysaccharide (LPS) to render macrophages nonspecifically tumoricidal (1-3). With the development of in vitro cloning techniques, it is now possible to examine directly the T lymphocytes which produce this and other lymphokines (4-9). Here we describe the phenotypes, antigen stimulation requirements and characteristics of lymphokine production by 72 clones generated against H-2, Mls, H-Y or Moloney leukemia virus (MoLV)-associated antigens.

MATERIALS AND METHODS

Derivation and maintenance of T cell clones (4,5,10; Kanagawa et al., submitted for publication) and assays for cytolytic activity (11), interleukin-2 (IL-2) (8) and interferon (IFN) (12) have been described elsewhere. Clonal supernatants were prepared by culturing 10^5 cloned cells with 5×10^6 irradiated (2000 rads) T cell-depleted spleen cells (or 10^6 irradiated MoLV-infected syngeneic spleen cells) in 1 ml for 24 hr.

MAF Assay: As described elsewhere (Kelso et al., submitted for publication), C57BL/6 macrophage monolayers were obtained by plating into flat bottomed microwells 2×10^5 peptone-induced peritoneal exudate cells or 5×10^4 bone marrow derived macrophages obtained from 6-8

days culture of bone marrow cells in L929 conditioned medium as a
source of macrophage colony stimulating factor (13). The macrophages
were incubated for 24 hours with clonal supernatants and LPS at a
dose which did not activate on its own (10 ng/ml for peritoneal
exudate cells, 100 ng/ml for bone marrow derived macrophages).
^{51}Cr-labeled P815 tumor cells (10^4 per well) were then added for
20 hours before measurement of ^{51}Cr release into the supernatant.
Background release in the absence of clonal supernatants was 25-35%
of maximum release with 0.5N HCl.

RESULTS AND DISCUSSION

        The capacity of supernatants from 2 antigen-stimulated cytolytic
T lymphocyte (CTL) clones to activate macrophages to lyse P815 tumor
cells is demonstrated in Figure 1. L3C5, a subclone of the L3 clone
(14), was derived by limiting dilution from a C57BL/6 anti-DBA/2

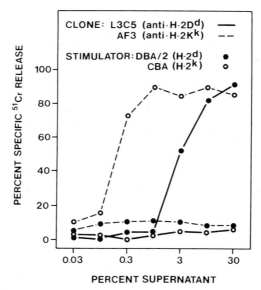

Fig. 1. Antigen specificity of MAF production by 2 cytolytic T lym-
         phocyte clones. Supernatants were collected from 24 hr
         cultures of the clones L3C5 (———) and AF5 (----) stimulated
         with irradiated T cell-depleted DBA/2 (●) or CBA/T6 (o)
         spleen cells. Clonal supernatants were assayed for MAF
         activity on C57BL/6 bone marrow derived macrophages as de-
         cribed in Methods.

mixed lymphocyte culture (MLC) and is specific for H-2D^d determinants. AF3 was cloned directly from B10.AQR anti-B10.A MLC by micromanipulation, and reacts to the K end of H-2^k, presumably H-2K^k. Incubation of these clones with irradiated T cell-depleted stimulator cells of the appropriate haplotype (DBA/2 for L3C5, CBA for AF3) resulted in release of a soluble activity which induced tumor cell lysis at dilutions as high as 1:30 and 1:300 respectively. No activity was detected in clonal supernatants from cloned cells or stimulators cultured alone (not shown) or from the clones when stimulated with cells of a third-party haplotype (Figure 1).

Several observations support the conclusion that the cloned cells release a factor which directly activates macrophages in this experimental system. Firstly, the use of bone marrow macrophages as effector cells, generally after 8 days culture when non-macrophage contamination is undetectable, indicates that both the activation and cytotoxic phases of the assay are independent of other cell types. Qualitatively similar results have been obtained using untreated or T cell-depleted adherent peritoneal exudate cells as effector cells. Secondly, evidence that the activity detected in this assay is produced by the cloned cells themselves came from experiments where clones were extensively depleted of stimulator cells by passage for 1 month in a source of IL-2 alone. Subsequent incubation of the clones with 2 µg/ml concanavalin A (Con A) resulted in MAF release to levels comparable with those produced in response to stimulator cells (9).

Results of the analysis of 72 T cell clones for cytolytic activity and their ability to produce the lymphokines MAF, interferon-γ and interleukin-2, are summarized in Table 1. Several conclusions can be drawn from this random screening of clones.

Firstly, MAF was produced by the majority of clones (68 out of 72) in both the allogeneic (H-2, Mls) and syngeneic H-2^b-restricted (H-Y, MoLV) antigenic systems. It should be noted that the phenotypes recorded here are stable properties of the clones. Thus, for example, clonal supernatants prepared from a given MAF positive clone 6 months apart contained comparable quantities of MAF, while MAF negative clones did not produce any detectable MAF activity when tested on several occasions. Furthermore, the description of a clone as MAF positive or MAF negative was not influenced by the choice of a 24 hour culture period for clonal supernatant production. Kinetic experiments with 4 MAF positive clones showed that activity was detectable within 3 hours and peaked within 6-20 hours after mixing with antigen, remaining at near-maximal levels throughout 7 days of culture; a MAF negative clone failed to produce detectable activity at any stage of culture.

Secondly, the ability of a clone to produce MAF was independent of its cytolytic activity or production of IL-2. Whereas there appear

Table 1.   Summary of Functional Activities of Clones Generated
           Against Different Antigens

Derivation of Clones	Specificity	Number positive/Number tested			
		CTL	MAF	IFN	IL-2
C57BL/6 anti-DBA/2[a]	H-2^d	7/9	9/9	9/9	2/9
C57BL/6 anti-DBA/2	Mls^a	0/2	2/2	2/2	2/2
BALB/c anti-DBA/2	Mls^a	0/3	2/3	2/3	3/3
B10.AQR anti-B10.A	H-2K^k	4/5	3/5	3/5	0/5
A.TH anti-A.TL	H-2I^k	0/3	3/3	3/3	2/3
C57BL/6 ♀ anti-♂	H-Y	25/25	24/25	24/25	0/25
C57BL/6 anti-MoLV	MoLV	13/25	25/25	25/25	9/25
Total		49/72	68/72	68/72	18/72

[a]The clones described in each line originated respectively from 7,
2, 2, 1, 2, 4 and 5 independent clonings.

to be only rare exceptions to the rule that IL-2 is produced by H-2I
or Mls-reactive Lyt-2 negative noncytolytic clones (15; Glasebrook
et al., submitted for publication), MAF is produced by most clones
of both the noncytolytic and cytolytic phenotypes. These observations
have now been extended to unprimed T lymphocytes cultured at limiting
dilution, where almost all Lyt-2 negative cells and about one third
of Lyt-2 positive cells which respond to an allogeneic stimulus have
been found to produce MAF (unpublished observations).

    Finally, whereas comparison of the lymphokines released by
various clones allowed the functional separation of MAF from IL-2
(and from granulocyte-macrophage colony stimulating activity (not
shown)), MAF production was not dissociated from IFN production for
any of the clones tested. In contrast with the results of Schultz
et al. (16) and Boraschi and Tagliabue (17), fibroblast-derived
IFN-β did not mimic MAF in this system. A recent report of biochemi-

cal separation of MAF from IFN in Con A-activated spleen cell supernatants (18) raises the interesting possibility that the MAF-IFN
correlation observed here reflects linked synthesis of two different
factors, or that they represent differently modified forms of a single
core molecule.

In summary, analysis of MAF production by T cell clones has
shown that (a) MAF is produced by the majority of both cytolytic and
noncytolytic clones, irrespective of their antigenic reactivity or
their ability to produce IL-2; (b) MAF and IFN production was correlated for all clones tested; and (c) induction of MAF release from
clones is antigen-specific (although antigen can be substituted by
Con A) and occurs within 3 hours of mixing with stimulator cells.
The observation that most cytolytic and noncytolytic clones produce
nonspecific mediators such as MAF and IFN, whose target populations
include nonlymphoid cells, points to a major role of these lymphocytes
in inflammatory responses in addition to their antigen specific
cytolytic and IL-2-producing functions.

REFERENCES

1. Weinberg, J. B., Chapman, H. A. Jr., and Hibbs, J. B. Jr., J.
      Immunol. 121:72, 1978.
2. Ruco, L. P., and Meltzer, M. S., J. Immunol. 121:2035, 1978.
3. Pace, J. L., and Russell, S. W., J. Immunol. 126:1863, 1981.
4. MacDonald, H. R., Cerottini, J.-C., Ryser, J. -E., Maryanski, J.
      L., Taswell, C., Widmer, M. B., and Brunner, K. T., Immunol.
      Rev. 51:93, 1980.
5. Glasebrook, A. L., Sarmiento, M., Loken, M. R., Dialynas, D. P.,
      Quintans, J., Eisenberg, L., Lutz, C.T., Wilde, D., and Fitch,
      F. W., Immunol. Rev. 54:225, 1981.
6. Nabel, G., Greenberger, J. S., Sakakeeny, M. A., and Cantor, H.,
      Proc. Natl. Acad. Sci. 78:1157, 1981.
7. Marcucci, F., Waller, M., Kirchner, H., and Krammer, P., Nature
      291:79, 1981.
8. Ely, J. M., Prystowsky, M. B., Eisenberg, L., Quintans, J., Gold-
      wasser, E., Glasebrook, A. L., and Fitch, F. W., J. Immunol.
      127:2345, 1981.
9. Glasebrook, A. L., Kelso, A., Zubler, R. H., Ely, J. M., Pry-
      stowsky, M. B., and Fitch, F. W., in "Isolation, Characteriza-
      tion, and Utilization of T Lymphocytes" (C. G. Fathman and
      F. W. Fitch, eds.), Academic Press, New York, in press.
10. Weiss, A., Brunner, K. T., MacDonald, H. R., and Cerottini, J.
      -C., J. Exp. Med. 152:1210, 1980.
11. Cerottini, J. -C., Engers, H. D., MacDonald, H. R., and Brunner,
      K. T., J. Exp. Med. 140:703, 1974.
12. Perussia, B., Mangoni, L., Engers, H. D., and Trinchieri, G., J.
      Immunol. 125:1589, 1980.
13. Meerpohl, H. -G., Lohmann-Matthes, M. -L., and Fischer, H., Eur.
      J. Immunol. 6:213, 1976.

14. Glasebrook A. L., and Fitch, F. W., Nature 278:171, 1979.
15. Widmer, M. B., and Bach, F. H., Nature 294:750, 1981.
16. Schultz, R. M., Papamatheakis, J. D., and Chirigos, M. A.,
        Science 197:674, 1977.
17. Boraschi, D., and Tagliabue, A., Eur. J. Immunol. 11:110, 1981.
18. Kniep, E. M., Domzig, W., Lohmann-Matthes, M. -L., and Kick-
        hofen, B., J. Immunol. 127:417, 1981.

DISCUSSION

RUSSELL: Reports are appearing that some T cell hybridomas may
produce either interferon or MAF. Your T lymphocyte clones were
either positive or negative for both. Do you have any explanation
for this apparent difference?

KELSO: At the moment, we're still open minded as to whether MAF and
interferon are actually the same molecule. If hybridomas mimic normal
lymphocytes and MAF and interferon are different, then an interesting
hypothesis is that the production of these two factors by normal
lymphocytes is somehow linked. Although they may actually be distinct
molecules, their production may go together.

IDENTIFICATION OF A T CELL HYBRIDOMA
WHICH PRODUCES EXTRAORDINARY QUANTITIES
OF MACROPHAGE ACTIVATING FACTOR

Robert D. Schreiber, Amnon Altman and David H. Katz

Department of Molecular Immunology
Research Institute of Scripps Clinic
La Jolla, California USA

INTRODUCTION

It is now generally accepted that under certain conditions the macrophage can express effector cell function toward a variety of neoplastic cells (1,2). Current evidence indicates that the macrophage tumoricidal response is initiated and/or augmented by a T cell product, denoted as macrophage activating factor or MAF (3-7). MAF is thought to alter or "prime" the macrophage in such a way as to make it responsive to a second or "triggering" signal which ultimately activates the tumoricidal potential of the cell (8-12).

The molecular identity of MAF and therefore the elucidation of its mechanism of action on macrophage populations has remained ill-defined. In part, this has been due to the limited amounts of MAF which could be produced under conventional conditions, amounts insufficient to achieve the full biochemical purification of this factor. In addition, because conventional culture supernatants of antigen- or mitogen-stimulated T cells normally contain a heterogeneous mixture of lymphokine activities which can effect macrophage function such as immune interferon (IFNγ) or migration inhibition factor (MIF), uncertainty exists as to the chemical identity of MAF.

In order to circumvent these problems, we initiated a series of experiments which we hoped would lead to the identification of a large scale source of MAF which could be used as starting material for the biochemical purification of this activity.

The present study documents the identification of a murine T cell hybridoma which secretes extraordinary quantities of a biologically

active factor that is functionally and biochemically identical to MAF
produced by conventional techniques. The identification of this
clone should prove to be invaluable in the ultimate purification of
MAF.

RESULTS

Identification of a T cell Hybridoma Clone which Produces MAF

        Four parental murine T cell hybridomas which had been constructed
by fusion of alloantigen-activated T cell blasts with the BW5147 T
cell lymphoma line (13) were tested for their ability to produce MAF.
The four cultures were chosen since they had previously been shown
to produce at least two other lymphokines, namely allogeneic effect
factor (13) and T cell growth factor (14). MAF was quantitated using
a ^{51}Cr-release assay which measured the MAF-dependent induction of
macrophage tumoricidal activity toward ^{51}Cr-labeled P815 mastocytoma
tumor cells (12). One unit of MAF is defined as that amount which
can activate 2×10^5 macrophages to effect 50% maximal ^{51}Cr-release
from 3×10^4 P815 targets. Only one of the hybridomas (number 24)
was found to produce MAF. Although hybridoma 24 did not produce
MAF as a constitutive product, it did secrete MAF in response to
stimulation with mitogens including Concanavalin A (Con A) and phytohemagglutinin. The amount of MAF activity which could be measured
in the stimulated culture supernatant of the uncloned hybridoma was
approximately 500 units/ml. This value represented only 20-30% that
found in conventional MAF preparations (2500 units/ml) which were
prepared by Con A stimulation of normal murine splenic cell cultures.

        In an attempt to identify a T cell hybridoma which could produce
higher amounts of MAF activity, the parental hybridoma was cloned by
limiting dilution. Twenty-seven cell colonies were obtained. None
of the clones secreted MAF constitutively. Figure 1 demonstrates
that 24 of the clones produced MAF when stimulated with 10 μg/ml
Con A. At least seven of the clones (A6, C7, D10, E3, G1, G5 and
H4) produced more MAF activity than the conventional MAF 8080 preparation depicted in the far right column on the figure. An additional
nine clones were found to produce greater than 200 units MAF/ml in
the initial screening but have not yet been fully quantitated by
titration and are denoted by the broken bars labeled N.T. >200.

        Of particular interest was clone G1 which produced extremely high
amounts of MAF activity but no detectable IFN activity. The supernatant from a stimulated culture of clone 24/G1, which was used in
Figure 1, displayed MAF activity (55,000 units/ml) 23 times greater
than that observed in the conventional MAF standard (2400 units/ml).
By stimulating the hybridoma at a higher cell density, we have been

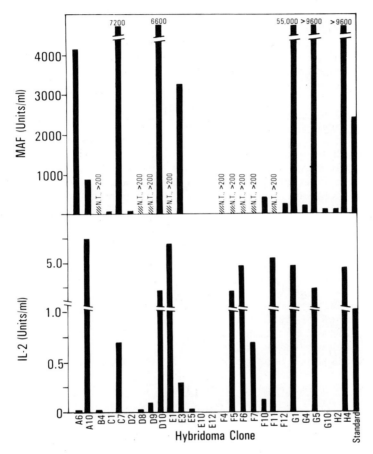

Fig. 1.   MAF and IL-2 production by clones of T cell hybridoma 24.
Five milliliter cultures of 27 T cell clones were stimulated
for 24 hours with 10 μg/ml Con A in DMEM containing 5% fetal
calf serum.   None of the clones produced MAF constitutively
but 15 of 27 produced IL-2 constitutively.   MAF and IL-2
activities of a conventional MAF standard prepared by Con
A stimulation of a normal murine splenic cell culture   are
shown by the vertical bars on the far right.

able to produce supernatants which display 100 times more MAF activity than conventional MAF preparations. The quantity of MAF activity that was observed in the respective supernatants appears to be a reflection of the actual MAF concentration. When hybridoma-derived and conventional MAF preparations were mixed at varying ratios, the amount of MAF activity measured in the mixtures closely approximated the theoretical values obtained by mathematical calculations based on the mixture composition. This observation ruled out the possibilities that the hybridoma supernatant contained an additional enhancing activity or that the conventional MAF preparation contained an additional inhibitory activity.

Demonstration of Independent Production of Distinct Lymphokines by T Cell Hybridoma Clones

Figure 1 also displays the interleukin 2 (IL-2) levels secreted by the individual T cell hybridoma clones. Unlike MAF, IL-2 was produced as a constitutive product by 15 out of 27 clones although only in low concentrations. Upon stimulation with Con A, 18 of the clones responded with IL-2 secretion. Ten clones (A10, C7, D10, E1, F5, F6, F11, G1, G5 and H4) produced more IL-2 than did normal mouse splenic cells stimulated under the same conditions.

A comparison of the amounts of MAF and IL-2 secreted by the various T cell hybridoma clones (Table 1) demonstrated that no correlation existed between the production of the two activities. For example, while clone A6 produced high amounts of MAF activity and only limited amounts of IL-2, clone A10 exhibited the opposite result. Clones G1, G5 and H4 produced levels of MAF which varied independently of IL-2. These results indicated that the discrete T cell clones produced individual repertoires of lymphokine activities.

Biochemical and Functional Comparisons of Hybridoma-Derived MAF and Conventional MAF

The results presented thus far indicated that we had identified a T cell hybridoma clone (24 G1) which produced an inordinately large amount of a factor which could activate macrophages to express tumoricidal activity. We next wished to confirm that this lymphokine was biochemically and functionally equivalent to MAF prepared by conventional techniques. Table 2 indicates that the two activities are identical by a variety of criteria. Hybridoma-derived and conventional MAF displayed identical pH and temperature sensitivities. Both preparations were stable to incubation at pH 7.0 or pH 5.0 but lost 85% of activity at or below pH 4.0. Both preparations, which also contained 5% fetal calf serum, were stable to incubation for 1 hour at $4^\circ C$, $37^\circ C$, or $56^\circ C$ but were totally inactivated following

treatment at 65°C for 1 hour. When prepared under serum free conditions, both MAF preparations were inactivated upon heating to 56°C.
Both hybridoma-derived MAF and conventional MAF required, to the same
extent, a second or triggering signal in order to induce full tumoricidal activity in macrophages. When subjected to molecular weight
analysis by gel filtration on Sephadex G100, both activities behaved

Table 1. Independent Production of Macrophage Activating Factor
          (MAF) and Interleukin 2 (IL-2) by Hybridoma 24 Clones

| Clone | (Lymphokine Activity (% of Standard) | |
	MAF	IL-2
A6	175	2
A10	33	675
E3	137	50
G1	2500	440
G4	4	0
G5	800	208
H4	800	440

Standard activity was produced by Con A stimulation of a normal
spleen cell culture.

as a single molecular species and displayed an apparent molecular
weight of 50,000 to 55,000 daltons. Forty to 60% of the applied
hybridoma and conventional MAF activities were recovered following
this separation step. The biochemical (15,16) and functional (8-12)
characteristics of our MAF preparations are also in agreement with
those reported for MAF by other laboratories.

Table 2.   Comparison of Biochemical and Functional Properties of
           Hybridoma-Derived MAF and Conventional MAF Preparations[a]

Property	T Cell Hybridoma MAF	Conventional Splenic MAF
pH sensitivity	85% inactivated at or below pH 4.0	85% inactivated at or below pH 4.0
Temperature sensitivity	Inactivated at 65°C (1 hour)	Inactivated at 65°C (1 hour)
Enhancement of activity by second signal	42-fold	50-fold
Molecular weight	55,000	50,000

[a]MAF prepared by Con A stimulation of either T cell hybridoma clone
24/G1 or normal murine splenic cell suspensions.

CONCLUSIONS

      The results presented in this report demonstrate that we have
identified a murine T cell hybridoma which produces large quantities
of MAF.   In addition, they also indicate the potential usefulness of
T cell hybridomas to produce distinct patterns of certain lymphokines
in the absence of selected other activities.   The identification
of hybridoma clone 24/G1 should prove to be an invaluable step in
providing the starting material needed for the biochemical purification
of MAF.   Once purified, MAF can conclusively be characterized with
respect to its biological activity and mechanism of action.

Acknowledgement

      This is publication number 2657 from the Research Institute of
Scripps Clinic.   R. D. Schreiber is a recipient of an American Heart
Association Established Investigatorship 77-202 and is supported by
United States Public Health Service Grant AI-17354 and a grant from
Eli Lilly and Company, Indiana.   A. Altman is a recipient of American
Cancer Society, Inc., Junior Faculty Research Award JRFA-17 and is
supported by USPHS Grants CA-24911 and CA-25803.   D. H. Katz is sup-
ported by USPHS Grant AI-07065 and CA-25803.

REFERENCES

1. Evans, R., and Alexander, P., _Nature_ 228:620, 1970.
2. Keller, R., _Lymphokines_ 3:283, 1981.
3. Evans, R., and Alexander, P., _Nature_ 236:168, 1972.
4. Lohmann-Matthes, M.-L., Ziegler, F. G., and Fischer, H., _Europ._ _J. Immunol._ 3:56, 1973.
5. Piessens, W. F., Churchill, W. J. Jr., and David, J. R., _J. Immunol._ 114:293, 1975.
6. Fidler, I. J., _J. Natl. Cancer Inst._ 55:1159, 1975.
7. Ruco, L. P., and, Meltzer, J. S., _J. Immunol._ 119:889, 1977.
8. Weinberg, J. B., Chapman, H. A. Jr., and Hibbs, J. B. Jr., _J. Immunol._ 121:72, 1978.
9. Ruco, L. P., and Meltzer, M. S., _Cell. Immunol._ 41:35, 1978.
10. Meltzer, M. S., _Lymphokines_ 3:319, 1981.
11. Pace, J. L., and Russell, S. W., _J. Immunol._ 126:1863, 1981.
12. Schreiber, R. D., Ziegler, H. K., Calamai, E., and Unanue, E. R., _Fed. Proc._ 40:1002, 1981.
13. Katz, D. H., Bechtold, T. E., and Altman, A., _J. Exp. Med._ 152: 956, 1980.
14. Altman, A., Sferruzza, A., Weiner, R. G., and Katz, D. H., _J._ _Immunol._ 128, 1982, in press.
15. Leonard, E. J., Ruco, L. P., and Meltzer, M. S., _Cell. Immunol._ 41:347, 1978.
16. Kniep, E. M., Domzig, W., Lohmann-Matthes, M.-L., and Kickhöfen, B., _J. Immunol._ 127:417, 1981.

DISCUSSION

UCCINI: We have found that MAF has specificity. Was your lymphokine specific for the mouse or will it work on human macrophages?

SCHREIBER: We haven't tried that yet.

UCCINI: It is difficult to produce MAF or gamma interferon without any stimulation. Do you know of any cell lines that produce these factors without stimulation?

SCHREIBER: We have found none, neither T cell lines nor hybridomas.

PICK: People using cytolytic assays have usurped the name MAF from the old literature. MAF was originally defined both in the guinea pig and in the mouse as a lymphokine which increased listeriocidal activity of macrophages. But bacteriological and cytocidal mechanisms may be completely different. I think there is a misleading trend appearing in the literature to indiscriminately use the same name for probably more than one biological entity. We should be careful in using the name MAF and when used, we should define it.

I would like to make a second comment. The interesting thing
about the requirement for the stimulation of your hybridoma is that
in some lines derived from human cancers, for instance the line derived from a patient with Sezary syndrome that produces MIF, you don't
need this second stimulus. However, there is also a line derived
from a human cancer which requires stimulation to produce leukocyte
inhibitory factor. Do you have a comment on the requirement for
stimulation in some cases and lack of it in others?

SCHREIBER: It may depend upon the biological activity of the specific
lymphokines. For instance, there are some T cell hybridomas which
produce allogeneic effect factor constitutively. Also, there are
some that produce interleukin 2. In fact, our hybridoma 24 produces
IL 2 constitutively. To date, however, we found no cell line which
can produce MAF or interferon as a constitutive product. I don't
know why unless these are generally such non-specific activating
factors that the regulation is at the level of secretion.

AUTOREGULATION OF MONONUCLEAR

PHAGOCYTE FUNCTION

Stephen W. Russell

Department of Comparative and Experimental Pathology
College of Veterinary Medicine
University of Florida, Gainesville, Florida 32610 USA

Among phagocytic cells the mononuclear phagocyte is particularly notable for the diversity of its functions. One explanation for such diversity is the fact that mononuclear phagocytes respond in a variety of ways to a number of different stimuli. Some of these stimuli can be produced by the mononuclear phagocytes themselves, raising the possibility that these cells are autoregulatory. It is my intention to consider this possibility here. Four classes of secretory products with autoregulatory potential will be considered, including proteinases, complement proteins, interferons, and prostaglandins.

PROTEINASES

Proteolytic enzymes were some of the earliest discovered secretory products of mononuclear phagocytes. Because of their ability to act extracellularly at physiological pH, neutral proteinases are among the most important. A variety of these has now been identified, including collagenase (1), elastase (2), plasminogen activator (3) and lysozyme (4). Many others have been recognized qualitatively but remain to be identified (5). A diverse range of stimuli will induce secretion of proteinase activity, including such things as phagocytosis, endotoxin or lymphokine (6,7). The lymphokine activity which stimulates macrophages to secrete elastase is a product of T lymphocytes. Evidence in support of this fact has recently been obtained from work performed with a cloned T cell hybridoma (8).

Aside from the obvious effect of substrate degradation, it would appear that proteinases free in the inflammatory milieu can directly influence the functions of mononuclear phagocytes. For example, exposure to trypsin, pronase, chymotrypsin or papain primes macro--

phages to respond to either particle ingestion or surface active
agents with an enhanced oxidative metabolic burst (9). Such an effect
could be beneficial if it either augmented the killing of microorganisms by oxidative pathways or contributed to extracellular killing
of tumor cells by activated macrophages (10,11). A second, rather
striking effect is the spreading of mononuclear phagocytes on a substratum that is induced by proteinases (12-14). Since these cells
move by crawling on tissue elements rather than by swimming through
interstitial fluids, an increased ability to attach to a substratum
may be relevant to more efficient functioning of macrophages at an
inflammatory site. Third, amplification of the inflammatory response
to a noxious stimulus is largely dependent on the production of pharma-
cologically active mediators, either by mononuclear phagocytes, other
cell types, or from plasma proteins. In their active form any of
these mediators potentially can affect mononuclear phagocytes. They
usually are produced or exist in an inactive form that must be cleaved
by some mechanism in order for them to have their effect. Proteinases
produced by mononuclear phagocytes have the capacity to mediate such
cleavage. For example, an acid proteinase isolated from macrophages
has been shown to cleave C5, also a product of macrophages, to produce
the chemotactically active fragment, C5a (5). A fourth function that
proteinases may have is down regulation of mononuclear phagocyte
function and the inflammatory response in general, either through
removal of receptors from cell surfaces or by degrading mediators.
As an example of this latter effect, a carboxypeptidase secreted by
stimulated guinea pig peritoneal macrophages has recently been reported by Kreuzpainter and his colleagues (16) to degrade the anaphyl-
atoxin, C3a, thereby eliminating its spasmogenic activity.

COMPLEMENT PROTEINS

     The classical and alternative pathways of complement activation
are the sources of a variety of mediators that are important in the
genesis of inflammatory responses. These two pathways relate to
each other through a common component, C3. We have already considered
that proteinases secreted at an inflammatory site by macrophages may
be of importance in activating components in the complement cascade.
It is also essential that we recognize that mononuclear phagocytes
synthesize and secrete many of the early components of both the classical and alternative pathways ((17) and Table 1) and that many of
the products of complement activation, especially of the early components, can directly stimulate macrophages and monocytes.

     With regard to production of complement components by mononuclear
phagocytes, it is becoming increasingly clear that differences exist
both on a population and individual cell basis. Thus, the amounts
and kinds of complement components produced by mononuclear phagocytes
at an inflammatory site could differ because of changes either in
their rates of synthesis (18), the amounts of biologically active

Table 1. Complement Proteins Synthesized by Mononuclear
          Phagocytes

C1	Factor B
C1q	Factor D
C2	Properdin
C3	C3b INA
C4	β1H
C5	

proteins that are secreted (19), the proportion of cells in the population that is synthesizing each of the various components (20-22),
or a combination of these possibilities. As an example of these
kinds of differences, Hartung et al. (page 525) show that differences in the level of activation influence the extent to which macrophage populations produce complement components C2, C3 and C4.

Secretion of complement proteins by mononuclear phagocytes may
be modulated in several ways. Feedback inhibition has been reported
for C4, and a lymphokine is purported to inhibit the synthesis of
C2 (24). Synthesis of C2 can be enhanced either by interaction of
pharmacologic agents with a specific nicotinic cholinergic receptor
on macrophages (25) or by exposing these cells to a lymphokine produced by T lymphocytes (26). It is not known how the stimulatory
lymphokine relates to the inhibitory one mentioned earlier, but that
it exists is clear from the analysis of the products of T lymphocyte
hybridomas (8) and clones (27). Interaction of antigen-antibody
complexes with mononuclear phagocytes increases protein synthesis
generally and, as a consequence, that of C2, C4, C3, C5, factor B,
properdin, C3b inactivator and β1H (28). When examined more specifically, it was concluded that interaction of complexes with Fc receptors is responsible for augmenting the synthesis of C2, while
interaction with C3b receptors inhibits synthesis of both C2 and
factor B (29).

Autoregulation could result from interaction of biologically
active complement fragments with either the cells that made them or,
in a broader sense, with other mononuclear phagocytes in the same
population. Many possibilities exist. For example, factor B, when

activated (Bb) and mediating its effect as a proteinase, causes mononuclear phagocytes to spread on their substratum (12,14). It is
probably for this reason that Bb also inhibits the motility of these
cells (30). The active fragment C5a enhances macrophage motility
and is chemotactic. In addition, it will induce macrophages to
release superoxide anion and prostaglandin E_2 (31), as well as to
secrete lysosomal hydrolases (32). C3b, the large activation fragment of C3, can also induce enzyme secretion by macrophages (33,34).
The fact that these cells secrete C3 and, in response to stimulation
by C3b, also secrete proteinases with the ability to cleave C3, suggests the potential for amplification. It has been postulated that
macrophages might be activated by C3b to kill tumor cells (35). The
effector was suggested to be C3a, generated perhaps by proteinases
secreted by the activated macrophages. This is certainly an attractive possibility. However, due to the fact that others have been
unable to obtain cytocidal effects with large amounts of highly purified C3a (36) the validity of this hypothesis must now be questioned.
Additional effects of C3b on mononuclear phagocytes include stimulation of a respiratory burst (37) as well as induction of prostaglandin
E and thromboxane B_2 release. This subject is discussed by Hadding
et al. on page 531.

INTERFERON

        Boraschi and her colleagues (page 519) describe effects that
fibroblast interferon has on macrophages. Mononuclear phagocytes
also produce interferon in response to a diverse range of stimuli.
There are two basic kinds of interferon referred to as type I and type
II. The former type contains alpha and beta subgroups made by leukocytes and fibroblasts, respectively. Type II, or gamma interferon,
is also known as immune interferon. It is made by antigen or mitogen
stimulated T lymphocytes. Type I and II interferons apparently interact with cells through different receptors (39) and are antigenically
distinct from each other (40). Of these two general types, inter-
feron(s) produced by macrophages falls into the type I group. Three
important characteristics it shares with this group are relative acid
stability, lability at 56°C and susceptibility to neutralization by
antibody made against fibroblast interferon (Table 2).

        The effects of interferon on mononuclear phagocytes have been
reviewed recently by Schultz (41). Of the effects he describes, most
of them can only be listed here. Included are enhancement of phagocytic activity (although not for all kinds of particles, e.g., infectious agents; see references 42-44), increased spreading on glass
or plastic, induction of growth suppression of intracellular microorganisms, enhanced antiviral activity, and elevation of levels of
cytoplasmic lactate dehydrogenase. Details on these activities can
be obtained from Schultz' review and the pertaining, original papers.
An additional effect recently reported by Lee and Epstein (45) is that

Table 2. Properties of Interferon (IFN) Produced by
          Mononuclear Phagocytes

Treatment	Effect	Source(s) of Macrophages
pH 2,1 day	No effect or slight loss of activity	Peritoneum Spleen
pH 2,5 days	50% activity lost	Bone marrow
56°C,30 min.	100% activity lost	Bone marrow Peritoneum Spleen
Trypsin	100% activity lost	Peritoneum
Antibody against L cell IFN	Activity neutralized	Bone marrow Peritoneum Spleen

Data from References 60-62.

of reversible interference with mononuclear phagocyte differentiation.
Of all of the effects that interferon has on mononuclear phagocytes,
perhaps the one that has attracted the most attention lately is its
apparent ability to activate macrophages to mediate tumor cell killing
or growth inhibition. Either type I or type II interferon added
directly to macrophage cultures has been shown to be effective (46).
Indirect support for the hypothesis that interferon is involved in
the process of activation stems from the finding that many activators
of macrophages also induce these cells to produce interferon (41).
Further evidence comes from neutralization experiments performed with
anti-interferon antibodies by Schultz and Chirigos (46). They showed
that (a) antibody against type I (L-cell) interferon would prevent
L-cell interferon from inducing cytotoxicity in macrophages; (b)
the antibody would prevent agents such as bacterial lipopolysaccharide, pyran copolymer or polyinosinic-polycytidylic acid from
activating macrophages; and (c) antibody directed against type I
interferon would not prevent activation induced by antigenically
distinct type II interferon. In addition to confirming that interferon can activate macrophages, these data clearly suggest that an

autoregulatory, positive feedback mechanism may be operative in the process of macrophage activation.

If one accepts that interferon can activate macrophages, the next consideration is whether or not it is the only activator. For example, what about the lymphokine that is found in culture super- nates of stimulated lymphocytes? Is it the same as, or different from, the immune (gamma) interferon that is invariably found in the same supernates? The fact that interferon activity shares properties and is produced coordinately with one of the other lymphokines that affects macrophages, migration inhibitory factor (MIF), was first noted by Youngner and Salvin (40). Shultz and Chirigos (47) followed with work showing that interferon and the lymphokine that is respon- sible for macrophage activation (MAF) are also strikingly similar. Experience in our own laboratory (unpublished) suggests that culture supernates rarely differ when the two activities are quantified, i.e., when one is high the other is too, and vice versa. A definitive answer as to whether MAF and gamma interferon activities are attribut- able to the same or different molecules probably awaits the analysis of supernates from cultures of cloned T lymphocytes or T cell hy- bridomas. The T lymphocyte clones described by Kelso and collabora- tors (page 493) either produced MAF and interferon or neither of these activities. A definitive answer as to whether MAF and gamma interferon activities are attributable to the same or different enti- ties probably awaits the isolation and characterization of the rele- vant molecules from cultures of cloned T lymphocytes or T cell hy- bridomas. The work of Kelso and her colleagues (page 493) and Schreiber et al. (page 499) reported here represents a step in this direction.

PROSTAGLANDINS

Prostaglandins are also secreted by macrophages in response to many different kinds of stimuli. Recent reviews by Gemsa (49) and by Morley (50) extensively cover the subject.

Prostaglandins are acidic lipid mediators that are produced through metabolism of arachidonic acid mobilized from cell membrane phospholipids subsequent to phospholipase (A_2 or C) activation. Pros- taglandins are produced via the cyclooxygenase pathway of arachidonic acid metabolism, while another whole set of mediators, e.g., the leukotrienes, are generated via the lipoxygenase pathway. Lipoxy- genase products produced by leukocytes recently have been reviewed (51) and will not be considered here.

Although there are exceptions, such as the augmentation of col- lagenase production by macrophages (52), as a generalization one can say that prostaglandins have an inhibitory effect on mononuclear phagocytes. For example, prostaglandin E is reported to inhibit the production of mononuclear phagocytes from bone marrow (53), to inhibit

phagocytosis (54), and to reduce motility (54). Schnyder et al.
report that PGE suppresses plasminogen activator secretion and the
elevation of various cell-associated enzyme activities (page 535)
PGE also appears to be involved in the negative control of macrophage activation for tumor cell killing (55,56). Although PGE is present in macrophage cultures in high concentration within hours of
applying an activating dose of lipopolysaccharide, it does not prevent
the development of cytolytic activity. Instead, it shuts it off once
it has developed (56). Introduction of indomethacin or other cyclooxygenase inhibitors prevents macrophages from synthesizing PGE and,
therefore, both augments and sustains the expression of cytolytic
activity. The mechanism underlying the inhibitory effect is, as yet,
unclear. When used in conjunction with activator, as little as 15
minutes of exposure to PGE_2 inhibited the mediation of killing; however, killing was not affected if macrophages were preincubated with
PGF_2 for 1 hour, washed, and then exposed to an activating concentration of lipopolysaccharide (56). Although one would expect intracellular elevation of cyclic AMP to be a part of the inhibitory
mechanism and there is experimental evidence in support of this contention (57) , more than a simple increase may be involved. Several
lines of evidence suggest this conclusion. For example, a lymphokine
activity in the supernatant medium of concanavalin A-stimulated spleen
cells interferes with the negative regulatory effect of PGE_2 but does
not quantitatively alter the intracellular elevation of cAMP in
lymphokine-treated macrophages, compared to controls (58). Preincubation of macrophages with PGE before exposure to activator, as
described above, would elevate intracellular levels of cAMP but it
does not inhibit cytolytic activity. Finally, PGI_2 (prostacyclin)
induces a quantitatively similar increase in intracellular levels of
cAMP, compared to PGE_2, but appears not to interfere with the expression of killing by activated macrophages, at least when used in a
single dose (Taffet, S M., Eurell, T. and Russell, S. W., unpub-
blished). If elevation of cAMP is important, it may be that localized
changes in its concentration, rather than whole cell increases, are
the critical factor.

CONCLUSION

     From these brief considerations, it is clear that autoregulation
is likely an important part of the biology of mononuclear phagocytes.
For example, Schultz (59) has hypothesized that interferons and E-type
prostaglandins may relate in Yin-Yang fashion to regulate macrophage
activation. Review of the preceding sections shows that such interplay could be extensive among the agents and their cleavage products
that have been considered here. However, it may be well to maintain
a healthy skepticism until it can be determined whether or not the
kinds of interplay that are observed in the relatively simple, static
conditions that characterize culture systems are also found in the
more complicated, constantly changing milieu of an inflammatory

lesion. This very important point, as well as a number of others, needs to be addressed before we will be certain that autoregulation is as important as it appears to be in the modulation of mononuclear phagocyte functions.

Acknowledgment

The author would like to thank Mrs. Judith McCallister for typing the manuscript. Supported in part by research grant CA 31199 from the National Cancer Institute.

REFERENCES

1.   Werb, Z., and Gordon, S., J. Exp. Med. 142:346, 1975.
2.   Werb, Z., and Gordon, S., J. Exp. Med. 142:361, 1975.
3.   Unkeless, J. C., Gordon, S., and Reich, E., J. Exp. Med. 139: 834, 1974.
4.   Gordon, S., Todd, J., and Cohn, Z. A., J. Exp. Med. 139:1228, 1974.
5.   Adams, D. O., Kao, K.-J., Farb, R., and Pizzo, S. V., J. Immunol. 124:293, 1980.
6.   Gordon, S., Fed. Proc. 36:2707, 1977.
7.   Page, R. S., Davies, P., and Allison, A. C., Int. Rev. Cytol. 52:119, 1978.
8.   Jones, C. M., Braatz, J. A., and Herberman, R. B., Nature 291: 502, 1981.
9.   Johnston, R. B. Jr., Chadwick, D. A., and Cohn, Z. A., J. Exp. Med. 153:1678, 1981.
10. Nathan, C. F., and Cohn, Z. A., J. Exp. Med. 154:1539, 1981.
11. Adams, D. O., Johnson, W. J., Fiorito, E., and Nathan, C. F., J. Immunol. 127:1973, 1981.
12. Gotze, O., Bianco, C., and Cohn, Z. A., J. Exp. Med., 149: 372, 1979.
13. Rabinovitch, M., and DeStefano, M. J., Exp. Cell Res. 77:323, 1973.
14. Sundsmo, J. S., and Gotze, O., Cell. Immunol. 52:1, 1979.
15. Synderman, R., Shin, H. S., and Dannenberg, A. M. Jr., J. Immunol. 109:896, 1972.
16. Kreuzpainter, G., Damerau, B., and Brade, V. Abstracts of the IXth International Complement Workshop, Molec. Immunol., in press, 1982.
17. Bentley, C., Zimmer, B., and Hadding, U., in "Lymphokines" (E. Pick and M. Landy, eds.), vol. 4, pp. 197-230, Academic Press, New York, 1981.
18. Cole, S. F., Mathews, W. J. Jr., Marino, J. T., Gash, D. J., and Colten, H. R., J. Immunol. 125:1120, 1980.
19. Oor, Y. M., Harris, D. E., Edelson, P. J., and Colten, H. R., J. Immunol. 124:2077, 1980.
20. Wyatt, H. V., Colten, H. R., and Borsos, T., J. Immunol. 108: 1609, 1972.

21. Barber, T. A., and Burkholder, P. M., J. Immunol. 120:716, 1978.
22. Auerbach, H. S., Lalande, M. E., Latt, S., and Colten, H. R. Abstracts of the IXth International Complement Workshop, Molec. Immunol., in press, 1982.
23. Matthews, W. J., Marino, J. T., Goldberger, G., Gash, D., and Colten, H. R., Fed. Proc. 38:1011, 1979.
24. Cambier, A., and Peltier, A. -P., Biomedicine. 28:332, 1978.
25. Whaley, K., Lappin, D., and Barkas, T., Nature 293:580, 1981.
26. Littman, B. H., and Ruddy, S., Cell. Immunol. 43:388, 1979.
27. Goldman, M. B., Prystowsky, M., Ely, J., Fitch, F. W., and Goldman, J. N., Abstracts of the IXth International Complement Workshop, Molec. Immunol., in press, 1982.
28. McPhaden, A., Lappin, D., and Whaley, K., Immunol. 44:193, 1981.
29. Whaley, K. and Lappen, D., Abstracts of the IXth International Complement Workshop, Molec. Immunol., in press, 1982.
30. Bianco, C., Gotze, D., and Cohn, Z. A., J. Immunol. 120:1765, 1978.
31. Kunkel, S. L., Plewa, M. C., Fantone, J. C., and Ward, P. A., Abstracts of the IXth International Complement Workshop, Molec. Immunol., in press, 1982.
32. McCarthy, K., and Henson, P. M., J. Immunol. 123:2511, 1979.
33. Schorlemmer, H. U., and Allison, A. C., Immunol. 31:781, 1976.
34. Schorlemmer, H. U., Bitter-Suermann, D., and Allison, A. C., Immunol. 32:929, 1977.
35. Ferluga, J., Schorlemmer, H. U., Baptista, L. C., and Allison, A. C., Clin. Exp. Immunol. 31:512, 1978.
36. Goodman, M. G., Weigle, W. O., and Hugli, T. E., Nature 283:78, 1980.
37. Schopf, R. E., Hammann, K. P., Scheiner, O., Lemmel, E.-M., and Dierich, M. P. Abstracts of the IXth International Complement Workshop, Molec. Immunol., in press, 1982.
38. Hadding, U. Hartung, H. P., Rasokat, H., and Gemsa, D. Abstracts of the IXth International Complement Workshop, Molec. Immunol., in press, 1982.
39. Branca, A. A., and Baglioni, C., Nature 294:768, 1981.
40. Youngner, J. S., and Salvin, S. B., J. Immunol. 111:1414, 1973.
41. Schultz, R. M., in "Lymphokine Reports" (E. Pick and M. Landy, eds.) vol. 1, pp. 63-97, Academic Press, New York, 1980.
42. Remington, J. S., and Merigan, T. C., Nature 226:361, 1970.
43. Kazar, J., Gillmore, J. D., and Gordon, F. B., Infec. Immun. 3:825, 1971.
44. Mizunoe, K., Hiraki, M., Nagano, Y. and Maehara, N., Japan. J. Microbiol. 19:235, 1975.
45. Lee, S. H. S., and Epstein, L. B., Cell. Immunol. 50:177, 1980.
46. Schultz, R. M., and Chirigos, M. A., Cell. Immunol. 48:52, 1979.
47. Schultz, R. M., and Chirigos, M. A., Cancer Res. 38:1003, 1978.
48. Ratliff, T. L., Thomasson, D. L., McCool, R. E., and Catalona, W. J., J. Reticuloendothel. Soc., in press, 1982.
49. Gemsa, D., in "Lymphokines" (E. Pick and M. Landy, eds.), vol. 4, pp. 335-375, Academic Press, New York, 1981.

50. Morley, J., in "Lymphokines" (E. Pick and M. Landy, eds.), vol. 4, pp. 377-394, Academic Press, New York, 1981.

51. Bokoch, G. M., Boeynaems, J. M., and Hubbard, W. C., in "Lympho-kines" (E. Pick and M. Landy, eds.), vol. 4, pp. 271-295, Academic Press, New York, 1981.

52. Wahl, L. M., Olsen, C. E., Sandberg, A. L., and Mergenhagen, S. E., Proc. Natl. Acad. Sci. USA 74:4955, 1977.

53. Pelus, L. M., Broxmeyer, H. E., Kurland, J. I., and Moore, M. A. S., J. Exp. Med. 150:277, 1979.

54. Oropeza-Rendon, R. L., Speth, V., Hiller, G., Weber, K., and Fischer, H., Exp. Cell Res. 119:365, 1979.

55. Schultz, R. M., Pavlidis, N. A., Stylos, W. A., and Chirigos, M. A., Science 202:320, 1978.

56. Taffet, S. M., and Russell, S. W., J. Immunol. 126:424, 1980.

57. Schultz, R. M., Pavlidis, N. A., Stoychkow, J. N., and Chirigos, M. A., Cell. Immunol. 42:71, 1979.

58. Taffet, S. M., Pace, J. L., and Russell, S. W., J. Immunol. 127: 121, 1981.

59. Schultz, R. M., Medical Hypoth. 6:831, 1980.

60. Blach-Olszewska, Z., and Cembrzynska-Nowak, M., Acta. Biol. Med. Germ. 38:765, 1979.

61. Havell, E. A., and Spitalny, G. L., Ann. N. Y. Acad. Sci. 350: 413, 1980.

62. Neumann, C., Macher, E., and Sorg, C., Immunobiol. 157:12, 1980.

DISCUSSION

BAGGIOLINI: One of my worries concerns all the possible pathophysiol-ogy involvements which one can ascribe to proteinases. The activation or de-activation of complement components by proteinases released from mononuclear phagocytes or polymorphonuclear leukocytes must be viewed with caution. Let's consider inactivation of anaphylatoxins. Anaphylatoxins will be bound to cell receptors and be engulfed very fast. So there is no need, in this particular case and in many others, to have an inactivation mechanism. Another thing is the activation of C3 or C5, which splits to give active moieties. Nobody has ever shown that this is something specific. It could just happen during the progressive aggregation of these molecules. I tend to believe that if you took an enzyme, such as neutrophil elastase, which is a very powerful and versatile enzyme, and a protein, really any protein, the protein would be cleaved. I think there is danger in involving these proteinases in all sorts of pathological situa-tions.

RUSSELL: I agree very much with you. If you go to the literature there are many such examples. What happens in a tissue culture system is certainly very different from what happens in a living animal. What is important is to recognize the potential that exists and then try and address some of the questions that you raised.

PICK: I would like to comment on the issue of the prostaglandins. We tend to think only about E type prostaglandins which are basically meant for export.   In order to do something to the cell, they have to act on a receptor.   Yet, we have a whole range of products that are derived from the lipoxygenase pathway.   Some are meant for export, but some have an intracellular regulatory function, which as far as we know do not require a receptor.   Since you cannot produce prostaglandins without activating phospholipase, there are a number of intermediary products which can also be extremely active such as phosphotidic acid.   I predict a time will come when perhaps more will be known about the regulatory intracellular functions in the macrophage of arachidonic acid products other than prostaglandins.

RUSSELL: The whole area of lipid mediators is a very exciting new area, not only in terms of mononuclear phagocyte biology, but also in terms of the inflammatory response in general.   We have overlooked many of these because of problems related to stability and identification but we are finally overcoming these technical hurdles.   In the next five years, I too predict that we are going to hear a great deal more about these mediators.

REGULATION OF MACROPHAGE

FUNCTIONS BY INTERFERON

Diana Boraschi,[1] Elena Pasqualetto,[2] Pietro Ghezzi,[2]
Mario Salmona,[2] John E. Niederhuber,[3]
and Aldo Tagliabue[1]

[1]Sclavo Research Center, Siena, Italy
[2]IRFMN Milan, Italy
[3]University of Michigan School, Ann Arbor
 Michigan, USA

Mononuclear phagocytes exert a major regulatory influence on many
lymphocyte functions, including antibody production, lymphocyte proliferation, and the production of lymphokines (reviewed in Reference
1). Highly suppressive macrophages can be induced in mice by injection of a variety of agents such as C. parvum, BCG and pyran
copolymer. Such in vivo stimulation concomitantly enhances macrophage
tumoricidal activity in vitro and this observation led to the belief
that a common activation mechanism underlay both functions (2-3).
Unstimulated peritoneal macrophages also express suppressor activity
(2,4) and low but significant levels of tumoricidal activity (5).
This "natural" cytotoxicity can be highly enhanced by in vitro exposure to lymphokines, interferons, bacterial endotoxins and other
substances (6-8). In vitro modulation of macrophage suppression
with the same agents that enhance tumoricidal activity would help
verify the hypothesis of a common regulation for the two activities.
We now summarize our recent observations on the possibility of distinguishing the mechanism of regulation of macrophage cytotoxicity
from that controlling suppression.

The effects of either fibroblast interferon (IFN) or the lymphokine macrophage activating factor (MAF) on macrophage mediated tumor
cytolysis and on suppression of lymphocyte proliferation is presented
in Table 1. Tumor cytolysis was measured by the release of tritiated
thymidine from prelabeled TU5 tumor cells after 48 h of cocultivation
with mouse peritoneal macrophages (A:T = 20:1) preexposed to the
activating stimuli in vitro (7). Suppressor activity was calculated
as reduction of thymidine incorporation by spleen cells pulsed with

ConA and then cultured with in vitro pretreated macrophages at a 1:10
ratio of macrophages to spleen cells (4). IFN significantly enhanced
tumor cytolysis, regardless of the time of macrophage exposure to
IFN. In contrast, MAF increased cytolysis only after 4 h of pretreatment, being ineffective at 20 h (7,9). The in vitro modulation of suppressor activity was different. Four hours of preexposure
to MAF or IFN did not alter macrophage suppressor activity. After
20 h of pretreatment, MAF-treated macrophages were as suppressive
as control cells while the same duration of exposure to IFN drasti-
cally reduced macrophage suppressor activity. Thus, by modulating
macrophage functions in vitro with either MAF or IFN it was possible
to show that the regulation of suppression is distinguishable from
that of cytolysis.

In an attempt to clarify the interactions between IFN and macrophages that led to a reduction of suppressor activity, we examined
the role of surface Ia molecules (see Reference 10). As shown in
Table 2, suppressor activity from C3H/HeN mice (H-2^k) was not affected by depletion of Ia positive macrophages by means of anti-I-A^k
and complement treatment. Pretreatment with IFN reduced the suppressor activity of control macrophages by about 50%. However, after
depletion of Ia positive macrophages this effect disappeared (only

Table 1.   Modulation of Macrophage Functions by Macro-
           phage Activating Factor (MAF) and Interferon
           (IFN)

Activity	Time of Exposure	Percent Activity Macrophages Exposed To		
		Medium	MAF	IFN
Tumor cytolysis	4 h	17	38*	38*
	20 h	15	16	37*
Suppression of lymphocyte proliferation	4 h	59	65	57
	20 h	43	37	2*

*P<0.01 vs. medium control

10% reduction in anti-I-A^k-treated macrophages). Macrophages treated
with the irrelevant antiserum and complement showed activity comparable to that of control macrophages treated with complement alone.
These data suggest that Ia negative macrophages are the suppressor
cells. In contrast, the modulation of suppression of IFN appears
to be mediated by Ia positive macrophages which in turn affect the
suppressor cell. A similar analysis was performed in parallel on
tumor cytolysis. Here, the effector cell was identified as an Ia
negative macrophage. At variance with the observations regarding
suppressor activity, the modulatory effect of IFN on cytocidal capacity did not depend upon the presence of Ia positive macrophages (11).
A clear dichotomy was evident between the modulatory effects of IFN
on suppressive and tumoricidal capacity of macrophages. Therefore
it is reasonable to hypothesize that distinct regulatory mechanisms
underlie these two functions.

Several soluble factors released by macrophages have been indicated as responsible for either their cytolytic or suppressor activity. Among these factors, oxygen intermediates are of special
interest (12,13). Accordingly, we investigated the ability of macrophages exposed for different times to either IFN or MAF to produce
two oxygen metabolites in response to opsonized zymosan. Release
of superoxide anion was measured as superoxide dismutase-inhibitable
reduction of ferricytochrone c (9), while hydrogen peroxide production
was quantitated as scopoletin oxidation in presence of horseradish
peroxidase (14). Data in Table 3 demonstrate a close parallelism

Table 2. Role of Ia Positive Macrophages in IFN
          Mediated Regulation of Suppressor Activity

Macrophages exposed to	Suppression as % of Control Macrophages Treated With		
	Complement only	anti-I-A^s + complement	anti-I-A^k + complement
Medium	100	99	104
Interferon	53	53	90*

*P < 0.01 vs. complement only
 Anti I-A^s = (ATH x B10.HTT) F$_1$ anti-A.TL serum
 Anti I-A^k = (B10.A x A.TL) F$_1$ anti-B10.HTT serum

between modulation of suppressor activity and of oxygen intermediate
production by macrophages exposed to IFN. In fact, neither superoxide
anion nor hydrogen peroxide release was affected by preexposure of
macrophages for 4 h to IFN or MAF in vitro. MAF was unable to affect
release of both superoxide anion and hydrogen peroxide after 20 h
of exposure. In contrast, production of both oxygen intermediates
was strongly reduced by exposure for 20 h to IFN. From these observa-
tions.we infer that the ability of IFN to reduce suppressor activity
may relate to its capacity to decrease oxygen intermediate release.
On the other hand, neither MAF- nor IFN- induced enhancement of cytolysis appears to correlate with production of superoxide anion and
hydrogen peroxide. Preliminary data on release of PGE_2, a substance
with a well-established primary role in macrophage mediated suppression (12), indicate that macrophages exposed to IFN for 20 h produce
much less PGE_2 than control cells (11).

In conclusion, it is possible to discriminate between mechanisms
of regulation of macrophage cytolytic and suppressor activity by the
use of IFN in vitro. Whereas IFN increased tumoricidal activity,
it dramatically reduced suppressor capacity. Different macrophage
subsets appear to be involved in the regulation of the two activities by IFN. Finally, analysis of the biochemical basis for the
IFN induced reduction of suppressor activity led to the conclusion
that macrophages exposed to IFN are less able to release suppressive
molecules, such as oxygen metabolites and prostaglandins.

Table 3. Production of Oxygen Intermediates by
          Macrophages

Oxygen Intermediate	Time of Exposure	Oxygen Intermediates % of Control Macrophages Exposed to		
		Medium	MAF	IFN
Superoxide anion	4 h	100	98	90
	20 h	100	112	17*
Hydrogen peroxide	4 h	100	94	97
	20 h	100	93	32*

*P<0.01 vs. medium control

Acknowledgement

        The authors thank Mr. Patrick Ronan for the preparation of anti-
sera. This work was partially supported by the Associazione Italiana
per la Ricerca sul Cancro and by the Italian National Research Council.

REFERENCES

1. Nelson, D. S., in "Immunobiology of the Macrophage" (D. S. Nelson,
        ed.), pp. 235-257, Academic Press, New York, 1976.
2. Baird, L. G., and Kaplan, A. M., Cell. Immunol. 28:22, 1978.
3. Kaplan, A. M., Morahan, P. S., and Regelson, W., J. Natl. Cancer
        Inst. 52:1919, 1974.
4. Boraschi, D., Soldateschi, D., and Tagliabue, A., Eur. J. Immunol.
        12:320, 1982.
5. Tagliabue, A., Mantovani, A., Kilgallen, M., Herberman, R. B.,
        and McCoy, J. L., J. Immunol. 122:2363, 1979.
6. Evans, R., and Alexander, P., Transplantation 12:227, 1971.
7. Boraschi, D., and Tagliabue, A., Eur. J. Immunol. 11:110, 1981.
8. Doe, W. F., and Henson, P. M., J. Exp. Med. 148:544, 1978.
9. Boraschi, D., Ghezzi, P., Salmona, M., and Tagliabue, A., Immuno-
        logy 45:621, 1982.
10. Niederhuber, J. E., Immunol. Rev. 40:28, 1978.
11. Boraschi, D., Pasqualetto, E., Ghezzi, P., Salmona, M.,
        Rotilio, D., Donati, M. B., and Tagliabue, A., in "Natural
        Cell-mediated Immunity" (R. B. Herberman, ed.), Academic
        Press, New York, in press.
12. Metzger, Z., Hoffeld, J. T., and Oppenheim, J. J., J. Immunol.
        124:983, 1980.
13. Nathan, C. F., Silverstein, S. C., Brukner, L. H., and Cohn,
        Z.A., J. Exp. Med. 149:100, 1979.
14. Nathan, C. F., and Root, R. K., J. Exp. Med. 146:1648, 1977.

DISCUSSION

UNKNOWN: What was the stimulus used to induce superoxide and hydrogen
peroxide production on one hand, and PGE production on the other?

BORASCHI: For prostaglandin production, we did not use any stimulation. We just put the cells in culture. As for superoxide and
hydrogen peroxide production, we used opsonized zymosan. The effect
of interferon is exactly the same.

BAGGIOLINI: You found a modest inhibition of superoxide production
and a rather striking inhibition of hydrogen peroxide production.

BORASCHI: In that particular experiment we did, but usually the depression is almost comparable.

PICK: What was the viability of your cells after 24 hours exposure to interferon?

BORASCHI: The same as in control cells.

PICK: It's common dogma in the superoxide field that resident mouse macrophages are poor producers of superoxide and almost ineffective as hydrogen peroxide producers. Most of this work was done with phorbol esters. Did you have any difficulty in inducing hydrogen peroxide?

BORASCHI: Yes, we had terrible difficulties in getting significant amounts of both hydrogen peroxide and superoxide. Our success has depended on the dose of zymosan we used in culture.

INFLUENCE OF MACROPHAGE ACTIVATION ON THE

SYNTHESIS OF COMPLEMENT COMPONENTS C2, C3, C4

Hans-Peter Hartung, Bernd Zanker
and Dieter Bitter-Suermann

Institute of Medical Microbiology
University of Mainz
Hochhaus Augustuspl.
6500 Mainz   F.R.G.

INTRODUCTION

Macrophages are a major site of complement synthesis. In the guinea pig production of complement components C1, C2, C4, C3, D, B, and P by peritoneal macrophages has been demonstrated (reviewed in Ref. 1). Whereas marked differences exist in biological activity between resident, elicited and activated macrophages (2), we investigated whether this holds true for the synthesis and secretion of complement components C2, C3, and C4.

MATERIALS AND METHODS

Collection of Cells and Preparation of Monolayers: Peritoneal macrophages were collected from untreated guinea pigs (300 - 500 g) or from animals injected with human serum albumin (0.5% in 25 ml saline) or C. parvum (14 mg in 2 ml saline) 4 or 7 days, respectively, before sacrifice. Monolayers were established by allowing 3 x 10^6 cells - in the absence of serum - to adhere to surfaces of plastic dishes (Becton and Dickinson, Heidelberg, 3.5 cm in diameter) for 2 hrs at 37°C, 5% CO_2. In order to remove nonadherent cells, monolayers were washed vigorously and 95% of the remaining cells were able to engulf latex particles. Less than 5% of the adherent cells were granulocytes on microscopic examination following differential staining. Monolayers were incubated in Dulbecco's modified Eagle's Medium without serum for another 12 hours. Viability of cells was determined by measuring release of lactate dehydrogenase (LDH) into culture supernatants. Supernatants were collected at

timed intervals and assayed for N-acetyl-β-D-glucosaminidase activity
(3) and for C2, C3, and C4. Actual cell numbers at the end of the
incubation period were determined by measuring the DNA content of the
monolayers (4).

   Assay for C2, C4, C3: C2 and C4 in supernatants were determined
by a hemolytic assay. Samples were incubated with sensitized erythrocytes in the presence of serum from guinea pigs genetically deficient
in C2 or C4 respectively (5,6), and hemoglobin release determined
photometrically (412 nm). C3 levels were determined by ELISA (7).
In brief, C3 containing samples were added to vials coated with anti
guinea pig C3 and after washing interacted with peroxidase labeled
anti guinea pig C3 IgG. The amount of labeled antibody bound was
determined by addition of substrate ABTS (2,2 azino-bi-3-ethyl-
benzthiazoline-sulfonic acid) (Serva, Heidelberg) and absorption at
414 nm compared with that of a standard guinea pig C3. The smallest
amount of C3 detectable was 10 pg.

RESULTS

   Synthesis and Secretion of C3: Resident macrophages cultured
over a period of 12 hours secreted about 70 ng $C3/10^6$ cells into the
supernatant (Figure 1). Macrophages elicited with human serum albumin
secreted 88 ng $C3/10^6$ cells and C. parvum activated macrophages 183
ng $C3/10^6$ cells. Secretion could be abolished by incubating the
cells in the presence of 0.1 µg/ml cycloheximide. There was no difference in C3 secretion whether medium was replaced every 3 hours
or not (data not shown).

   Secretion of Hemolytically Active C2 and C4: Determination of
functional C2 and C4 in culture supernatants showed marked differences
between resident, elicited and activated macrophages (Figure 1).
C. parvum activated macrophages secreted 4 times as much hemolytically
active C2 and twice as much C4 as resident cells. Intermediate
values of both components were obtained from cells elicited with human
serum albumin.

   Kinetics: The time course of C3 secretion by resident elicited
and activated macrophages over a period of 12 hours is presented also
in Figure 1. Whereas resident cells showed a decline in secretion,
both elicited and activated macrophages displayed nearly constant
rates of C3 secretion. Secretion of hemolytically active components
C2 and C4 also proceeded at nearly constant rates over a period of
12 hours (data not shown).

   Enzyme Secretion: In order to correlate synthesis of complement
components with another indicator of macrophage activation, we studied
synthesis and secretion of the lysosomal enzyme N-acetyl-β-D-glucosaminidase. The degree of activation was reflected by the amount of

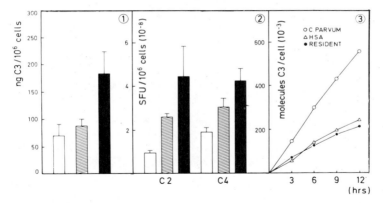

Fig. 1.   Secretion of complement components by resident, elicited,
and activated guinea pig macrophages. Panel 1:   secretion
of C3 by macrophages cultured for 12 hours.   C3 was measured
by ELISA.   Open column: resident macrophages; hatched
column: macrophages elicited by injection of human serum
albumin; dark column: C. parvum activated macrophages.
Panel 2:   Concomitant secretion of hemolytically active
C2 and C4.   Results are reported as site forming units (SFU)/
10^6 cells/12 hours.   Panel 3:   Kinetics of C3 secretion calculated per cell.

enzyme released into the supernatants inasmuch as in C. parvum induced
macrophages, total enzyme content was more than 7 times that of resident cells, and the amount released 5 times as high (Table 1).

DISCUSSION

     Although a number of reports present evidence for macrophages
being major producers of complement components, little attention
has been paid to whether complement synthesis in macrophages can be
influenced by inflammatory stimuli.   We have determined that resident
guinea pig peritoneal macrophages secrete about 70 ng C3/10^6 cells/12
hours.   This amount is 100 - 300 times higher than the production of
any other complement protein (1), thus reflecting the situation in
plasma where C3 is the most abundant component.   Further, we have
demonstrated that complement synthesis by macrophages is dependent
upon degree of macrophage activation.   Thus, release of C3 and hemolytically active C2 and C4 rose 2 - 4 fold in C. parvum activated
macrophages when compared to resident ones.   Intermediate values
were found in albumin elicited macrophages.   Thus, there is a stepwise

Table 1. Synthesis and Release of N-acetyl-β-D-Glucosaminidase
          by Resident, Elicited and Activated Macrophages

Type of Macrophage	Total Activity	Supernatant Activity
Resident	780 ± 30	200 ± 50
Elicited by human serum albumin	3010 ± 800	450 ± 140
C. parvum	5790 ± 580	1100 ± 180

Total enzyme activity (intracellular + extracellular) and enzyme
released into the supernatant were determined after 12 hours of
culture and are reported in mU/10^6 cells/12 hours. Data are
given as mean ± standard deviation on 15 experiments.

increase in complement production by resident, elicited and activated
macrophages. These changes in complement secretion parallel similar
alterations in lysosomal enzyme content and release.

    When studying the percentage of macrophages actually synthesizing
C2 and C4, Cole et al. (8) found no difference between resident and
starch elicited macrophages. They concluded that enhanced synthesis
of C2 and C4 was due to increased production per single cell. Since
it is not currently possible to perform similar hemolytic plaque
assays for C3, we do not know whether the increased production by
activated macrophages is caused by increased synthesis per cell or
by recruitment of more cells sharing in the generation of C3.

    In conclusion, we have quantitatively estimated C3 secretion by
guinea pig peritoneal macrophages and demonstrated for the first time
that secretion of C3, the central protein of the complement activation
pathway, is markedly enhanced in activated macrophages. Inflammatory
and immunological stimuli cause macrophages to secrete increased
amounts of C2, C3, C4 thereby augmenting local availability of complement in the face of accelerated turnover at sites of inflammation.

Acknowledgement

    This work was supported by a grant of the Stiftung Volkswagenwerk.

REFERENCES

1. Bentley, C., Zimmer, B., and Hadding, U., in "Lymphokines" Vol 4
      (E. Pick, ed.), pp. 197-239, Academic Press, New York, 1981.
2. Karnovsky, M. L., and Lazdins, J. K., J. Immunol. 121:809, 1978.
3. Woollen, J. W., Heyworth, R., and Walker, P. G., Biochem. J. 78:
      111, 1961.
4. Einstein, L. P., Schneeberger, E. E., and Colten, H. R., J. Exp.
      Med. 143:114, 1976.
5. Bitter-Suermann, D., Hoffmann, T., Burger, R., and Hadding, U.,
      J. Immunol. 127:608, 1981.
6. Gaither, T. A., Alling, D. W., and Frank, M. M., J. Immunol. 113:
      574, 1974.
7. Zimmer, B., Hartung, H. P., Scharfenberger, G., Bitter-Suermann,
      D., and Hadding, U., submitted for publication.
8. Cole, F.S., Matthews, W. G., Marino, J. T., Crash, D. J., and
      Colten, H. R., J. Immunol. 125:1120, 1980.

DISCUSSION

SCHREIBER: Since both C2 and C4 are labile proteins and once cleaved
no longer have any hemolytic activity, have you correlated their
antigeneity with function?

HARTUNG: No. Our test basically depends upon the antigenic properties of C3. The advantage of the conventional hemolytic assay is
that it reflects functional activity but it is less sensitive by
about a hundred fold. Your point is well made for we don't know if
the C3 secreted into the supernatants is actually functional. We
are now developing ELISA assays using monoclonal antibodies with
specificities for the various split products of C3. From data in
the literature, C3 released from starch induced guinea pig peritoneal
macrophages was mainly in the native form in supernatants. We didn't
do tests based on the antigenic properties of C2 or C4.

SCHREIBER: Were you able to examine any of these released proteins,
for the possibility that they're being released in their pro form,
that is pro-C4 which is a single chain molecule instead of three,
or pro-C3 which is a single chain molecule instead of two?

HARTUNG: No.

BITTER-SUERMANN: If we extensively concentrate the supernatants of
our macrophage cultures, we get C3 hemolytic activity in contrast to
the human system where you get no activity.

PERITONEAL MACROPHAGES RELEASE
PROSTAGLANDIN E AND THROMBOXANE
B_2 IN RESPONSE TO C3b

Ulrich Hadding, Hans-Peter Hartung, Heinrich Rasokat
and Diethard Gemsa

Inst. for Medical Microbiology
Univ. Mainz
Hochhaus Augustuspl. D-6500 Mainz, F.R.G.

Inst. of Immunology
Univ. Heidelberg
F.R.G.

INTRODUCTION

Macrophages have surface receptors for C3b, the larger product cleaved from complement component C3 upon activation of both the classical and alternative complement pathways (1). We have investigated whether or not C3b stimulates macrophages to release prostaglandin E (PGE) and thromboxane B_2 (TXB_2), metabolites of arachidonic acid produced in the cyclooxygenase pathway (2).

MATERIALS AND METHODS

C3b was prepared from purified guinea pig C3 (3) by interacting C3 with the enzymic complex cobra venom factor-bound, activated complement component B. C3b was separated from C3a by gel-filtration and from the enzymic complex by immunoabsorption.

Peritoneal macrophages were collected from outbred guinea pigs (300 - 500 g) 4 days after injection of human serum albumin (25 ml 0.5% in saline). 1.5×10^6 cells were allowed to adhere to 16 mm culture wells in the absence of serum. After 2.5 hrs nonadherent cells were removed by vigorous washing, stimuli added to a final volume of 1.5 ml/well of Dulbecco's modified Eagle's medium without serum, and incubated for 18 hrs. Viability of the cells was tested by measuring lactate dehydrogenase released into the culture supernatants (4).

PGE and TXB_2 were determined by radioimmunoassay (5).

RESULTS

C3b stimulated the release of PGE in a dose dependent manner (Table 1). The effect of C3b was comparable to that of zymosan, a well-known inducer of prostaglandin synthesis. The stimulatory effect of C3b was abrogated when indomethacin was added to the cultures. C3b also induced the release of significant amounts of TXB_2.

When TXB_2 levels were determined in supernatants collected at timed intervals following addition of 80 μg/ml C3b, it was found that TXB_2 release rose above spontaneous secretion at 2 hrs (2.2 ng/ml), increased at 6 hours to 10.8 ng/ml and approached a plateau after 12 hours of 18 ng/ml (after 18 hours, 19.8 ng/ml).

In order to demonstrate that the stimulatory action was due to C3b, we passed C3b down an immunoabsorbent column (CnBr activated Sepharose 4B   Pharmacia, Uppsala, to which monoclonal anti-C3 IgG was bound) and measured PGE and TXB_2 release from cultures exposed

Table 1.   Influence of C3b on Release of Prostaglandin E
            by Macrophages

Stimulant	PGE ng/ml
Medium Only	0.4
C3b 9 μg/ml	1.5
19 μg/ml	3.1
38 μg/ml	4.5
75 μg/ml	6.5
C3b (75 μg/ml) + Indomethacin (0.5 μg/ml)	0.0

Zymosan (50 μg) released 5 ng PGE/ml.

to the effluent.   Immunoabsorption of C3b almost completely abolished
the stimulating effect.

DISCUSSION

     Macrophages metabolize arachidonic acid by various pathways to
prostaglandins, thromboxanes and leukotrienes (6).   A number of inflammatory stimuli elicit release of prostaglandins from macrophages
(7).   Gemsa et al., while studying the modulation of phagocytosis-
induced liberation of prostaglandins, found that bacteria coated with
IgG plus complement were the most efficient particles to evoke PGE
release from macrophages (8).

     Our findings demonstrate that fluid phase C3b causes macrophages
to produce PGE and TXB$_2$.   Other biological effects of C3b include
immune adherence, opsonization and secretion of lysosomal enzymes
(reviewed in Ref. 9).   The latter action of C3b may be important in
the pathogenesis of chronic inflammation (10).   Since prostaglandins
are involved in inflammatory responses (11), the results presented
here indicate an additional pathway by which C3b may play a pathogenic role in inflammation.

REFERENCES

1.   Griffin, F. M., Bianco, C., and Silverstein, S. C., J. Exp. Med.,
        144:1269, 1975.
2.   Samuelsson, B., Goldyne, M., Granström, E., Hamberg, M., Hammar-
        ström, S., and Malsten, C., Ann. Rev. Biochem. 47:997, 1978.
3.   Meuer, S., Becker, S., Hadding, W., Bitter-Suermann, D., Immuno-
        biol. 154:135, 1978.
4.   Wroblewski, F., and LaDue, J. S., Proc. Soc. Exp. Biol. Med.,
        90:210, 1955.
5.   Grimm, W., Seitz, M., Kirchner, H., and Gemsa, D., Cell Immunol.
        40:419, 1978.
6.   Stenson, W. F., and Parker, C. W., J. Immunol. 125:1, 1980.
7.   Gemsa, D., in: "Lymphokines" Vol. 4 (E. Pick, ed.), pp. 335-375,
        Academic Press, New York, 1981.
8.   Gemsa, D., Seitz, M., Menzel, J., Grimm, W., Kramer, W., and
        Till, G., Adv. Exp. Med. Biol.114:421, 1979.
9.   Hadding, U., Agents and Actions 7:24, 1980.
10. Schorlemmer, H. U., Allison, A. C., Immunology 31:781, 1976.
11. Kuehl, F. A. Jr., and Egan, R. W., Science 210:978, 1980.

DISCUSSION

BAGGIOLINI: This is a very interesting finding because it's the
only one I know where you have no stimulation of the oxidative burst

yet stimulation of PGE release by a phagocyte. These findings bring into discussion the action of C3b as an activator of macrophages.

DAVIES: Some years ago, Stossal looked at the opsonically active form of C3, but I don't think it's been heard of again. The suggestion was made that it was the inactivated form C3bI, which was the opsonically active form distinct from C3b. I wonder if you looked at C3bI and if that might stimulate the formation of toxic oxygen metabolites?

HADDING: I don't know if C3bI is the opsonized form. You can prepare particles carrying C3b but only in the absence of serum. In serum, the C3b splits further. In the absence of serum, you are not working in a physiological environment. We started with a pure component in which there was no further modification of the C3b. We purposely used an enzyme which would only split at one point and nothing more. We tried to have a stimulus where we could say, "We know what we have." So, I cannot tell what would happen if we prepared C3bI. I don't know if it would behave similarly.

OTTENDORFER: I'd like to comment on the question put forward by Dr. Baggiolini. There are reports which show differential effects of stimulators of the oxidative burst and the release of arachidonic acid cyclo-oxygenase products. They've found that some stimulants induce an oxidative burst but fail to produce prostaglandin release.

DIERICH: Human C3b prepared with trypsin acts on human monocytes to increase their chemiluminescence. Just to make it very clear, the human situation differs from that in guinea pigs. In guinea pigs, C3b prepared with cobra venom factor B causes no increase in guinea pig macrophage chemiluminescence.

PICK: In a situation where a stimulant is clearly a prostaglandin or thromboxane synthesis inducer but in which there is no oxidative burst, it is perhaps wise to try high concentrations of indomethacin or ETYA to close both the cyclo-oxygenase and lipoxygenase pathways. I would predict that then there would be an increase in the concentration of non-oxidized fatty acid, which may be the real inducer of the oxidative burst. It is a paradox that the more arachidonic acid flows into the oxidative pathway, the less is the chance for a burst induction. As far as Dr. Dierich's comment, I think it's dangerous to use chemiluminescence as an absolute indicator of the burst because there are so many possible interpretations. I would not use it unless you have a clear inhibition by some enzyme or anti-oxidants. I think it's safe to use known products such as Dr. Hadding has used.

PROSTAGLANDIN E_2 IS A FEED-BACK

REGULATOR OF MACROPHAGE ACTIVATION

Jörg Schnyder, Beatrice Dewald and Marco Baggiolini

Wander Research Institute, a Sandoz research unit
Wander Ltd., P. O. Box 2747
CH-3001 Berne, Switzerland

INTRODUCTION

Phagocytosis is a main function of macrophages and a powerful stimulus of macrophage differentiation (1). When quiescent macrophages are induced to phagocytose, a number of immediate responses are observed: (a) respiratory burst with release of superoxide and H_2O_2; (b) release of arachidonic acid oxygenation products (3) and (c) secretion of lysosomal hydrolases (1,4). The delayed consequence of phagocytosis is a differentiation or activation process characterized by an increased production of cellular and export proteins that result in cell growth and the expression of novel secretory functions.

Prostaglandin E_2 (PGE$_2$) is the main cyclooxygenase product released by macrophages which engage in phagocytosis. Since E-type prostaglandins were reported to modify a variety of lymphocyte, mononuclear phagocyte and osteoclast functions, we have tested the effects of PGE$_2$ and of PG synthesis inhibition on a number of biochemical correlates of macrophage activation.

MATERIALS AND METHODS

Quiescent macrophages were obtained by peritoneal lavage from non-treated male mice and were cultured according to standard techniques (5). Adherent cells were stimulated by zymosan phagocytosis as described in a previous paper (1) in the presence of indomethacin and/or PGE$_2$ in various concentrations. Established methods were used for the assay of lactate dehydrogenase (5), β-glucuronidase (5), plasminogen activator (5), alkaline phosphodiesterase I (6), and hexose monophosphate (HMP) shunt activity (7). PG production was measured by a radioimmunoassay for E-type prostaglandins (Miles Laboratories, Inc.).

RESULTS

Table 1 shows that blockade of PG production or addition of PGE_2 has no effect on the immediate-onset responses to phagocytosis and does not alter HMP shunt activity and β-glucuronidase release in resting (i.e. non-phagocytosing) cells. Therefore, one can conclude that the respiratory burst and the secretion of lysosomal hydrolases do not depend on the levels of cyclooxygenase products which the macrophages release upon stimulation.

Quiescent macrophages which are kept in culture for several days spread on the surface of the culture dishes and enlarge slightly, as shown by a moderate increase in the cellular contents of enzymes like β-glucuronidase and alkaline phosphodiesterase I. Such changes are more pronounced when the same cells are stimulated by phagocytosis. The effect of the phagocytic stimulus is best shown by plasminogen activator secretion which is detectable from about day 3 on in stimulated cultures, but remains very low in the nonstimulated controls. As shown in Table 2, these changes are amplified both in phagocytosing and non-phagocytosing cells when indomethacin is present in the culture medium at concentrations which block PG production. On the other hand, again in both cases, secretion of plasminogen activator and the rise in cellular β-glucuronidase and alkaline phosphodiesterase I are prevented by addition of PGE_2, in the presence and absence of indomethacin.

DISCUSSION

This study shows that inhibition of cyclooxygenase facilitates macrophage activation while addition of PGE_2 prevents or reverses this process. These effects suggest that PGE_2 acts as an inhibitory regulator of macrophages. PGE_2 is released by the macrophage in response to stimuli leading to activation and appears to exert an inhibitory control over the activation process. It is conceivable that other cyclooxygenase products released by macrophages, i.e. PGI_2 (3) and TXA_2 (8) act similarly. Ref. 9 presents a complete account of this work.

Table 1.   Effect of Cyclooxygenase Inhibition by Indo-
           methacin and PGE_2 Addition on Phagocytosis-
           Stimulated Macrophages.   Immediate-Onset
           Responses

Phago-cytosis	Additions indo-meth-acin μM	PGE_2 μM	Hexose monophosphate shunt nmoles/10^6 cells	PGE_2 production ng/10^6 cells	β-glucuroni-dase secretion mU/10^6 cells
no	–	–	2.06	24.5	0.45
no	1.0	–	2.04	3.5	0.47
no	–	0.1	2.09	15.6	0.41
yes	–	–	35.6	102.9	3.08
yes	1.0	–	44.1	3.2	3.14
yes	–	0.1	34.1	–	3.34
yes	1.0	0.1	39.9	–	3.04

Phagocytosis was initiated by addition of zymosan (8 particles
per cell) at time 0 and eliminated by washing after 1 hour.
Indomethacin and PGE_2 were added 10 min before zymosan and were
then present throughout the experiment.   Hexose monophosphate
shunt activity was measured by the liberation of $^{14}CO_2$ from
^{14}C-1-glucose during phagocytosis or the corresponding control
period.   PGE and β-Glucuronidase were measured on day 2.

Table 2. Effect of Cyclooxygenase Inhibition by Indo-
          methacin and PGE$_2$ Addition on Phagocytosis-
          Stimulated Macrophages. Delayed Responses.

Phago-cytosis	Additions indo-meth-acin μM	PGE$_2$ μM	Plasminogen activator secretion U/10^6 cells	Cellular β-glucuronidase mU/10^6 cells	Cellular alkaline phospho-diesterase I mU/10^6 cells
no	–	–	55	2.2	2.6
no	1.0	–	179	2.6	3.3
no	–	0.1	0	1.5	1.4
no	–	1.0	2	1.2	0.9
yes	–	–	467	3.9	5.0
yes	0.1	–	1186	6.7	6.8
yes	1.0	–	621	6.2	6.1
yes	–	0.1	408	3.6	4.1
yes	–	1.0	72	2.0	3.0
yes	1.0	1.0	74	2.0	3.2

Conditions as in Table 1. The delayed response is represented by
the amount of plasminogen activator secreted between day 7 and 10
and by the cellular levels of β-glucuronidase and alkaline phosphodiesterase I on day 10. On day 0, both of these levels were
1.5 mU/10^6 cells.

REFERENCES

1. Schnyder, J., and Baggiolini, M., J. Exp. Med. 148:1449, 1978.
2. Badwey, J. A., and Karnovsky, M. L., Ann. Rev. Biochem. 49:695,
        1980.
3. Davies, P., Bonney, R. J., Humes, J. L., and Kuehl, Jr., F. A., in
        "Mononuclear Phagocytes. Functional Aspects, Part II" (R. van
        Furth, ed.), pp. 1317-1345, Martinus Nijhoff Publishers, The
        Hague/Boston/London, 1980.
4. Davies, P., Page, R. C., and Allison, A. C., J. Exp. Med. 139:
        1262, 1974.
5. Schnyder, J., and Baggiolini, M., J. Exp. Med. 148:435, 1978.
6. Edelson, P. J., and Erbs, C., J. Exp. Med. 147:77, 1978.
7. Schnyder, J., and Baggiolini, M., Proc. Natl. Acad. Sci. USA 77:
        414, 1980.
8. Brune, K., Glatt, M., Kälin, H., and Peskar, B. A., Nature
        274:261, 1978.
9. Schnyder, J., Dewald, B., and Baggiolini, M., Prostaglandins
        22:411, 1981.

DISCUSSION

LAZDIN: It was presented in poster session that, during activation
for cytotoxicity with thioglycollate induced macrophages, there is
a decrease in the production of plasminogen activator. In your
opinion, what is the relevance of plasminogen activator as a marker
for macrophage activation?

SCHNYDER: Thioglycollate elicited macrophages produce a large amount
of plasminogen activator, but they produce no hydrogen peroxide.
And that is the reason why they are not cytotoxic.

LAZDIN: No. What I meant was that when thioglycollate induced
macrophages are activated by MAF or by LPS to become cytotoxic, the
rate of production of plasminogen activator decreases rather than
increases. So it seems that there is an inverse relationship between
plasminogen activator production and activation of macrophages for
cytotoxicity.

SCHNYDER: What was the concentration of LPS? When we add LPS to our
cultures, the cells are not happy and leak LDH and cause secretion
of lysozyme.

BAGGIOLINI: May I interrupt? In think I'm partly responsible for
the impression that LPS depresses plasminogen activator secretion.
In many systems, like non-elicited macrophages or starch or thioglycollate elicited macrophages one can monitor in culture the state
of activation of the macrophages by looking at plasminogen activator
and other enzymes. This doesn't mean that plasminogen activator

secretion is always high in the case of macrophage activation. Indeed, we had one example where this was not the case, and this is what you're talking about.

DAVIES: I think it's a question of terminology. "Activation" should be used only for cells capable of increased bactericidal or cytocidal activities. All other things are perhaps stimulated or elicited activities.

SCHNYDER: We used activation in a biochemical rather than functional sense.

RUSSELL: This is a confusing field and we tend to make it more so by using the term activation so indiscriminately. It has recently been suggested that we should say "activation for..." and then define whatever the specific function is that we are talking about. In this case, it would be activation for plasminogen activator secretion.

CRUCHAUD: Have you tried any other non-steroidal inflammatory drug, because indomethacin could have a phosphodiesterase effect.

SCHNYDER: We have tested a battery of cyclo-oxygenase inhibitors.

CRUCHARD: Have you looked for elastase or collagenase in your supernatants? If we add PGE_2 we can increase collagenase production.

SCHNYDER: Collagenase is not always secreted like plasminogen activator. Plasminogen activator was increased and collagenase was not. Thus, the enzymes need not be secreted concurrently.

PICK: Collagenase is an exceptional situation. It's probably the only enzyme whose secretion is stimulated by PGE_2.

SCHNYDER: If you have cells making collagenase and prostaglandin and if you add indomethacin to the culture, you decrease collagenase secretion. If you now add prostaglandin, you can restore the original level.

SECTION 8

ANTIGEN PRESENTING MACROPHAGES

MECHANISMS CONTROLLING DIFFERENTIATION AND

FUNCTION OF ANTIGEN-PRESENTING MACROPHAGES

M. Feldman,[1] E. Tzehoval,[1] Y. Ron,[1] P. De Baetselier,[3]
M. Fridkin,[2] and S. Segal[1]

[1]Department of Cell Biology, The Weizmann Institute
of Science, Rehovot 76100, Israel
[2]Department of Organic Chemistry, The Weizmann Institute of Science, Rehovot 76100, Israel
[3]Free University of Brussels, Brussels, Belgium

Mechanisms controlling the differentiation and function of antigen-presenting macrophages were studied using two distinct approaches. One asked either what are the genetic determinants and their gene-products which have to be expressed on antigen-presenting cells in order to enable macrophages to present antigen to lymphocytes, or alternatively what are the exo-cellular signals which induce the expression of these functional gene products.

We shall confine ourselves to a discussion of some of the inducing exo-cellular factors, which either trigger the differentiation of a macrophage into a differentiated, mature antigen-presenting cell, or activate the immunogenic function of the mature cell.

Considering the activation process of the differentiated macrophage, one can guess the identity of some of the signals according to the cell surface receptors which characterize the macrophage membrane. Thus, the existence of Fc-receptors suggested that Ig molecules might function as regulating signals for macrophage function. The simplistic approach attributed to these molecules an "opsonizing" effect, increasing the phagocytic or pinocytic activity of macrophages. With regard to phagocytosis, there is no doubt that Ig molecules may increase particle uptake. Yet, studies initiated by Victor Najjar indicated that the effect is actually not mediated by the intact Ig molecule. Rather, from the Ig bound to the Fc receptor, a tetrapeptide (Thr, Lys, Pro, Arg), tuftsin, is cleaved off, and it is this peptide which mediates the phagocytic process (1-3).

Since antigen-presenting cells are also equipped with Fc receptors, and since antigen uptake by such cells seems a prerequisite for
antigen presentation, we carried out experiments to test whether
tuftsin activates antigen presentation to lymphocytes, and whether
it acts by increasing the antigen load within the immunogenic macrophage (4).

TUFTSIN ACTIVATES THE IMMUNOGENIC FUNCTION OF MACROPHAGES

We used a system in which a monolayer of in vitro antigen-pulsed
peritoneal macrophages presented antigen (KLH) to virgin spleen lymphocytes (5). Such spleen cells, having interacted with the antigen-
presenting cells for 18 h, were exposed to x-irradiation to prevent
their replication and were then injected into the footpad of a syngeneic mouse. In the adjacent popliteal lymph nodes, they subsequently recruit effector T lymphocytes. The latter were then
measured in vitro for thymidine uptake in specific response to KLH
(5). We found that this response was significantly augmented when
the introduction of KLH to the macrophages was concomitant with
treatment of the macrophages with synthetic tuftsin (4). Yet it
appeared to us that this augmentation could not be attributed to
increased pinocytosis, i.e., to an increase in antigen load within
the antigen-presenting cell, because synthetic analogs of tuftsin,
such as (Ala')tuftsin, H-Ala-Lys-Pro-Arg-OH or the quintapeptide
Thr-Lys-Pro-Arg-Gly manifested an even greater potentiating immunogenic effect than that obtained by tuftsin, yet these peptides had
no effect on phagocytosis. We therefore suggested that tuftsin
activates the immunogenic function of macrophages but not by increasing antigen uptake (4).

Comparing the effect of various synthetic tetrapeptide analogs
of tuftsin which have a suppressing effect on antigen presentation,
it appeared that although the entire active tetrapeptide is most
probably recognized by the tuftsin receptor on the macrophage cell
surface, it is the Pro-Arg that represents the sequence which triggers
the intracellular processes in the antigen-pulsed cells which end in
an immunogenic signal fired by the macrophage to the spleen lymphocytes (4).

THE SPLEEN EMITS AN INDUCING SIGNAL TO THE ANTIGEN-PRESENTING CELLS

Compared to peritoneal macrophages, the incidence of Ia-positive
macrophages in the spleen is significantly higher. Since these are
the macrophages which are engaged in antigen presentation, it seemed
to us of interest to test whether the spleen exerts a controlling
effect on the antigen-presenting macrophages of the peritoneum.
Furthermore, one of the two enzymes which cleaves off the tuftsin
tetrapeptide from the CH_2 portion of the Ig is located in the spleen

while the other enzyme, the leukokininase, is located on the macrophage cell surface.   We therefore carried out a series of experiments
testing the capacity of thioglycollate-induced peritoneal macrophages
from splenectomized donors to induce in spleen cell populations from
normal donors the generation of antigen-specific initiator lymphocytes. We found that such peritoneal macrophages from donors splenectomized 3-4 months earlier were impaired, compared to macrophages
from normal donors (6). In fact, a similar impairment was found when
KLH-pulsed macrophages were tested for their capacity to induce a
lymphoproliferative response by presenting the antigen to KLH-primed
T lymphocytes. Thus, both the primary and the secondary stimulation
with antigen via peritoneal macrophages obtained from splenectomized
animals were significantly impaired, compared to macrophages from
sham operated mice (6). The impairment in the immunogenic function
of these macrophages could not be attributed to an impaired antigen
uptake by the antigen-presenting cells, since the labeled KLH in
macrophages from antigen-pulsed splenectomized donors was similar
to that measured in antigen-pulsed macrophages from normal animals.
This could not reflect "surface-absorbed" antigen, since similar
results were obtained following treatment of the macrophages with
trypsin. It thus appears that the peritoneal macrophages from splenectomized donors take up an amount of antigen similar to that taken
up by macrophages from normal donors, yet their capacity to present
antigen to lymphocytes is significantly impaired (6). It is impossible at this stage to attribute this impairment to a defined factor
within the spleen. Yet, it is clear that the spleen exerts a regulating effect on nonspecific macrophages. In fact, although the
uptake of KLH by peritoneal macrophages was not affected by splenectomy, uptake of Ig-coated micrococci was reduced (6). Whether this
reduction reflected an impairment relevant to the immunogenic defect,
remains an open question. One possibility, however, is noteworthy:
if the lack of tuftsin is a factor, then it cannot be attributed to
a lack of the spleen enzyme that splits the carboxy terminal of the
tetrapeptides, since macrophages from splenectomized donors fed with
micrococci treated with Ig from normal donors (in which the spleen
enzyme could have functioned) showed a reduced level of phagocytosis.
A decreased activity of the macrophage leukokininase could account
for the lack of tuftsin cleavage on the macrophage cell surface but
then one assumes that the activity of a macrophage-associated enzyme
is controlled by the spleen. No other observations support this
notion.

THE THYMUS CONTROLS THE DIFFERENTIATION OF THE ANTIGEN-PRESENTING
SUBSET OF MACROPHAGES

     Do the antigen-presenting peritoneal macrophages constitute a
defined subpopulation of cells, distinct in structural, functional
and developmental properties? To approach this question we adopted
the discontinuous BSA gradient for fractionating peritoneal macro-

phages (7). We found that at the interfaces of the various density
layers of BSA, there was a segregation of the macrophage population
into subsets with different morphological properties (8). We then
tested the capacity of fractions 1 to 4, representing cells accumulated at the 8/11, 11/15, 15/20 and 20/30% BSA interfaces, respectively, to present KLH antigen to spleen lymphocytes. Monolayers of
macrophages were prepared from each of these fractions, the monolayers were then pulsed with KLH, and spleen cells were seeded on
top of these monolayers for 18 h of interaction; they were then
collected, irradiated and injected to syngeneic mice. Six days later,
lymphocytes recruited in the regional lymph nodes were assayed for
specific response to KLH.

The results indicated that cells from fraction 1, at the 8/11%
interface, were the most potent in antigen-presenting capacity. These
cells also showed the highest density of Ia molecules (8). On the
other hand, macrophages of fraction 3 were found to be devoid of Ia,
and also inactive in antigen presentation. This fraction was, however, the most active in opsonin-dependent phagocytosis. Thus, it
appears that antigen presentation and phagocytosis are two distinct
functions performed by distinct subsets of macrophages.

We found previously that macrophages from nude or thymectomized
mice are impaired with respect to their capacity to present antigen
to naive lymphocytes (9). When we subjected peritoneal lymphocytes
to BSA fractionation, we found that fraction 1, the most immunogenic
fraction, was missing in the peritoneum of nude mice. Fraction 3,
which is the least immunogenic, yet is highly phagocytic, constituted
a great part of the nude peritoneal macrophage population (8).

To further demonstrate that the thymus is indeed involved in
controlling the differentiation of the antigen-presenting subset of
peritoneal macrophages, we tested the capacity of thymic cells to
restore the defective immunogenic capacity of macrophages from athymic
mice, and to induce the appearance of the "missing" fraction. We
found that the intraperitoneal inoculation into nude mice of hydro-
cortisone-resistant thymocytes from normal donors completely restored
the otherwise impaired capacity of nude peritoneal macrophages to
present antigen to unprimed lymphocytes. Light scatter analysis of
the peritoneal cells indicated that the initial different scatter of
the peritoneal population of the nude mice had become identical to
that of the heterozygous control mice (8).

It should be noted finally that the impairment in antigen presentation of peritoneal macrophages from nude donors applies to the
primary antigenic stimulation. We found that macrophages from nude
mice which were unable to trigger naive lymphocytes could present
antigen (KLH) to primed T cells. This may indicate basic differences
between the control of primary and secondary presentation of antigen
to lymphocytes. One should consider the following alternatives:

(a) The differences manifest only quantitative differences in the level of immunogenic signals required for stimulating unprimed compared to primed lymphocytes. (b) Macrophages engaged in antigen presentation to unprimed lymphocytes constitute a different subpopulation from those which stimulate memory lymphocytes. (c) The lymphocytes receiving the immunogenic signal in a primary response constitute a fundamentally different population from the primed T cells responding in a secondary stimulation. Macrophages from nude mice can stimulate the latter, but not the other type of lymphocytes.

SUMMARY

We have briefly reviewed our studies on the mechanisms controlling the differentiation and activation of peritoneal antigen-presenting cells. We demonstrated that the peritoneal population is composed of two main subsets of cells, only one of which participates actively in primary antigen presentation. The latter is missing in athymic mice and seems to differentiate under the influence of the shortlived, cortisone-resistant subpopulation of thymocytes. The maturation of the peritoneal macrophages is subjected also to an additional inducing effect, that of the spleen. Macrophages from splenectomized donors are impaired both with respect to antigen presentation to naive and to primed lymphocytes, and with respect to phagocytosis of "opsonized" bacteria. The mature antigen-presenting cell is subjected to activating signals deriving from the Fc-bound Ig molecule. This is mediated via a tetrapeptide, tuftsin, which is cleaved off the CH_2 portion of the Ig and activates the immunogenic effect of the antigen-pulsed macrophage.

Acknowledgement

Supported by Grant No. 2645 from the U.S.-Israel Binational Science Foundation, Jerusalem.

REFERENCES

1. Najjar, V. A., and Nishioka, K., Nature 228:672, 1970.
2. Nishioka, K., Constantopoulos, A., Satoh, P. S., and Najjar, V. A., Bioch. Biophys. Res. Comm. 47:172, 1972.
3. Fridkin, M., Stabinsky, Y., Zakuth, V., and Spirer, Z., Bioch. Biophys. Acta 496:203, 1977.
4. Tzehoval, E., Segal, S., Stabinsky, Y., Fridkin, M., Spirer, Z., and Feldman, M., Proc. Nat. Acad. Sci. USA 75:3400, 1978.
5. Steinman, L., Tzehoval, E., Cohen, I. R., Segal, S., and Glickman, E., Eur. J. Immunol. 8:29, 1978.
6. Ron, Y., De Baetselier, P., Feldman, M., and Segal, S., Eur. J. Immunol. 11:608, 1981.
7. Rice, S. G., and Fishman, M., Cell. Immunol. 11:130, 1974.

8. Tzehoval, E., De Baetselier, P., Feldman, M., and Segal, S., Eur.
      J. Immunol. 11:323, 1981.
9. Tzehoval, E., Segal, S., and Feldman, M., Proc. Nat. Acad. Sci.
      USA 76:4056, 1979.

DISCUSSION

UNANUE: We find normal numbers of Ia positive macrophages in the
spleen and peritoneal cavity of athymic mice. Further, T cells
are able to respond to the macrophages bearing antigen. Thus our
results are quite different from yours. I think the difference involves the kind of assays that we each use. The macrophages from the
thymic versus athymic animal are probably different. For instance,
macrophages from athymic mice are more activated cells since nude
mice are subject to infection. I'm concerned about your system of
measuring T cell proliferation for it contains a number of unknown
variables. Perhaps in your system there is a dysfunction of the
macrophage but in a more direct type of assay, such as we are using,
it is not evident. Therefore, I'm questioning whether you directly
measure antigen presentation or something else.

FELDMAN: The assay which we use for measuring proliferating T cells
is a very direct assay, based on thymidine incorporation. The induction phase of the system may appear rather complicated, but we feel
that we imitate in a clear way the stages which an inductive process
must follow in vivo. We take a monolayer of macrophages and pulse
with antigen. We then seed the spleen cells as initiators. They
interact as well as with T cells and the latter ones are the ones
which we measure.

KENYON: In our studies with nude mice, we used paired litter mates
as controls. We saw no real difference between them. If anything,
antigen presentation was better with the nude mouse.

FELDMAN: I repeat. Antigen presentation which we have measured was
impaired in the nude mouse. It was impaired in neonatally thymectomized animals. It was impaired in adult thymectomized mice 2 months
after thymectomy.

CHARACTERISTICS OF AN IA$^+$

ANTIGEN-PRESENTING TUMOR CELL LINE

D. A. Cohen, L. A. Smith and A. M. Kaplan

Department of Surgery
Virginia Commonwealth University
Medical College of Virginia
Richmond, Virginia   23298 USA

INTRODUCTION

The controlling elements responsible for regulating Ag-specific proliferation appear to be associated with the gene products of the I region within the major histocompatibility complex. Numerous studies (1-5) have indicated that for efficient T-cell proliferation to occur, Ag must be presented to the T cell on the surface of an Ag-presenting cell (APC) in either direct or indirect association with Ia molecules expressed by the APC. A complication in studying the nature of the relationship of Ag to Ia has been that the APC population represents a small and variable population and is very likely comprised of several cell types. The ability to achieve homogeneous populations in sufficient quantities for molecular studies has been hampered by the fact that APCs in the peripheral organs (macrophages or dendritic cells) appear to be terminally differentiated and, as such, are relatively resistant to most cloning techniques. An alternative approach for deriving stable clones of APCs is to develop tumor cell lines which express antigen-presenting capacity; these cell lines can then be used as homogeneous populations to evaluate the interaction of Ag and Ia molecules on the surface of the APC. We have recently described several adherent tumor cell lines which possess morphological and functional characteristics of splenic dendritic cells, including constitutive expression of surface Ia molecules (6). We describe in this report that these cell lines (P388AD) have the ability to present Ag to primed T cells which have been depleted of endogenous APC.

METHODS

Cell Cultures: The procedures for the isolation of P388AD from the P388 leukemia have been described in detail elsewhere (6). Briefly, the P388 leukemia (nonadherent) was cultured in vitro under conditions which selected for adherent variants. The adherent cultures were cloned by limiting dilution and several adherent clones (P388AD) were selected for study. Nonadherent clones of the P388 leukemia were also obtained by limiting dilution and were designated P388NA.

Immunofluorescence: Surface Ags were detected by indirect immunofluorescence as previously described (6). Alloantisera anti-I-A.11,16 and anti-H-2D^d were obtained from the National Institutes of Health (NAIAD), Bethesda, MD. Monoclonal antibodies (Ab), anti-Thy-1.1 and M1/70 (anti-Mac-1) were obtained from the Salk Institute. The following monoclonal reagents were a generous gift from Dr. Noel Warner (Becton Dickinson and Co., Mountain View, CA): anti-Ly 1, Ly 2, ThB, E2, DNL 1.9, and Ly 9.1.

Receptors and Enzymes: The procedures for detection of surface receptors for the Fc portion of Ab and for the C3b fragment of complement and for detection of the enzymes peroxidase and nonspecific esterase were described previously (6).

Functional Assays: Procedures for phagocytosis, primary mixed lymphocyte reactions (MLR) and primary syngeneic mixed lymphocyte reactions (SMLR) have been described elsewhere (6). Secretion of interleukin-1 (IL-1) following stimulation with lipopolysaccharide (LPS) was determined by Dr. Stephanie Vogel (Uniformed Services, University of the Health Sciences, Bethesda, MD) utilizing a PHA-induced thymocyte proliferation assay (7). The ability of trinitrophenyl (TNP)-modified clones to induce hapten-specific cytolytic T cells (CTL) was performed essentially as described by Shearer, et al. (8) except that TNP-modified clones, syngeneic to the responding T-cell population, were used as stimulators and/or targets. Presentation of Ag to primed T cells was determined in a proliferative assay as follows. DBA/2 female mice were immunized subcutaneously into the footpads and base of the tail with 100 μg/site of purified turkey gamma globulin (TGG) emulsified 1:2 in complete Freund's adjuvant. Seven to 14 days later, draining lymph nodes were removed and T cells were enriched on nylon-wool columns and incubated at 4°C for 30 min in a 1:10 dilution of anti-Ia^d antiserum (B10.LG x A.TFR-4 anti-B10.D2) followed by a 30 min incubation at 37°C in prescreened rabbit complement (1:10) (low-tox M, Accurate Chemical Corp., Westbury, NY). This T-cell population was considered devoid of endogenous APC since no response to soluble TGG was detected unless unprimed, splenic accessory cells were added. Five x 10^5 T cells were then cultured

in microtiter plates for 4 days with various concentrations of mitomycin C-treated spleen cells or clones which were preincubated with
TGG (5 mg for 60 min at 37° C) and extensively washed (Ag-pulsed
clones). The extent of proliferation was determined with an 18 h
pulse of tritiated thymidine (2 μCi/well). Proliferation of T cells
cultured with Ag-pulsed clones was considered Ag-specific if the CPM
were greater than that seen with unpulsed clones.

RESULTS AND DISCUSSION

        Previous studies (6) with clones of P388AD (adherent) and P388NA
(non-adherent) indicated that distinct differences existed between
adherent and nonadherent clones, in spite of the fact that both were
derived from the same parental nonadherent tumor, P388 leukemia.
A comparison of P388AD and P388NA has been made with respect to
surface Ags, surface receptors, cytoplasmic enzymes and several immunological functions and is summarized in Table 1. H-2 Ags of the
D region and I-A subregion are expressed in all clones tested, adherent or nonadherent. The expression of I region gene products has
been verified by several techniques (immunofluorescence, complement-
mediated cytotoxicity and quantitative absorption) and with several
monoclonal reagents and alloantisera. The presence of Ia, as well
as the absence of phagocytic activity and Fc and C3b receptors, clearly distinguishes P388AD from the well-characterized macrophage tumor
line, P388D1. In an attempt to further identify the origin of these
tumors, we obtained a series of monoclonal reagents from Dr. Noel
Warner which could be used to screen for surface markers on T cells,
B cells, macrophages or their precursors. The derivation and specificities of these reagents have been described in detail by Ledbetter
and Herzenberg (10). Of the seven additional monoclonal reagents
tested, only Thy 1, Ly 1 and DNL 1.9 Ags were detected. Expression
of DNL 1.9 (expressed on B cells and pre-B cells) is consistent with
a B-cell origin of the P388 leukemia; however, simultaneous expression
of Thy 1, Ly 1 and Mac-1 in P388AD confuses any simple interpretation
as to the origin of the tumor. Although Thy 1 appears inconsistent
with a myeloid/monocyte origin, expression of this T cell Ag has also
been found in another myeloid/monocyte tumor, WEHI-3 (11). In addition, the expression of Ly 1 (a T-cell marker) is also found on some
B-cell tumors (12) and although Ly 1 is expressed on both P388AD and
P388NA, expression on P388AD is diminished when compared to P388NA.
The data from membrane Ag analysis suggest a possible pre-B-cell
origin for both P388NA and P388AD (Ly1⁺, DNL 1.9⁺, FcR⁻, C3bR⁻ sIg⁻);
however, when one looks at the expression of Mac-1 and Ia as well
as cellular morphology in P388AD, this cell line appears more like
an immature macrophage or dendritic cell (Ia⁺, Mac-1⁺, Fc⁻, C3b⁻,
nonphagocytic). Clearly, more work is necessary before we know the
true origin of this tumor.

        Many studies have suggested that expression of Ia Ags is a prerequisite for induction of the primary MLR and SMLR (13-16). P388AD but not P388NA stimulated unprimed allogeneic (C57B1/6) or syngeneic (DBA/2) T cells during mixed culture. The nonadherent P388NA expressed equivalent amounts of membrane Ia but failed to trigger T-cell proliferation. Warner et al. (11) similarly found a lack of correlation between Ia expression and T-cell stimulation when they tested Ia$^+$ B-cell lymphomas for their ability to stimulate an MLR.

Stimulation of T-cell proliferation by an Ia$^+$ inducer cell apparently requires at least two separate signals from the inducer cell: recognition of Ia molecules and release of a nonspecific soluble factor, IL-1 (17,18). In an attempt to distinguish between the stimulatory capacities of our Ia$^+$ P388AD and P388NA clones, we determined the ability of the clones to synthesize IL-1 after stimulation by LPS. All clones of P388AD synthesized IL-1, whereas all P388NA clones were negative for IL-1 synthesis. Thus, it would appear that IL-1 synthesis may be a necessary requirement for induction of the MLR and SMLR by P388AD.

Since clones of adherent P388AD possess two necessary components of APCs, we felt it necessary to assess whether the clones would present Ag to primed T lymphocytes. To deplete all endogenous APCs in the T-cell population so as to be certain that Ag, which may be shed by the tumor cell clones, is not inadvertently presented by contaminating APC (19), nylon-wool-enriched T cells derived from draining lymph nodes of TGG-primed DBA/2 mice were treated with anti-Ia^d and complement to eliminate endogenous APC. T cells treated in this manner failed to proliferate in response to soluble TGG unless splenic filler cells were added. Data presented in Figure 1 demonstrate that Ag-pulsed P388AD.2 but not NA clones presented TGG to primed T cells nearly as well after anti-Ia$^+$ C treatment as before treatment. The only evidence that we have obtained suggesting that P388NA has any Ag-presenting ability has been the capacity to induce TNP-specific CTL in vitro in DBA/2 spleen cells. In those experiments in which both TNP-modified P388AD and TNP-modified P388NA induced CTL, whole spleen cell preparations were used, thus the opportunity for cooperation with endogenous accessory cells existed and may account for the presentation of TNP by the nonadherent P388NA.

These data suggest that the adherent Ia$^+$ clones of P388AD are cell lines which have the capacity to present Ag to primed T cells. The presentation of Ag by P388AD tumor cells appears to be independent of any endogenous APC in the T-cell preparations and may be a function of surface Ia expression and IL-1 secretion by the clones. As such, we feel that these cell lines, P388AD, may be a useful tool for the analysis of Ag-Ia interactions on the surface of an APC.

Table 1. Cell Characteristics of P388AD and P388NA Tumor
Cell Lines

Characteristcs	P388AD	P388NA
Surface Antigens:		
I-A	+	+
H2-D	+	+
Mac-1 (macrophage)	+	−
Thy 1 (T cell)	+	+
Ly 1 (T-cell subset)	50% weak	+
Ly 2 (T-cell subset)	−	−
ThB (thymocyte, B-cell subsets)	−	−
E2 (macrophage, B-cell subsets)	−	−
DNL 1.9 (B and pre-B cells)	+	+
Ly 9.1 (T cell, B cell)	−	−
sIg	−	−
Surface Receptors:		
Fc Receptor	−	−
C3b Receptor	−	−
Enzymes:		
Peroxidase	−	−
Esterase	+	+/−
Functions:		
Phagocytosis	−	−
MLR Stimulation	+	−
Syngeneic MLR Stimulation	+	−
IL-1 Synthesis	+	−
TNP-Specific CTL Induction	+	+
Antigen Presentation	+	−

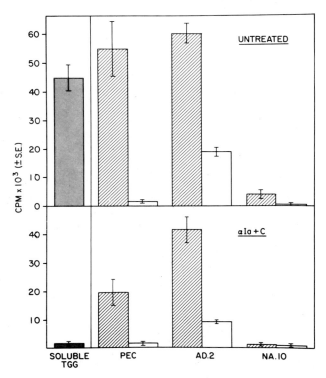

Fig. 1.   Presentation of TGG to primed T cells by antigen-pulsed
          P388AD.2.   Immune DBA/2 lymph node cells (see text) were
          enriched for T cells on nylon-wool columns and either
          treated with anti-Ia^d and complement (lower panel) or un-
          treated (upper panel).   Controls for anti-Ia^d alone and
          complement alone were comparable to untreated cells and
          are omitted from the figure.   The immune T cells were then
          stimulated in microtiter wells with TGG alone (▨) or
          washed DBA/2 peritoneal exudate cells, P388AD.2 or P388NA.10
          previously incubated with (▨) or without TGG (☐).   The bars
          represent ± SEM.

REFERENCES

1. Benacerraf, B., and Germain, R. N., Immunol. Rev. 38:70, 1978.
2. Rosenthal, A. S., and Shevach, E. M., J. Exp. Med. 138:1194, 1973.
3. Lee, K. C., and Wong, M., J. Immunol. 125:86, 1980.
4. Schwartz, R. H., Yano, A., and Paul, W. E., Immunol. Rev. 40:153, 1978.
5. Benacerraf, B., J. Immunol. 120:1809, 1978.
6. Cohen, D. A., and Kaplan, A. M., J. Exp. Med. 154:1881, 1981.
7. Mizel, S. B., Oppenheim, J. J., and Rosenstreich, D. L., J. Immunol. 120:1497, 1978.
8. Shearer, G. M., Eur. J. Immunol. 4:527, 1974.
9. Julius, M. H., Simpson, E., and Herzenberg, L. A., Eur. J. Immunol. 3:645, 1973.
10. Ledbetter, J. A., and Herzenberg, L. A., Immunol. Rev. 47:63, 1979.
11. Warner, N. L., Daley, M. J., Richey, J., and Spellman, C., Immunol. Rev. 48:197, 1979.
12. Lanier, L. L., Warner, N. L., Ledbetter, J. A., and Herzenberg, L. A., J. Immunol. 127:1691, 1981.
13. Minami, M., and Shreffler, D. C., J. Immunol. 126:1774, 1981.
14. Steinman, R. M., and Witner, M. D., Proc. Natl. Acad. Sci. USA 75:5137, 1978.
15. Glimcher, L. H., Longo, D. L., Green, I., and Schwartz, R. H., J. Exp. Med. 154:1652, 1981.
16. Yamashita, U., and Shevach, E. M., J. Immunol. 124:1773, 1980.
17. Farrar, W. L., Mizel, S. B., and Farrar, J. J., J. Immunol. 124:1773, 1980.
18. Germain, R., J. Immunol. 127:1964, 1981.
19. Schwartz, R. H., Kim, K. J., Asofsky, R., and Paul, W. E., in "Macrophage Regulation of Immunity" (E. R. Ynanur and A. S. Rosenthal, eds.), pp. 277-284, Academic Press, New York, 1980.

DISCUSSION

ERB: If you take a non-adherent cell line and add interleukin 1, do you get antigen presentation?

KAPLAN: Yes, but we've just done one experiment. We recently got some human interleukin 1 from Larry Lachman, and it converts the non-adherent cells into antigen presenting cells at least in the mixed lymphocyte reaction.

FELDMAN: It reminds me of old experiments of Ethan Shevach in the guinea pig. He claimed that when he tested Ia positive leukemic cells, just by virtue of them having Ia, they presented antigen. If the TNP antigen was chemically bound to them, he got negative results. His conclusion was that Ia is necessary but not sufficient to present antigen.

KAPLAN: That's similar to the data obtained by Noel Warner with some of the WEHI lines and others.

FIDLER: You made the comment that the cells reverted from adherent to non-adherent. Did you mean that? The word reverted implies a certain movement and suggests that being adherent is a more progressive state than being non-adherent.

KAPLAN: Not necessarily.

FIDLER: If you put TPA on the non-adherent cells will they become adherent?

KAPLAN: Interestingly, if we put TPA on the non-adherent cells they become adherent and they begin to express Fc receptors and to a lesser extent C3b receptors. If we put TPA on the adherent cells, they do not change very much and they do not express Fc receptors or C3b receptors.

BELLER: In terms of these cells presenting antigen by themselves, have you used a more conventional procedure for obtaining macrophage depleted lymph node cultures? Clearly when you use anti-Ia and complement, you will deplete the antigen presenting cells, but you will not get rid of all macrophages. The possibility remains that you have a collaboration between the Ia negative macrophage and the P-388 adherent cell.

KAPLAN: We're very concerned about that possibility because during a syngeneic or allogeneic mixed lymphocyte reaction lymphokines are produced which could be converting the Ia negative macrophages to Ia positive macrophages. We've tried removing the macrophages but we haven't been able to remove them to the extent we get absolutely no response to soluble turkey gamma globulin. That's really what we have to be able to do as a negative control for the experiment. Our other approach is to assay antigen presentation to a series of T cell lines in which the problem of contaminating macrophages is eliminated.

ANTIGEN PRESENTATION IN VITRO BY A
MURINE MACROPHAGE CELL LINE

Yolande Buchmüller, Jacques Mauël
and Giampietro Corradin

University of Lausanne
Institute of Biochemistry
1066 Epalinges, Switzerland

INTRODUCTION

Macrophages have been shown to play a central role in the elicitation of immune responses. One of their multiple functions is the presentation of antigen in a proper form to naive or immune T cells. The biochemical steps involved in this process are not well understood. The availability of a macrophage cell line capable of supporting an antigen-driven T cell proliferation would certainly aid in the elucidation of the biochemical and cellular requirements for antigen presentation. To this purpose, the murine macrophage cell line IC-21 was tested to see whether or not it was capable of presenting soluble and particulate antigens.

MATERIALS AND METHODS

The IC-21 macrophage cell line was originally obtained by transformation of C57BL/6 murine peritoneal macrophages by Simian virus 40 (1) and exhibits several macrophage properties, including adherence, Fc receptor function, phagocytosis, response to migration inhibitory factor, killing of tumors and parasites upon activation, and regulation of a primary antibody response in vitro (2,3,4). Several antigens were used in these studies: beef and tuna cytochrome c, apocytochrome c, apocytochrome c derivatized with a 2,4-dinitro-5-aminophenyl group, native or denatured ovalbumin and Leishmania tropica parasites.

T cells were derived from lymph node cells of C57BL/6 and BDF_1 mice, immunized 7 days previously by subcutaneous injection at the

base of the tail with antigen in complete Freund's adjuvant (5).
Draining lymph nodes were disrupted, and the cells were washed and
distributed into culture flasks, at a concentration of 4×10^6 cells/
ml in 5 ml of Click's or Iscove's medium supplemented with 5% fetal
calf serum or 0.5% normal mouse serum, together with the appropriate
antigen. The resulting blasts were either maintained in culture by
weekly additions of fresh serum-containing medium, fresh antigen and
20×10^6 irradiated syngeneic spleen cells (6) or purified over a
Ficoll or Percoll gradient and placed in culture for one or two days
in the presence of conditioned medium as source of T cell growth
factor. Alternatively, cells were kept in the primary culture for
9-10 days and then used as such. These sources of T cell blasts were
strictly antigen and macrophage dependent (6).

To assay for antigen presentation by the IC-21 macrophage cell
line, cultured T cell blasts were centrifuged, resuspended in fresh
medium and distributed at a concentration of 10^4 cells/0.2 ml of
medium in microtiter plates, in the presence of 10^4 mitomycin-treated
IC-21 macrophages or 10^6 irradiated spleen cells and the appropriate
antigen. After 3 days of culture, 1 µCi of methyl-^{3}H-thymidine was
added to each well, and thymidine incorporation was measured 24 h
later.

RESULTS

As shown in Table 1, comparable thymidine incorporation was
observed when either irradiated spleen cells or mitomycin-treated
IC-21 macrophages were added to ovalbumin specific BDF$_1$ cells or
Leishmania specific C57BL/6 T cell blasts in the presence of the
specific antigen. Negligible proliferation was observed in the
absence of specific antigen, or in the absence of irradiated spleen
cells or mitomycin-treated IC-21 macrophages; this indicates that
proliferation of antigen-specific T cells is macrophage dependent.
Similar and consistant results were also obtained with T cells specific for all the antigens tested. Depending upon the experiment,
either the irradiated spleen cells or the IC-21 cell line appeared
to give a more optimal response. These variations are probably due
to the fact that conditions for optimal proliferation, that is the
number of accessory cells, may vary slightly from experiment to
experiment. For instance, in some experiments, T cell blast proliferation was better with 2×10^4 than with 10^4 IC-21 macrophages.
However, these results clearly indicated that the IC-21 macrophage
cell line is capable of processing and presenting the antigen to
specific T cells.

Finally, experiments were performed to determine whether the
antigen presenting capacity of IC-21 macrophages was specifically
inhibited by a monoclonal anti-Ia antiserum. As shown in Table 2,
proliferation of ovalbumin specific T cells was almost totally

Table 1. Proliferative Response of Antigen Specific T
          Cells in the Presence of IC-21 Macrophages

Cells	CPM $\pm$ S.D. x 10^3	
	BDF$_1$ T cells	C57BL/6 cells
T cells + Ag	1.6 $\pm$ 0.3	0.3 $\pm$ 0.1
T cells + IC-21	1.1 $\pm$ 0.2	0.3 $\pm$ 0.1
T cells + IC-21 + Ag	8.9 $\pm$ 1.4	17.6 $\pm$ 2.1
T cells + C57BL/6 spleen	0.5 $\pm$ 0.2	0.9 $\pm$ 0.2
T cells + C57BL/6 spleen + Ag	5.2 $\pm$ 1.1	22.7 $\pm$ 2.0

Cell cultures contained 10^4 T cells blasts, 10^4 IC-21 macrophages or 0.5 x 10^6 spleen cells. BDF$_1$ T cell blasts were
specific for apocytochrome c and were kept in culture for
60 days.   C57BL/6 T cell blasts were specific for Leishmania
tropica parasite and were kept in culture for 8 days.

inhibited by a monoclonal anti-Ia[b,s] antibody (7) when added to cultures containing IC-21 macrophages. The presence of Ia molecules
on the surface of IC-21 macrophages was detected by using a bio-
tinylated monoclonal anti-Ia reagent and FITC avidin. These reagents
were prepared to control for the presence of specific Fc receptors.
For this purpose, 7% of normal mouse serum, 7% normal rat serum and
an excess of a non relevant monoclonal antibody of the same class
as the specific anti-Ia reagent were added prior to addition of the
specific anti-Ia antibody.

CONCLUSIONS

     All the results shown here would strongly indicate that the IC-21
macrophage line is capable of specifically presenting antigen to T
cells. On the other hand, formal proof of the antigen presenting
capacity of this cell line has not been presented in this report
since it could be argued that residual macrophages were present in

Table 2. Inhibition of a Proliferative Response by
          Monoclonal Anti-Ia Antibody

Cells	CPM $\pm$ S.D. x 10^3
T cells + ovalbumin	0.6 $\pm$ 0.4
T cells + IC-21	0.3 $\pm$ 0.1
T cells + IC-21 + ovalbumin	15.9 $\pm$ 0.6
T cells + IC-21 + ovalbumin + anti-Ia^{b,s}	2.5 $\pm$ 1.8

10^4 T cell blasts specific for ovalbumin were used.
Cells were stimulated in culture for 5 days, purified
on a Percoll gradient and cultured for 1 day in a conditioned medium as source of T cell growth factor.
Monoclonal anti-Ia was directed against determinants
encoded by I-A locus of b and s haplotypes (7) and
kindly provided by Dr. A. Williams.

all the preparations of T cell blasts tested. Experiments are
currently under way to obtain specific T cells devoid of any residual
macrophages.

REFERENCES

1. Mauel, J., and Defendi, V., J. Exp. Med. 134:336, 1971.
2. Morahan, P. S., J. Reticuloendothel. Soc. 27:223, 1980.
3. Stewart, C. C., in "Macrophage Regulation of Immunity" (E. R.
       Unanue and A. S. Rosenthal, eds), pp. 455-476, Academic Press,
       New York, 1980.
4. Mocarelli, P., Pelmer, J., and Defendi, V., Immunol. Commun. 2:
       441, 1973.
5. Corradin, G., Etlinger, H. M., and Chiller, J. M., J. Immunol.
       119:1048, 1977.
6. Schrier, R. D., Skidmore, B. J., Kurnick, J. T., Goldstine, S. N.,
       and Chiller, J. M., J. Immunol. 123:2525, 1979.
7. McMaster, W. R., and Williams, A. F., Immunol. Rev. 47:117, 1979.

DISCUSSION

ERB: Have you used antigen pulsed IC-21 cell line macrophages as well?

CORRADIN: We used them a couple of times but pulsing antigen did not work as well as if antigen was left in the medium throughout the culture period.   At the time we did these experiments with pulsed antigen, we used Leishmania as the antigen, so maybe there was a problem.

Ir GENE REGULATION OF T CELL PROLIFERATION: REQUIREMENT FOR Ia AND ANTIGEN EXPRESSION ON THE SAME ACCESSORY CELL

Lawrence B. Schook*, Darryl A. Campbell and
John E. Niederhuber

Department of Microbiology and Immunology and
Division of Surgical Oncology
University of Michigan Medical Center
Ann Arbor, Michigan   48109 USA

INTRODUCTION

We and others have shown that treatment of antigen presenting cells with alloantiserum restricted to the I-A subregion inhibits their ability to present the terpolymer GAT to immune T cells (1-3). The inhibition is independent of complement and antiserum recognizing other H-21 subregions has no effect (Figure 1). These results have been interpreted to suggest that anti I-A antibodies mask the Ia antigen (Ir gene product) and, thus, block the T cell receptor(s) from recognizing either Ir gene product and specific antigen in complex or separately (4).

This study examines two other possible explanations. The first possibility being that anti-Ia antibody treatment of antigen presenting cells induces or enhances suppression or secondly that two antigen presenting cells may be required -- one expressing Ir gene product and function and a second presenting only the specific antigen. While suppression would be induced by anti-Ia treatment, the second alternative would be inhibited by specific anti-Ia antibodies.

This report evaluates these two potential explanations using purified splenic macrophages pretreated with anti I-A serum prior to $Glu^{60}Ala^{30}Try^{10}$ (GAT) pulsing and the addition of GAT primed T cells.

---

* Present address: Department of Microbiology and Immunology, Box 678, Medical College of Virginia, Richmond, VA   23298.

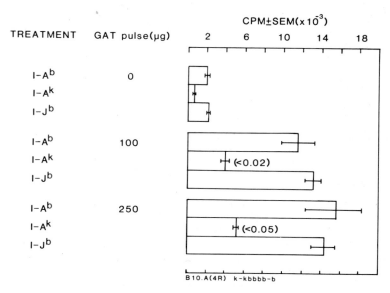

Fig. 1.   Effect of H-21 restricted alloantisera on antigen presenta-
          tion by B10.A (4R) splenic macrophages. Splenic macro-
          phages were prepared as previously described (6,7), incubated
          with anti-Ia sera for 30 minutes, washed, and pulsed with
          GAT. The sera for all experiments described were: anti
          I-A^b, (A X B10.D2) anti-B10.A(5R) absorbed with EL-4 leukemic
          cells; anti I-A^k, (A.TH X B10.A(5R)) anti-A.TL; and anti
          I-J^b, B10.A(5R) anti-B10.A(3R). The purified lymph node T
          cells (4 x 10^5/well) were added and the cultures were incu-
          bated for 96 hours. The cultures were pulsed with 1 μCi
          ^{3}H-thymidine 18 hours prior to harvesting. The P values
          were determined by comparing the results obtained after
          treating cultures with anti I-A^k to anti I-A^b(inappropriate).

In order to determine which of the restrictions discussed above is
operative, different antigen presenting cells were added to the
anti I-A treated splenic macrophages and T cell cultures. The following antigen presenting cells were used to reconstitute the inhibited proliferative response: (a) GAT-pulsed purified splenic
macrophages; (b) GAT-pulsed purified splenic macrophages treated
with anti I-A and complement; and (c) purified splenic macrophages.

RESULTS AND DISCUSSION

This report evaluates the role of Ir gene products expressed by splenic macrophages and their association with GAT in the presentation of antigen to immune lymphocytes. To determine whether two antigen presenting cells - one presenting antigen and one expressing Ir gene function - were involved in a proliferative response, purified splenic macrophages were added to the cultures. These purified splenic macrophages have been shown to be comprised of approximately equal numbers of Ia bearing and non-Ia bearing antigen presenting cells (4).

Treatment of B10.A splenic macrophages with alloantiserum restricted to the I-A^k subregion resulted in significant inhibition of the T cell response (Figure 2). Addition of increasing numbers of purified splenic macrophages to these cultures did not result in T cell proliferation. Thus, the addition of Ia bearing antigen

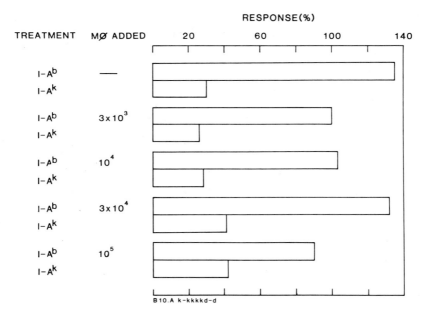

Fig. 2.   Addition of increasing numbers of antigen presenting cells for reconstituting an anti-Ia inhibited proliferative response. Purified splenic macrophages (6,7) were added to the anti-Ia treated cultures prepared as described in Figure 1.   The medium control responses declined with addition of increasing numbers of antigen presenting cells (from 22,797 CPM when no cells were added to 6,513 CPM when 10^5/antigen presenting cells were added).

presenting cells cannot provide Ir function to cultures containing
GAT pulsed antigen presenting cells treated with anti-Ia antibodies.
This would suggest that the T cell receptor(s) must recognize both
Ir gene product (function) and the specific antigens, e. g. GAT,
on the same accessory cell. One can also rule out that an insufficient number of Ia bearing cells were added, since high numbers
of splenic macrophages (10^5/culture) were added, until non-specific
suppression occurred.

The possibility of inducing a suppressive state by anti I-A
serum treatment of splenic macrophages was addressed in a similar
manner. In these experiments, various purified splenic macrophages
were added to the cultures (Figure 3) to restore T cell responsiveness. Three populations were used: (a) splenic macrophages, as a
source of Ia bearing cells (Ia^+); (b) splenic macrophages pulsed
with GAT, as a source of Ia bearing cells expressing membrane-bound
antigen (Ia^+ GAT); and (c) complement pulsed with GAT, as a source
of non-Ia bearing cells expressing membrane bound antigen (Ia^- GAT).

If suppression was responsible for the decrease in antigen-
induced T cell proliferation none of these added preparations should

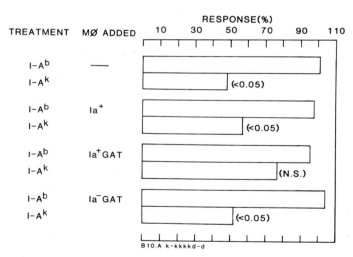

Fig. 3. Addition of various antigen presenting cells to reconstitute
an anti-Ia inhibited proliferative response. The populations
of GAT pulsed and/or anti I-A^k + complement treated antigen
presenting cells (10^4/well) were added to cultures treated
as described above. The medium control responses for each
group were 12,928-16,248 CPM.

result in a significant T cell response. However, if masking of the
Ir gene product alone or in association with processed antigen was
the actual event being observed, then the antigen presenting cells
(Ia^+ GAT) should result in T cell proliferation. In cultures which
received either the Ia^+ or the Ia^- GAT splenic macrophages, no
significant increase in T cell responsiveness occurred. However,
anti-Ia antibody blocked monolayers and T cell cultures which received
Ia bearing macrophages with membrane associated GAT (Ia^+ GAT) did
have a significant T cell proliferation.

SUMMARY

These results demonstrate that only splenic macrophages expressing both Ir products (function) and the antigen (GAT) were able
to reconstitute the T cell response of cultures in which the splenic
macrophage monolayer had been treated with appropriate anti I-A antibodies. This inhibition was not a result of induced suppression
caused by anti-Ia treatment but rather a masking of Ia antigen which
prevents the T cell receptor(s) from recognizing either the Ir gene
product and specific antigen in complex or separately.

Acknowledgement

This work was supported in part by a grant from the National
Institutes of Health (CA-25052-03AI). L. B. S. was a recipient of
post-doctoral research fellowships awarded by the National Institute
of Health (CA-06268-03) and the Elsa Pardee Foundation.

REFERENCES

1. Schook, L. B., and Niederhuber, J. E., Fed. Proc. 40:4138, 1968.
2. Schook, L. B., and Niederhuber, J. E., in "Heterogeneity of Mono-
        nuclear Phagocytes" (O. Förster and M. Landy, ed.), pp. 208-
        213, Academic Press, New York, 1981.
3. Schwartz, R. H., Yano, A., and Paul, W. E., Immunol. Rev. 40:
        153, 1981.
4. Unanue, E. R., and Rosenthal, A. S., ed., in "Macrophage Regula-
        tion of Immunity", Academic Press, New York.
5. Cowing, C., Schwartz, B. D., and Dickler, H., J. Immunol. 120: .
        378, 1978.
6. Niederhuber, J. E., Allen, P., and Mayo, L., J. Immunol. 122:
        342, 1979.
7. Niederhuber, J. E., and Allen, P., J. Exp. Med. 151:1103, 1980.

DISCUSSION

FELDMAN: What will antibodies to the GAT determinant do when applied to these macrophages? Old experiments in guinea pigs showed that anti I-A antibodies blocked the immunogeneic response but anti antigen at least 3 hours after the interaction between the macrophage and the antigen did not block initiation of proliferation. Have you tried this technique?

SCHOOK: I really cannot address that point since we have not done kinetic experiments.

DORIA: You have presented evidence that pretreatment of macrophages with anti I-J does not prevent GAT antigen presentation. How do you reconcile this result with previous results from your own laboratory showing that such treatment prevents the antibody response in vitro?

SCHOOK: I wish I could. In other experiments, Niederhuber showed that if you took splenic macrophages and treated them with anti I-A or anti I-J without complement, blocking of the primary in vitro antibody response occured with antisera to the I-J subregion and not with antisera to the I-A subregion. Obviously this is a converse. At this point, it is premature for us to speculate other than perhaps to say that it may be related to differences between induction of a primary antibody response and a secondary proliferative response. However, you really can't compare the two experimental protocols.

STEVENSON: You mentioned that some monoclonal antibodies against I-A also block the reaction. Do all monoclonal antibodies against I-A block in your system?

SCHOOK: Only the appropriate monoclonal antibody.

STEVENSON: Would the monoclonals against the framework antigens of I-A all block, or do they have to be against the polymorphic areas?

SCHOOK: You obviously work in the human system. I don't think that we can draw parallels between the human and the mouse at this time. The monoclonal antibody that we use in these experiments recognizes specificity Ia.17, which appears to map to a beta chain that's encoded within the I-A subregion.

BIOCHEMICAL EVIDENCE FOR MULTIPLE Ia MOLECULES

Chella S. David, William P. Lafuse and Michele Pierres

Department of Immunology, Mayo Clinic and Medical School,
Rochester, Minnesota   55905 USA and Centre d' Immunologie
Inserm, Marseille, France

INTRODUCTION

Both Ia antigens and Ir genes map to the I-A and I-E subregions
of the mouse H-2 gene complex. The I-A subregion codes genes for
three polypeptide chains, designated A_α, A_β, and E_β while the I-E
subregion maps the gene for E_α (1). The map position of a third
invariant Ia polypeptide chain designated I is not known. Thus,
the I-A molecule and I-E molecule are formed by the non-covalent
binding of the three chains, α (34,000 M.W.), β (28,000 M.W.) and I
(31,000 M.W.). Recent studies using mutant strains and monoclonal
antibodies have clearly shown that Ia antigens are the immune response
gene products. The Ia molecules are involved in generating histo-
incompatibility, alloreactivity, autoreactivity, as well as generating immune response to a variety of foreign antigens. Studies
using alloreactive and antigen-specific T-cell clones have shown
that Ia molecules most probably mediate immune activity by presentation of antigens to T-cells (2). Yet, it is not clear how the Ia
molecules interact specifically with a variety of antigens under Ir
gene control. One possibility is that Ia molecules have several
antigenic sites which interact with different antigens. Recent
studies have shown that there may be multiple epitopes of similar Ia
antigens which are spatially distributed through the Ia molecule
(3). Another possibility is that there are multiple Ia molecules,
each of which can interact specifically with different antigens.
To investigate this question, we have begun the biochemical study of
Ia molecules using monoclonal antibodies. In this report, we will
present preliminary evidence that there are multiple I-A and I-E
molecules (4).

RESULTS AND DISCUSSION

Multiple I-E Molecules

Monoclonal antibodies used in this study are described in Table
1. Antigenic extracts of B10.A ($A^k E^k$) were pretreated with excess
13-4 (anti-E_α), 17-3-3 (anti-E_β) or H9-14.8 (anti-E^k_β) monoclonal
antibodies. When the B10.A extract was pretreated with excess 13-4
monoclonal antibody and then tested with an anti-Ia.7 alloantisera
(D2.GD x B6)F_1anti-B10.HTG), a substantial proportion of the I-E
molecules remained. When 17-3-3 monoclonal antibody was used in the
pretreatment step, 25% of the I-E molecules precipitated by the anti-
Ia.7 alloantiserum remained. When H9-14.8 was used in the pretreatment step, all of the I-E molecules precipitated by the anti-Ia.7
alloantiserum were removed. These results suggest that 13-4 and 17-
3-3 monoclonal antibodies recognize subpopulations of I-E^k molecules
while H9-14.8 is expressed on all molecules. To determine if 17-3-3
and 13-4 monoclonal antibodies recognize the same I-E molecule,
reciprocal sequential precipitations with these monoclonal antibodies
were done (Figure 1). When 13-4 was used in the pretreatment step,
it only slightly reduced the molecules precipitated by 17-3-3 and
H9-14.8 monoclonals.Similarly, pretreatment with 17-3-3 only slightly
reduced the molecules precipitated by the 13-4 and H9-14.8. These
results show that some of the I-E molecules express either 13-4 or
17-3-3 determinants while other I-E molecules express both determinants. Identical results were also obtained in reciprocal sequential
precipitations of 17-3-3 and 13-4 monoclonal antibodies with B10.A(3R)
($A^b E^k$), B10.A(5R) ($A^b E^k$) and B10.S (9R) ($A^s Exyk$) extracts.

Table 1. Description of Monoclonal Antibodies

Monoclonal	Specificity	Region	Chain	Source
13-4	Ia.7	I-E	E_α	L. Herzenberg
17-3-3	Ia.22	I-E	$E_\beta^{k,s,b,r}$	D. Sachs
H9-14.8		I-E	E_β^k (private)	M. Pierres
17-227	Ia.15	I-A	$A_\alpha^{b,d,k}$	G. Hammerling
H9-15.4	Ia.7,	I-A,I-E	$A_\alpha^{b,q}\ E_\alpha$	M. Pierres
H39-459	Ia.7,	I-A,I-E	$A_\alpha^{b,q}\ E_\alpha$	M. Pierres

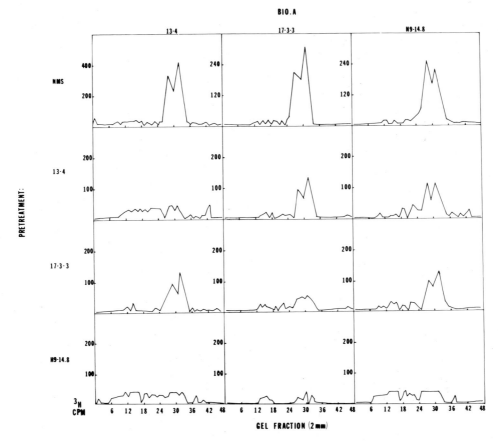

Fig. 1.   Sequential immunoprecipitation analysis of I-E molecules
          from B10.A.   Aliquots were pretreated with excess NMS, 13-4,
          17-3-3, or H9-14.8 monoclonal antibody.   After removal of
          immune complexes with S. aureus, the extracts were divided
          and subjected to second immunoprecipitations with the same
          monoclonal antibodies.

Isoelectric Focusing Comparisons of I-E Molecules

     To confirm the results obtained by sequential precipitations,
isoelectric focusing patterns of I-E molecules precipitated by 13-4
and 17-3-3 were compared (Figure 2).   13-4 and 17-3-3 positive I-E
molecules were immunoprecipitated from a B10.A(3R) ($A^b E^k$) extract
and run on one-dimensional isoelectric focusing gels.   The gels were

fractionated into 1 mm fractions and the pH gradient of each gel
determined with a pH microelectrode. The ampholine mixture of 1:1
biolyte 5–7 and pharmalyte 4–6.5 was chosen to give a pH gradient of
7 to 4.5. Although the pH gradient varied slightly from gel to gel,
the isoelectric pH of various peaks was consistently the same from
gel to gel. Previous 2-D electrophoresis studies and isoelectric
focusing analysis of isolated alpha and beta chains has shown that
E_α chains appear on isoelectric focusing gels as a complex set of
peaks with isoelectric points near pH 5.0 while A_e beta chains are
more basic (pH 6.7–5.4). The E_α peaks precipitated with 17–3–3 gave
a complex set of four peaks at pH 5.2–4.9 in contrast to the 13–4

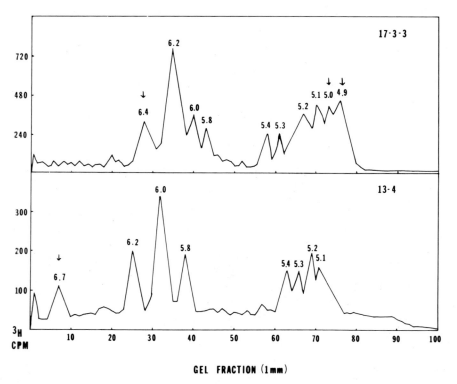

Fig. 2.   Analysis of I–E molecules precipitated from B10.A(3R) with
           17–3–3 and 13–4 monoclonal antibodies and run on isoelectric
           focusing gels. Gels were fractionated into 1 mm fractions
           with the Gilson gel fractionator, the pH of each fraction
           was determined with a pH microelectrode, and the radioactiv-
           ity in each fraction determined. Arrows indicate A_e and E_α
           peaks unique to 13–4 or 17–3–3.

monoclonal antibody which precipitated two E_α peaks with isoelectric points 5.2 and 5.1. This suggests that 17-3-3 monoclonal antibody precipitates E_α chains (pH 5.0 and 4.9) not precipitated by 13-4. A_e beta chains precipitated with 13-4 and 17-3-3 monoclonal antibodies focused at pH's of 6.7-5.8. Three A_e peaks (pH 6.2, 6.0. 5.8) were precipitated by both 13-4 and 17-3-3 while 17-3-3 monoclonal antibody precipitated a unique A_e beta peak at pH 6.4. 13-4 monoclonal antibody precipitated a unique A_e peak at pH 6.7. Two peaks at 5.4 and 5.3 precipitated by both 13-4 and 17-3-3 could be either the E_α chains or A_e chains.

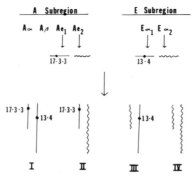

Fig. 3.   Model of multiple I-E molecules defined by monoclonal anti-
          bodies 17-3-3 and 13-4.

Our studies suggest that there may be at least four I-E molecules (Figure 3). An important question is how many A_e and A_α chains exist. One possibility is that each A_e peak present on the isoelectric focusing gels represents a single A_e beta chain. If this is true, there would be a minimum of five A_e beta chains precipitated by 13-4 and 17-3-3 monoclonal antibodies with four A_e beta chains bearing 17-3-3 and one A_e beta chain that does not express 17-3-3. If a similar number of E_α polypeptide chains exist and there is free association between all of the A_e beta polypeptide chains and E_α chains, 25 separate I-E molecules could be formed. If each of these I-E molecules interact with different antigens, then the specificity seen in immune response might be explained. However, it is also possible that some of the A_e peaks may be due to differences in complexity of the carbohydrate portions of the A_e chains. Some of the heterogeneity present in the E_α peaks is clearly due to differences in glycosylation. Cullen, et al. (5) recently showed that B cells and spleen adherent cells differ in A_α and E_α polypeptide chains. B cells were found to express extra alpha bands in IEF gels that are not found in the adherent cell population. These extra alpha bands expressed on B cells were removed with neuraminidase digestion suggesting that they were due to alpha chains containing a large quantity of N-acetyl neuraminic acid. Interestingly, Cullen, et al. (5) also found that the neuraminidase digestion had no effect on the beta chain IEF bands. This suggests that different beta IEF peaks might be due to differences in protein structure and not carbohydrate structure.

Crossreaction of Anti-I-E^k Monoclonal Antibodies with Subsets of I-A^b Molecules

Monoclonal antibodies, H9-15.4 and H39-459 were derived from an A.TH anti-A.TL immunization. When tested on I^k haplotypes, both antibodies reacted with the I-E molecule on the basis of direct binding of ^{125}I antibody to spleen cells, direct immunoprecipitation of Ia molecules from radiolabeled LPS spleen blasts as well as sequential immunoprecipitations. Surprisingly, H9-15.4 and H39-459 were also positive against the I-E negative haplotypes, H-2^b and H-2^q. To investigate the question of whether these monoclonal antibodies cross-react with I-A molecules of the b and q haplotypes, sequential precipitation analysis was undertaken. B10 (H-2^b) Ia extract was pretreated with anti-A^b alloantiserum (B10.MBR x A)F_1 anti-B10.A(5R). This pretreatment removed all of the Ia molecules detected by H9-15.4 and H39-459, confirming the crossreaction with the I-A^b molecule. When a recombinant strain, B10.A(3R) (A^{b}E^k) was tested, pretreatment with either anti-A^b alloserum or anti-Ia.7 alloserum did not completely remove Ia molecules precipitated by the monoclonal antibodies. This suggests that in this recombinant, H9-15.4 and H39-459 reacted with both I-A and I-E molecules.

To examine the possibility that these monoclonal antibodies may be differentiating subsets of I-A^b molecules, sequential precipitations were done (Figure 4). Besides H9-15.4 and H39-459, a third monoclonal anti-A^b antibody, 17-227 which detects the Ia.15 specificity was also included. B10 Ia extract was pretreated with either excess H9-15.4, H39-459, or 17-227, and then tested for remaining I-A^b molecules with the alloantiserum. With each of the monoclonal antibodies, pretreatment only partially removed the I-A^b molecules precipitated with the alloantiserum. This study suggests that these monoclonal antibodies detect subsets of A^b molecules. Pretreatment with excess 17-227 monoclonal antibody had little or no effect on immunoprecipitation of I-A^b molecules by H39-459 and H9-15.4. Likewise, pretreatment with H9-15.4 or H39-459 monoclonal antibody did not remove molecules detected by 17-227. This result sug-

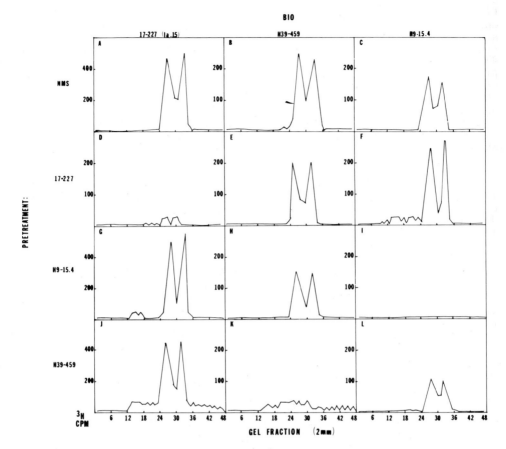

Fig. 4.   Sequential analysis of I-A^b molecules. Aliquots of a B10
          extract were pretreated with excess NMS, 17-227, H9-15.4,
          and H39-459 monoclonal antibody and a second immunoprecipita-
          tion done with same monoclonals.

gests that 17-227 precipitates a subset of I-A^b molecules not detected
by H9-15.4 and H39-459. Although H9-15.4 and H39-459 showed the same
H-2 strain distribution, they differ when compared by sequential
precipitation. Pretreatment with H9-15.4 only partially removed Ia
molecules detected by H39-459. Similarly, pretreatment with H39-459
only removed some of the Ia molecules precipitated by H9-15.4. This
indicates that some I-A^b molecules detected by H9-15.4 and H39-459
express only H9-15.4 or H39-459 determinants, while other I-A^b molecules express both determinants.

To explain these results suggesting multiple I-A and I-E molecules, we propose that in addition to gene duplication of ancestral genes to yield the I-A and I-E molecules, additional gene duplications occurred during the evolution forming multiple copies of genes encoding the A_α, A_β, E_β, and E_α Ia polypeptide chains. After gene duplication, diversification through mutations would have occurred resulting in determinants being expressed on some of the multiple alpha and beta polypeptide chains and absent on others. These events probably occurred early in the evolution as a recent study by Hurley, et. al. (6) suggests that in man multiple HLA-DR molecules exist. The presence of multiple I-A molecules may also explain some of the specificity seen in the immune response to various antigens.

SUMMARY

Sequential immunoprecipitation and isoelectric focusing studies using anti-Ia monoclonal antibodies have shown multiple I-A and I-E Ia molecules. These studies suggest that the genes coding for Ia polypeptide chains A_α, A_β, E_β and E_α might have undergone further duplication and mutated. This would explain the diversity seen with Ia molecules in their ability to "present" numerous self and non-self antigens for generation of immune response.

Acknowledgement

The studies reported herein were supported by NIH grants CA 24473, AI 14764 and by funds from the Mayo Foundation. The authors thank Suresh Savarirayan and his associates for superb mouse husbandry, Paula Corser for excellent technical assistance and Mary Steege for skillful preparation of the manuscript.

REFERENCES

1. Jones, P. P., Murphy, D. B., McDevitt, H. O., J. Exp. Med. 148: 925, 1978.
2. Fathman, C. G., Kimoto, M., Melvold, R. W., David, C. S., Proc. Natl. Acad. Sci. USA 78:1853, 1981.
3. Pierres, M., Kourilsky, F. M., Rebovan, J. P., Dosseto, M., Caillol, D., Eur. J. Immunol. 10:950, 1980.
4. Lafuse, W. P., Corser, P. S., David, C. S., Immunogenetics, in press.
5. Cullen, S. E., Kindle, C. S., Schreffler, D. C., Cowing, C., J. Immunol. 127:1478, 1981.

DISCUSSION

FELDMAN: We know that low responders to an antigen such as insulin can be made to respond if the insulin is presented slightly dif-

ferently, as for instance with adjuvants. To what extent is it possible to determine which kind of new I-A determinants are being expressed following treatment with adjuvant like molecules?

DAVID: In such cases, I don't think any new I-A molecules are being
generated. When treated with adjuvant, the antigen determinant which
cannot associate with Ia by itself can do so now. In this way, insulin plus adjuvant is now able to be presented.

FELDMAN: Since your whole argument is based on precipitation, how
efficient is your precipitation technique? Have you tried to recycle
one monoclonal 2 or 3 times? Most people who have tried precipitation
know that on one cycle you often don't bring down everything.

DAVID: We have close to 50 monoclonals against I-A of mouse, and
some of these monoclonals showed a similar kind of removal and others
did not. We have tried combining two different monoclonals. For
example, we would take two distinct monoclonals both of which precipitated the same molecules and use them in two different steps. They
still did not remove all the molecules, so most probably they are not
the same molecule. However, the possibility remains that the molecules are identical except for glycosylation or something similar.

MANN: Do you have any evidence that the allo determinants are on
alpha chains?

DAVID: With the I-E subregion, I think we have very good evidence
that Ia.7 of the E molecule is on the alpha chain and that Ia.22 is
most probably on the beta chain. Two other specificities have been
identified: Ia 47 and 48 are both on the beta chain. With the A
molecule, whether or not specificity 17 and 15 are on the beta or
alpha chain has still not been proven. John Freld told me recently
that he has developed a method by which he can now separate the chains
without using the Ia reactivity. So maybe we will be able to see
where the specificities are located.

MANN: In the human system, we found the MB and MT determinants are
most probably on the heavy chain. And the DR determinants are on
the light chains.

WILLIAMSON: Do your two monoclonals independently cap I-A on the
surface of the cell?

DAVID: We don't do capping studies. I don't like them.

WILLIAMSON: But capping would be a test to see if there are two sorts
of molecules outside the cell. It would be more definitive than
precipitation.

DAVID: I don't know. Capping cells can create other problems too.

Ia DETERMINANTS ON MACROPHAGES:

SIGNIFICANCE AND ROLE IN THE IMMUNE RESPONSE

Peter Erb, Angelika C. Stern and Michael J. Cecka

Institute for Microbiology
University of Basel
Basel, Switzerland

INTRODUCTION

Since the discovery that macrophages are essential for the resistance to intracellular infection, these mononuclear cells have become recognized as being important not only in nonspecific but also in specific immunity. There is now general agreement that the induction of any immune response is dependent on macrophages. However, besides macrophages, other cells which are not considered to belong to the macrophage lineage are also involved in the induction of at least some immune processes. Thus, these "inducer" cells are heterogeneous and the use of the term "accessory cells" instead of macrophages .seems to be more appropriate to describe these cells. However, as most of the evidence provided in this paper directly concerns the classical macrophage the term "macrophage" will be used, keeping in mind that the functional activities described will not necessarily be restricted to this particular cell type.

EXPRESSION OF I REGION ASSOCIATED ANTIGEN (Ia) ON MACROPHAGES

In 1973, Rosenthal and Shevach (1) showed that for antigen-specific proliferation of guinea pig T cells macrophages and T cells had to be identical at the major histocompatibility complex (MHC) in order to cooperate successfully. This important observation was confirmed by us in a different system, the induction of antigen-specific T helper cells in vitro (2,3). Thus, only macrophages identical at the I-A subregion of the H-2 complex induced or restimulated T helper cells. Since then, the genetic restriction of T cell–macrophage interactions in various systems has been firmly established (4-9). In 1975, we also demonstrated that antigen-incubated macrophages released a mediator, called genetically related macrophage factor (GRF) which induced T helper cells in the absence of additional macrophages and antigen (2,10). GRF was characterized serologically and was found to consist of a complex of macrophage derived Ia products and immunogenic fragments (11). Some of the major features of GRF are listed in Table 1. GRF was thus the first demonstration that macrophage derived Ia is not only important in the activation of T cells, but also that T cells recognize antigen only in the context of Ia. Recently, Puri and Lonai described a molecule very similar to GRF, which they termed IAC (Ia-associated antigen complex) (12).

Many investigators have looked at Ia products on macrophages using various techniques such as immunofluorescence, antibody and complement (C') mediated cytotoxicity, electron microscopic visualization of biosynthetic labeling. Ia antigens coded for by genes of the I-A and/or I-E/C subregion have been detected on macrophages or macrophage-like cells obtained from various anatomical sites (13-24). However, the various macrophage populations differ in their expression of Ia antigen (Table 2), eg. peritoneal exudate macrophages only contain a small proportion of Ia positive (Ia+) cells, while the percentage of Ia+ monocytes or macrophages in the spleen or thymus is much higher. It is now clear that only Ia+ macrophages express functional accessory cell activity. Thus, treatment of macrophages with the appropriate anti-Ia sera and C' abolishes their T cell

Table 1. Properties of Genetically Related Macrophage Factor (GRF)

Source	Peritoneal exudate cells Bone marrow derived macrophages Dendritic cells incubated with antigen for 4 days
Effect	Induces antigen-specific T helper cells in vitro and in vivo Restimulates helper memory cells Activity is genetically restricted
Nature	Ia (coded by I-A subregion) -antigen complex Molecular weight 60,000 to 75,000 daltons Sensitive to trypsin, papain Adsorbed by anti-Ia (I-A) immunoadsorbents Adsorbed by anti-antigen immunoadsorbents
Target	Binds to Lyl+,2,3+ and Lyl+,2,3-

Table 2.   Proportion of Ia+ Accessory Cells of Macrophage and
           Non-macrophage Origin

Accessory Cell	Percent Ia Positive	References
Peritoneal exudate macrophages	5 - 20	13,14
Splenic macrophages	40 - 60	15
Alveolar macrophages	5 - 80[a]	16,17
Thymic macrophages	40 - 60	18
Monocytes in spleen or peri- toneal exudate	30 - 50	19
Bone marrow derived macrophages	20 - 80[b]	20
Langerhans cells	100	21,22
Kupffer cells	50 - 80	23
Interdigitating cells	100	24
Dendritic cells	100	25
P cells	100	26
Tumor lines[c]	100	27,28,29

[a]Weinberg and Unanue (16) only found 5% Ia positive cells, while
Lipscomb et al. (17) found 80% Ia positive cells.
[b]Bone marrow derived macrophages were cultured in vitro: percentage Ia positive cells depends on the in vitro culture
period and the source of colony stimulating factor used.
[c]B cell lymphomas, leukemias, SV_{40} transformed.

activating property. Moreover, separation of macrophages into an Ia+
and an Ia- fraction showed that only the Ia+ cells activated T cells
to become helper cells. A typical example using bone marrow derived
macrophages is shown in Figure 1. However, as mentioned before, there
is accessory cell heterogeneity, i.e. accessory cell function is not

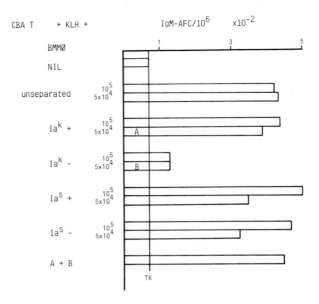

Fig. 1. Ia positive bone marrow derived macrophages (BMMφ) induce
         KLH-specific T helper cells. BMMφ from CBA mice were cul-
         tured with CSF for 7 days and treated with either anti-Ia[k]
         or anti-Ia[s] (both sera provided by the Transplantation Im-
         munology Branch, NIAID, Bethesda) and then with FITC-con-
         jugated rabbit anti-mouse immunoglobulin (see Ref. 47).
         The anti-Ia treated cells were then separated into stained
         and unstained cells using a fluorescence activated cell
         sorter (FACS-II). Two concentrations of unseparated and
         separated BMMφ were then incubated with 3×10^6 macrophage
         depleted normal CBA T cells and 0.1 µg KLH. After 4 days
         the cells were harvested, counted and 5×10^4 live cells added
         to 5×10^6 CBA spleen cells (as a source of B cells) and TNP-
         KLH (0.05 µg). Four days later the IgM anti-DNP response
         (TNP and DNP cross-react at the antibody level) was deter-
         mined. As controls, T cells were incubated wth KLH (NIL)
         and spleen cells were incubated with TNP-KLH alone (TK).
         AFC = antibody forming cells.

exclusively restricted to macrophage or macrophage-like cells. Other
cells such as dendritic cells (25), persisting cells (26) or some
tumor cell lines (27-29) can also express accessory cell function
(see Table 2) though it is not yet clear whether their activity is
restricted to the activation of certain T cell functions.

REGULATION OF Ia EXPRESSION OF MACROPHAGES

Under in vitro culture conditions Ia+ splenic, thymic or peritoneal exudate macrophages become Ia- within a short time, i.e. one
day (15,18,30). However, both synthesis as well as expression of Ia
is resumed by these macrophages provided they are exposed to phagocytic stimuli such as bacteria (Listeria, C. parvum), latex, zymosan,
opsonized sheep red blood cells, antigen-antibody complexes, or even
antigen(e.g. keyhole limpet hemocyanin (KLH)) alone (30). If these
stimuli are added at the beginning of the culture period, they prevent
the loss of the Ia expression for a prolonged time or even enhance
Ia expression (30). These in vitro results confirmed in vivo experiments done by the same group and others, in which mice injected with
Listeria, KLH or BCG showed a marked increase of Ia+ macrophages in
the peritoneal exudate (31,32).

Besides phagocytic stimuli, lymphokines obtained from Con A
activated spleen cells (30), from Trypanosoma cruzi immune spleen
cells (34) or from Mycobacteria or Listeria immune T cells (35,36)
also enhance the synthesis and expression of Ia antigens on macrophages in vitro as well as in vivo demonstrating a T cell dependent
regulation of the synthesis and expression of Ia on macrophages.

We have verified this observation by comparing the helper T
cell activating property of adherent peritoneal macrophages which
were kept in culture for 3 days and stimulated by various means.
Thus, starch induced peritoneal exudate cells were cultured in microtiter plates (10^5/well) for 72 hours. After 48 hours,to some adherent
peritoneal exudate cells Con A, LPS, KLH, spleen cells, nylon wool
purified T cells, or B cells (anti-Thy 1+C' treated spleen cells,
10^6) with or without Con A or LPS were added and incubated for a
further 24 hours. The wells of the plates were then carefully washed
to remove all non-adherent cells and KLH-primed T cells were added
with or without KLH. After 4 days in culture, the T cells were
harvested and their helper activity was measured in a cooperation
system in which $5x10^4$ T cells and $4x10^5$ B cells (DNP-primed) were
incubated with TNP-KLH. After 5 days the IgG anti-TNP response was
determined. A typical result is shown in Figure 2. Peritoneal exudate cells kept in culture for 24 hours were still highly active in
T helper cell restimulation, while the activity of older peritoneal
exudate cells (48,72 hours) was very low. Adding KLH, Con A or LPS
for 24 hours to the 72 hour peritoneal exudate cells only marginally
improved their activity. However, adding spleen cells or T cells
together with Con A for 24 hours dramatically increased the T helper
cell activating capacity of the 72 hour old peritoneal exudate cells.
The same cells alone or incubated with LPS had no effect. B cells
and Con A added for 24 hours also increased the activity of the 72
hour peritoneal exudate cells, but this effect can be most likely
attributed to contaminating T cells. B cells alone or incubated with
LPS had no effect. The results demonstrate that 'old' peritoneal

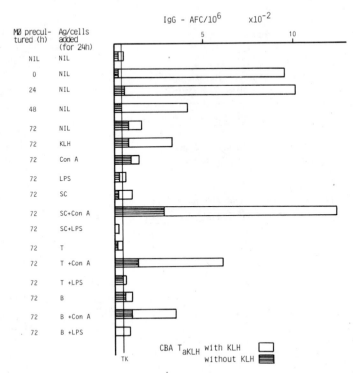

Fig. 2.   Transient accessory cell function of peritoneal exudate
          cells. Starch induced peritoneal exudate cells (1x10^5)
          were plated into microtiter wells and cultured for 24, 48,
          or 72 hours. The 72 hour cells were washed after 48 hours
          and then incubated for an additional 24 hours with either
          the indicated antigen (KLH, 10 µg; Con A, 10 µg; or LPS,
          50 µg) or 10^6 cells (spleen cells abbreviated SC; nylon wool
          purified T cells; or B cells obtained as anti-Thy 1 plus
          complement treated spleen cells). All wells were washed
          to remove non-adherent cells, antigen and mitogen. There-
          after 8x10^5 KLH-primed CBA T cells were added and incubated
          for 4 days with or without KLH (2 µg). The T cells were
          then harvested and 5x10^4 T cells added to 4x10^5 DNP-primed
          B cells and incubated with TNP-KLH (0.005 µg). Five days
          later the IgG anti-TDP response was determined. For controls
          see Figure 1. AFC = antibody forming cells.

exudate cells which were not functional regain accessory cell function
only if preincubated with Con A activated spleen or T cells, but not
if preincubated with antigens or mitogens alone. In preliminary
experiments, the supernatant of Con A activated spleen cells or T
cells had the same effect as the cells, i.e. reactivated the accessory
cell function of old peritoneal exudate cells. Taken together, the
expression of Ia on macrophages kept in vitro (and also in vivo) is
only a transient event and this correlates with the functional capacity of these macrophages. However, there is an exception to the rule.
The percentage of Ia+ bone marrow derived macrophages increases with
the time in culture, but this increase does not correlate with their
functional T helper cell inducing capacity (Stern et al. manuscript
submitted). Thus, the percentage of Ia+ bone marrow derived macrophages increases from about 10-20% on day 3 up to 80% on day 15 and
remains constant for several weeks. In contrast, the T helper cell
inducing capacity of bone marrow derived macrophages is highest between 3 and 10 days and then declines. Clearly, bone marrow derived
macrophages cannot directly be compared with peritoneal exudate cells
or splenic macrophages, because bone marrow derived macrophages divide
and require colony stimulating factor for their growth. Moreover, the
source of colony stimulating factor (CSF) determines the amount of
Ia+ bone marrow derived macrophages obtained in vitro. Lung conditioned medium induces up to 80% Ia+ bone marrow derived macrophages
(20) while L cell conditioned medium only induces between 20 and 40%
Ia+ cells (37,38). Most likely, lung conditioned medium contains
lymphokines which enhance the percentage of Ia+ cells. Nevertheless,
the bone marrow derived macrophage experiment is an example of the
inability of Ia+ macrophages to function as accessory cells for a
certain T cell function and thus demonstrates that expression of Ia
on macrophages is not necessarily linked to their functional capacity. It is open and not yet tested whether these Ia+ 'old' bone
marrow derived macrophages are capable of inducing other T cell functions, e.g. T cell proliferation, DTH, or whether their functional
activity is more dependent on their differentiation stage than on
the expression of Ia.

    Investigating the Ia+ and Ia- population of peritoneal exudate
cells or splenic macrophages, it was found that the Ia+ macrophages
can convert to Ia- and, if appropriately stimulated, back to Ia+ again.
Whether the original Ia- macrophage population can also convert into
Ia+ cells seems to be dependent on the method of stimulation. Phagocytic stimuli cannot convert Ia- into Ia+ macrophages (30), while
lymphokines obtained from Con A activated spleen cells can (33). It
is not clear whether each macrophage is equipped to synthesize and
express Ia, whether the synthesis and expression of Ia is dependent
on the differentiation stage of the macrophage or whether subpopulations of Ia+ and Ia- macrophages exist. The conversion of Ia- into
Ia+ macrophages, a process which does not seem to be dependent on
a proliferative expansion of Ia- precursor cells (33), favors subpopulations more than differentiation dependent Ia expression.

ROLE OF Ia ON MACROPHAGES

The exact role of Ia products on macrophages is still not clear.
It is most likely that Ia on macrophages is involved in antigen presentation to T cells. T cell activation only occurs if antigen is
presented in context of Ia, be it on the surface of macrophages or
by way of soluble mediators (GRF, IAC). T cell activation can be
blocked by appropriate anti-Ia sera which act at the macrophage
level. Using Ir gene controlled antigens, it has been shown that
responder T cells are only activated by responder but not by non-
responder macrophages suggesting that Ia might be the mediator of Ir
gene function (39). There is more direct evidence for such a correlation. Rosenwasser and Huber (40) showed that the Ir gene control of
the proliferative T cell response to bovine insulin was associated
with the expression of Ia.W39 (a private specificity of I-A^b) by
antigen presenting cells. More recently, Lin et al. used a mouse
strain with a mutation in I-A (probably A$_\beta$ gene) to demonstrate that
this mutation resulted in a selective immune response defect to bovine
insulin but not to (T,G)-A-L or collagen at the antigen presenting
macrophage (41). This observation is best explained by a direct
association of the A$_\beta$ polypeptide chain of the Ia^b molecule with the
expression of the immune responsiveness to bovine insulin. Thus,
both studies suggest that the Ia molecules at the macrophage level
are the relevant membrane structures coded for by Ir genes and that
non-responsiveness of the animal might be at least partly due to an
Ir gene defect at the level of the antigen presenting macrophage.
However, other explanations of this phenomenon must also be considered. It is conceivable that Ir genes are not expressed in macrophages, but at another site, e.g. T cells. The failure to generate
a T cell response could then be explained by the lack of appropriate
recognition sites on T cells. Alternatively, Ir gene defects might
only be a quantitative phenomenon, i.e. a matter of affinity (42),
or might reflect a regulatory imbalance between help and suppression
(43). Indeed, the concept of Ir gene expression at the level of the
antigen presenting macrophage has been challenged by recent observations that under certain circumstances macrophages from low responder
animals can present antigen (44-46). If this is proven correct, it
would not necessarily speak against an antigen presenting function
of Ia but severely limit the possibility that Ia is the mediator
of Ir gene function.

Acknowledgement

This work was supported in part by the NIAID (grant 1R01 AI
14513-03) and the Juvenile Diabetes Foundation (grant 85R058).

REFERENCES

1. Rosenthal, A. S., and Shevach, E. M., J. Exp. Med. 138:1194, 1973.

2. Erb, P., and Feldmann, M., Nature 254:352, 1975.
3. Erb, P., and Feldmann, M., J. Exp. Med. 142:460, 1975.
4. Miller, J. F. A. P., Vadas, M. A., Whitelaw, A., and Gamble, J., Proc. Natl. Acad. Sci. USA 72:5039, 1975.
5. Kappler, J. W., and Marrack, P. C., Nature 262:797, 1976.
6. Pierce, C. W., Kapp, J. A., and Benacerraf, B., J. Exp. Med. 144: 371, 1976.
7. Thomas, D. W., and Shevach, E. M., J. Exp. Med. 144:1236, 1976.
8. Yano, A., Schwartz, R. H., and Paul, W. E., J. Exp. Med. 146:828, 1977.
9. Farr, A. G., Dorf, M. E., and Unanue, E. R., Proc. Natl. Acad. Sci. USA 74:3542, 1977.
10. Erb, P., and Feldmann, M., Eur. J. Immunol. 5:759, 1975.
11. Erb, P., Feldmann, M., and Hogg, N., Eur. J. Immunol. 6:365, 1976.
12. Puri, J., and Lonai, P., Eur. J. Immunol. 10:273, 1980.
13. Unanue, E. R., Dorf, M. E., David, C. S., and Benacerraf, B., Proc. Natl. Acad. Sci. USA 71:5014, 1974.
14. Schwartz, R. H., Dickler, H. B., Sachs, D. H., and Schwartz, B. D., Scand. J. Immunol. 5:731, 1976.
15. Cowing, C., Schwartz, B. D., and Dickler, H. B., J. Immunol. 120:378, 1978.
16. Weinberg, D. S., and Unanue, E. R., J. Immunol. 126:794, 1981.
17. Lipscomb, M. F., Toews, G. B., Lyons, C. R., and Uhr, J. W., J. Immunol. 126:286, 1981.
18. Beller, D. I., and Unanue, E. R., J. Immunol. 124:1433, 1980.
19. Henry, C., Goodman, J. R., Chan, E., Kimura, J., Lucas, A., and Wofsy, L., J. Reticuloendothelial Soc. 26:787, 1979.
20. Stern, A. C., Erb, P., and Gisler, R. H., J. Immunol. 123:612, 1979.
21. Frelinger, J. G., Wettstein, P. J., Frelinger, J. A., and Hood, L., Immunogenetics 6:125, 1978.
22. Stingl, G., Katz, S. I., Abelson, L. D., and Mann, D. L., J. Immunol. 120:661, 1978.
23. Richman, L. K., Klingenstein, R. J., Richman, J. A., Strober, W., and Berzofsky, J. A., J. Immunol. 123:2602, 1979.
24. Hoffmann-Fezer, G., Götze, D., Rodt, H., and Thierfelder, S., Immunogenetics 6:367, 1978.
25. Steinman, R. M., Kaplan, G., Witmer, M., and Cohn, Z. A., J. Exp. Med. 149:1, 1979.
26. Schrader, J. W., and Nossal, G. J. V., Immunol. Rev. 53:61, 1980.
27. McKean, D. J., Infante, A. J., Nilson, A., Kimoto, M., Fathman, C. G., Walker, E., and Warner, N., J. Exp. Med. 154:1419, 1981.
28. Cohen, D. A., and Kaplan, A. M., J. Exp. Med. 154:1881, 1981.
29. Buchmüeller, Y., Mauël, J., and Corradin, G., in "Macrophages and Natural Killer Cells: Regulation and Function" (S. Normann and E. Sorkin, eds.), pp. 557-561, Plenum Press, New York, 1982.
30. Beller, D. I., and Unanue, E. R., J. Immunol. 126:263, 1981.

31. Beller, D. I., Kiely, J. M., and Unanue, E. R., J. Immunol. 124: 1426, 1980.
32. Ezekowitz, A. B., Austyn, J., Stahl, P. D., and Gordon, S., J. Exp. Med. 154:60, 1981.
33. Steeg, P. S., Morre, R. N., and Oppenheim, J. J., J. Exp. Med. 152:1734, 1980.
34. Steinman, R. M., Nogueira, N., Witmer, M. D., Tydings, J. D., and Mellman, I. S., J. Exp. Med. 152:148, 1980.
35. Nussenzweig, M. C., Steinman, R. M., Gutchinov, B., and Cohn, Z. A., J. Exp. Med. 152:1070, 1980.
36. Scher, M. G., Beller, D. I., and Unanue, E. R., J. Exp. Med. 152:1684, 1980.
37. Lee, K. C., and Wong, M., J. Immunol. 125:86, 1980.
38. Mottram, P. L., and Miller, J. F. A. P., Eur. J. Immunol. 10: 165, 1980.
39. Rosenthal, A. S., Barcinski, M. A., and Blake, J. T., Nature 267:156, 1977.
40. Rosenwasser, L. J., and Huber, B. T., J. Exp. Med. 153:1113, 1981.
41. Lin, C. C., Rosenthal, A. S., Passmore, H. C., and Hansen, T. H., Proc. Natl. Acad. Sci. USA 78:6406, 1981.
42. Grossman, Z., and Cohen, I. R., Eur. J. Immunol. 10:633, 1980.
43. Araneo, B. A., Jowell, R. L., and Sercarz, E. E., J. Immunol. 123:961, 1979.
44. Stötter, H., Imm, A., Meyer-Delius, M., and Rüde, E., J. Immunol. 127:8, 1981.
45. Ishii, N., Baxevanis, C. N., Nagy, Z. A., and Klein, J., J. Exp. Med. 154:978, 1981.
46. Kimoto, M., Krenz, T. J., and Fathman, C. G., J. Exp. Med. 154: 883, 1981.
47. Erb, P., Stern, A. S., Alkan, S. S., Studer, S., Zoumbou, E., and Gisler, R. H., J. Immunol. 125:2504, 1980.

DISCUSSION

CORRADIN: Would you comment on the idea that Ia antigens are in part carbohydrate and that Ir genes code for glycosyl transferases.

ERB: That remains a possibility. So far we have no evidence against the proposition but also very little in support of it.

UNANUE: I wish to comment on your very interesting observation that bone marrow derived macrophages after 28 days in culture have Ia surface determinants but present antigen weakly. One possible explanation is that the day 28 population of macrophages contains inhibitory cells. This suggestion could be tested by mixing day 28 macrophages with fresh ones and see if they interfere with antigen presentation. Lung conditioned media is a very strong activating media and there is an inverse relationship between the degree of activation and

the secretion of interleukin 1. As the macrophages become more
activated, they secrete less interleukin 1. Thus, a second possible
explanation for your results is that the 28 day macrophages do not
secrete a key molecule such as interleukin 1. A third possibility
is that these activated macrophages do not take up the antigen. I
wonder if you have evaluated any of these possibilities.

I wish to add a further comment. When we looked at certain
populations of macrophages, we found a dichotomy between their capacity to bind T cells, their capacity to produce interleukin 1, and their
capacity to induce T cell proliferation. For example, thioglycollate
elicited macrophages are very, very poor producers of interleukin 1
compared to peptone elicited macrophages. In your situation wherein
you activated your macrophages by prolonged culture in lung conditioned media, I wondered whether or not you may have created a defect
of one sort or another in the macrophage.

ERB: Actually, we looked for macrophage suppression with mixing
experiments and were not able to demonstrate suppression by these
old macrophages. Further, they seemed to be as active in phagocytosis
as younger macrophages. Thus, there was no difference in phagocytosis. Concerning interleukin 1 production, I don't know. We will
have to test it.

VON BLOMBERG: I was very pleased with your conclusion that not all
Ia positive macrophages were stimulatory because it was our conclusion
as well. Did you test the lymphokine modulated macrophages? These
were macrophages which were initially Ia positive, became Ia negative
after culture and regained Ia upon addition of lymphokine. Were these
macrophages which now had enhanced Ia expression due to lymphokine
more stimulatory than your original cell population?

ERB: The experiments to which you refer were not done by myself.
As near as I can remember, they were not more stimulatory. They just
regained accessory cell function.

VON BLOMBERG: But they were much more Ia positive.

BELLER: I don't think there is any particular reason to assume a
stochiometric relationship between presenting cell function and Ia
antigen density. While Ia is necessary, it is not sufficient.
Antigen presentation is an extremely complex process and you really
can't expect to see a direct relationship between the amount of Ia
on a cell and its ability to present antigen to T cells.

ERB: I concur based upon our experiments with bone marrow macrophages. Day 7 bone marrow macrophages are almost 50% Ia positive
and younger ones are much less so being about 10 to 20% Ia positive.
However, the same number of cells are needed to effect T cell function. We can not use less of day 7 macrophages than of day 3.

TOKUNAGA: Where does Ia come from on the T cells?

ERB: At least in humans, it seems that the T cells produce their own Ia.   In mouse, it might be as well.

REGULATION OF MACROPHAGE Ia

EXPRESSION IN VIVO AND IN VITRO

David I. Beller and Emil R. Unanue

Department of Pathology
Harvard Medical School
Boston, Massachusetts   02115 USA

The expression of Ia antigens on macrophages is critical for
their function as antigen-presenting cells (1,2). Since the ability
of individual macrophages to synthesize and express Ia is short-lived
(3,4), the control mechanisms regulating these events become of major
importance. We have recently shown that, in vivo, the phenotype of
the exudate macrophage can be regulated by immunologic stimuli, such
as administration of hemocyanin, sheep erythrocytes, and especially
the intracellular bacterium Listeria monocytogenes (5). With intraperitoneal injection of Listeria, the level of local Ia-bearing macrophages (as determined by immunofluorescence) can reach 90% or more,
while in normal mice, or mice given inflammatory stimuli such as
peptone, it is only about 5 to 10%. All experiments reported here
employed a monoclonal antibody (clone 10-2.16) recognizing specificities 17 of the I-A subregion of fkr and s haplotypes. Using
another monoclonal antibody (against I-E/C) as well as A.TH anti
A.TL antibodies, we have found that the response to modulation of
I-A subregion determinants is characteristic of Ia antigens in general
(4,7). This amplification of the Ia-positive subset was shown, by
adoptive transfer, to require the function of activated T cells.
Moreover, when peritoneal exudate cells from Listeria-immunized mice
were cultured with heat-killed Listeria, they produced a soluble
mediator which, when injected intraperitoneally into normal mice,
induced a similar, selective augmentation of the Ia-positive population (6). The T cell origin of this activity has been confirmed by
its production by T cell lines and hybridomas. It is not genetically
restricted in its action, has a molecular weight of approximately
50,000 daltons, and is referred to as macrophage (Ia-positive)-
recruiting factor (MIRF).

In evaluating the mode of action of MIRF in vivo, we sought

591

to distinguish between the possibilities of conversion of resident
peritoneal macrophages from Ia-negative to Ia-positive in situ, as
opposed to the "recruitment" of a distinct target cell population
via the circulation.   These alternatives were tested by evaluating
the effects of irradiation in vivo (7).   Table 1 reveals that lethal
irradiation of mice prior to injection of immune T cells and killed
Listeria, or MIRF, totally abrogated the response to the immune
stimuli.   Additionally, reconstitution of irradiated mice with bone
marrow substantially restored their responsiveness.   Since irradiation
suppresses the immigration of phagocytes into the exudate, and the
subsequent transfer of bone marrow restores this process (8,9), it
appears that the bone marrow or circulating monocyte is the preferential target of MIRF in vivo.   The selective increase in Ia-bearing
macrophages in the exudate thus appears to be due primarily to an
accumulation of this induced population at the site of stimulation,
rather than the induction of Ia on resident macrophages.   Moreover,
once induced by infection with live Listeria, by transfer of immune
T cells with killed bacteria, or by administration of MIRF, the
maintenance of this expanded Ia-positive population was also radiation
sensitive (4,7).   Thus, both induction and maintenance are dynamic
processes requiring a viable stem cell source.

In vitro experiments also support the idea that the resident
macrophage is not the preferential target of the T cell lymphokine.
For the demonstration of Ia induction in vitro, we used a more potent
source of the lymphokine, the culture medium conditioned by the
continuous growth of Listeria-reactive T cell lines.   Figure 1 shows
the kinetics of response of resident cells compared to those elicited
by peptone or Listeria.   The time at which Ia was expressed was
indicative of the manner in which the macrophages had been elicited.
Moreover, as mentioned before, the ability to express (and synthesize
(4)) Ia was rapidly lost, even in the presence of T cell lymphokine.
Only after the characteristic refractory period was Ia actually induced on these cells.

This refractory period could not be explained by the lack of a
receptor or other target site on the macrophage membrane which might
need to be induced in culture.   Both resident and peptone-elicited
macrophages could be triggered by a two-hour exposure to the lymphokine given within the first twenty-four hours of culture (Figure 2).
After washing and reculture, the cells went on to express Ia with the
same kinetics and to the same extent as those parallel cultures
exposed to the lymphokine continuously.   Thus, in the presence of
the T cell lymphokine, all macrophage populations rapidly become
committed to express Ia.   What distinguishes the different populations
is the relative proportion of inducible cells as well as the duration
of the refractory period.

Table 1. Effect of Lethal Irradiation and Bone Marrow Reconstitution on the Ability of Immune and Non-Immune Stimuli to Generate a Peritoneal Exudate

Stimulus	X-ray	Bone Marrow	% I-A+	PEC/Mouse (x 10^{-6})	I-A+ (x 10^{-5})	I-A- (x10^{-5})
MIRF	–	–	58.5±11.8	7.05±0.44	33.1 ± 8.6	23.2± 6.5
T cells and Listeria	–	–	73.6± 7.5	8.08±2.6	43.3 ±10.2	16.7± 9.8
T cells	–	–	17.3± 2.3	3.97±2.3	4.5 ± 2.5	22.3±14.1
Oil	–	–	12.4± 2.4	8.68±0.95	7.4 ± 1.1	52.7± 6.5
None	–	–	9.9± 1.6	3.05±0.61	2.1 ± 0.6	18.7± 3.6
MIRF	+	–	4.7± 1.4	0.15±0.10	0.05± 0.03	1.0± 0.77
T cells and Listeria	+	–	4.4± 0.32	0.19±0.07	0.05± 0.03	1.2± 0.49
Oil	+	–	3.0± 1.2	0.23±0.07	0.05± 0.01	1.6± 0.52
None	+	–	1.6± 0.46	0.24±0.04	0.03± 0.01	1.5± 0.21
MIRF	+	+	37.6± 6.7	3.0 ±0.86	9.1 ± 2.8	15.4± 6.6
T cells and Listeria	+	+	55.5± 4.7	3.1 ±0.71	13.2 ± 3.2	10.5± 1.8
Oil	+	+	10.5± 3.8	3.5 ±0.56	2.6 ± 0.70	23.1± 5.1
None	+	+	7.2± 2.9	0.66±0.13	0.33± 0.10	4.5± 1.2

On Day 0, some mice received 900 R irradiation, and selected groups were reconstituted with 6 x 10^7 syngeneic bone marrow cells injected intravenously fifteen to twenty hours later. The different stimuli were administered intraperitoneally beginning on Day 7, and peritoneal exudate cells (PEC) were harvested and evaluated on Day 10. Average bone marrow cell counts (x 10^{-6}) on Day 10 were: Normal mice: 32.5±0.64; irradiated mice: 0.29±0.01; irradiated/reconstituted mice: 14.5±0.41.

Finally, the functional activity of macrophages that had been induced to express Ia was assessed in antigen presentation to the Listeria-immune T cell line (Figure 3). Macrophages cultured for six to seven days in medium and which had become totally Ia-negative were no longer able to stimulate the antigen-dependent proliferation of these T cells. However, macrophages cultured for this time period

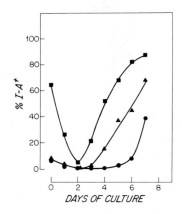

Fig. 1.   Kinetics of induction of I-A on different macrophage populations. Peritoneal exudate cells were elicited by Listeria
followed one week later by peptone (■), peptone alone,
(▲), or no stimulus (resident; ●). Macrophages were purified by adherence to glass coverslips and cultured in RPMI
medium ± supernatant from continuous Listeria-immune T cell
lines (10% v/v). I-A was determined by staining with fluorescent antibodies. In the absence of supernatant, all
cultures showed less than 2% I-A-positive cells after Day 3.

in the T cell lymphokine and on which Ia had been induced, stimulated T cell proliferation to the same extent as fresh macrophages. Thus, the spontaneous transition from Ia-positive to Ia-negative, and the induced expression of Ia, correlate extremely well with the functional competence of these macrophages in inducing T cell responsiveness to complex, multideterminant antigens.

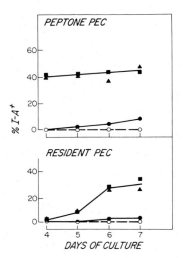

Fig. 2.   Induction of I-A expression by brief exposure to T cell line
          supernatant.   Peptone-elicited (top panel) or resident
          (bottom panel) macrophages were exposed to the active super-
          natant for the initial twenty-four hours of culture (▲)
          or only the first (●), or last (■) two hours of this
          twenty-four-hour period.   Control cultures were kept in
          medium (○).   There is a marked increase in the inducibility
          of macrophages after overnight culture.

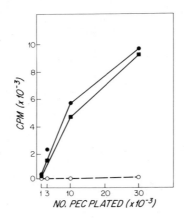

Fig. 3.   Antigen-presenting function of macrophages induced to express
          I-A determinants. Peptone-elicited macrophages were cultured
          in medium (○) or 10% T cell line supernatant (●) for seven
          days and their ability to stimulate antigen-dependent pro-
          liferation of the Listeria-immune T cell line compared to
          freshly isolated macrophages (■). After pulsing with
          antigen (heat-killed Listeria) macrophages were irradiated
          and then 5×10^4 immune T cells added. H^3-thymidine incor-
          poration was assessed during the third day of culture. Un-
          stimulated macrophages lost both I-A and antigen-presenting
          function, while induced macrophages displayed the acquisition
          of both activities, to an extent comparable with that of
          fresh macrophages. T cells exposed to continuous antigen
          showed no proliferation above background, suggesting an
          absolute dependence on exogenous accessory cells.

REFERENCES

1.   Yamashita, U., and Shevach, E., J. Immunol. 119:1584, 1977.
2.   Longo, D., and Schwartz, R., J. Exp. Med. 151:1452, 1980.
3.   Steinman, R., Roqueira, N., Witmer, M., Tydings, J., and Mellman,
     S., J. Exp. Med. 152:1248, 1980.
4.   Beller, D. I., and Unanue, E. R., J. Immunol. 126:263, 1981.
5.   Beller, D. I., Kiely, J.-M., and Unanue, E. R., J. Immunol. 124:
     1426, 1980.
6.   Scher, M. G., Beller, D. I., and Unanue, E. R., J. Exp. Med. 152:
     1684, 1980.
7.   Scher, M. G., Unanue, E. R., and Beller, D. I., J. Immunol. 128:
     447, 1982.
8.   Volkman, A., and Gowans, J., Brit. J. Exp. Pathol. 46:50, 1965.
9.   Volkman, A., and Gowans, J., Brit. J. Exp. Pathol. 46:62, 1965.

DISCUSSION

STEVENSON: Some members of J. Oppenheim's laboratory have demonstrated an effect of gamma interferon on macrophages. Could interferon be one of the specific lymphokines operative in recruiting or inducing Ia positive macrophages?

BELLER: I think it's premature to try to ascribe a molecular identity to MIRF. In our hands, beta interferon has absolutely no effect on Ia induction in vivo or in vitro. Several other laboratories also have found that beta interferon is without effect. This contrasts to Oppenheim's finding that it has significant activity. The question remains whether or not MIRF is really gamma interferon. Oppenheim has blocked lymphokine induced Ia expression by using anti-gamma interferon antibodies. Unfortunately, there is no hybridoma at the moment against gamma interferon. Thus, he has used an antiserum directed against Con A spleen cell supernatants which have interferon activity but of course also various lymphokine activities. So what he really has is an anti-lymphokine antibody which you'd expect to have blocking activity. While he has tried to purify the Ia inducing lymphokine, the progress so far does not allow a definitive conclusion about the molecular or biological nature of the effector molecule. I wouldn't rule out gamma interferon because his data are certainly suggestive, but I think it's too early to know.

VAN MAARSSEVEEN: Did you say that after 6 days of culture the resident cells adapted so that they have Ia?

BELLER: In the presence of lymphokine, yes.

VAN MAARSSEVEEN: Could it be that you have selected for the Ia positive cells because your original resident population was not pure? There are some exudate macrophages in any unstimulated cavity. If the resident cells do not proliferate as readily as exudate macrophages, then after some days in culture there would be more exudate macrophages.

BELLER: If your point is that there is selective proliferation, I can tell you that the data are exactly the same if you irradiate the cultures. There is no requirement for proliferation. Further, the total cell number and total cell protein in these cultures is constant over 7 days.

SCHOOK: I can't see how anti-Ia plus complement treatment of bone marrow cells really shows that you have one macrophage lineage. This experiment shows that you have a stem cell that doesn't express Ia, but it does not rule out separate stem cells for Ia positive and Ia negative macrophages.

BELLER: The experiment shows that there is not a stem cell that already possesses Ia. If there are discrete stem cells and two lineages, they both would have to be Ia negative, and at some time, one of them would be induced to express Ia. So they would both be derived from a similarly Ia negative cell but not necessarily the same cell. On the other hand, we know that all bone marrow derived macrophage stem cells can be induced to express Ia in culture. So, as far as we know, there doesn't seem to be a phagocyte population that's refractory to induction. The implication is that there's a single lineage of phagocytes within which all macrophages can be induced to express Ia.

LIEBOLD: You nicely presented the up regulation of a positive macrophage T cell interaction but could you briefly elaborate on the down regulation? How do the I-J molecules come into the game and how do you stop the process?

BELLER: I'm not sure of the role of I-J molecules; we haven't looked at them. Down regulation may proceed in a number of ways. From David Snyder's work in our laboratory, it appears to be an autoregulation by the macrophage via prostaglandin production. If you culture macrophages, induce Ia with lymphokine and then add prostaglandins, there is a much more rapid loss of Ia than in the absence of prostaglandins. Similarly, if you examine fresh cells in culture, you notice a rapid loss of Ia with a half life of about 12 to 24 hours. In the presence of indomethacin, the half life is significantly extended. Thus prostaglandins and specifically PGE2 may down regulate the system.

LIEBOLD: So down regulation works on the macrophages as well as on the T cells.

BELLER: That's true. Prostaglandins also inhibit T cell proliferation.

FELDMAN: It's absolutely clear that lymphokines can induce the phenotypic expression of Ia on macrophages, yet we do not know whether these lymphokines act as physiological regulators in vivo. I think it's absolutely urgent to get monoclonal antibodies to these lymphokines, to inject them in vivo and to see whether they block any of these phenomena.

BELLER: Such an experiment would be difficult to do because the basal level of Ia is independent of T cell function, at least in our hands.

FELDMAN: So we shall never know whether or not these lymphokines function physiologically?

BELLER: The point is that in the absence of lymphokines, you're not going to have an animal whose macrophages are non-functional. What you could assess is whether or not such antibodies could block enhanced function or the increased expression of Ia in response to immunologic stimuli.

TOKUNAGA: How do T cells recognize Ia antigen?

BELLER: I really wish I knew how to answer that question.

Ia AND MACROPHAGES IN ALLOPROLIFERATION

AND ALLOCYTOTOXICITY

D. R. Katz, A. A. Czitrom, Marc Feldmann and
G. H. Sunshine

I.C.R.F., Tumour Immunology Unit
Department of Zoology
University College London
Gower Street, London, WCIE 6BT, U.K.

Current experiments on primary allogeneic responses in our laboratory are based on the observation that the prototype macrophage washed out of the peritoneal cavity does not stimulate allogeneic T cell proliferation in vitro (1) but does stimulate T helper cell induction (2) and cloned antigen specific T helper cells (3). The dendritic cell which clearly differs from a conventional macrophage is exceptional in its efficiency in primary mixed lymphocyte culture (MLC) stimulation (1). Another contrast of major immunologic relevance between these two cell types is that peritoneal macrophages do not express Ia (or Class II MHC products) on their surface to any great extent (usual range 5-10% specific fluorescence) whereas dendritic cells are strongly Ia positive (70-80%).

We have compared other accessory cells with these two standardised cell types in Ia expression and in MLC stimulation and have examined the effect of anti Ia antisera in this system. We have compared different accessory cells in the generation of cytotoxic T cells and examined the role of Ia in this assay. We have examined some of the influences which affect Ia expression in accessory cells.

CELL TYPES

The different types of accessory cells used are shown in Table 1. All cells were derived from mouse spleen, lymph node and peritoneal cavity. Those from solid tissue fell into two groups: those derived on the commencement day of the experiment, and those which required preparation starting one day previously. Both B and T cells were derived on the day of the experiment.

601

Table 1. Summary of Different Types of Accessory Cells

Cell Type	Method of Isolation	Ia Expression
Spleen cells	Cell suspension by crushing	+++
Spleenic accessory cells	Spleen cells adherent at 2h to a fibronectin coated plate	++
Dendritic cells (18h non-adherent, Fc negative)	Low density cells (10/23%) interface of discontinous albumin gradient. Adherent at 2h and non-adherent at 18h	++++
18h non-adherent Fc positive	Low density cells adherent at 2h and non-adherent at 18h	+++
18h adherent cells	Low density cells which are adherent at 2h and at 18h	++
Peritoneal cells	Cells washed out of ·unstimulated peritoneal cavity	±
B cells	Spleen cells which express surface Ig	++
T cells	Spleen cells which do not adhere to fibronectin plates and which pass through nylon wool and/or G10 columns	±

Ia expression was assessed by indirect immunofluorescence, with monoclonal anti I-A^k and I-E^k antibodies (courtesy of Dr. G. Hammerling). Control cells were from a H-2^d haplotype. Ia expression was graded as follows: ± = <10%; + = 10=20%; ++ = 20-40%; +++ = 40-60%; ++++ = >60%.

Ia EXPRESSION

Table 1 compares Ia expression on the accessory cells used in these experiments. The whole spleen population includes Ia on B cells, macrophages, dendritic cells, and endothelium. The adherent

fractions (at both 2 and 18 h) were removed from the culture vessel
by treatment with 3 mM ethylenediamine tetra-acetic acid. When these
cells were examined in situ there were generally more cells which
expressed Ia on their surfaces. The dendritic cell population was
the most strongly Ia positive but the 18 h non-adherent Fc positive
cells also expressed Ia even if corrected for non-specific Fc binding
as in a radioimmunoassay (data not shown).

     The 18 h fractions all showed a higher percentage of Ia positive
cells than the populations derived on the same day as the experiment.
Several variables introduced during the preparation of these cells
were examined to see if they could account for this difference. Two
representative examples are shown in Table 2. The exposure to albumin
in the initial separation step is not the most important factor since
there was a range of Ia expression within an individual gradient.
Furthermore, time in culture may influence Ia expression, but this
is not inevitable. Peritoneal cells lost their Ia after 24 hours
in vitro. Spleen adherent cells maintained for 24 hours after removal
of non-adherent cells still showed the same range of Ia expression.
However, after exposure to UV light spleenic adherent cells lost
this staining. We have found the following factors to influence Ia
expression on accessory cells: cell type, other cell surface antigens, nature of antecedent stimulation, time in vitro, treatment in
vitro, and time in cell cycle.

PRIMARY MIXED LYMPHOCYTE CULTURES (MLC)

     Primary MLC were examined by comparing the capacity of accessory
cells from one strain of mouse to stimulate either syngeneic or allogeneic T cells. The method used was a 4 day I-125-iododeoxyuridine
incorporation system. A representative experiment is shown in Table
3. Dendritic cells were the most potent stimulators, and peritoneal
cells were not effective. Similarly, B cells did not induce alloproliferation but 18 h non-adherent positive and spleenic accessory
cells did (1). The inhibitory effect of anti I region antisera are
shown in Table 4. Dendritic cell stimulation is inhibited by appropriate anti Ia antisera but not by an inappropriate antibody.

     In vivo, immunologic stimulation of peritoneal cells has been
shown to influence Ia (4). We have used a wide range of peritoneal
stimulants to see (a) if we could increase Ia expression on peritoneal
cells and (b) if this increase was reflected in augmented allogeneic
stimulatory capacity. The results are shown in Table 5, which show
that it is possible to render peritoneal cells more strongly Ia
positive but this does not necessarily coincide with increased allo-
stimulation (1).

Table 2.   Effect of Albumin Gradients and Time in Culture
           on Ia Expression

Cell Type	Ia Expression
Albumin Gradient	
10/23% interface (A)	++
23/26% interface (B)	+
26/29% interface (C)	±
29/35% interface (D)	±
Time in Culture	
2 hour spleen adherent cells	++
2 hour spleen adherent cells maintained for 24 hours	++
Peritoneal cells	±
Peritoneal cells maintained for 24 hours	−

Cells from spleen were spun on a discontinuous gradient.
Ia expression was determined and graded as described in
Table 1.

IN VITRO ALLO-CYTOTOXIC T CELLS

    Different irradiated stimulator populations were examined for
their capacity to generate allo-cytotoxic T cells in vitro.   After
5 days in culture, the cytotoxic cells were assayed on ^{51}Cr labeled
48 hr concanavalin A-induced blasts cells from the stimulator strain
of mouse.   Syngeneic control stimulator cells did not generate allo-
cytotoxic T cells.   The results are expressed as percentage specific
chromium release at different effector : target ratios.   Table 6.
shows that dendritic cells and 18 h non-adherent Fc positive cells
were potent stimulators but peritoneal cells were not.   Dendritic
cell stimulation was abolished by adding the appropriate antisera.

Table 3.   Effect of Stimulator Cells on the Mixed Lymphocyte
Culture

| Stimulator cells (B10G) | | Responder cells (purified T cells) | |
| | | Syngeneic | Allogeneic |
Type	Number	(B10.G)	(B10A)
Spleen	3×10^5	216	18,828
Peritoneal cells	10^4	403	1,652
Spleenic accessory cells	10^4	313	11,242
Dendritic cells	10^4	450	15,830

Stimulator cells were irradiated with 2000 rads.   Responder
cell data are reported in counts per minute in a 4 day I-125-
iododeoxyuridine assay.

Similar results were obtained when the mouse strains differed only
at the K or D end of the MHC (B10A anti B10AQR and B10T 6R anti B10G
respectively) even though the Class II MHC products on these cells
are identical (5).

CONCLUSIONS

These findings show that Class II products are important in the
induction of both alloproliferation and allocytotoxicity. There is
a correlation between Ia expression and primary allostimulatory
capacity with dendritic cells and peritoneal cells at two extremes and
with other cell types ranging between these two. Anti Ia antisera
inhibit the primary MLC and also the generation of allo-cytotoxic T
cells even when the stimulator population is relatively enriched for
Ia expressing cells and when the Ia itself is syngeneic rather than
acting as an alloantigen. However, this correlation is not invariable
since B cells fail to induce alloproliferative responses and since
increased Class II product expression on peritoneal cells is not
always associated with increased allostimulatory capacity. Moreover,
the susceptibility of Ia expression to a variety of influences in the
culture system suggests that it is a relatively unstable component of
the stimulation in these types of assay.

Table 4.   Inhibition of Dendritic Cell Stimulation of Mixed
           Lymphocyte Cultures by Anti Ia Antisera

| | | | Responder cells (purified T cells) | |
| | | | Syngeneic | Allogeneic |
Type	Number	Antisera	(ATL)	(ASW)
Spleen	3×10^5	-	2167	15395
	10^4	-	2576	3058
Peritoneal cells	10^4	-	3081	2747
Dendritic cells	10^4	-	2873	11961
	10^4	anti- Ia^k	236	791
	10^4	anti- Ia^b	3624	11849

Stimulator cells (ATL)

Stimulator cells were irradiated with 2000 rads. Responder
cell data are reported in counts per minute in a 4 day I-125-
iododeoxyuridine assay. Antiseara was prepared as follows:
Anti Ia^k (ATH anti ATL) and anti Ia^b (B10.A x A anti C57 B1/6)
(Courtesy of Dr. I.F.C. McKenzie).

     Three hypothetical explanations for these findings are of inter-
est. One is that there is a difference between Ia on one cell and
Ia on another. There is recent evidence that this may be the difference between spleenic accessory cells and B cells (6). Second,
it may require interleukin (IL-1) production as well as Ia to induce
stimulation. The antibody studies illustrate that synergy may be
necessary. Thirdly, dendritic cells may lack an inhibitory mechanism
which blocks Ia intrinsically, and this allows them to induce un-
restrained proliferation and allo-cytotoxic T cells in these assays.

Acknowledgement

     Thanks are due to Susan Edwards, Marjorie Liddel and Allison
Kingsbury for expert technical assistance and to Philippa Wells for
typing. DRK and GHS are ICRF research fellows and AAC holds a centen-
nial fellowship from the Canadian MRC.

Table 5.   Induction of Ia Expression on Peritoneal Cells and
           Absence of Effect on Mixed Lymphocyte Cultures

			Responder cells (Purified T cells)	
Stimulator cells (CBA)				
			Syngeneic	Allogeneic
Type	Number	Ia Expression	(CBA)	(Balb)
Dendritic cells	1×10^4	7.6	1717	12231
	5×10^3		1264	5021
Peritoneal cells	1×10^4	1.3	2101	4483
	5×10^3		2122	3621
+2% starch	1×10^4	5.2	743	1485
	5×10^3		1454	2446
+ pristeine	1×10^4	1.0	1101	1729
	5×10^3		1413	2170

Stimulator cells were irradiated with 2000 rads. Ia expression is reported in % specific binding in radioimmunoassay.
Responder cells data are reported in counts per minute in a
4 day I-125-iododeoxyuridine assay.

Table 6.   Effect of Stimulator Cells on Capacity to Generate
           Cytotoxic T Cells

Stimulator cells		Anti- serum Against	Responder T cell	% specific Cr release E/T ratios of		
Type	Number			30:1	10:1	3:1
				(B10A targets)		
B10A spleen	6×10^5	–	B10G	55	45	24
	5×10^3	–		14	4	0
B10A dendritic	5×10^3	–	B10G	70	56	29
B1018 non- adherent Fc positive	5×10^3	–	B10G	66	55	38
B10A peritoneal	5×10^3	–	B10G	3	0	0
				(B10S targets)		
B10S spleen	6×10^5	–	B10AQR	82	63	35
	5×10^3	–	B10AQR	7	3	2
B10S dendritic	5×10^3	–	B10AQR	58	32	17
	5×10^3	Ia^s	B10AQR	4	0	0
	5×10^3	Ia^k	B10AQR	27	10	0

Stimulator cells were irradiated at 2000 rads.   E/T = effector
to target cell ratio.

REFERENCES

1. Sunshine, G. H., Katz, D. R., Czitrom, A.A., Eur. J. Immunol.
       12:9, 1982.
2. Erb, P., and Feldmann, J. Exp. Med. 142:460, 1975.
3. Schreier, M. H., Iscove, N., Tees, R., Aarden, L., and von
       Boehmer, H., Immunol. Rev. 51:315, 1980.
4. Beller, D. I., Kiely, J. M., and Unanue, E., J. Immunol. 124:
       1426, 1980.
5. Czitron, A. A., Katz, D. R., Sunshine, G. H., Immunology, 1982,
       in press.
6. Cullen, S. E., Kindle, C. S., Shreffler, D. C., and Corning, C.,
       J. Immunol. 127:1478, 1981.

DISCUSSION

AHMANN: Are the responder cells whole spleen cells or purified T cells?

KATZ: Purified T cells.

AHMANN: If you take a purified T cell population, they are accessory cell dependent. So when you add back the stimulator cells, they provide not only stimulation, but accessory cell function. If you use whole spleen cells as your responder population, you can find B cells as good stimulators as well as splenic adherent cells. Are you really asking if it's an accessory cell phenomenon or is it actually the Ia stimulatory determinant that you're looking at?

KATZ: You suggest that the feeder like function of accessory cells can be separated from its Ia stimulatory function. Since we have purified the population, we would argue that we are providing both from the same cell.

SECTION 9

ROLE OF MACROPHAGES IN HUMAN DISEASE

FIBRONECTIN: A MEDIATOR OF PHAGOCYTOSIS

AND ITS POTENTIAL IN TREATING SEPTIC SHOCK

Frank A. Blumenstock and Thomas M. Saba

Department of Physiology
Albany Medical College of Union University
Albany, New York   12208 USA

Clearance of certain types of particulate matter from the circulation by the cells of the RES is dependent upon the presence of specific proteins in the plasma which support the clearance process by opsonization (1,2,3). One protein actively involved in mediating RES clearance of non-bacterial particulates is a large plasma glycoprotein referred to as α_2 surface binding (SB) glycoprotein (4,5). This protein, which has been known by a number of terms, was originally demonstrated by its depletion from the plasma of animals subjected to RES blockade with gelatinized colloid. In these studies, the in vitro support of hepatic macrophage uptake of colloid by plasma from blockaded animals was directly proportional to the deficit in in vivo RE clearance observed following intravascular colloid injection (6). The in vitro activity of this opsonizing protein was demonstrated to be heparin dependent, cryoprecipitable from plasma in the presence of heparin and relatively sensitive to heat denaturation (7,8). Early biochemical characterization of this particular opsonin demonstrated that it was non-immunoglobulin in nature and independent of the complement system (9,10). Further, it was probably identical to a plasma factor which had been described earlier as being functional in the RES clearance of gelatinized radiogold (11).

With the demonstration that this serum factor was involved in RES clearance, a variety of studies implicated it in RES alterations consequent to disease processes (12). Early studies emphasized its potential importance in RE function during tumor growth. Animals with transplanted tumors demonstrated cyclic alterations in RES function that correlated closely with the opsonic activity of the plasma (13). Whereas RES clearance is important in host defense (14-16), these studies suggested that part of the host defense process against tumors was related to the presence of opsonic substances in the

613

plasma. Other studies demonstrated that intravascular tumor cell
injection resulted in RES blockade accompanied by deficits in humoral
support of in vitro RES phagocytosis (17). Similar to animal studies,
patients with cancer have diminished plasma opsonic activity when
compared to normal controls (18). In experimental animals, surgical
trauma, which depressed RE clearance function and plasma opsonic
activity, enhanced host sensitivity to intravenous tumor cell injection (19). Studies on the etiology of RES depression following traumatic injury have revealed that such injury causes RES function to
decline in association with decreased humoral support for in vitro
hepatic phagocytosis (20). In traumatized man, all patients demonstrated a significant depression of plasma opsonin levels immediately
post-injury and those patients who survived normalized their plasma
opsonic levels while non-survivors who suffered from multiple organ
failure and sepsis did not (21). Thus, a variety of indirect evidence
suggests that plasma factor(s) that support RES clearance function
are important in the host response to injury, shock and tumor growth
in both man and experimental animals.

    Purification of the plasma factor from human and animal sources
revealed it to be a high molecular weight glycoprotein (Mr=440,000)
consisting of two identical or nearly identical subunits of Mr=220,000
(22,23,24). It had an electrophoretic migration equivalent to α_2
globulin. With the development of monospecific antibody to the protein, it was possible to measure its concentration in the plasma of
rats (400-600 µg/ml) and in man (330+25 µg/ml) (24.25). By immunoassay, we confirmed that RE blockade in rats resulted in a plasma
deficit in the protein (24) and that post-traumatic RE dysfunction
in man and rats was also associated with a depression in circulating
plasma opsonin levels (21,26). Further, intravenous administration
of antibody to the opsonic protein increased the sensitivity of rats
to traumatic injury (26) and injection of semi-purified opsonic
protein fractions inhibited tumor growth in animals (27). Thus, the
biochemical purification of the molecule responsible for macrophage
recognition of gelatin-coated colloid allowed for a more in-depth
study of its mechanism of function and importance in host defense.

OPSONIC PROTEIN, COLD INSOLUBLE GLOBULIN AND FIBRONECTIN

    Cold insoluble globulin (CIg) was first identified by Morrison
et al. in 1948 (28). This protein was present in fibrinogen fractions
of human plasma prepared by cryoprecipitation. It resisted purification and was originally considered to be a protein species somehow
related to fibrinogen. This concept was shown to be false when it
was demonstrated that CIg was distinct from fibrinogen; for instance,
it remained unclottable in the presence of fibrinogen under certain
conditions (29). Subsequently purified in 1970 from human plasma by
Mosesson and co-workers (30), CIg was demonstrated to have a molecular
weight of 440,000 with a subunit molecular weight of 220,000 (31).

Further characterization identified it as an α_2 globulin with an affinity for heparin (31). In the presence of CIg, fibrinogen was cryoprecipitable but in its absence, fibrinogen remained soluble (31). At this time, its biological function was unknown but it was hypothesized that it somehow was involved in the coagulation process since it was so closely associated with fibrinogen. Indeed, subsequent investigations revealed that it could be incorporated into the stabilized fibrin clot by activated plasma transglutaminase (Factor XIIIa) (32). Cold insoluble globulin was demonstrated by immunoassay (33) to be depressed in the plasma following elective surgery and to be a significant contaminant in the plasma cryoprecipitate used for Factor VIII replacement therapy (31). We were struck by the similarities in structure between plasma CIg and the opsonic protein. We have been able to demonstrate that the two proteins have antigenic identity, identical molecular weights in both the native and reduced form, and identical electrophoretic mobilities (4,5,25).

The discovery of the common identity of opsonic α_2SB glycoprotein and CIg finally revealed a biological role for CIg: that is, the opsonization of particulates and augmentation of phagocytosis. Since it had already been demonstrated that critically injured patients suffering from sepsis and multiple organ failure had circulating opsonin deficiency, the discovery of the identity of the opsonic protein and CIg provided the opportunity to study the effect of reversing the opsonic deficiency by infusion of plasma cryoprecipitate whose CIg content was 10-fold that of plasma (34). Administration of 10 units of fresh cryoprecipitate to a series of septic patients with deficits by immunoassay and bioassay in circulating opsonic α_2SB glycoprotein levels reversed both the immunoassay and bioassay deficiencies restoring RE function as measured in vitro (34,35). The pulmonary function of these patients appeared to improve and there was a reversal in the septic state as demonstrated by an increase in alertness, a normalization of leukocyte levels and arterial pulse, an improvement in various hemodynamic parameters, negative blood cultures and an alleviation of the febrile state (34).

Clinical evaluation of cardiovascular, pulmonary and renal functions in such patients suggested that reversal of opsonic α_2SB glycoprotein deficiency improved organ function as reflected by ventilation-perfusion balance, peripheral vascular hemodynamics and oxygen consumption, and improved glomerular filtration rate (36). These original observations are now being reevaluated in our institution in a prospective double blind study utilizing cryoextracted plasma as a control substance.

Independently cell biologists studying alterations in cell surface proteins during oncogenic transformation described the existence of a large, fibrous protein on the external cell surface which appeared to participate in cell-cell interaction and cell adhesion to a substratum (37). This important cell surface protein, which is

lost upon oncogenic transformation in tissue culture, is antigenically
related to plasma CIg (38), and thus to the opsonic protein. This
adhesive glycoprotein enhances the adherence of fibroblasts to a
collagenous substratum and has been named cell surface or tissue
fibronectin. It has been demonstrated in vivo to mediate attachment
of mesenchymally derived cells to collagen. Although this cell surface form of fibronectin is antigenically related to plasma fibronectin (CIg, opsonic α_2SB glycoprotein) there are subtle differences in the two forms of the protein, especially with respect
to their relative solubility. The plasma form is more soluble at
physiological pH and also is slightly smaller, having a molecular
weight 10,000 less than the cell surface form of the molecule (39).
There may be slight differences in amino acid sequences since monoclonal antibodies can be generated which recognize only one form of
the molecule (40). Functionally, fibronectin may (a) be involved in
RES clearance of blood-borne particulates, (b) wound healing (fibronectin has been shown to be chemotactic for fibroblasts (41) and
possibly for macrophages), and (c) cellular organization at the tissue
level including adherence of vascular endothelium to basement membrane (36).

FIBRONECTIN: STRUCTURE AND FUNCTIONAL PROPERTIES

     The demonstration that fibronectin in its soluble form in the
plasma participates in RES clearance and that another form of the
molecule is important for cellular attachment and adherence, stimulated a whole series of investigations directed at understanding
the extent of its functions in vivo and how these functions relate
to its structure. Both forms of the molecule possess unique domains
that have high affinities for heparin, actin, Staph. aureus, collagen
(or gelatin) and fibrin (42).

     The opsonic properties of the protein in the in vitro liver
slice assay have been shown to depend upon heparin (7). Heparin will
also amplify the phagocyte uptake of gelatinized particles by rat
and mouse peritoneal macrophage monolayers in vitro (43,44). The
heparin binding domain has been identified as being the same domain
involved in fibronectin binding to fibroblasts (42). The high affinity for heparin and the enhanced macrophage uptake of fibronectin
opsonized particles in the presence of heparin raises the possibility
that heparin functions as a co-factor for the opsonic activity of the
molecule by enhancing the binding of opsonized particles to the
macrophage cell surface.

     The actin binding site is on the same structural domain as the
high affinity site for fibrin and gelatin, and therefore fibronectin
may be involved in the clearance of actin from the circulation following tissue damage due to traumatic injury or sepsis (45).

Fibronectin possesses two domains for binding fibrin: one high affinity binding site which is on the same domain as the gelatin binding site, and another low affinity binding site on the N terminal domain that is involved in transglutaminase mediated covalent cross-linking of fibronectin to fibrin and collagen (42). Fibronectin has been demonstrated to participate in RES clearance of fibrin micro-aggregates but it may also maintain the solubility of fibrin monomers since fibronectin inhibits fibrin-fibrin interactions and fibrin-collagen interaction in vitro (46). It is interesting to note that fibronectin not only inhibits the interaction of fibrin with itself and with collagen, but it also resolubilizes polymerized fibrin monomers in vitro (46). This property of fibronectin may be important in disseminated intravascular coagulation (DIC). Several studies have documented decreased plasma fibronectin levels in this syndrome (47,48). It has been postulated that fibrin microaggregates are opsonized by plasma fibronectin and cleared by the RES. However, an alternative possibility exists. Tissue damage may consume plasma fibronectin with concomittant activation of the coagulation cascade generating circulating fibrin monomer. In the presence of decreased levels of fibronectin, the circulating fibrin monomers self-associate, forming microaggregates which embolize and occlude the microcircula-tion. They are not cleared by the RES, again due to decreased plasma fibronectin levels. Part of the improvement following cryoprecipitate therapy observed in patients suffering from multiple organ failure in association with post-injury sepsis could be attributed to this mech-anism. Therefore, the infused plasma fibronectin could enhance RES clearance of circulating microparticulates, inhibit formation of new microparticulates and allow fibrinolysis to degrade previously en-trapped fibrin in the microcirculation.

Fibronectin has also been demonstrated to bind Staph. aureus via the N terminal domain and thus augment neutrophil phagocytosis (49) of pathogenic strains of this bacterium commonly responsible for post-burn septicemia. Since fibronectin deficiency in burn patients precedes the development of sepsis (50,51), it is possible that fibronectin may play a role in antibacterial resistance as well as RES clearance of products of tissue trauma and bacterial sepsis.

It is evident that fibronectin may function in a variety of ways to enhance host defense processes and to maintain organ structure and function. The new rapid immunoturbidimetric assay (52) may be a valuable means to evaluate fibronectin in the clinical setting and to delineate the importance of fibronectin in various disease states. While collectively the findings summarized here strongly support the therapeutic use of opsonically active fibronectin in the treat-ment of septic injured patients, we emphasize that additional con-trolled clinical studies are needed to critically test this hypothe-sis.

Acknowledgement

        Supported by NIH Grants GM-21447 and AI-17635. Clinical studies
were performed in the Albany Trauma Center, GM-15426. The authors
wish to acknowledge the secretarial assistance of Mrs. Maureen Davis
for her invaluable assistance in the preparation of this manuscript.

REFERENCES

1. Saba, T. M., Arch. Int. Med., 126:1031, 1970.
2. Wright, A. B., and Douglas, S. R., Proc. Roy. Soc. 73:128,
        1904.
3. Stossel, T. P., New Eng. J. Med. 290:717, 1974.
4. Blumenstock, F. A., Saba, T. M., Weber, P., and Laffin, R., J.
        Reticuloendothelial Soc. 23:35, 1977.
5. Blumenstock, F. A., and Saba, T. M., Adv. Shock Res. 2:44, 1979.
6. Saba, T. M., and Di Luzio, N. R., Amer. J. Physiol. 216:197, 1969.
7. Saba, T. M., Filkins, J. P., and Di Luzio, N. R., J. Reticulo-
        endothelial Soc. 3:398, 1966.
8. Di Luzio, N. R., McNamee, R., Olcay, I., Kitahama, A., and Miller,
        R. H., Proc. Soc. Exp. Biol. Med. 145:311, 1974.
9. Allen, C., Saba, T. M., and Molnar, J., J. Reticuloendothelial
        Soc. 13:410, 1973.
10. McLean, S., Siheel, J., Molnar, J., Allen, C., and Sabet, T.,
        J. Reticuloendothelial Soc. 19:127, 1976.
11. Murray, I. M., Amer. J. Physiol. 204:655, 1963.
12. Saba, T. M., in "Medicine in Transition",(E. P. Cohen, ed.), pp.
        297-332, Univ. of Illinois Press, Chicago, 1981.
13. Saba, T. M., and Antikatzides, T. G., Brit. J. Cancer 32:471,
        1975.
14. Old, L. J., Clarke, D. A., Benacerraf, B., and Goldsmith, M.,
        Ann. N. Y. Acad. Sci. 88:264, 1960.
15. Zweifach, B. W., Ann. N. Y. Acad. Sci. 88:203, 1960.
16. Biozzi, G., Halpern, B. N.,Benacerraf, B. and Stiffel, C., in
        "Physiopathology of the Reticuloendothelial System" (B. N.
        Halpern, ed.), pp. 204-225, Charles C. Thomas, Springfield,
        IL., 1957.
17. Di Luzio, N. R., Miller, E., McNamee, R., and Pisano, J. C.,
        J. Reticuloendothelial Soc. 11:186, 1972.
18. Pisano, J. C., Jackson, J. P., DiLuzio, N. R., and Ichinose, H.,
        Cancer Res. 32:11, 1972.
19. Saba, T. M., and Antikatzides, T. G., Brit. J. Cancer 34:381,
        1976.
20. Kaplan, J. E., and Saba, T. M., Am. J. Physiol. 230:7, 1976.
21. Scovill, W. A., Saba, T. M., Kaplan, J. E., Bernard, H. R.,
        and Powers, S. R., J. Surg. Res. 22:709, 1977.
22. Blumenstock F. A., Weber, P., and Saba, T. M., J. Biol. Chem.
        252:7156, 1977.
23. Blumenstock, F. A., Saba, T. M., Weber, P., and Laffin, R., J.
        Biol. Chem. 253:4387, 1978.

24. Blumenstock, F. A., Weber, P., Saba, T. M., and Laffin, R., _Amer._
     J. Physiol. 232:R80, 1977.
25. Saba, T.M., Blumenstock F. A., Weber, P., and Kaplan, J. E.,
     Ann. N. Y. Acad. Sci. 312:43, 1978.
26. Kaplan, J. E., Saba, T. M., and Cho, E., _Circ. Shock_ 3:203, 1976.
27. Saba, T. M., and Cho, E., _J. Reticuloendothelial Soc._ 22:583,
     1977.
28. Morrison, P. R., Edsall, J. T., and Miller, S. G., _J. Amer. Chem._
     Soc. 70:3103, 1948.
29. Polara, B., Thesis 63-7948. University of Minnesota, University
     Microfilms, Ann Arbor, MI, 1963.
30. Mosesson, M. W., and Umfleet, R. A., _J. Biol. Chem._ 245:5278,
     1970.
31. Mosesson, M. W., _Ann. N. Y. Acad. Sci._ 312:11, 1978.
32. Mosher, D. F., _Ann. N. Y. Acad. Sci._ 312:38, 1978.
33. Aronsen, K. F., Ekelund, G., Kindmark, C.-O., and Laurell, C. B.,
     Scand. J. Clin. Lab. Invest. 29:Suppl. 124:127, 1972.
34. Saba, T. M., Blumenstock, F. A., Scovill, W. A., and Bernard,
     H. R., _Science_ 201:622, 1978.
35. Scovill, W. A., Saba, T. M., Blumenstock, F. A., Bernard, H. R.,
     and Powers, S. R., _Annals of Surg._ 188:521, 1978.
36. Saba, T. M., and Jaffe, E., _Am. J. Med._ 68:577, 1980.
37. Yamada, K. M., and Olden, K., _Nature_ 275:179, 1978.
38. Ruoslahti, E., and Vaheri, A., _J. Exp. Med._ 141:497, 1975.
39. Yamada, K. M., and Kennedy, D. W., _J. Cell Biol._ 80:492, 1979.
40. Atherton, P. T., and Hynes, R. O., _Cell_ 25:133, 1981.
41. Postlethwaite, A. E., Keski-Oja, J., Balian, G., and Kang, A. H.,
     J. Exp. Med. 153:494, 1981.
42. Engel, J., Odermatt, E., Engel, A., Madra, J. A., Furthmayer, H.,
     Rohde, H., and Timple, R., _J. Molec. Biol._ 150:97, 1981.
43. van de Water, L., Schroeder, S., Crenshaw, E. B., and Hynes,
     R. O., _J. Cell Biol._ 90:32, 1981.
44. Blumenstock, F. A., Saba, T. M., Roccario, E., Cho, E., and
     Kaplan, J. E., _J. Reticuloendothelial Soc._ 30:61, 1981.
45. Estes, J. E., Dillon, B. C., and Saba, T. M., _J. Reticuloendo-_
     thelial Soc. Abstracts of the 18th Annual National Meetings,
     p. 13A, 1981.
46. Kaplan, J. E., Snedeker, P. W., _J. Lab. Clin. Med._ 96:1054, 1980.
47. Mosher, D. F., and Williams, E. M., _J. Lab. Clin. Med._ 91:729,
     1978.
48. Mosher, D. F., _Thromb. Res._ 9:37, 1976.
49. Proctor, R. A., Pendergast, E., and Mosher, D. F., _Clin. Res._
     27:650, 1979.
50. Lanser, M. E., and Saba, T. M., _J. Reticuloendothelial Soc._ 30:
     415, 1981.
51. Brodin, B., Bergham, L., Frigerg-Nielson, S., Nordstrom, H., and
     Schildt, B., _Excerpta Medica_, 7th World Congress of Anesthe-
     siology, Hamburg, p. 504, 1980.
52. Saba, T. M., Albert, W. H., Blumenstock, F. A., Evanega, G.,
     Staehler, F., and Cho, E., _J. Lab. Clin. Med._ 90:473, 1981.

DISCUSSION

VAN FURTH: You mentioned a rapid method for determining fibronectin levels. What is a good rapid method?

BLUMENSTOCK: Boehhringer-Mannheim sells a turbidometric assay which we have used satisfactorily. The values we get correlate very well with those determined by rocket immunoassay. The advantages are that you can do it manually with a spectophotometer and get answers within 20 minutes.

VAN FURTH: What is the mechanism for the decrease in soluble fibronection in certain diseases?

BLUMENSTOCK: Concerning the mechanism of decrease following traumatic injury, there's evidence of actual sequestration at the site of injury. These studies have been done with ^{125}I-labeled fibronectin. Following surgical incision, there's a localization of fibronectin at the area of injury as you might expect, since exposed collagen binds fibronectin. So, sequestration is one reason for consumption. Secondly, following injury there is the release of debris into the plasma and this consumes fibronectin. In addition, aggregates of fibrin bind fibronectin, are cleared, and this depletes circulating fibronectin. Also, damaged platelets similarly consume fibronectin, as does circulating collagen. Finally, in some patients, there also may be a defect in synthesis. We really don't know what causes fibronectin depletion but these are good possibilities.

VAN FURTH: Where is it produced?

BLUMENSTOCK: There are two forms of fibronectin: cell surface form and the plasma form. I think that's been well shown. There is some evidence that the plasma form might be synthesized in the liver, but we really don't know. Macrophages can synthesize fibronectin but the level of synthesis is not high. Depressed levels such as those associated with RE blockade, come back quite rapidly within 6 hours. We have some evidence in rats that the half-time of fibronectin in the circulation is very short, probably less than 20 hours, which is really fast clearance for a plasma protein. Thus, macrophages are probably not the major site of synthesis or the only one. When we injected splenium[75] methionine, it was incorporated into the protein. We produced RE blockade and there was a slight increase in the synthetic rate following the blockade. But we still don't know where it came from.

ISLIKER: The amount of fibronectin is very high, in fact 300 micrograms per ml. What is the lowest level you have seen in diseased patients?

BLUMENSTOCK: In patients with disseminated intravascular coagulation,

it may go down to 0, and commonly we see levels as low as 20 or
30 μg/ml.

ISLIKER: An additional means by which fibronectin might be depressed
in disease is that immune complexes activate complement and then
Clq binds the fibronectin.

KENYON: When you have lysis of a fibrin clot, is the fibronectin also
lysed?

BLUMENSTOCK: The fibronectin is not lysed. We did our experiments
in the presence of a protease inhibitor.

UNKNOWN: To what extent is fibronectin an acute phase protein?

BLUMENSTOCK: I do not believe that it's a classic acute phase protein. Acute phase proteins are barely detectable in the normal state,
but appear in very high levels during an inflammatory response. There
is some evidence that fibronectin does rise after inflammation, but
it's already there at high levels and just goes up a little higher.

LAZDIN: In your opinion, is so-called reticuloendothelial blockade
always due to the depletion of fibronectin?

BLUMENSTOCK: I don't think that all RE blockade is due to the consumption of fibronectin. The classical means of inducing blockade
was to use colloids that were gelatin stabilized. As such, fibronectin depletion may have been critical to the blockade. Curiously,
following RE blockade, there's an increased sensitivity of these
animals not only to trauma but also to tumor growth, hemorrhagic
shock, and endotoxin shock.

FIBRONECTIN BINDS TO CHARGE-MODIFIED PROTEINS

Matti Vuento,[1] Mirja Korkolainen,[1] and Ulf-Håkan Stenman[2]

[1]Department of Biochemistry, University of Helsinki, Unioninkatu 35, SF-00170 Helsinki 17, Finland
[2]Department of Obstetrics and Gynaecology, University Central Hospital, Helsinki, Finland

INTRODUCTION

Binding of fibronectin to a variety of macromolecular ligands has been demonstrated in vitro. These include proteins such as collagen, fibrin, actin, Clq and fibronectin itself, but also glycosaminoglycans, DNA and polyamines. Furthermore, fibronectin interacts with the surface of animal cells and bacteria (reviewed in Ref. 1). Some of these interactions contribute to the organization of extracellular matrixes in cell cultures and in connective tissue. The binding of fibronectin to bacteria, collagen, fibrin and actin probably are important for the role of circulatory fibronectin as an opsonin. In many cases, fibronectin appears to bind denatured or modified ligands more readily than native ones. Thus binding of fibronectin to collagen (2) and actin (3) is enhanced by denaturation of these proteins. Fibronectin binds more efficiently to fibrin than to its parent molecular form fibrinogen (4). Conformational factors may also play a role in the binding of fibronectin to Clq (5). We have tested the effect of chemical modification on the interaction of fibronectin with several serum proteins. Modification of carboxyl groups of lysozyme, albumin and IgG caused these proteins to avidly bind to fibronectin. This modification also enhanced the self-association of fibronectin.

MATERIALS AND METHODS

Fibronectin was purified from plasma as described (6) and labeled with ^{125}I by lactoperoxidase-catalyzed iodination (7). Human

IgG and albumin were from Kabi, Stockholm, Sweden. Other proteins
were from Sigma, St. Louis, MO., USA. The proteins were immobilized
by coupling to cyanogen bromide activated Sepharose (6). Modification
of carboxyl groups was performed with 1-ethyl-3-(3-dimethylamino-
propyl) carbodiimide at pH 4.75 in the presence of 1 M glycine ethyl
ester (8). Amino groups were modified with succinic anhydride (9).
Reduction and alkylation were performed as described (10). Heat-
denaturation of Sepharose-linked proteins was done at 75°C for 30
minutes. Full details of these procedures will be published elsewhere (11). Affinity chromatography experiments were carried out
using columns (5 ml) of carboxyl-modified, Sepharose-linked fibronectin, IgG, albumin or lysozyme. The columns were equilibrated with
Tris-buffered saline, pH 7.5. Pooled human serum (2.5 ml) was applied
to the columns, which were then washed with the equilibrating buffer
(30 ml) and eluted with 8M urea in the equilibrating buffer. SDS-
PAGE was carried out in vertical gel slabs (6% acrylamide) using the
buffer system of Weber & Osborn (12). Binding of fibronectin to
various proteins was studied by incubating labeled fibronectin with
aliquots (20 μl) of Sepharose-linked proteins. Bound fibronectin
was separated by centrifugation. Bovine serum albumin was added to
the incubation buffers to prevent binding of labeled fibronectin
to tube walls.

RESULTS AND DISCUSSION

     In preliminary experiments, a number of immobilized proteins
were treated with carboxyl-modifying reagents. After this treatment
the following proteins bound iodinated fibronectin: fibronectin,
IgG, albumin, lysozyme, and cytochrome c. However, chicken ovalbumin,
human α_1-acid glycoprotein and soybean trypsin inhibitor did not
bind fibronectin before or after treatment. Albumin, IgG and fibronectin were selected for further studies as major plasma proteins.
Results from binding experiments are shown in Table 1. Labeled
fibronectin bound to immobilized native fibronectin, but not to native
albumin or IgG. This result is in line with the self-association
tendency of fibronectin observed earlier (13). Carboxyl-modification
of immobilized fibronectin enhanced the binding of labeled fibronectin. Immobilized IgG and albumin also effectively bound labeled
fibronectin after this modification. Succinylation, reduction and
alkylation or heat-denaturation did not have such an effect.

     To study whether or not the interaction of fibronectin with
carboxyl-modified proteins was strong enough to take place under
physiological conditions, we performed affinity chromatography experiments (Figure 1). When serum was passed through columns of
carboxyl-modified proteins linked to Sepharose immunoreactive fibronectin was depleted, and it was present in fractions eluted from the
columns with urea (not shown). Protein eluted from the columns was
analyzed by SDS-PAGE (Figure 2). Polypeptide bands with Mr of 220,000

Table 1. Binding of ^{125}I-labeled Fibronectin to Various Proteins
          Coupled to Agarose

Protein coupled to Sepharose 4B	Modification	^{125}I-fibronectin Bound to 20 µl of Gel (% of Applied Radioactivity)
None	None	2
	Carbodiimide	2
Fibronectin	None	25
	Succinylation	6
	Reduction and alkylation	17
	Heat-denaturation	3
	Carbodiimide	80
IgG	None	7
	Succinylation	4
	Reduction and alkylation	9
	Heat-denaturation	9
	Carbodiimide	80
Albumin	None	4
	Succinylation	3
	Reduction and alkylation	4
	Heat-denaturation	6
	Carbodiimide	45

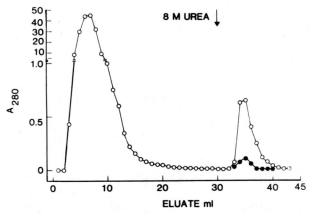

Fig. 1. Affinity chromatography of human serum on carboxyl-modified
         IgG linked to Sepharose. Part of the elution curve obtained
         with nonmodified IgG-Sepharose is shown as filled circles.

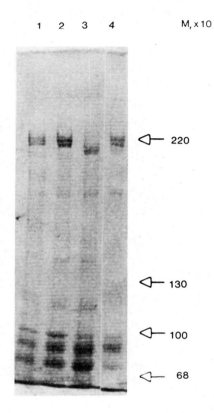

Fig. 2.   Dodecylsulfate polyacrylamide gel electrophoresis.   Electro-
          phoretic analysis of protein material bound from serum to
          columns of carboxyl-modified IgG (1), lysozyme (2), albumin
          (3) and fibronectin (4).   The migration is from top to
          bottom.   Migration positions of molecular weight markers
          are shown by arrows.

are seen in the stained gels.   These bands were identified as fibronectin, because they were not present if fibronectin had been removed
from serum with gelatin-Sepharose (6).   An additional set of polypeptide bands with Mr 68,000-100,000 was consistently present (Figure 2).

    Modification of carboxyl groups by the carbodiimide reaction
used in this study leads to a reduction of negatively charged groups
of the modified proteins.   Since fibronectin is anionic at neutral
pH (14), its binding to carboxyl-modified proteins could be due to

electrostatic attraction. On the other hand, drastic changes in the surface charge of the modified proteins could give rise to conformational alterations with an exposure of hydrophobic regions, to which fibronectin may bind. In any case, the interaction of fibronectin with carboxyl-modified proteins is strong enough to take place at physiological ionic strength and pH and in the presence of other serum components. Although no experimental evidence is available at present on the biological significance of this new phenomenon, we would like to propose that the ability of fibronectin to recognize modified proteins could be important for the opsonin function of fibronectin. The negative charge of proteins could be reduced in vitro e.g. by proteolysis. Fibronectin could bind to such modified proteins and facilitate their uptake by phagocytic cells.

Acknowledgement

This work was supported by the Research Council for Natural Sciences, Academy of Finland.

REFERENCES

1. Vaheri, A., Keski-Oja, K., Vartio, T., Alitalo, K., Hedman, K., and Kurkinen, M., in "Gene Families of Collagen and Other Proteins" (Prockop and Champe, eds.), pp. 161-178, Elsevier, Amsterdam, 1980.
2. Engvall, E., Ruoslahti, E., and Miller, E. J., J. Exp. Med. 147: 1584, 1978.
3. Koteliansky, V. E., Glukhova, M. A., Morozkin, A. D., Musatov, A. P., Shirinsky, V. P., Tskhovrebova, L. A., and Smirnov, V. N., FEBS Lett. 133:31, 1981.
4. Stemberger, A., and Hörmann, H., Hoppe-Seyler's Z. Physiol. Chem. 357, 1003, 1976.
5. Menzel, E. J., Smolen, J. S., Liotta, L.., and Reid, K. B., FEBS Lett. 129:188, 1981.
6. Vuento, M., and Vaheri, A., Biochem. J. 183:331, 1979.
7. Marchalonis, J. J., Biochem. J. 113:299, 1969.
8. Carraway, K. L., and Koshland, D. E. Jr., Methods in Enzymology 25:616, 1972.
9. Klotz, I. M., Methods in Enzymology 11:576, 1967.
10. Vuento, M., Salonen, E., Salminen, K., Pasanen, M., and Stenman, U.-H., Biochem. J. 191:719, 1980.
11. Vuento, M., Korkolainen, M., and Stenman, U.-H., submitted for publication.
12. Weber, K., and Osborn, M., J. Biol. Chem. 244:4406, 1969.
13. Vuento, M., Vartio, T., Saraste, M., von Bonsdorff, C.-H., and Vaheri, A., Eur. J. Biochem. 105:33, 1980.
14. Vuento, M., Wrann, M., and Ruoslahti, E., FEBS Lett. 82:227, 1977.

DISCUSSION

BLUMENSTOCK: Denatured proteins bind fibronectin and such binding should mediate their removal. Have you determined if fibronectin mediates macrophage uptake of charge modified proteins?

VUENTO: Not yet.

KENYON: What was the exact form of fibronectin used in your experiment?

VUENTO: The fibronectin was isolated from plasma and the soluble, native fibronectin was labeled with iodine. The other proteins were coupled to agarose. We measured the binding of native fibronectin to the modified protein gels.

KENYON: Whenever we have attempted to label fibronectin, even using the lacto-peroxidase method, the molecule breaks down. Thus, it is difficult to ascertain what you're dealing with.

VUENTO: We used lacto-peroxidase catalyzed iodination under conditions wherein only one atom of iodine was incorporated into the molecule. In about 50 iodinations, we didn't see any significant break down of the protein.

ISLIKER: If you aggregate albumin, it will bind Clq and Clq will bind the fibronectin similar to modified IgG. I wonder whether this whole binding phenomenon might not be mediated by complement?

VUENTO: You may have a point. We didn't check our preparations for aggregation.

RES-SATURATION WITH ACCUMULATION OF HEMOGLOBIN-HAPTOGLOBIN COMPLEXES IN PLASMA. A METHOD FOR SIMULTANEOUS QUANTITATION OF FREE HAPTOGLOBIN AND HEMOGLOBIN-HAPTOGLOBIN COMPLEXES

Ivan Brandslund,[1] and Sven-Erik Svehag[2]

[1]Department of Clinical Chemistry, Odense University
Hospital, DK-5000 Odense C, Denmark
[2]Institute of Medical Microbiology, Odense University
DK-5000 Odense C, Denmark

INTRODUCTION

The presence of visible amounts of hemoglobin in plasma is usually regarded as a certain sign of acute intravascular hemolysis. When hemoglobin (Hb) is released intravascularly it combines irreversibly with haptoglobin (Hp). One molecule Hp (MW 98-114 KD) binds up to 4 molecules of Hb (MW 16.2 KD) as 2 $\alpha\beta$ units. These HbHp complexes have molecular weights ranging from 114,200 daltons when Hp is unsaturated to 162,800 daltons when Hp is fully saturated with four Hb molecules (assuming the genetic type of Hp with a molecular weight of 98,000 daltons). One g of Hp binds up to about 0.65 g Hb.

The HbHp complex is rapidly removed by the reticuloendothelial system by first order kinetics with half-life of 8-12 minutes when the clearance system is unsaturated, or by zero order kinetics and an elimination rate of 130 mg Hb per hour per liter of plasma when the system is saturated (1). When the circulating Hp is rapidly saturated with Hb, the HbHp complexes are removed within 10 hours and the Hp concentration will reach zero and remain near zero values as long as hemolysis continues (2). If the amount of Hb released exceeds the binding capacity of Hp, the heme part of Hb will bind to hemopexin and the plasma hemopexin concentration will decrease and stabilize at a level reflecting the severity of the hemolysis (3).

RESULTS

In two patients suffering from penicillin-induced serum sickness-like adverse reactions, high and rising plasma concentrations of Hb

were measured for several days (Figure 1) (4). The Hb concentrations
corresponded to known maximal Hp binding capacity, and Hb and Hp were
shown to be present in plasma as HbHp complexes, as evidenced by
gel chromatography and quantitative precipitation of Hb using antibody
against Hp (5). Parallel determinations of hemopexin (Figure 1)
showed that the hemolysis exceeded Hp binding capacity. As the HbHp
accumulated, hemopexin concentrations decreased (Figure 1, patient 1).
The data suggested that HbHp were being removed by a mechanism different than hemopexin. This has been a matter of dispute (5,6,7).
We have proposed that this type of HbHp accumulation reflects a
saturation of blockade of the RES (4,9,10) in conditions where hemolysis is pronounced. The Hb release was caused by a Coombs' negative
microangiopathic hemolysis secondary to a diffuse intravascular

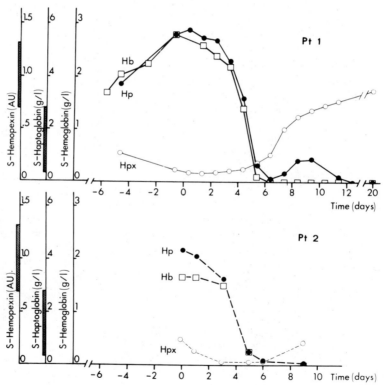

Fig. 1.   Changes in serum hemoglobin (Hb, □), haptoglobin (Hp, ●)
           and hemopexin (Hpx, ○) levels in two patients with peni-
           cillin-induced hemolytic uremic syndrome and accumulation
           of HbHp complexes. Penicillin treatment was discontinued
           on day 0.

coagulation as a part of a hemolytic uremia-like syndrome (11), and
not by the well known Coombs' positive penicillin-induced immune
hemolysis.

From a third patient who developed penicillin-induced HbHp accumulation, serial serum samples had been frozen as a routine procedure. When the patient showed a rise in temperature after 3 weeks
of parenteral penicillin treatment, daily serum samples were collected. The patient's plasma gradually became deeply red colored.
The serum samples were examined by double-decker rocket immunoelectrophoresis (12). It proved possible to separate and quantify HbHp
complexes in the bottom gel containing anti-Hb antibodies (Dako,
Copenhagen, Denmark) and free Hp in the upper gel containing anti
-Hp antibodies (Dako) (Figure 2, bottom part). Conventional rocket
immunoelectrophoresis was used to measure Hp using anti-Hp antibodies
as shown in the upper part of Figure 2. A steady increase in total
Hp (Hp + HbHp) was seen during the first symptoms of the adverse
reaction (specimens 16-21). This increase was real, as a saturation
of Hp with Hb does not influence the height of the Hp rockets. The
increase in total Hp was caused by an accumulation of HbHp consequent
to intravascular hemolysis. These results indicate that the penicillin reaction was followed by an inhibition of HbHp clearance,
while the clearance of hemopexin was not affected. The inhibited
HbHp clearance seems to be due to a general clearance defect, as
evidenced by the concomitant fall to low levels of plasma fibronectin,
a protein known to participate in the removal of waste products from
the circulation (specimens 12-19 in Figure 3) (13). That the HbHp
accumulation reflects a general RES saturation is indicated not only
by the low levels of fibronectin but also by the serious clinical
state of the patients, resulting in a reversible syndrome of respiratory distress, renal insufficiency, classical complement activation,
consumption coagulopathy and multiorgan failure (11). We have found
also unusually high plasma concentrations of fibrinogen degradation
products. These are believed to be cleared by the RES (14) and may
involve fibronectin.

How penicillin provokes this situation is not clear. Two alternative mechanisms seem possible, one being the result of immune complex formation secondary to the interaction of penicillin metabolites
with certain plasma proteins, the other being modification of plasma
proteins leading to their removal via fibronectin and hence fibronectin consumption and RES-saturation.

We believe the HbHp accumulation is a general phenomenon associated with an overloading of the RES and a consumption of fibronectin. We have observed the occurrence of HbHp in patients with
severe burns (5), various intoxications (5,15), trauma and septicemia
(e.g. specimens 9 to 11 in Figure 2). We have not yet sufficient
quantitative measurements to be able to conclude whether the occurrence of HbHp in these conditions represents an acute, transient state
or a gradual continued accumulation as in the penicillin reactions.

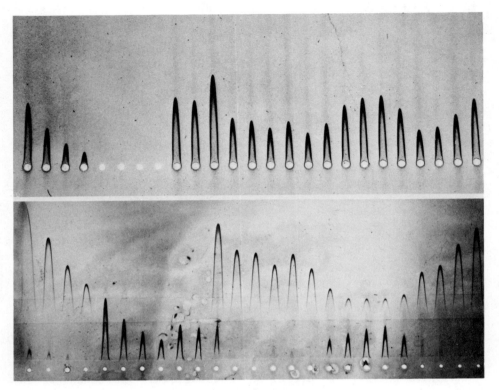

Fig. 2.   Accumulation of Hemoglobin-Haptoglobin complexes (HbHp) in
           a patient's serum during the early phase of a penicillin-
           induced adverse reaction. Upper part: Quantitation of total
           Hp (free Hp + HbHp) by anti-Hp antibodies. Specimens 1-4 are
           Hp standards 2.6, 1.3, 0.65 and 0.32 g/l, specimens 9-25 are
           patient serum samples from days -29, -28, -26, -16, -12, -10,
           -6, -3, -2, -1, 0, +1, +2, +4, +5, +8, +16 in relation to
           discontinuation of penicillin treatment. Lower part: Simul-
           taneous determination of HbHp complexes and free Hp by use of
           anti-Hb in the bottom gel (precipitates HbHp) and anti-Hp in
           the top gel, precipitating only free Hp, as HbHp has been re-
           tained in the bottom gel. Specimens 1-4 are Hp standards
           5.2, 2.6, 1.04 and 0.52 g/l. 5-8 are HbHp-standards contain-
           ing 5.2, 2.6, 1.04 and 0.52 g Hp per liter. Specimens 9-25
           are the same serum samples as above. Specimens 9-11 are from
           the acute bacteremic phase of endocarditis, 12-15 from the
           three weeks asymptomatic phase after institution of peni-
           cillin treatment, 16-21 represent daily tests during six days
           of the penicillin reaction, 22-25 daily values from the con-
           valescence period. Penicillin treatment was discontinued
           between samples 19 and 20.

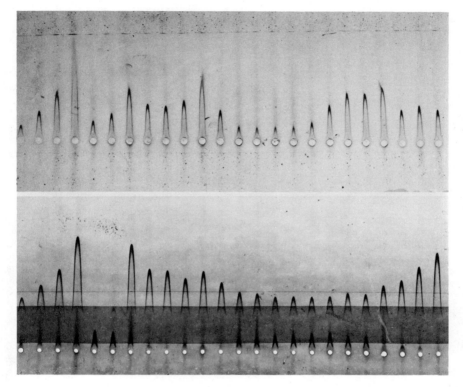

Fig. 3.   Fibronectin concentrations in serum from the same patient
          as in Figure 2. Note the abrupt fall in fibronectin (upper
          part) at the time of HbHp accumulation (lower part). HbHp
          disappeared and fibronectin concentrations normalized after
          discontinuation of penicillin treatment between samples
          17 (day 0) and 18. Samples 1-4 in the upper part of the
          figure are fibronectin standards containing 25%, 50%, 100%
          and 200% of normal values. Samples 5-24 are serum samples
          from days -29, -28, -26, -16, -12, -10, -6, -3, -2, -2, -1,
          -1, 0, +1, +2, +2, +4, +5, +8, +16. Samples 1-4 (bottom)
          are Hp standards: 0.52, 1.3, 2.6 and 5.2 g/1.

REFERENCES

1. Laurell, C. B., and Nyman, M., Blood 12:493, 1957.
2. Nyman, M., Scand. J. Clin. Lab. Invest. 11, suppl. 39:1, 1959.
3. Müller-Eberhard, U., New England J. Med. 283:1091, 1970.
4. Brandslund, I., and Hyltoft Petersen, P., in "Comptes Rendus
       du 4eColloque de Pont-a-Mousson de Biologie Prospective"
       (G. Siest and Galteau, M.-M., eds.), pp. 246-248, Masson,
       Paris, 1979.
5. Brandslund, I., Hyltoft Petersen, P., Brinkløv, M. M., Klint
       Anderson, P., and Parlev, E., Scand. J. Clin. Lab. Invest.,
       in press.
6. Hersko, C., Cook, J. D., and Finch, C. A., J. Lab. Clin. Med.
       80:624, 1972.
7. Liem, H. H., Biochem. Biophys. Acta 343:546, 1974.
8. Wada, T.,Ohara, H., Watanabe, K., Kinoshita, H., and Yachi, A.,
       J. Reticuloendothelial Soc. 8:185, 1970.
9. Brandslund, I., and Hyltoft Petersen, P., Lancet II:633, 1979.
10. Brandslund, I., and Hyltoft Petersen, P., Lancet I:103, 1980.
11. Brandslund, I., Hyltoft Petersen, P., Strunge, P., Hole, P., and
       Worth, V., Haemostasis 9:193, 1980.
12. Brandslund, I., Siersted, H. C., Svehag, S.-E., and Teisner, B.,
       J. Immunol. Methods 44:63, 1981.
13. Saba, T. M., and Jaffe, E., Am. J. Med. 80:577, 1980.
14. Walsh, R. T., and Barnhart, M. I., Thromb. Diath. Haemorrh.
       suppl. 36:83, 1969.
15. Brinkløv, M. M., Andersen, P. K., Nielsen, P. S., Brandslund,
       I., Hole, P., and Larsen, O., Uqeskr. Laeg 141:2606, 1979.

DISCUSSION

BLUMENSTOCK: I noticed that your plasma samples were very red and
very cloudy.

BRANDSLUND: As a matter of fact, the plasma was filled with cell
aggregations, namely thrombocytes, and these aggregates disappeared
concurrently with the hemoglobin/haptoglobin complexes after dis-
continuation of penicillin.

GORDON: Is there any direct evidence that the hemoglobin/haptoglobin
complexes go into macrophages and not into other hepatic cells?

BRANDSLUND: No.

α_2M-PROTEINASE COMPLEXES ARE TAKEN UP

BY MACROPHAGES DURING JOINT INFLAMMATION

T. L. Vischer, E. Flory and K. Muirden

Division of Rheumatology, Hôpital Cantonal
Universitaire, Geneva and Department of Medicine
University of Melbourne, Australia

During inflammation, proteinases are released by inflammatory and tissue cells. The activation of the complement, the coagulation and the fibrinolytic systems produces additional proteinases. Active proteinases become inactivated by various inhibitors in the circulation and tissues. Among the plasma proteinase inhibitors, alpha$_1$-antiproteinase and α_2-macroglobulin (α_2M) are most abundant (1). The α_2M-proteinase complexes are removed from the circulation by fixed tissue macrophages in the liver and spleen and broken down (2,3). In the experiments summarized in this paper, we have looked for the presence of α_2M-proteinase complexes in phagocytic cells of synovial tissues and exudates from patients with joint diseases. In parallel, we established that only α_2M-proteinase complexes and not free α_2M or proteinases are taken up by phagocytic cells.

UPTAKE OF α_2M-TRYPSIN COMPLEXES BY MACROPHAGES

Both peptone-induced and non-elicited mouse peritoneal macrophages were exposed at room temperature in RPMI 1640 medium to human α_2M, trypsin or α_2M-trypsin complexes for 30 minutes. After washing, cytocentrifuge smears were prepared and stained with FITC-conjugated antibodies against α_2M or trypsin. The macrophages exposed to α_2M-trypsin complexes, but not those exposed to either native α_2M or trypsin, contained small cytoplasmic inclusions positive for α_2M or trypsin. From the results summarized in Table 1, we can conclude that macrophages take up only α_2M-trypsin complexes but not free α_2M. Similar results have been reported previously (4-6). Neutrophil leukocytes containing α_2M-trypsin complexes were not observed during these experiments. A detailed account of this work has been reported in Ref. 7.

635

In addition, cultured synovial lining cells obtained from biopsies or cadaver joints (8,9) were used for similar experiments. They were exposed to normal human plasma with trypsin or plasmin added to form spontaneous α_2M-proteinase complexes. Inclusions with α_2M could be seen by the immunofluorescence method when the cells were incubated at 23°C, but not when the incubation was done at 4°C and in the presence of iodoacetamid, or in the absence of proteases. Details of these experiments can be found in Ref. 10. Synovial cells are thus similar to macrophages in regard to uptake of α_2M-proteinase complexes.

α_2-MACROGLOBULIN-TRYPSIN COMPLEXES IN CELLS AND TISSUES FROM HUMAN JOINTS

We first prepared cytocentrifuge smears of cells obtained from various synovial fluids aspirated from joints of patients with rheumatoid arthritis (N = 9), ankylosing spondylitis (N = 5), chondrocalcinosis (N = 2) or osteoarthrosis (N = 10). Using an FITC-conjugated anti-human α_2M antibody, cytoplasmic inclusions similar to those found in the previous in vitro experiments could be seen in monocytes, but not in granulocytes. The more inflammatory fluids, defined arbitrarily by protein content and number of white blood cells per mm^3, had a mean of 31% of monocytes positive for α_2M whereas the less inflammatory fluids had only 8% positive. However, there was no direct correlation with cell number and protein content. The occurrence of α_2M inclusions had no relation to the clinical diagnosis of the patients providing the synovial fluid. Details are reported in Ref. 7.

Table 1.   Endocytosis of α_2M-trypsin Complexes by Mouse Macrophages

Stimulants Added to Macrophages	Fluorescent Markers	
	FITC-Rabbit Anti-Human α_2M	FITC-Rabbit Anti-Human Trypsin
None	negative	negative
Trypsin	negative	negative
Native α_2M	negative	negative
α_2M-trypsin complexes	strongly positive	positive

Synovial membranes from 10 patients with rheumatoid arthritis, 8 with osteoarthritis and 2 with normal joints were examined by immunofluorescence for α_2M after fixation with cold ethanol : granular α_2M deposits were seen mainly in synovial lining cells but also in perivascular macrophage-like cells and in a few stromal cells. The intensity of α_2M staining increased in the more inflamed tissues (RA) whereas less inflamed tissues contained few or no deposits. Deposits of immunoglobulins and the C3 complement component were also more frequent in the more inflamed tissues (10). Although fibroblasts can take up free α_2M, we think that the inclusions at least in the synovial lining cells and the macrophage-like cells in the deeper layer correspond to α_2M-proteinase complexes.

CONCLUSION

These experiments provide evidence that some proteinases liberated during joint inflammation are complexed with α_2M and taken up by macrophages and phagocytic synovial cells, both in tissues and exudates. In vitro, the complexes disappear within 12 hours (7,10). This uptake by macrophages corresponds to a mechanism of removal and destruction of α_2M-proteinase complexes. Recently we have shown that endocytosis of α_2M-proteinase complexes can activate resident murine macrophages to produce and secrete other neutral proteinases, which could play a role in inflammation (11). Thus, the disposal of α_2M-proteinase complexes by macrophages might be another way of perpetuating chronic arthritis. An immunoregulatory role has been attributed to α_2M- proteinase complexes in relation to this effect on macrophages (12). This might have implications for the deranged immune mechanisms in rheumatoid arthritis.

Acknowledgement

This work was supported by grants from the Swiss National Fund, the Swiss Federal Hygiene Office, the Australian Arthritis and Rheumatism Foundation, and the Arthritis and Rheumatism Association of Victoria.

REFERENCES

1. Heimburger, N., in "Proteases and Biological Control" (E. Reich, D. B. Rifkin, E. Shaw, eds.), pp. 367-386, Cold Spring Harbor Laboratory, Cold Spring Harbor, 1975.
2. Ohlsson, K., Acta Physiol. Scand. 81:269, 1971.
3. Ohlsson, K., and Laurell, C. B., Clin. Sci. Mol. Med. 51:87, 1976.
4. Dolovich, J., Debanne, M. T., and Bell, R., Am. Rev. Resp. Dis. 112:521, 1975.
5. Debanne, M. T., Bell, R., and Dolovich, J., Biochim. Biophys. Acta 411:295, 1975.

6. Debanne, M. T., Bell, R., and Dolovich, J., Biochim. Biophys.
       Acta 428:466, 1976.
7. Flory, E., and Vischer, T. L., Rheumatol. Int. 1:61, 1981.
8. Fraser, J. R. E., and Catt, K. S., Lancet 30:1437, 1961.
9. Clarris, B. J., Fraser, J. R. E., Moran, C. J., and Muirden, K.
       D., Ann. Rheum. Dis. 36:293, 1977.
10. Flory, E. D., Clarris, B. J., and Muirden, K. D., Ann. Rheum.
       Dis., 1982, in press.
11. Vischer, T. L., and Berger, D., J. Reticuloendothelial Soc.
       28:427, 1980.
12. Hubbard, W. J., Hess, A. D., Hsia, S., and Amos, D. B., J. Im-
       munol. 126:292, 1981.

DISCUSSION

VICTOR: Rheumatoid synovium is a very difficult tissue for immunofluorescence analysis because of problems associated with rheumatoid
factor. Did you use any particular technique to avoid these problems?

VISCHER: We used both normal rabbit anti-α_2M antibody and Fab fragments. But there was still quite a bit of background.

BORTH: Everyone here knows that α_2M protease complexes are taken up
by the RES. How did you discern cells which secrete α_2-macroglobulin
from cells which take up the complexes?

VISCHER: I think that's an important problem. Many cells have been
shown more or less conclusively to produce α_2-macroglobulin. I think
it's a question of sensitivity. With immunofluorescence, we have
never seen a macrophage or fibroblast react that had not been exposed
to serum α_2-macroglobulin or its complexes. Endogenous synthesis
usually resulted in a very homogenous stain whereas the particulate
staining was probably due to endocytosis.

BORTH: You found fewer positive cells in osteoarthritis than in
rheumatoid arthritis?

VISCHER: Yes. But it was not related to diagnosis but to the extent
of inflammation. In osteoarthritis or chondrocalcinosis, we found
more positive cells the greater the extent of inflammation.

VICTOR: How can you tell if complexes are inside the cell and not
on the surface of the cell?

VISCHER: With the macrophages we did surface staining and didn't
see them on the surface.

VICTOR: Were the specimens fixed cells?

VISCHER: Yes.

GORDON: Did you say that there are non-macrophages in the synovial lining that also take up the complexes?

VISCHER: It's difficult to say because the cells differentiate in culture and they all were cultured for a few days.

RICHTER: Was α_2-macroglobulin deposited at sites of fibrin deposition?

VISCHER: There was no clear correlation with fibrin.

SPLENIC FUNCTION IN PRIMARY GLOMERULONEPHRITIS

Susan Lawrence, B. A. Pussell and J. A. Charlesworth

Department of Nephrology
Prince Henry Hospital, Sydney and
The Renal Unit
Wollongong Hospital, Wollongong, NSW, Australia

INTRODUCTION

Recent studies have shown that the tissue distribution and deposition of immune complexes may be altered by activation or saturation of the reticuloendothelial system (RES) (1-3). RES hypofunction has been demonstrated in patients with systemic lupus erythematosus (4), systemic vasculitis (5) and mixed cryoglobulinemia with nephritis (6). These authors suggested that defective RES function may have contributed to the pathogenesis of the nephritis found in these putative immune complex-mediated diseases.

The present study was aimed at measuring splenic function in primary glomerulonephritis by using the clearance of altered, radio-labeled, autologous red blood cells (RBC). Where possible, double-isotope studies were used to examine the clearance of erythrocytes that had been damaged by two different methods. In vitro studies were also undertaken to define the site of uptake of the altered red cells and to examine the function of human macrophages in the presence of preformed immune complexes and serum from patients with various forms of primary glomerulonephritis.

METHODS

Selection of Patients: Patients with a diagnosis of primary glomerulonephritis based on clinical, histological, biochemical and serological criteria were studied. All had serum creatinine, serum albumin, 24 hour protein excretion, full blood count, complement profile (CH50,C4,C3,C1q) and C1q binding assay for immune complexes

641

measured on the day of the study, using standard techniques. Anti-glomerular basement membrane antibody was measured by a solid phase radioimmunoassay (7).

Disease categories were: mesangial IgA nephropathy (12 patients); acute post-infectious glomerulonephritis (4 patients); membrano-proliferative glomerulonephritis (6 patients); focal glomerulosclerosis (4 patients); membranous glomerulonephritis (3 patients); minimal change glomerulonephritis (1 patient) and anti-glomerular basement membrane disease (1 patient). Serial studies were performed in 3 patients - 1 each with minimal change glomerulonephritis, acute glomerulonephritis with crescents and anti-glomerular basement membrane disease. Control studies were performed in 11 healthy volunteers.

Preparation of Erythrocytes: Red cells were radiolabeled with either sodium pertechnetate (^{99m}Tc) or sodium chromate (^{51}Cr) using standard techniques (4,5) and damaged by one of three methods: a) thermal stress (8), b) chemical alteration with the sulfhydryl inhibitor N-ethylmaleimide (NEM, Calbiochem-Behring, USA) or c) coating with antibody to Rhesus D antigen (10).

Clearance Studies: The clearance studies were carried out according to the protocol of Lockwood et al. (5). In order to examine the clearance of erythrocytes that had been damaged by different methods, seventeen double-isotope studies were performed in 13 individuals: 7 with heat-damaged and NEM-treated cells and 10 with antibody-coated and NEM-treated cells.

In Vitro Macrophage Studies: Peritoneal macrophages were harvested from patients on uncomplicated maintenance peritoneal dialysis. The cells were separated by a standard technique and seeded onto flat-bottomed microtiter wells (Nunc, Denmark) (about 5×10^5/well). After 1 hour incubation at 37°C the non-adherent cells were removed by washing and the adherent macrophages examined for viability by trypan blue dye exclusion. The adherent macrophages were then incubated for one hour at 37°C with fresh medium containing 20 µl of various test substances and/or 200 µl antibody-coated erythrocytes (EA). The test reagents were IgG-coated latex particles (Calbiochem-Behring, USA), carbon particles (Norit-A), pooled normal human sera, serum from patients with glomerulonephritis, preformed immune complexes (BSA/rabbit anti-BSA prepared in 5-fold antigen excess), sheep EA, human EA (10) and NEM-treated human red cells (9). The cells were again washed, fixed, stained and examined by light microscopy. Macrophage function was assessed by the ability of these cells to bind and phagocytose red cells in the presence of medium alone, preformed immune complexes, serum, inert particulate matter or Fc-bearing material (i.e. IgG-coated latex particles).

Three low-power fields were examined and the percentage of macro-

phages forming EA rosettes calculated. The number of erythrocytes
either adherent to or ingested by the macrophages was also noted.
Results were expressed as the mean of at least 3 separate experiments
performed in duplicate.

RESULTS

     In 6 healthy volunteers, the half-life of heat damaged erythro-
cytes ranged between 10.5 and 16.5 minutes. Seventeen studies were
performed in 14 patients with primary glomerulonephritis. The clearance half life was abnormal in 5/7 patients with IgA nephropathy,
2/2 with focal glomerulosclerosis, and 1/1 with membrano-proliferative
glomerulonephritis. One patient with minimal change glomerulonephritis was studied serially and the clearance time returned to
normal with clinical remission following steroid therapy (half life
of 23.5, 15.5, and 15 minutes). In contrast to these abnormal clearances, 2 patients with acute post-streptococcal glomerulonephritis
and 1 patient with anti-glomerular basement membrane disease had
normal red cell clearances.

     Thirty-five studies were performed in 28 individuals using NEM-
treated erythrocytes. In 5 healthy volunteers, clearance half life
ranged between 10 and 22.5 minutes. Abnormal clearance was detected
in 6/8 patients with IgA nephropathy, 4/6 with membrano-proliferative
glomerulonephritis, 2/2 with focal glomerulosclerosis, 2/3 with
membranous glomerulonephritis, 1/2 with minimal change glomerulonephritis, 0/1 with acute glomerulonephritis and 3/4 with anti-
glomerular-basement membrane disease. This latter patient was studied
serially at weekly intervals and, following immunosuppressive therapy
and vigorous plasma exchange, clearance half life improved accompanied
by a fall in anti-glomerular-basement membrane antibodies and serum
creatinine (Figure 1). One patient with acute crescentic glomerulonephritis (without linear immunofluorescence in renal biopsy or circulating antibody to glomerular basement membranes) also was studied
serially before and after treatment with immunosuppressive drugs.
Although the clearance half life remained abnormal, there was a
definite improvement with therapy corresponding to the improved
renal function and clinical condition of the patient (half life of
90, 87, and 34 minutes with serum creatinine of 0.45, 0.38, and
0.11 mM/L respectively). One patient who had undergone splenectomy
previously had a prolonged erythrocyte clearance time and served as
a positive control.

     Ten studies were performed using antibody coated erythrocytes
together with NEM-treated cells in double isotope labeled clearance
studies. Individuals whose red cells lacked Rh D antigen or who
had undergone splenectomy had a clearance half life greater than
120 minutes. Abnormal clearance times were found in 2/2 patients
with membranous glomerulonephritis, 1/1 with minimal change glomerulo-

nephritis, and 1/1 with membrano-proliferative glomerulonephritis.
In serial studies performed on the latter patient, the clearance half
life improved in response to therapy. In 5/9 double-isotope studies,
abnormalities in clearance of antibody-coated and NEM-treated cells
occurred concurrently. In one healthy individual who was Rh D negative, the NEM-treated erythrocyte survival was normal (10.5 minutes)

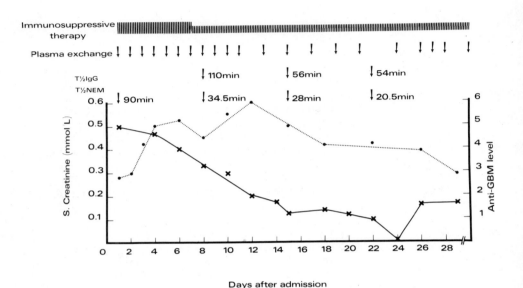

Fig. 1.   Clinical course of a patient with anti-glomerular basement
          membrane disease. The fall in antibody level was accompanied
          by improved renal function and splenic clearance of both
          NEM-treated and antibody-coated red cells. (●) serum
          creatinine; (x) anti-glomerular basement membrane antibody
          level.

          $T\frac{1}{2}$NEM = clearance half life of NEM-treated erythrocytes
                    (normal range 10-22.5 minutes).
          $T\frac{1}{2}$IgG = clearance half life of antibody-coated erythrocytes
                    (normal range <61 minutes.

while the antibody-coated clearance time was grossly abnormal exceed-
ing 120 minutes (Figure 2). No correlation could be demonstrated
between abnormalities of splenic function as measured by any of these
three techniques and the levels of complement binding immune complexes, proteinuria, serum albumin, renal function or complement
abnormalities.

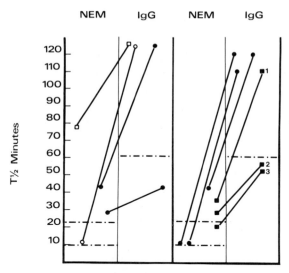

Fig. 2. Results of simultaneous, double isotope labeled red cell clearance studies using NEM-treated and antibody-coated erythrocytes.

□ Control study in a subject post-splenectomy
○ Control study in Rh D negative individual
● Patients with glomerulonephritis
■ 1,2,3 serial studies in patient with anti-glomerular basement membrane disease

The results of the in vitro experiments are presented in Table 1. These studies indicate the presence of at least 2 separate types of receptor on the surface of human peritoneal macrophages: an Fc receptor and a non-Fc receptor. The phagocytic function of each of these receptors could be blocked selectively. In preliminary experiments, we have demonstrated that diluted serum from a patient with anti-glomerular basement membrane disease inhibited the binding of inert particles and Fc bearing materials, suggesting that both receptor types were blocked.

CONCLUSION

Our studies have demonstrated that splenic function is impaired in the majority of patients with primary glomerulonephritis and that in certain types of glomerulonephritis improvement in splenic function accompanies clinical remission. Other workers (5,12) have reported similar findings in patients with systemic vasculitis and rheumatoid arthritis. Unlike these authors, however, we were unable to show

Table 1. In Vitro Experiments with Human Peripheral Macrophages

Preparation	Cells Positive %	EA per Cell
Sheep EA	0 – 80	4 – 8
+ IC, NHS or anti-GBM serum	0 – 15	1 – 2
+ NHS 1/20 or BSA	40 – 75	3 – 5
+ anti-GBM serum 1/20	10 – 30	1 – 2
+ IgG-coated Latex beads	40 – 50	2 – 3
+ carbon particles	60 – 80	3 – 5
Human EA	40 – 50	2 – 3
+ IC, NHS or anti-GBM serum	10 – 15	1 – 2
+ NHS 1/20	40 – 60	2 – 3
+ anti-GBM serum 1/20	10 – 15	1 – 2
+ IgG-coated Latex beads	10 – 15	1 – 2
+ carbon particles	40 – 50	2 – 3
Human NEM-RBC	35 – 40	3 – 4
+ IC, NHS or IgG-coated Latex beads	30 – 40	2 – 3
+ carbon particles	10 – 15	1 – 2
+ anti-GBM serum 1/20	10 – 15	1 – 2
Macrophages + carbon alone	90 – 95	Ingested carbon +++
Macrophages + IgG latex beads alone	90 – 95	Ingested latex +++

Abbreviations: EA=antibody-coated erythrocytes; IC=immune complexes of BSA/rabbit anti-BSA in 5-fold antigen excess; BSA= bovine serum albumin; NEM-RBC=N-ethylmaleimide-treated red blood cells; NHS=pooled normal human serum; anti-GBM=serum from patient with antibody to glomerular basement membrane.

any significant correlation between red cell clearance times and other parameters of disease activity, e.g. Clq binding immune complexes. In individual subjects studied serially, the clearance half life varied as the disease process remitted or relapsed.

Acknowledgement

This work was supported by the National Health and Medical Research Council of Australia.

REFERENCES

1. Atkinson, J. P., and Frank, M. M., J. Clin. Invest. 53:1742. 1974.
2. Ford, P. M., Brit. J. Exp. Path. 56:523, 1975.
3. Barcelli, U., Rademacher, R., Ooi, Y. M., and Ooi, B. S., J. Clin. Invest. 67:20, 1981.
4. Frank, M. M., Hamburger, M. I., Lawley, T. J., Kimberly, R. P., Platz, P. H., New Engl. J. Med. 300:518, 1979.
5. Lockwood, C. M., Worlledge, S., Nicholas, A., Cotton, C., Peters, D. K., New Engl. J. Med. 300:524, 1979a.
6. Hamburger, M. I., Gorevic, P. D., Lawley, T. J., Franklin, E. C., Frank, M. M., Trans. Assoc. Am. Phy. 92:104, 1979.
7. Lockwood, C. M., Amos, N., and Peters, D. K., Kidney Int. 16:93, 1979b.
8. Crome, P., and Mollison, P. L., Brit. J. Haemat. 10:137, 1964.
9. Jacob, H. S., and Jandl, J. M., J. Clin. Invest. 41:1514, 1962.
10. Jaffe, C. J., Vierling, J. M., Jones, E. A., Lawley, T. J., Frank, M. M., J. Clin. Invest. 62:1069, 1978.
11. Thompson, R. A., and Rowe, D. S., Immunology 14:745, 1968.
12. Williams, B. D., Lockwood, C. M., Pussell, B. A., and Cotton, C., Lancet, i:1311, 1979.

DISCUSSION

BRADFIELD: Previous studies from Hammersmith Hospital imply that the reduced clearance by the spleen in certain diseases is due to an overloading with immune complexes. But some of the diseases which you showed are not thought to be mediated by immune complexes, such as minimal change anti-basement membrane disease. Did you look for a correlation between the presence of immune complexes and reduced vascular clearance in your patients? If it's not due to overloading by immune complexes, what do you hypothesize is the relationship between lowered splenic clearance and the disease process?

PUSSELL: Our first hypothesis was that an excessive amount of immune complexes or overload did account for the abnormal deposition of immune complexes within the glomerulus. However, immune complexes in primary glomerulonephritis were not often found. There was no correlation between red cell clearance and the presence of circulating

immune complexes either by correlation coefficient or by a Chi square
type analysis. The presence of immune complexes was examined using
3 assays: a Clq binding assay, a monoclonal rheumatoid factor binding
assay and a Raji cell binding assay.

RICHTER: Recently, the Hammersmith group have shown that uptake of
heat damaged red blood cells is actually an indication of splenic
blood flow. To what extent did hemodynamic alterations occur in
your patients? Further, since there is a strong association between
abnormal splenic uptake and DRW 3 positive patients, did you tissue
type some of your patients?

PUSSELL: We did not tissue type our patients. With respect to the
question about splenic blood flow, there was no reason to believe
that hemodynamic alterations existed in our patients. There was no
correlation with serum albumin which you might have expected if there
were hemodynamic changes. Some patients had very low serum albumin
levels and normal clearance and some with low serum albumin had
abnormal clearances.

IMMUNOPATHOLOGY OF PULMONARY GRANULOMAS

Quentin N. Myrvik

Department of Microbiology and Immunology
Bowman Gray School of Medicine
Winston-Salem, North Carolina   27103 USA

INTRODUCTION

Granulomatous-type inflammations can be divided into two general
categories: (a) nonimmunologic or foreign body type and (b) immunologic or hypersensitivity type. The foreign body type granulomas
are readily produced experimentally with substances like carrageenan,
plastic beads, bentonite, talc and certain microbial cell-wall products, such as mycobacterial cord factor. The major principle involved
in these types of granulomas is that the irritant is usually water
insoluble and slowly digestible or nondigestible. Accordingly, it
represents the basic cellular response of chronic inflammation which
is dominated by macrophage mobilization to the site of the irritant.
As a rule, the lymphocyte response is minimal and the level of macrophage activation is limited. The characteristic response in this
case is that repeated injections of these types of irritants produce
the same time course in the evolution of the lesion (1).

In contrast, immunologic or hypersensitivity type granulomas
characteristically show an accelerated response upon subsequent injection of the eliciting substances. The hypersensitivity granulomas
which are common in diseases like tuberculosis, histoplasmosis,
leprosy and blastomycosis exhibit lesions which can vary considerably
in size. The classic immunologic or hypersensitivity type granuloma
is the tubercle in tuberculosis (Figure 1). This lesion represents
a highly organized immune structure consisting of large numbers of
macrophages that congregate around the infectious (antigen) focus
together with a rather striking lymphocyte mantle which is located
circumferentially around the periphery of the granuloma. Generally
speaking, the macrophages beneath the lymphocyte mantle exhibit a
progressive state of activation from the exterior to the interior.

649

Macrophages become highly activated during the development of the
granuloma and take on the characteristic, morphologic appearance of
epithelioid cells. Giant cells are also a common feature of this
granuloma. In the above diseases, the granulomas commonly develop
central necrosis that may lead to massive lung destruction and cavity
formation (Figure 2).

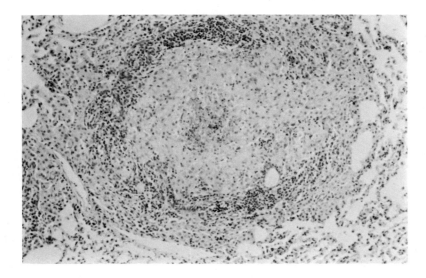

Fig. 1.   Tubercle elicited by BCG in BCG-vaccinated rabbits. Rabbits
          were sensitized s.c. with 200 μg of heat-killed BCG in 0.2
          ml light mineral oil. Three weeks later, the tuberculin
          positive animals were given 5 mg of heat-killed BCG by the
          intratracheal route. The animals were sacrificed 3 weeks
          later.

     Present data strongly support the concept that classical hyper-
sensitivity granulomas are mediated by immune or sensitized T cells
which, in turn, liberate an array of lymphokine mediators that orches-
trate the accumulation of macrophages and their activation. This
review will briefly summarize some of the concepts that relate to
the tissue-damaging events of infectious hypersensitivity type granuloma as well as their regulation.

POSSIBLE CHARACTERISTICS RELATED TO NECROSIS

Macrophage Turnover Rate

The work of Spector and colleagues (2,3,4), have clearly established that the lifespan of macrophages in hypersensitivity type
granulomas are markedly shortened. As a general rule, the survival
time of macrophages in hypersensitivity type granulomas range from
one to three weeks, whereas in foreign body granulomas the macrophage
lifespan can be as long as three months. This marked difference in
lifespan is most likely related to the level of macrophage activation
and the degree of host cell-antigen interactions.

Focal Deposition of Antigen

It is axiomatic that the chronic nature of granulomatous lesions
depend on persistence of antigen. Accordingly, a depot of insoluble
antigen becomes the organizational center of the granulomatous reaction. The purpose of this immune structure must be to wall off and
destroy the offending agent. In this regard, we have demonstrated
in our laboratory that cross-linked tuberculoprotein can essentially
reproduce the cellular events that occur when tubercle bacilli are

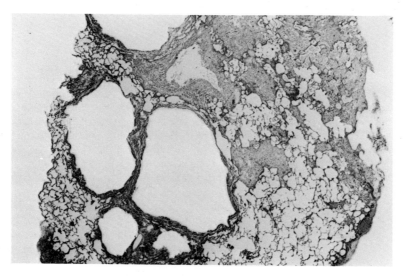

Fig. 2.   Cavity formation in a rabbit tubercle elicited by BCG.
          Animals were treated as in Figure 1 except two cycles of
          sensitization were used. The rabbits were sacrificed 6
          weeks after the challenge dose was administered.

employed to elicit allergic pulmonary granulomas in a tuberculin
hypersensitive animal (5). In contrast, soluble tuberculoprotein
elicits a diffuse mononuclear response rather than the development
of discrete granulomatous lesions.

Phagocytosis-induced Injury

One of the major mechanisms that could explain a shortened lifespan of activated macrophages in granulomas is based on the observation that highly activated macrophages undergo a phagocytosis-induced
death when they ingest certain particulate materials (6). This has
been demonstrated with various mycobacteria as well as zymosan. In
this case, the oxygen-generated metabolic products (superoxide anion,
hydrogen peroxide, etc.) probably contribute to the early demise
of activated macrophages. This event is directly associated with
immune macrophages that exhibit a marked burst in the hexose monophosphate shunt upon triggering with zymosan. Since the antigen
depot is in the center of the lesion and coincides with the location
of necrosis, the synchronous death of macrophages in the center of
the granuloma appears to be triggered by the particulate antigen or
organisms. We have observed that in the case of the experimental
sarcoid-type of granuloma (2-3 weeks after a single i.v. injection
of BCG), necrosis can be triggered in vivo by the i.v. or intra-
tracheal injection of zymosan (unpublished). This lends support to
the concept that phagocytosis-induced injury may be one of the important initiating events responsible for focal necrosis in tubercle-
type granulomas.

Is the Intercellular Matrix Different in Necrotizing Allergic Granulomas?

One of the remarkable features of the tubercle-type granulomas
is the nature and the highly organized pattern in which macrophages
accumulate around the infectious or antigen focus. Studies from
our laboratory indicate that a macrophage agglutinating factor (MAgF)
accumulates in the lung during delayed hypersensitivity reactions
(7). This agglutination is readily evident in the alveoli 48 hours
after challenge of a tuberculin positive animal (Figure 3). The MAgF
complex, which is a potent agglutinating factor for alveolar macrophages, contains hyaluronic acid and other protein constituents including fibronectin (8; unpublished). It is possible that the inter-
cellular matrix in allergic granulomas is different from that in the
sarcoid-type and foreign body granulomas because a comparable MAgF
component is not found in lavage fluids. Accordingly, if a special
matrix could prevent macrophage movement in the tubercle-type allergic
granulomas, the focal injury of macrophages might be intensified.
In contrast, it is possible that a less adhesive matrix might occur
in the sarcoid-type or foreign body granulomas which would allow more

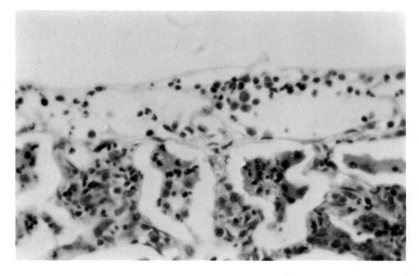

Fig. 3. Agglutination of macrophages in alveoli of tuberculin positive rabbits 48 hours after challenge dose was administered.

macrophage mobility during its evolution. This would promote dis-
persal of cells destined to succumb and prevent large numbers of
macrophages dying en masse in a single focus.

Correlation of Necrosis with Delayed Hypersensitivity

The observations of Yamamura et al. (9) convincingly indicated
that hypersensitivity is correlated with central necrosis of tuber-
cles. Accordingly, if a tuberculin-positive animal is desensitized
and challenged, the animal will develop granulomas but without classical central necrosis. This observation suggests that the circulating T cells which mediate dermal delayed hypersensitivity contribute
somehow to the necrosis of the center of tubercle type granulomas.
It is possible that the blastogenic lymphocytes liberate cytotoxic
lymphokines as a consequence of antigen administration. Alternatively, these T cells could "overactivate" the first macrophages
that arrive at the site of antigen challenge.

It is of special interest that following desensitization, strong
MIF activity is still exhibited by immune alveolar cells, whereas
blastogenic activity is very low. A highly relevant question that
remains unanswered pertains to whether blastogenic T cells represent
a separate subset of T cells and whether this subset of T cells
somehow triggers tissue damage in the lung during elicitation of
tubercle-type granulomas that undergo central necrosis.

REGULATORY MECHANISMS IN IMMUNE GRANULOMAS

Macrophage Suppression of Blastogenesis

It has been reported that alveolar macrophages can suppress mitogen-induced blastogenesis of lung lymphocytes (10,11). The suppression appears to be mediated by PGE_2 (12). We have also observed a suppression of T cell blastogenesis of alveolar cells from tuberculin positive animals incubated with PPD. However, MIF positive activity is not affected. This suggests that macrophages progressively down regulate blastogenic events of T cells but not the production of MIF as the macrophages increasingly dominate the lesions. This is compatible with the concept that MIF producing lymphocytes are largely responsible for cell-mediated immunity and circulating blastogenic T cells are largely responsible for dermal hypersensitivity. Accordingly, selective suppression by macrophages of blastogenesis would retain the mobilization and activation of macrophages but simultaneously regulate the cytotoxic potential of blastogenic T cells.

Suppression of Blastogenesis and MIF Production by T Suppressor Cells

It has been observed by Doughty and Phillips (13) employing the schistosome egg granuloma model that late in the course of the granuloma, namely about 20 weeks, T suppressor cells produce significant suppression of both blastogenesis and MIF production. This appears to represent a down regulation in those instances where there may be a strong and sustained granulomatous response over an extended period. This appears to be different from the macrophage type suppression which seems to be selective only for blastogenic T cells.

ACCUMULATION OF IMMUNE T CELLS IN THE LUNG AND THEIR POTENTIAL TO MEDIATE PULMONARY GRANULOMATOUS RESPONSES

Antigen Driven Homing of Immune T Cells in the Lung

Recent evidence suggests that T cells which mediate macrophage mobilization and activation can home to tissues and become noncirulating memory cells (14). Evidence for this largely comes from the analysis of transfer experiments at different intervals after immunization. For example, animals immune to Listeria monocytogenes cannot transfer immunity (parabiosis) at 26 days after immunization. The suggestion is that immune T cells have entered the tissues and have become noncirculating. This concept is also supported by observations that deposition of antigen in the lung can selectively sequester specific T cells in the lung (15). There is also indirect evidence that when animals are desensitized to an accelerated pulmonary granulomatous response, the desensitization process tends to drive immune T cells to the lung (unpublished). This observation can explain why there is no correlation between the granuloma index and the level of dermal hypersensitivity. In this regard, it is of special interest that if animals are given 400 rads of total body radiation they can mount a normal accelerated macrophage response in the lung, provided the lungs are shielded during the radiation treatment. This further supports the idea that the lung can function and express cell-mediated immunity with resident immune T cells. It seems likely that during desensitization procedures, circulating T cells can be driven to tissues and that while an animal may express anergy to a dermal skin test, the animal is still competent to express a cell mediated immune reaction in a local organ like the lung. Furthermore, if such immune T cells present in the lung preferentially produce mediators of cell mediated immunity, this expression of immunity would tend to avoid excessive activities of circulating blastogenic T cells which, as mentioned earlier, might have the potential to produce tissue damage. This observation emphasizes the point that a state of dermal anergy is not necessarily equivalent to lack of a potential cell mediated immune expression in a local organ like the lung.

DISCUSSION

     The immune granuloma can be looked upon as the ultimate immune structure to confront facultative intracellular parasites that are difficult to kill and digest which naturally results in chronic and persistent infections. The remarkable walling-off of an infectious focus represents a highly specialized and unique host response to these types of infections. While the blastogenic T cell population may contribute importantly to this expression of immunity, since it provides a means to expand the clone of specifically immune T cells, it is possible that an oversupply of circulating blastogenic T cells could set the stage for exaggerated tissue damage in the lung. It is therefore of potential importance that macrophages, which accumulate in the immune granuloma, are capable of selectively suppressing blastogenesis but not MIF production. This provides a unique set of circumstances because it puts a damper on potential tissue-damaging blastogenic T cells but allows the cell mediated immune lymphokine production system to remain intact.

     One of the major unanswered questions is whether the T cells that mediate dermal delayed hypersensitivity reactions are different from the T cells that mediate cell-mediated immunity. This is an important point because there has been a tendency to equate dermal delayed hypersensitivity with a state of cell-mediated immunity even though some investigators have claimed to find a dichotomy. In this regard, it is not certain whether MIF producing cells are derived from a down regulated blastogenic population. This possibility may obtain when macrophages accumulate and suppress blastogenesis. One of the basic and most important questions to be answered relates to whether two distinct subsets are operating in these two reactions.

It would appear that monoclonal antibody reagents can ultimately answer this question.

Acknowledgement

Support from HL 16769 is acknowledged.

REFERENCES

1. Boros, Dov L., in "Granulomatous Inflammations" (P. Kallos, ed.), Progress in Allergy, Vol. 24, pp. 183-267, S. Karger, Basel, 1978.
2. Spector, W. G., and Lykke, A. W. J., J. Path. Bact. 92:163, 1966.
3. Spector, W. G., and Ryan, G. B., in "Mononuclear Phagocytes" (van Furth, ed.), Davis, Philadelphia, 1970.
4. Spector, W. G., Lykke, A. W. J., and Willoughby, D. A., J. Path. Bact. 93:101, 1967.
5. McGee, M. P., and Myrvik, Q. N., J. Reticuloendothelial Soc. 22: 253, 1978.
6. McGee, M. P., and Myrvik, Q. N., Infect. Immun. 26:910, 1979.
7. Galindo, B., Myrvik, Q. N., Love, S. H., J. Reticuloendothelial Soc. 18:295. 1975.
8. Love, S. H., Shannon, B. T., Myrvik, Q. N., and Lynn, W. S., J. Reticuloendothelial Soc. 25:269, 1979.
9. Yamamura, T., Ogawa, Y., Maeda, H., and Yamamura, Y., Amer. Rev. Resp. Dis. 109:594, 1974.
10. Herscowitz, H. B., Conrad, R. E., and Penline, K. J., Adv. Exp. Med. Biol. 121A:459, 1980.
11. Ansfield, M. J., Kaltreider, H. B., Caldwell, J. L., and Herskowitz, F. N., J. Immunol. 122:542, 1979.
12. Demenkoff, J. H., Ansfield, M. J., Kaltreider, H. B., and Adam, E., J. Immunol. 124:1365, 1980.
13. Doughty, B. L., and Phillips, S., J. Immunol. 128:37, 1982.
14. Jungi, T. W., J. Reticuloendothelial Soc. 30:33, 1981.
15. Lipscomb, M. F., Myons, C. R., O'Hara, R. M Jr., and Stein-Streilein, J., J. Immunol. 128:111, 1982.

DISCUSSION

KENYON: In reference to delayed hypersensitivity causing a necrotic focus at the center of the granuloma, were you implying that you can get a positive delayed hypersensitive reaction with other kinds of antigen-induced granuloma? Or were you referring specifically to BCG?

MYRVIK: The model that we have used almost exclusively has been with BCG. The principles that relate to other infectious disease granulomas are probably the same. We have not worked with berryllium. As I recall, central necrosis is not a characteristic feature of berryllium induced granulomas.

KENYON: You said that if zymosan or another antigen is given that death occurs. Are you referring to death of the macrophage?

MYRVIK: Yes. I am referring to phagocytosis-induced death of the macrophage. This can be demonstrated in two ways. One can inject zymosan intratracheally followed by viability counts on the lavaged populations or you can inject zymosan intravenously and elicit foci of necrosis in the granulomatous lung. I'm cautious in drawing a firm parallelism here because zymosan is a complex biological material. For instance, it is an excellent activator of complement. We need another substance that could trigger the HMS shunt that wouldn't have the additional complications of zymosan, such as activation of complement by the alternate pathway.

VAN FURTH: Do you think it is the phagocytic induction of oxidative metabolites surrounding the macrophage that kills them or is it specific for zymosan?

MYRVIK: Since unrelated viable bacteria, dead bacteria or cell walls trigger this shunt, it is not peculiar for zymosan. For example, Microbacterium phlei or Corynebacterium parvum induce an equal or better oxidative burst than BCG; obviously, it is not an antigen specific event. Since catalase produces partial protection for the HMS-related death of macrophages, we postulate that oxidative metabolities are involved.

VAN FURTH: Recent studies and new calculations have shown that in normal steady state conditions in mice, the turnover of macrophages in the lung is about 7 days. I know we initially reported that it was much longer, but it's 7 days. If you look at the data of the late Prof. Spector, I think there is an even longer life span of the macrophages in the lung during granulomas and not a shorter one.

MYRVIK: In the carrageenan granulomas studied by Prof. Spector, he had survival times up to three months. So if you have a non-immunologic granuloma, you might indeed have a longer survival than in the steady state. However, Spector found that the survival times of macrophages in hypersensitivity granulomas were as short as 2 weeks. Granulomatous inflammation could have a trapping effect perturbing normal clearance of alveolar macrophages via the muco-ciliary escalator.

AN INVESTIGATION ABOUT CELL PROLIFERATION IN LYMPH NODE GRANULOMAS OBTAINED FROM SARCOIDOSIS PATIENTS

A. van Maarsseveen,[1] R. van der Gaag,[1] and J. Stam[2]

[1]Department of Pathology, Free University Hospital Amsterdam, The Netherlands
[2]Department of Pulmonology, Free University Hospital Amsterdam, The Netherlands

INTRODUCTION

The general concept of the mononuclear phagocytic system is that stem cells differentiate to promonocytes to monocytes which are released from the bone marrow into the circulation to eventually enter the tissues as macrophages (1,2). If a granuloma is present, these cells differentiate into epithelioid cells and/or multinucleated giant cells (1). The development and function of granulomas in pathology is reviewed in references 3-5. Spector described high turnover granulomas characterized by cell infiltration and extensive local proliferation, and low turnover granulomas in which these events are much less (6,7). The toxicity of the granuloma inducing agent determined what kind of granuloma was produced. It is generally agreed that granulomas in sarcoidosis are high turnover granulomas (1,7). In an earlier report (8), we described in these granulomas a very low number of mitotic figures. In this study, we measured local cell metabolism in lymph node granuloma slices with radio-nucleotides.

MATERIALS AND METHODS

Mediastinal lymph nodes were taken clinically and sliced into uniform 500 micron thick sections (9). The slices were put onto grids, stretched and incubated in RPMI-1640 bicarbonate medium containing 20% pooled inactivated human serum and L-glutamine (0.3 mg/ml). Each slice was incubated for three hours with ^{3}H-Thymidine

(3HTdR, spec.act.25 Ci/mmol) or ^{3}H-Uridine (3HUdR, spec.act. 30 Ci/
mmol), at a concentration of 4 μCi/ml medium. Thereafter, the grids
were washed three times and cultured for another hour with unlabeled
nucleotide (0.05 mmol). They were then washed three times, fixed in
neutral buffered formalin and embedded in paraffin. Five micron
thick sections were made perpendicular to the slice, after which
autoradiography was performed (10). The sections were stored for
three weeks at 4°C, developed and stained with hematoxylin and eosin.
The epithelioid cells were examined for grains overlaying the nucleus,
as an index of radionucleotide incorporation. Serial sections of
the slices yielded information about the total degree of activity
in one granuloma. The labeling index was calculated as follows:

$$\frac{\text{number labeled epithelioid cells}}{\text{number labeled + unlabeled epithelioid cells}} \times 100$$

Non-sarcoid granulomas were examined as controls. These were
produced by injecting sensitized guinea pigs with Freund's complete
adjuvant (0.05 ml) intravenously. After 5 days, the granulomatous
pulmonary tissue (11) was examined. All sections from patients with
sarcoidosis showed characteristic epithelioid cell granulomas without
necrosis. Very few labeled epithelioid cells were present in the
granulomas after exposure to 3HTdR, in contrast to the surrounding
lymphoid tissue (Table 1).

Using serial sections, we calculated the total number of labeled
cells per granuloma. In most granulomas, the number of proliferating
cells was very low. In only one granuloma was there a relatively
high number of labeled epithelioid cells (Table 2). Labeled multinucleated giant cells were not observed.

The same technique when applied to granulomatous pulmonary tissue
from the guinea pig revealed a high number of DNA synthesizing cells.
(Table 3). This observation has been reported (7,12,13).

In contrast to these results with 3HTdR, sections exposed to
3HUdR revealed labeled cells in nearly all granulomas. To verify
if the rate of RNA synthesis present in one granuloma section is
representative for the whole granuloma, serial sections were evaluated. When one section disclosed a low labeling activity, the
same was observed in the other parts of the granuloma (Figure 1).
When a high labeling index was observed in one area of a granuloma,
the same index was present also in the other parts of the granuloma.

CONCLUSION

In this study, the intact granulomatous tissue from lymph nodes

Table 1.   Quantitation of Lymph Node Granulomas of Sarcoidosis
           Patients in which no Tritiated Thymidine Labeled
           Epithelioid Cells were Observed

	Granulomas	
Patient	Number Examined	% With No Labeled Epithe- lioid Cells
1	76	92.1
2	52	86.6
3	76	86.9
4	94	95.6
5	56	96.4
6	84	96.4
7	31	93.4
8	10	0.0

obtained from proven sarcoidosis patients was investigated. We used
serial sections to deduce events taking place in one entire granuloma.
Proliferation of cells was measured by 3HTdR incorporation and RNA
dependent protein synthesis was measured by 3HUdR incorporation as
evaluated by autoradiograms applied to slices of fresh lymphoid
tissue.   The relationship between the rate of incorporation in vivo
and in vitro has already been established during short-time incuba-
tions (14,15). Our specimens cultured for several hours kept their
normal morphologic characteristics.

        Experimental pulmonary granulomas from guinea pigs revealed a
high degree of labeling, consistent with reported observations that
Freund's adjuvant induced granulomas are high turnover granulomas (7,
16). However, sections of lymph nodes from patients with sarcoidosis
generally showed a very low degree of proliferating epithelioid cells.

Table 2.   Determination of Tritiated Thymidine Labeling in 6
            Lymph Node Granulomas Obtained from Two Patients with
            Sarcoidosis

| | Epithelioid Cells | | | |
| | Patient 3 | | Patient 5 | |
Granuloma	Number	% Labeled	Number	% Labeled
1	293	3.8	70	9.4
2	334	3.7	496	3.0
3	244	1.2	94	1.1
4	289	0.7	178	0.5
5	454	0.4	270	0.0
6	60	0.0	226	0.0

The total number of epithelioid cells in each granuloma counted
in serial sections.   The percentage of labeled epithelioid cells
was determined on that same granuloma.

But serial sections revealed some granulomas with a high labeling
index, comparable with data reported from animal experiments (7,12,
13).   In contrast to some reports suggestive of high turnover in
sarcoid granulomas (1,6,7), we concluded that sarcoidosis mostly is
characterized by low turnover granulomas.   This conclusion is based
on low local DNA synthesis.   With our slice technique, we could show
RNA synthesis in epithelioid cells in intact granulomas.   Different
granulomas synthesize RNA at significantly different levels. We concluded that granulomas of sarcoidosis patients are mostly low turnover, low proliferating granulomas, suggesting a granuloma inducing
agent of low toxicity (5,7).

Table 3. Percentage of Tritiated Thymidine Incorporation into
Experimental High Turnover Granulomas from Guinea Pigs

Cell Type	% Labeled Cells
All inflammatory cells	4.0±0.8
Macrophages	75.3±5.5
Epithelioid cells	8.6±3.4
Lymphocytes	8.2±3.4

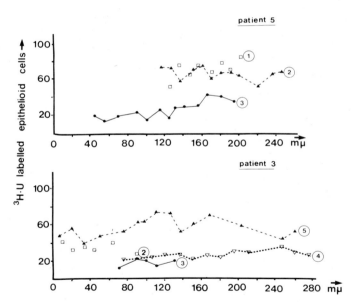

Fig. 1. The degree of tritiated uridine labeled epithelioid cells
present in sarcoid granulomas from two lymph nodes evaluated
by serial sections. Granuloma ③ < ② and granuloma ④
< ② or ⑤ .

Acknowledgement

We thank Prof. Dr. W. Maassen and Dr. D. Grechuchna of the
Rührlandklinik, Essen, W. Germany for clinical histories and tissue;
Mrs. Ph. Broekhuyzen for technical assistance and Mrs. K. Horta for
typing the manuscript. This study was supported by a grant of the
Koninklijke Nederlandse Centrale Vereniging tot bestrijding der
Tuberculose.

REFERENCES

1. Epstein, W. L., Pathobiol. Ann., 7:1, 1977.
2. van Furth, R., in "Mononuclear Phagocytes", p. 151, Blackwell
    Scientific Publication, Oxford, 1970.
3. Anderson, W. A. D., Pathology, 6th ed., The C. V. Mosby Company,
    St. Louis, MO., 1971.
4. Boros, D. L., Ann. New York Acad. Sci. 278:36, 1976.
5. Warren, K. S., Ann. New York. Acad. Sci. 278:7, 1976.
6. Ryan, G. B., and Spector, W. G., J. Path. 99:139, 1969.
7. Spector, W. G., Ann. New York Acad. Sci. 278:3, 1976.
8. van Maarsseveen, A. C. M. Th., Veldhuizen, R. W., Alons, C. L.,
    Stam, J., and Mullink, H., 1982, in preparation.
9. Stadie, W. C., and Riggs, B. C., J. Biol. Chem. 3:687, 1944.
10. Caro, L. G., van Tubergen, R. P., and Kolb, J. A., J. Cell Biol.
    15:173, 1962.
11. Galbenu, P., Roels, H., and Willems, J., Exp. Cell Biol. 46:355,
    1978.
12. Papadimitriou, J. M., and Spector, W. G., J. Path. 105:187, 1971.
13. Wolfart, W., Beitr. Path. Anat. 129:436, 1963.
14. Willems, J., Galand, P., and Chretien, J., Lab. Invest. 23:6,
    1970.
15. Luke, D. A., Virchows Arch. B. (Cell Pathol.) 29:343, 1979.
16. Papadimitriou, J. N., and Spector, W. G., J. Path. 106:37, 1972.

DISCUSSION

MYRVIK: Would you comment on the low and high turnover of the cells
in the sarcoid granuloma compared to the steady state? Dr. Hunninghake has lavaged a lot of sarcoid patients. Occasionally he gets
macrophages in the lavage that are relatively immature yet show a
high degree of activation. But apparently most patients don't show
that. I'm wondering if you could relate your observations to the
immunologic state of the sarcoid?

VAN MAARSSEVEEN: I don't know if our granulomas are characterized
by activation of the macrophage. Early, there is an enormous infiltration of mononuclear cells because in these granulomas there are

probably toxic agents.   The macrophages come in and they die.   So,
you've got a recruitment of cells and a high turnover granuloma.   When
the agents' toxicity is very low,you don't need so many macrophages
because they are ingesting and containing the agents.

MYRVIK: Presumably this could oscillate during the course of illness.

VAN MAARSSEVEEN: Yes.

FUNCTIONAL ASSESSMENT OF ALVEOLAR MACROPHAGES IN

ALLERGIC ASTHMATIC PATIENTS

P. Godard,[1] M. Damon,[2] J. Chaintreuil,[2] O. Flandre,[2]
A. Crastes de Paulet,[2] and F. B. Michel[1]

[1]Clinique des Maladies Respiratoires, Hôpital
l'Aiguelongue, avenue du Major Flandre, 34059
Montpellier Cédex, France
[2]Inserm U 58 et Laboratoire de Biochimie A (Pr.
A. Crastes de Paulet), Faculté de Médecine, 34060
Montpellier Cédex, France

INTRODUCTION

Alveolar macrophages play a critical role in host defense against infections and other noxious environmental agents, both by phagocytosis of inhaled particles and by the elaboration of a number of specific mediators which may affect functions of other cell types (1,2).

Alveolar macrophages are the principal resident cells obtained by broncho-alveolar lavage and it has been suggested that they play a role in bronchial asthma (3). We have studied alveolar macrophages of allergic asthmatic patients and our data showed an impaired functional activity of these cells.

MATERIALS AND METHODS

Patients consisted of 19 individuals with allergic bronchial asthma (aged 18 to 62 years, mean = 25) and 7 healthy volunteers (aged 20 to 35 years, mean = 28). All medications were discontinued 8 days prior to the study. Broncho-alveolar lavage was performed by instillation of 300 to 400 ml of 0.9% saline at room temperature into a subsegmental bronchus followed by gentle aspiration (4,5). Prior to culture, total cell counts were determined in a hemocytometer and differential cell counts made from smears stained with Giemsa and toluidine blue.

Alveolar macrophages were purified by adherence over 2 hours
and cultured for an additional 24 hours. The percentage of macrophages in the adherent cell population was assessed by two methods:
phagocytosis of latex particles and nonspecific esterase staining.
The functional status of alveolar macrophages was evaluated by three
methods: (a) viability as assessed by dye exclusion, (b) zymosan
phagocytosis as assessed by the percentage of cells which had ingested
2- or more particles (zymosan particles at a concentration of 50 μg/ml)
and (c) generation of three lipid mediators of inflammation: PGE_2,
$PGF_{2\alpha}$ and TxB_2. These prostaglandins were measured by radioimmunoassay. The release of prostaglandins and thromboxane into the culture
medium was studied with (a) resting cells, (b) cells stimulated by
zymosan (50 μg/ml in 15 cases, and 25 μg/ml and 50 μg/ml in two
cases), and (c) cells stimulated by specific allergen (Dermatophag-
oides pteronyssinus).

RESULTS

The cell yield per ml of recovered lavage fluid was statistically
equivalent for controls and patients with asthma. The differential
count for macrophages, lymphocytes and neutrophils was also equivalent
in the two groups, whereas the eosinophil count was higher in allergic
asthmatics than in control subjects (3.9 $\pm$ 1.6% versus 0.4 $\pm$ 0.3%;
$p < 0.05$).

Macrophage viability was significantly lower in cell cultures
from asthmatics (77 $\pm$ 3% versus 86 $\pm$ 3%; $p < 0.05$). Zymosan phagocytosis was 66 $\pm$ 10% in cultures from control subjects and was significantly lower in cultures from asthmatics (40 $\pm$ 5%; $p < 0.05$).
Unstimulated macrophages from control and asthmatic patients released
equivalent quantities of PGE_2 , $PGF_{2\alpha}$ and TxB_2. Zymosan (50 μg/ml)
phagocytosis induced a 2 to 3 fold increase in the release of each
prostaglandin from macrophages of normal subjects ($p < 0.01$) and a
1 to 2 fold increase from macrophages of allergic asthmatic patients
($p < 0.01$) (Table 1). In the two patients, the zymosan-induced
increase of prostaglandin release was not dependent upon the amount
of zymosan in the culture medium (Table 2).

In four patients, alveolar macrophages were stimulated by the
addition of the specific allergen D. pteronyssinus, Laboratory Stal-
lergenes, France. Release of TxB_2 in ng/ml/10^6 cells was for unstimulated cells (26.6) and for stimulated cells with 10 ng/ml allergen
(50.6), 100 ng/ml allergen (56.0); 500 ng/ml allergen (40.6) and 2.5
μg/ml allergen (30.2).

DISCUSSION

Three mechanisms have been proposed to explain the pathophysi-
ology of bronchial asthma: (a) bronchial hyperreactivity (6); (b)

Table 1.   Release of PGE_2, $PGF_{2\alpha}$ and TxB_2 from Alveolar
Macrophages of Normal and Asthmatic Patients

Measurement	Control Subjects (n = 7)	Asthmatic Patients (n = 15)
Resting Macrophages		
PGE_2	9.3 + 5.7	6.7 + 2.3
$PGF_{2\alpha}$	8.2 + 2.9	8.5 + 1.9
TxB_2	27.2 + 8.7	29.0 + 7.2
Stimulated Macrophages		
PGE_2	18.1 + 9.3	12.0 + 2.6
$PGF_{2\alpha}$	20.8 + 4.4	13.2 + 1.8
TxB_2	91.7 + 23.6	63.9 + 7.7

Data are reported in $ng/10^6$ cells.   Macrophages were stimulated
with zymosan (50 µg/ml).

autonomous nervous system imbalance (7); (c) local release of chemical
mediators (8).   The latter has been suggested because the' broncho-
alveolar cells (obtained by lavage in animals and human beings) are
able to release chemical mediators under specific stimulation (8,9).
Among these cells, there are several lines of evidence   that implicate
alveolar macrophages as important: (a) alveolar macrophages have surface receptors for IgE (10,11); (b) alveolar macrophages from asthmatic allergic patients can be activated by addition of anti-IgE or
the specific allergen to release lysosomal enzymes (12), platelet
activating factor (13), superoxide (12) and prostaglandins; and (c)
alveolar macrophages are able to limit the concentration of SRS-A
available to respiratory smooth muscle (14).

Our data show that the viability and functional activity of
alveolar macrophages are impaired in asthmatic subjects.   The high
eosinophil count found in lavage fluid could impair macrophage viability since the major basic protein of the eosinophil has some cytotoxic actions (15).   The consequence could be an impaired regulation
of mediator release in the broncho-alveolar lumen of allergic asthmatic patients.

Table 2.   Release of PGE$_2$ from Alveolar Macrophages
During Zymosan Phagocytosis

	Unstimulated Cells	Stimulated Cells (Zymosan)	
		25 µg/ml	50 µg/ml
Patient 1	14.2	84.8	24.5
Patient 2	5.4	14.0	10.3

Data are reported in ng/10^6 cells.

Acknowledgement

This work was supported by grant from Le Fond Special du Comité National de Lutte Contre la Tuberculose et le Conseil Scientifique de la Faculté de Médecine de Montpellier.

REFERENCES

1. Voisin, C., in "Allergologie" (J. Charpin, ed.), pp. 76-79, Flammarion, Paris, 1980.
2. Green, G. M., Jakab, G. J., Low, R. B., Davis, G. S., Am. Rev. Resp. Dis. 115:479, 1977.
3. Arnoux, B., Caeiro, M. H., Cerrina, J., Benveniste, J., Respiration 42 (suppl. 1): 4, 1981.
4. Hunninghake, G. W., Gadek, J. E., Kawanami, O., Ferrans, V. J., Crystal, R. G., Am. J. Path. 97:149, 1979.
5. Godard, P., Aubas, P., Calvayrac, P., Taïb, J., Michel, F. B., Nouv. Pres. Méd. 10:3141, 1981.
6. Orehek, J., Gayrard, P., Smith, A. P., Grimaud, C., Charpin, J., Am. Rev. Resp. Dis. 115:937, 1977.
7. Nadel, J., in "Bronchial Asthma. Mechanisms and Therapeutics" (E. B. Weiss, M. S. Siegel, ed.) pp. 155-162, Little Brown Co., Boston, 1976.
8. Patterson, R., McKenna, J. M., Suszko, I. M., Solliday, N. H., Pruzansky, J. J., Roberts, M., Kehoe, T. J., J. Clin. Invest. 59:217, 1977.
9. Patterson, R., Harris, K. E., Suszko, I. M., Roberts, M., J. Clin. Invest. 57:586, 1976.
10. Anderson, C. L., Spiegelberg, H. L., J. Immunol. 126:2470, 1981.

11. Capron, A., Joseph, M., Dessaint, J. P., Capron, M., Tonnel,
    A. B., Respiration 42 (suppl. 1):20, 1981.
12. Joseph, M., Tonnel, A. B., Capron, A., Voisin, C., Clin. Exp.
    Immunol. 40:416, 1980.
13. Arnoux, B., Duval, D., Benveniste, J., Eur. J. Clin. Invest.
    10:437, 1980.
14. Patterson, N. A., Graig, I. D., J. Allergy Clin. Immunol. 67:
    435, 1981.
15. Frigas, D., Loegering, D. A., Solley, G. O., Farrow, G. M.,
    Gleich, G., Mayo Clin. Proc. 56:345, 1981.

DISCUSSION

BAGGIOLINI: You have shown that macrophages from asthmatic people
produce less arachidonic acid metabolites either at rest or after
stimulation. You have shown also that these cells inactivate slow
reactive substances (SRS-A). With this observation, you contradict
some of the ideas which many people suggest as mechanisms explaining
why asthmatics are more prone to bronchospasm. For example, thromboxane A2 is bronchospastic and of course SRS-A is bronchospastic.
How do you reconcile your data with these mechanisms?

GODARD: I can't explain it. Others have shown that some leukotrienes
are produced by human alveolar macrophages.

BAGGIOLINI: There is no doubt about the production of leukotrienes
by alveolar macrophages but in your patients you had also an enhancement of inactivation and that's what disturbed me.

GODARD: The inactivation of SRS-A by alveolar macrophages has been
published. It is not my work.

WILLIAMS: Can you be confident that you're lavaging the same parts
of the bronchial tree in your asthmatic as in your normal group of
patients. In the asthmatic, there is likely to be a lot of bronchospasm preventing access to the alveoli. You may just be dealing with
dead cells.

GODARD: All the cells had good viability.

WILLIAMS: But when you lavaged the patients, were you sure that
your fluid was getting to the same part of the bronchial tree?

GODARD: Oh yes, I think so.

MYRVIK: What is known about the Fc receptor on macrophages for IgE?

GODARD: I know of one group who is working on that but I don't know
the answer.

SECTION 10

ACTIVATION AND EFFECTOR MECHANISMS AGAINST MICROBES AND TUMORS

MACROPHAGE ACTIVATION AND EFFECTOR

MECHANISMS AGAINST MICROBES

Jacques Mauel

WHO Immunology Research and Training Centre
Institute of Biochemistry
1066 Epalinges, Switzerland

INTRODUCTION

This review is concerned with some of the mechanisms whereby macrophages destroy intracellular microorganisms, as well as with processes allowing pathogens to elude the microbicidal machinery directed at them. Considerable insight has been gained recently in this area, thanks particularly to detailed investigations into the interaction of macrophages with certain intracellular protozoa. Thus, the various points to be discussed will be illustrated by examples taken from the study of the behavior in macrophages of both prokaryotic and eukaryotic parasites, including Toxoplasma, spp., Trypanosoma spp. and Leishmania spp. The latter three species of microorganisms are pathogenic to man; such studies thus represent more than a fascinating academic exercise, as they are expected to provide fresh leads for the formulation of new therapeutic approaches.

INTERNALIZATION OF MICROORGANISMS BY MACROPHAGES: MORPHOLOGICAL OBSERVATIONS

Destruction of microorganisms by macrophages requires binding and then internalization of the former by the phagocytes. Little is known of the factors involved in these processes. Surface structures of microorganisms can activate complement by the alternative and, presumably, by the classical pathways (1,2) in the absence of antibodies, thus generating cleavage products that may act as chemoattractants for macrophages (3). Binding of antibodies and of components of complement to particulate antigens increases their susceptibility to ingestion by monocytes through attachment to relevant receptor structures on the surface of the phagocytic cells. Micro-

675

organisms can be ingested in the apparent absence of such factors
and it has been suggested that they may interact with mononuclear
phagocytes via lectin-like structures on the latter cells (4,5) and
sugars acting as ligands on the surface of the microorganisms (6,7).
Engulfment then proceeds via the sequential circumferential interaction between ligands and receptors (the "zipper" mechanism) as
described by Griffin and coworkers (8). Integrity of the macrophage
cytoskeletal networks is essential to the process of phagocytosis,
as ingestion of protozoa such as Leishmania (9,10) or Trypanosoma
cruzi (11) is inhibited by the microfilament-disorganizing agent
Cytochalasin B.

     Once intracellular, the microorganism will be enclosed in a
vacuole whose membrane derives from the plasma membrane of the host
cell. Depending on the host species and, possibly on the stage in
development of the microorganism, the microbe-containing phagosome
may vary in shape from a "loose" type in which an "empty" space can
be distinguished around the parasite by electron microscopy (12),
to a "tight-fitting" one in which the vacuolar membrane is closely
apposed to that of the microbe (13). The reasons for these differences and their relevance to the eventual fate of the microorganism
are uncertain.

     By use of appropriate cytological markers, it can be shown that
microbe-containing phagosomes normally fuse with primary and secondary
lysosomes, which contribute hydrolytic enzymes required for the
degradation of foreign material. Depending on the microorganism under
study, phagosome-lysosome fusion may or may not be indicative of
eventual destruction of the microbe. In any case, digestion by lysosomal enzymes appears to constitute a late step in the process of
elimination of the internalized microorganism, which requires a preliminary "lethal hit" presumably mediated by toxic products of the
host cell oxidative machinery.

INTERNALIZATION OF MICROORGANISMS BY MACROPHAGES: BIOCHEMICAL EVENTS

     Ingestion of microorganisms by macrophages (obtained usually
from the peritoneal cavity of experimental animals) may induce in the
latter cells a metabolic burst characterized by increased oxygen
uptake and the production of highly reactive oxygen metabolites (14).
For instance, phagocytosis of promastigotes of Leishmania tropica
and Leishmania donovani by resident murine peritoneal macrophages
(15), and of Toxoplasma gondii by human monocytes grown from the
peripheral blood (16) has been shown to induce the production of
superoxide and to stimulate the hexose monophosphate shunt. Ingestion
of Mycobacterium microti by mouse macrophages similarly stimulates
the release of hydrogen peroxide from the phagocytes.

     Phagocytosis-triggered oxidative events are particularly marked

when macrophages originate from animals that have been infected by
certain pathogens, or when the test cells have been elicited by intraperitoneal inoculation of inflammatory agents (17), pointing to a
remarkable degree of plasticity in the metabolic response of these
cells. A similar stimulation of oxidative metabolism can be induced
in the absence of phagocytic activity, when macrophages are exposed
to membrane-active agents. Whether differences in biochemical and
functional tests between macrophages elicited or activated in various
ways are truly qualitative, such as resulting from the emergence of
different cell populations, or whether they merely constitute a
reflection of the strength of the stimulus and of the sensitivity of
the tests is unclear at the present time (18).

     Little is understood of the precise mechanisms by which super-
oxide and other oxygen intermediates are generated as result of phagocytosis. Moreover, macrophages appear to be heterogeneous in this
respect: when peritoneal macrophages of various sizes were examined
histochemically after triggering with the membrane-active agent
phorbol-myristate-acetate, different cell populations could be distinguished on the basis of the site of hydrogen peroxide production
(whether localized within vacuoles, or on the plasmalemmal membrane)
(19). The first step in the phagocytosis-driven metabolic chain
appears to be the stimulation of a membrane-bound oxidase accepting
a reduced pyridine nucleotide (presumably NADPH) as an electron donor
(20,21), and catalyzing the univalent reduction of molecular oxygen
to superoxide. This enzymatic machinery would presumably become
incorporated in the phagosomal membrane, where its presence would
insure that lethal oxygen derivatives are delivered in the vicinity
of the microorganism to be destroyed. Superoxide itself appears to
display little microbicidal activity (22). However, scavengers of
this molecule, such as cytochrome c, strongly impair the capacity
of macrophages to destroy intracellular microorganisms. Thus, the
importance of superoxide in the antimicrobial mechanisms of phagocytes
presumably relates to its capacity to generate other oxygen derivatives. These are produced by the dismutation of superoxide (O_2^-) to
yield hydrogen peroxide (H_2O_2); formation of the hydroxyl radical
($HO^\cdot$) and of singlet oxygen ($^1O_2$) may then proceed by interaction of
the former two molecular species (14). Of importance might be the
recent identification of a very-low-potential cytochrome b (23) present
specifically in the membrane of phagocytes (24). This compound is
also incorporated into phagocytic vacuoles (25) where it may participate in the generation of toxic oxygen derivatives. The observation
that this cytochrome is missing(or functionally abnormal) in neutrophils from patients with chronic granulomatous disease (26) suggests
that it constitutes an important component of the microbicidal oxidase
complex of phagocytic cells.

     Induction of superoxide production in human neutrophils by chemo-
tactic peptides (but not by phorbol esters) is stimulated by extracellular calcium (27), and treatment of guinea-pig leukocytes by the

calcium ionophore A 23187 (together with calcium) enhances oxygen
uptake and glucose consumption through the hexose monophosphate
shunt (28). These findings suggest that transmembrane flux of
calcium and/or other ions play an important role in the induction
of the respiratory burst. Arachidonic acid metabolism through the
lipoxygenase pathway has been implicated also in the production of
reactive oxygen species by thioglycollate-elicited murine macrophages
exposed to a phagocytic stimulus (29).

MACROPHAGE ACTIVATION: ITS ROLE AND MECHANISMS

     Certain microorganisms have acquired the capacity to survive
within mononuclear phagocytes either by gaining access to host cells
without stimulating a respiratory burst or by avoiding or counter-
acting the effects of lysosomal enzymes. Destruction of these invad-
ers will require further stimulation of the oxidative machinery described above. Thus macrophage activation can be viewed as a means
of reactivating biochemical pathways that are concerned with intracellular killing, when phagocytosis alone has failed to evoke them
at a sufficient level.

     Macrophage "activation" can be analyzed at the biochemical,
morphological and functional levels, and there exists some controversy
as to the rightful utilization of the word (18). In the following
discussion, the term activation will be used in its restricted meaning
of an increased capacity to destroy or otherwise inhibit certain
microorganisms (whether intra- or extra-cellular) and certain types
of non-microbial target cells. In this sense, activation of macrophages can be induced by a variety of stimuli including microbial
products, for instance endotoxin (30), natural or synthetic polymers
(31,32), immunoglobulin complexes (33) and various other biological
substances (reviewed in Ref. 34). Importantly, activation of macrophages appears to constitute a normal effector phase of the cellular
immune response, involving products of specifically sensitized T
lymphocytes as shown both in vivo (35) and in vitro (36). Immunological activation appears to be mediated by soluble factors secreted
by the stimulated lymphoid cells (37). As a consequence, although
induction of the activated state is immunologically specific in that
it results from an antigen-specific immune response, its expression
is non-specific (38). Once activated, macrophages may destroy microorganisms or other targets that are immunologically unrelated to the
activating stimulus.

     Activated macrophages are characterized biochemically by changes
in the content or pattern of secretion and activities of various
enzymatic and other markers (18). Of particular relevance to their
increased microbicidal capacity is the observation that they are
metabolically far more responsive, upon phagocytic stimulation or
triggering by membrane-active agents, than their "normal" counter-

parts. However, phagocytosis is not a necessary prerequisite for induction of the metabolic burst in activated cells. Thus, incubation of macrophages with activating lymphokines stimulates the hexose monophosphate shunt, superoxide production and other biochemical markers of the activated state without further phagocytic stimuli (39,40).

CORRELATION BETWEEN INTRACELLULAR KILLING AND OXIDATIVE EVENTS IN MACROPHAGES

Several types of evidence indicate that killing of intracellular microorganisms is closely dependent on the oxidative metabolic activity of the phagocytes, as indicated below:

(a) Intracellular destruction of microbes correlates with stimulation of the respiratory burst and conversely, failure to trigger such metabolic activity is linked with intracellular survival. Thus, killing of T. gondii by human monocytes is accompanied by superoxide production, whereas mouse macrophages fail to respond metabolically to the same stimulus and allow the parasite to survive intracellularly (16). The same host cells, however, destroy L. tropica and L. donovani, which do trigger superoxide and hydrogen peroxide production upon phagocytosis (15). Mutant phagocytic cell lines that fail to respond to membrane-active agents by a metabolic burst also lack the capacity to destroy the above organisms (41). Induction of intracellular killing of Leishmania enriettii by incubation of parasitized macrophages with appropriate lymphokines is accompanied by a respiratory burst (40), and hydrogen peroxide production by various classes of activated mouse macrophages closely correlates with the capacity of these cells to kill T. cruzi (42).

(b) Incubation of macrophages with certain scavengers of oxygen metabolites favors intracellular survival and conversely, stimulation of the production or utilization of such intermediates enhances killing. Catalase, benzoate and histidine remove hydrogen peroxide, hydroxyl radicals and singlet oxygen, respectively, and inhibit the capacity of activated macrophages to kill toxoplasmas (43). The same effect is obtained by incubation of L. enriettii-infected cells with aminotriazole, an inhibitor of hydrogen peroxide metabolizing enzymes (40). Conversely, supplying exogenous peroxidase to subliminally-activated macrophages enhances the microbicidal activity of the latter cells (40).

(c) Microorganisms are rapidly destroyed in vitro when incubated in cell-free preparations containing hydrogen peroxide or a hydrogen peroxide generating system, peroxidase, and a halide (40,44); similar results can be achieved with a superoxide generating mixture (45). In this case also, appropriate scavengers protect the microorgansism from the lethal action of oxygen intermediates.

SURVIVAL MECHANISMS OF INTRACELLULAR MICROORGANISMS

A number of microorganisms are able to survive and multiply within macrophages. Their capacity to resist the microbicidal processes of mononuclear phagocytes is undoubtedly a major factor in their pathogenicity. Two categories of mechanisms appear to have evolved by intracellular pathogens to promote their survival within the intracellular environment: (a) they have learned to resist or circumvent the lytic machinery of lysosomes; and (b) more importantly perhaps, they may be internalized by phagocytes without stimulating a respiratory burst.

Intracellular Survival of Microbes: Morphological Observations

Ultrastructural studies of the relationship between macrophages and intracellular microbes have uncovered at least three types of mechanisms whereby parasites may avoid the harmful effects of lysosomal constituents.

(a) Inhibition of the fusion of phagosomes with lysosomes. Both Mycobacterium tuberculosis (46) and T. gondii (47) infecting mouse macrophages are found in phagocytic vesicles that do not fuse with surrounding lysosomes. Ultrastructural observations suggest that T. gondii can alter the properties of the vacuole in such a way that strips of endoplasmic reticulum and mitochondria are attracted towards its immediate surroundings, thus providing a protective coat that prohibits access of the lysosomes (47). This behavior depends on the integrity of the parasite: dead or antibody-coated toxoplasmas are unable to inhibit phagosome-lysosome fusion and are thus digested (47,48). Similar observations have been reported for M. tuberculosis (46,49).

(b) Resistance to lysosomal constituents. Certain microorganisms appear to thrive in the theoretically unhospitable phagolysosome. For instance, vacuoles harboring L. donovani or Leishmania mexicana in macrophages from man, hamster or mouse, are observed to fuse with lysosomes with no apparent harmful effects towards the parasites (12,50,51). The mechanisms of such resistance are unknown. The intravacuolar pH of Leishmania-infected macrophages is hardly modified by infection (52). Inert, non-digestible material such as the waxy capsule surrounding M. lepraemurium and other mycobacteria may protect these microorganisms from attack by bacteriolytic lysosomal enzymes (53). Alternatively, intracellular pathogens may release substances capable of inhibiting lysosomal activity. In this context, Leishmania parasites have been shown to synthesize a carbohydrate-rich substance, the excretory factor (54) that displays anti-β-galactosidase activity (55). This material is negatively charged, as are other moderators of lysosomal function (56). That such a compound may protect parasites against intracellular degradation is suggested by the demonstration that growth of Leishmania in other-

wise non-permissive host macrophages is considerably stimulated by
the addition of exogenous excretory factor (57). In the same line,
infection of macrophages in vitro by L. donovani has been shown to
promote the intracellular survival of another microorganism, Lepto-
monas costoris, suggesting an impairment of the digestive mechanisms
of the phagocytes (58).

        (c) Escape from phagolysosomes. As shown by ultrastructural
investigations, certain microorganisms appear capable of escaping
from the unhospitable intravacuolar spaces, to reach the cytoplasmic
matrix where their survival is unimpaired. So do trypomastigotes
of T. cruzi in mouse macrophages (11,59). This may represent a
crucial survival advantage, as parasites that remain confined within
the vacuolar compartment are destroyed. Mycobacterium leprae can
similarly be detected in extravacuolar locations (60). It is noteworthy that when T. cruzi is phagocytized by activated macrophages
the parasites are destroyed in phagolysosomes (61), suggesting that
they are killed by the metabolically more active phagocytes before
they had a chance to reach a safer intracellular location.

Intracellular Survival of Microbes: Biochemical Aspects

        Normal human monocytes and activated human or murine macrophages
destroy or inhibit the proliferation of T. gondii, whereas non-activated macrophages are unable to do so. Defective intracellular killing appears to depend on the capacity of the parasite to become internalized without stimulating a metabolic burst (16). This behavior
may be linked to characteristics of the parasite surface or to the
type of membrane receptors involved in the internalization process.
Indeed, coating the parasite with specific polyclonal (48,62) or
monoclonal (63) antibody renders it fully susceptible to intracellular
destruction as well as allows a phagocytosis induced oxidative burst
to be generated (16).

        It is important to note that microorganisms may succeed in taking
residency within phagocytic cells even under situations of unimpaired
metabolic activity. For instance, engulfment of culture promastigotes
of L. tropica and L. donovani by mouse peritoneal macrophages in vitro
is accompanied by the generation by the phagocytes of superoxide and
hydrogen peroxide, the latter product being toxic for the parasites.
Accordingly, most of the ingested organisms are destroyed intracellularly within 18 hours (15). A certain number of parasites are
observed to survive. However, they transform to and multiply as
amastigotes, a developmental form presumably better adapted to intracellular survival (64). Their destruction will depend on the capacity
of macrophages to become activated as result of stimulation by the
immune system or by other processes.

Parasite Survival in Activated Macrophages

        In vitro studies indicate that the susceptibility of microorgan-

isms to intracellular destruction by activated macrophages varies
considerably, depending on the origin and level of activation of the
host cell. Indeed, a parallel can sometimes be established between
the capacity of activated macrophages from a given species to destroy
a microorganism in vitro (or lack of it) and susceptibility of the
host to infection by this microorganism. For instance, guinea pigs
are very susceptible to infection by Leishmania enriettii; this correlates with the observation that activated guinea pig macrophages
fail to destroy the microorganism in vitro (65). Conversely, mice
are completely resistant to infection by this parasite, which correlates with the capacity of activated mouse macrophages to kill this
microorganism (37,66). A similar parallel can be found in mice
infected by L. tropica, where genetically determined variations in
resistance or susceptibility can be detected (67). In this case also,
in vitro studies have shown that macrophages from resistant strains
are more efficient at killing the parasite than are macrophages from
susceptible strains (68).

    The reasons are not clear for the varying susceptibilities of
different microorganisms to intracellular destruction by activated
macrophages from different sources. Resistance of parasites to the
cytolytic processes of macrophages may reflect in part the endowment
of the parasites as scavengers of oxygen metabolites (15). However,
a given microbe (for instance, L. enriettii) may resist destruction
in activated macrophages from a certain source (e.g. the guinea pig),
but be extremely susceptible to killing by activated cells from
another origin (e.g. the mouse). This suggests that either activated
macrophages differ qualitatively and/or quantitatively with respect
to their capacity to generate microbicidal products, or that other
factors in the intracellular environment unrelated to cytotoxic pathways play a crucial role in determining the susceptibility of microorganisms to intracellular destruction.

SUMMARY

    The term activation is used to designate biochemical and func-
tional changes that are induced in macrophages by a variety of stimuli, including interaction with microbial products, synthetic substances, immunoglobulins of different classes, and factors released
by lymphocytes. The changes observed comprise an increased capacity
to destroy intracellular microorganisms and non-microbial target cells
as well as the stimulation of biochemical pathways leading to the
release of enzymes and the generation of various toxic compounds.
Activation may thus be viewed as a process aimed at recalling those
metabolic functions that are necessary for killing, when phagocytosis
has failed to evoke them.

    The increased microbicidal capacity of activated macrophages
is linked to the production of oxygen intermediates, as illustrated

by the study of macrophage toxicity for certain intracellular proto-
zoan parasites. Scavengers of oxygen metabolites inhibit parasite
killing in macrophages; on the contrary, agents that stimulate the
production or utilization of such intermediates enhance the microbicidal effect of phagocytes.

Several mechanisms enable microorganisms to survive within macrophages. In some instances, intracellular survival appears to depend
on the capacity of microorganisms to be endocytized without awakening
the host cell oxidative machinery. In addition, the endowment of
microorganisms in endogenous enzymatic scavengers of oxygen metabolites may play a role in promoting intracellular survival. These
and other mechanisms, such as the property to avoid the harmful effects of lysosomal constituents by inhibiting phagosome-lysosome
fusion, or by releasing agents that block the lysosomal enzymatic
machinery, may explain why certain microbes are able to survive within
activated macrophages.

REFERENCES

1. Santoro, F., Bernal, J., and Capron, A., Acta Tropica 36:5, 1979.
2. Betz, S. J., and Isliker, H., J. Immunol. 127:1748, 1981.
3. Synderman, R., Shin, H. S., and Dannenberg, A. M., J. Immunol.
    109:896, 1972.
4. Kolb, H., and Kolb-Bachofen, V., Bioch. Biophy. Res. Comm. 85:
    678, 1978.
5. Glass, E., Stewart, J., and Weir, D. M., Immunology 44:529, 1981.
6. Weir, D. M., and Ogmundsdottir, H. M., in "Mononuclear Phagocytes:
    Functional Aspects", Part I (R. van Furth, ed.), pp. 865-881,
    Martinus Nijhoff Publishers, The Hague, 1980.
7. Dawidowicz, K., Hernandez, A. G., Infante, R. B., and Convit, J.,
    Nature 256:47, 1975.
8. Griffin, F. M., Griffin, J. A., Leider, J. E., and Silverstein,
    S. C., J. Exp. Med. 142:1263, 1975.
9. Alexander, J., J. Protozool. 22:237, 1975.
10. Ardehali, S. M., Khoubyar, K., and Rezai, H. R., Acta Tropica
    36:15, 1979.
11. Nogueira, N., and Cohn, Z. A., J. Exp. Med. 143:1402, 1976.
12. Chang, K. P., and Dwyer, D. M., J. Exp. Med. 147:515, 1978.
13. D'Arcy Hart, P., and Young, M. R., Exp. Cell Res. 118:365, 1979.
14. Klebanoff, S. J., in "Mononuclear Phagocytes: Functional
    Aspects:, Part II (R. van Furth, ed.), pp. 1105-1137,
    Martinus Nijhoff Publishers, The Hague, 1980.
15. Murray, H. W., J. Exp. Med. 153:1302, 1981.
16. Wilson, C. B., Tsai, V., and Remington, J. S., J. Exp. Med. 151:
    328, 1980.
17. Johnston, R. B., Chadwick, D. A., and Pabst, M. J., in "Mono-
    nuclear Phagocytes: Functional Aspects", Part II (R. van
    Furth, ed.), pp. 1143-1158, Martinus Nijhoff Publishers, The
    Hague, 1980.

18. Karnovsky, M. L., and Lazdins, J. K., J. Immunol. 121:809, 1978.
19. Karnovsky, M. L., Bawdey, J., Briggs, R., Karnovsky, M. J., and Lazdins, L. J., in "Mononuclear Phagocytes; Functional Aspects", Part II (R. van Furth, ed.), pp. 1495-1511, Martinus Nijhoff Publishers, The Hague, 1980.
20. Forman, H. J., Nelson, J., and Fisher, A. B., J. Biol. Chemistry 255:9879, 1980.
21. Bellavite, P., Breton, G., Dri, P., and Soranzo, R., J. Reticuloendothelial Soc. 29:47, 1981.
22. Babior, B. M., New Engl. J. Med. 298:659, 1978.
23. Cross, A. R., Jones, O. T. G., Harper, A. M., and Segal, A. W., Biochem. J. 194:599, 1981.
24. Segal, A. W., Garcia, R., Goldstone, A. H., Biochem. J. 196:363, 1981.
25. Segal, A. W., and Jones, O. T. G., Nature 276:515, 1978.
26. Segal, A. W., and Jones, O. T. G., FEBBS Lett. 110:111, 1980.
27. Lehmeyer, J. E., Snyderman, R., and Johnston, R. B., Blood 54: 35, 1979.
28. Romeo, D., Zabucchi, G., Miani, N., and Rossi, F., Nature 253:544, 1975.
29. Smith, R. L., and Weideman, M. J., Biochem. Biophys. Res. Comm. 97:973, 1980.
30. Alexander, P., and Evans, R., Nature (New Biol.) 232:76, 1971.
31. Cook, J. A., Holbrook, T. W., Parker, B. M., J. Reticuloendothelial Soc. 27:567, 1980.
32. Kaplan, A. M., Morahan, P. S., and Regelson, W., J. Natl. Cancer Inst. 52:1919, 1974.
33. Capron, A., Dessaint, J. P., Joseph, M., Rousseaux, R., Capron, M., and Bazin, H., Eur. J. Immunol. 7:315, 1977.
34. Ogmundsdottir, H. M., and Weir, D. M., Clin. Exp. Immunol. 40: 223, 1980.
35. Lane, F. C., and Unanue, E. R., J. Exp. Med. 135:1104, 1972.
36. Mauel, J., Behin, R., and Louis, J., Exptl. Parasitol. 52:331, 1981.
37. Buchmüller, Y., and Mauel, J., J. Exp. Med. 150:359, 1979.
38. Mackaness, G. B., J. Exp. Med. 120:105, 1964.
39. Nathan, C. F., Karnovsky, M. L., and David, J. R., J. Exp. Med. 133:1356, 1971.
40. Buchmüller, Y., and Mauel, J., J. Reticuloendothelial Soc. 29: 181, 1981.
41. Murray, H. W., J. Exp. Med. 153:1690, 1981.
42. Nathan, C., Nogueira, N., Juangbhanich, C. W., Ellis, J., and Cohn, Z. A., J. Exp. Med. 149:1056, 1979.
43. Murray, H. W., Juangbhanich, C. W., Nathan, C. F., and Cohn, Z. A., J. Exp. Med. 150:950, 1979.
44. Klebanoff, S. J., J. Exp. Med. 126:1063, 1967.
45. Murray, H. W., and Cohn, Z. A., J. Exp. Med. 150:938, 1979.
46. Armstrong, J. A., and D'Arcy Hart,P., J. Exp. Med. 134:713, 1971.
47. Jones, T. C., and Hirsch, H. G., J. Exp. Med. 136:1173, 1972.
48. Jones, T. C., Len, L., and Hirsch, H. G., J. Exp. Med. 141:466, 1975.

49. Armstrong, J. A., and D'Arcy Hart, P., J. Exp. Med. 142:1, 1975.
50. Alexander, J., and Vickerman, K., J. Protozool. 22:502, 1975.
51. Berman, J. D., Dwyer, D. M., and Wyler, D. J., Infect. Immunity 26:375, 1979.
52. Chang, K. P., in "The Host-Invader Interplay" (H. van den Bossche, ed.), pp. 231-234, Elsevier/North Holland, Amsterdam, 1980.
53. Draper, P., and Rees, R. J. W., Nature (Lond.) 228:860, 1970.
54. Slutzky, G. M., and Greenblatt, C. L., Bioch. Medicine 21:70, 1979.
55. El-On, J., Bradley, D. J., and Freeman, J. C., Exptl. Parasitol. 49:167, 1980.
56. Avila, J. L., and Convit, J., Biochem. J. 160:129, 1976.
57. Handman, E., and Greenblatt, C. L., Z. Parasitenkd. 53:143, 1977.
58. Kutish, G. F., and Janovy, J., J. Parasitol. 67:457, 1981.
59. Tanowitz, H., Wittner, M., Kress, Y., and Bloom, B. R., Am. J. Trop. Med. Hyg. 24:25, 1975.
60. Evans, M. J., and Levy, L., Infect. Immunity 5:238, 1972.
61. Kress, Y., Bloom, B. R., Wittner, M., Rowen, A., and Tanowitz, H., Nature (Lond.) 257:394, 1975.
62. Anderson, S. E., and Remington, J. S., J. Exp. Med. 139:1154, 1974.
63. Sethi, K. K., Eudo, T., and Brandis, H., Immunology Letters 2: 343, 1981.
64. Lewis, D. H., and Peters, W., Ann. Trop. Med. Parasitol. 71: 295, 1977.
65. Mauel, J., Behin, R., Biroum-Noerjasin, and Rowe, D. S., Clin. Exp. Immunol. 20:339, 1975.
66. Mauel, J., Buchmüller, Y., and Behin, R., J. Exp. Med. 148:393, 1978.
67. Preston, P. M., and Dumonde, D. C., in "Immunology of Parasitic Infections" (S. Cohen and E. Sadun, eds.), pp. 167-202, Blackwell Scientfic Publications, Oxford, 1976.
68. Behin, R., Mauel, J., and Sordat, B., Exptl. Parasitol. 48:81, 1979.

DISCUSSION

SCHNYDER: In your experiments with horseradish peroxidase, do you think this enzyme works from outside or inside the cell? Would you get the same effect using horseradish peroxidase together with your microbes?

MAUEL: The experiments are done as follows. We first infect the macrophage with parasites in the absence of peroxidase. Then we activate the infected macrophage with lymphokines in the presence or absence of peroxidase. If we add peroxidase without the activating stimulus, nothing happens to the parasite. You need both induction of a metabolic response by the macrophage and the utilization of some metabolic product, presumably hydrogen peroxide, by peroxidase to increase killing.

SCHNYDER: Do you think that horseradish peroxidase is taken up in the same vacuole as the microbes?

MAUEL: We should have done that experiment. According to the litera- ture, it will end up in the same vacuole as the parasite.

BRAY: Will macrophages from DBA/2 mice, which apparently can't be activated to kill Leishmania, do so in the presence of prostaglandin inhibitors?

MAUEL: We haven't tried prostaglandin inhibitors. However, the DBA/2 macrophages, which we can't activate to kill Leishmania tropica, can be activated to kill Leishmania enriettia, a guinea pig parasite.

UNKNOWN: Do you find Mycobacteria leprae outside the phagosomes?

MAUEL: Yes. I know of one report where Mycobacteria leprae were found in the cytoplasm. This work was done with armadillos.*

BRAY: Are there microorganisms that will kill the macrophage?

MAUEL: Certainly, if the microorganism grows too fast, they will kill the macrophages.

UNKNOWN: Many substances which act as activators are, in fact, toxic. When one is dealing with obligate intracellular organisms whose reproduction is dependent upon a healthy macrophage, some of the things we call activators may be substances that interfere with macrophage metabolism and thereby inhibit parasite survival.

---

*Marchiondo, A. A., Smith, J. H., and File, S. K., Naturally occur- ring leprosy-like disease of wild armadillos: ultrastructure of lepromatous lesions, J. Reticuloendothelial Soc. 27:311, 1980.

RELEASE OF SUPEROXIDE AND HYDROGEN PEROXIDE FROM GUINEA-PIG

ALVEOLAR MACROPHAGES DURING PHAGOCYTOSIS OF MYCOBACTERIUM BOVIS BCG

P. S. Jackett, P. W. Andrew and D. B. Lowrie

M. R. C. Unit for Laboratory Studies of Tuberculosis
Royal Postgraduate Medical School
London W12 OHS   U.K.

INTRODUCTION

Experimental evidence is accumulating to suggest that macrophages can kill tubercle bacilli by producing hydrogen peroxide. For example, among strains of Mycobacterium tuberculosis virulence correlates with hydrogen peroxide resistance (1,2) and peroxide-susceptible mutants are killed more readily than parent strains in the lungs of guinea-pigs, particularly when the availability of peroxide from macrophages increases with the onset of acquired immunity (3,4). Furthermore, killing of Mycobacterium microti by immunologically-activated mouse peritoneal macrophages is prevented by exogenous catalase (5). This study characterizes hydrogen peroxide release from guinea-pig alveolar macrophages during phagocytosis of Mycobacterium bovis BCG.

METHODS

Macrophages were obtained by pulmonary lavage of normal guinea-pigs or guinea-pigs that had been vaccinated 3 weeks earlier by intravenous injection of M. bovis BCG in Freund's incomplete adjuvant. The lavage fluid was medium 199 containing 25 mM-HEPES buffer pH 7.3 and 5 units of heparin per ml. The cells were allowed to adhere to plastic petri dishes for 1 hour then maintained for a further 17 hours in medium 199 containing 10% newborn calf serum before use, at which time no polymorphonuclear leukocytes were detectable by Giemsa staining and each monolayer contained about 10^6 macrophages by DNA estimation (6).

M. bovis BCG was a thoroughly washed suspension from 7 days

687

culture in 7H9 liquid medium (2) containing (5,6-^{3}H)uridine. The bacilli were either alive or heat-killed (80°C, 30 minutes) and either unopsonized or opsonized with serum from BCG-vaccinated guinea-pigs (37°C, 30 minutes) before final suspension in Hanks' balanced salt solution buffered with 25mM-HEPES (pH 7.3).

To measure release of superoxide and hydrogen peroxide during phagocytosis, rinsed monolayers were incubated with bacterial suspension in the presence of cytochrome c (1 mg/ml) or p-OH-phenylacetic acid (2.4 mM) and horseradish peroxidase (5 µg/ml) as previously described (4).

RESULTS

Release of hydrogen peroxide was directly proportional to the number of bacilli that became associated with macrophages during 1 hour of phagocytosis when bacilli were presented to macrophages at ratios of up to 40:1 (Figure 1).

Uptake of bacilli was linear for about 1 hour and linear release of superoxide and hydrogen peroxide followed after a short lag or acceleration period (Figure 2). Both phagocytosis and reduced oxygen release stopped after about 90 minutes. Without phagocytic stimulation, superoxide and hydrogen peroxide release were minimal.

Table 1 shows that although macrophages from vaccinated animals took up about twice as many bacilli in an hour as did macrophages from normal animals, hydrogen peroxide release per bacillus taken up was not different. Furthermore, although opsonization doubled the number of bacilli taken up there was no effect on hydrogen peroxide release per bacillus. The viability of the bacilli affected neither uptake of the bacilli nor hydrogen peroxide release.

DISCUSSION

Phagocytosis of BCG was an effective stimulus of hydrogen peroxide release, presumably by stimulating a system that reduces oxygen via superoxide which then dismutates to hydrogen peroxide. The cessation of superoxide and hydrogen peroxide release that coincided with the end of bacterial uptake does not preclude the continued liberation of these products into the phagocytic vacuole within the macrophage. Indeed there is evidence that peroxide-dependent killing of M. microti in mouse macrophages can continue for at least 24 hours after phagocytosis (5). The acceleration phase at the start of superoxide and hydrogen peroxide release may largely reflect the increasing stimulation through increasing contact as attachment and phagocytic

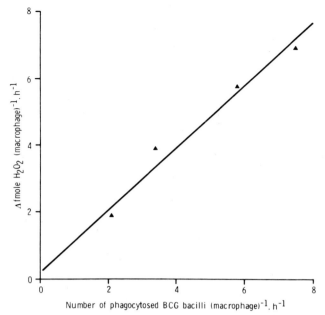

Fig. 1.   The amount of hydrogen peroxide released from guinea pig
          alveolar macrophages during phagocytosis of M. bovis BCG
          in relation to the number of bacilli taken up. Monolayers
          from a BCG-vaccinated animal were incubated at 37°C for 1
          hour with opsonized live BCG at bacillus:cell ratios of 10,
          20, 30 and 40:1.

engulfment proceed since little delay was seen with the soluble
stimulus phorbol myristate acetate (unpublished data).

     The absence of a difference in responsiveness to BCG between
macrophages from normal and BCG-vaccinated animals was notable since
not only were the lungs of the vaccinated animals granulomatous
yielding 2 to 10 times more macrophages but also the macrophages were
twice as active in phagocytosis. This contrasts with our findings
for alveolar macrophages from guinea pigs responding to an intravenous
challenge infection with M. tuberculosis after intraperitoneal plus
intramuscular vaccination: these macrophages were both more phagocytic and released more superoxide and hydrogen peroxide upon phagocytic stimulation within a few days of the challenge infection, unlike
macrophages from control (infected but non-vaccinated) animals (4).
This suggests that an immunological component of the inflammatory
reaction was necessary for an increased hydrogen peroxide response
to phagocytic stimulation and was deficient in the reaction evoked
here by intravenous Freund's adjuvant plus BCG.

Table 1. Release of Hydrogen Peroxide (H_2O_2) by Alveolar Macrophages from Normal and BCG-vaccinated Guinea pigs in Response to Phagocytosis of BCG in vitro

Stimulus	Normal Guinea pigs		Vaccinated Guinea pigs	
	Bacilli per Macrophage	Increased H_2O_2 Release in fmole/Bacillus/Hour	Bacilli per Macrophage	Increased H_2O_2 Release in fmole/Bacillus/Hour
Dead BCG	2.6	0.16	5.8	0.32
Live BCG	2.5	0.14	6.6	0.12
Opsonized dead BCG	6.8	0.14	12.2	0.11
Opsonized live BCG	6.1	0.15	12.4	0.24

Uptake of (^{3}H)-labeled BCG and the amount of H_2O_2 released were measured after 1 hour of phagocytosis. Data were averaged from 5 independent experiments with cells from normal guinea pigs and 8 with cells from vaccinated animals. Values for H_2O_2 release from phagocytosing macrophages were corrected for release from control macrophages and differences in bacillus uptake.

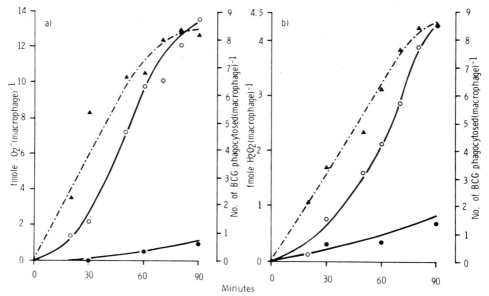

Fig. 2.   The kinetics of release of superoxide (a) and hydrogen per-
          oxide (b) during phagocytosis. Monolayers from a vaccinated
          animal were incubated at 37°C with opsonized live BCG at a
          bacillus:cell ratio of 30:1 (o——o) or without bacilli
          (●——●).   The kinetics of uptake of bacilli are also shown
          (▲——▲).

Since opsonization of tubercle bacilli failed to enhance hydrogen
peroxide release per bacillus above that observed with unopsonized
bacilli, the tubercle bacillus may be an inherently potent stimulus.

Tubercle bacilli, including BCG, inhibit fusion of lysosomes
with phagosomes and this ability is overcome if the bacilli are either
killed or antibody-coated before phagocytosis (7,8). Since neither
killing nor antibody-coating affected the hydrogen peroxide release
per bacillus taken up it is evident that the two macrophage responses,
phagolysosome formation and release of hydrogen peroxide, are substantially independent.

REFERENCES

1. Mitchison, D. A., Selkon, J. B., and Lloyd, J., J. Path. Bact.
    86:377, 1963.

2. Jackett, P. S., Aber, V. R., and Lowrie, D. B., J. Gen. Micro-
        biol. 104:37, 1968.
3. Jackett, P. S., Aber, V. R., Mitchison, D. A., and Lowrie, D. B.,
        Br. J. Exp. Path. 62:34, 1981.
4. Jackett, P. S., Andrew, P. W., Aber, V. R., and Lowrie, D. B.,
        Br. J. Exp. Path. 62:419, 1981.
5. Walker, L., and Lowrie, D. B., Nature 293:69, 1981.
6. Labarca, C., and Paigen, K., Anal. Biochem. 102:344, 1980.
7. Armstrong, J. A., and Hart, P.D'A., J. Exp. Med. 142:1, 1975.
8. Lowrie, D. B., Jackett, P. S., Aber, V. R., and Carrol, M. E. W.,
        in "Mononuclear Phagocytes: Functional Aspects" (R. van Furth,
        ed.), pp. 1057-1075, Martinus Nijhoff, The Hague, Boston,
        London, 1980.

DISCUSSION

UNKNOWN: You mentioned that Mycobacteria vary in their sensitivity
to hydrogen peroxide. Is there a spectrum of sensitivity, from Mycobacterium smegmatis to those strains used for vaccination to the
pathogenic strains?

JACKETT: We've never looked at any of the non-pathogenic, rapidly
growing Mycobacteria.

UNKNOWN: Is BCG more or less sensitive than Mycobacterium bovis or
Mycobacterium tuberculosis?

JACKETT: BCG is as resistant to hydrogen peroxide as the most
virulent tubercle bacillus. I don't believe that resistance to
hydrogen peroxide is the only thing that accounts for virulence.
Conversely, you can have low virulence without having susceptibility
to hydrogen peroxide.

PICK: I remember a paper published about 10 years ago, before knowledge of the oxidative burst, in which rabbits were immunized with
BCG and then their macrophages were used to prepare an organic extract. The extract was found to be a semi-potent killer of Mycobacteria. The interpretation was absent at that time. I would like
to suggest that, in some cases, the actual killer of the pathogen
is a lipid peroxide produced in the course of the oxidative burst.
Do you have any information on the sensitivity of Mycobacteria to
lipid peroxide?

JACKETT: Hydrogen peroxide by itself is not adequate to kill Mycobacteria directly. One possibility is that it gets into the Myco-

bacteria and attacks some vital biochemical machinery. Another possibility is that it generates intermediate toxic materials, such as
lipid peroxides. It doesn't have to be the final effector.

MAUEL: Is there a correlation between level of hydrogen peroxide
generated from macrophages of BCG immune animals when they phagocytize
Mycobacteria and intracellular killing of the microorganism?

JACKETT: It's quite remarkable, but the amount of hydrogen peroxide
released per bacillus taken up by a normal guinea pig macrophage
is identical to that released from a macrophage obtained from a
Freund's adjuvant vaccinated guinea pig. Although the vaccinated
animals have granulomas in the lung with more macrophages, they do
not release more hydrogen peroxide per bacillus engulfed than do
normal macrophages. That contrasts with what we find when we vac-
cinate the guinea pig in a more conventional manner without Freund's
adjuvant. The macrophages obtained from the challenged and infected
lung in these immune animals are activated and release more hydrogen
peroxide per bacillus engulfed. In other words, it seems that you
need a more potent immunological stimulus than the inflammatory
stimulus provided by Freund's adjuvant alone.

METCALF: You're measuring hydrogen peroxide released from a cell,
but surely it's the concentration within the cell and around the
organism that is critical. Is there no way to measure such hydrogen
peroxide?

JACKETT: I don't know of a way of measuring the concentration within
the phagolysosome. If anyone does, I'd be most interested because
that is the crucial question.

STIMULATION BY VARIOUS LECTINS OF INTRACELLULAR

KILLING OF MICRO-ORGANISMS BY PHAGOCYTIC CELLS

P. C. J. Leijh, Th. L. van Zwet and R. van Furth

Department of Infectious Diseases
University Hospital
2333 AA Leiden
The Netherlands

INTRODUCTION

The ingestion of bacteria by phagocytes is accompanied by a metabolic burst during which oxygen is consumed and converted into superoxide anion and hydrogen peroxide. Studies performed in the last decade have shown a correlation between the generation of these oxygen intermediates and the microbicidal and tumoricidal activity of the cells (1-3). Recently, we described the extracellular requirement for serum factors, especially IgG and C3b, by monocytes and granulocytes to obtain maximal intracellular killing of micro-organisms (4-6).

To obtain insight into the mechanism of this extracellular stimulation and to correlate this stimulation with the functioning of the oxygen dependent microbicidal systems, we investigated the effect of various plant lectins on intracellular killing.

MATERIALS AND METHODS

Lectins: These were obtained as follows: Concanavalin A (Con A, Miles-Yeda, Israel), wheatgerm (Sigma Chemical Co., St. Louis, Mo.), peanut lectin (Sigma), lentil lectin (Pharmacia, Uppsala, Sweden), phytohemagglutinin (PHA, Wellcome Netherlands, Amsterdam, The Netherlands), soybean lectin (Pharmacia), and succinyl-concanavalin A (Industrie Biologique Francaise, Clichy, France).

Leukocytes and Serum: Monocytes were collected by differential centrifugation of blood of healthy donors on Ficoll-Hypaque according to Boyum (4). The interphase layer was washed four times with phosphate buffered saline (PBS) containing 0.5 U heparin/ml, and suspended in Hanks' balanced salt solution (HBSS) with 0.1% (w/v) gelatin to a concentration of 1 x 10^7 monocytes/ml. The percentage of granulocytes in this suspension was less than 2.

Granulocytes were obtained by dextran sedimentation of the Ficoll-Hypaque pellet to remove the erythrocytes (6), and resuspended in HBSS with gelatin to a concentration of 10^7 cells/ml. In this suspension more than 98% of the cells were granulocytes.

Serum from the blood of healthy donors with blood group AB was prepared by clotting blood for 1 hr at room temperature, and centrifuging the clot for 20 min at 1100 g. Serum was stored in 2-ml aliquots at -70°C.

Micro-organisms: S. aureus were cultured overnight in Nutrient Broth no. 2, harvested by centrifugation at 1500 g, washed twice with PBS, and suspended in HBSS with gelatin to a concentration of 10^7 bacteria/ml. S. aureus were opsonized by incubation of 5 x 10^6 bacteria/ml with 10% (v/v) AB serum for 25 min at 37°C under slow rotation. The suspension was then cooled to 4°C, and the bacteria harvested by centrifugation, washed with ice-cold HBSS, and resuspended to a concentration of 10^7/ml.

Phagocytosis and Killing of S. aureus: Phagocytosis of S. aureus by monocytes and granulocytes was measured as a decrease in the number of viable extracellular bacteria during incubation of 5 x 10^6 monocytes or granulocytes/ml and 5 x 10^6 bacteria/ml in the presence or absence of serum or the lectin under study for 60 min at 37°C under slow rotation (4 rpm). The number of viable extracellular bacteria was determined microbiologically in the supernatant after differential centrifugation for 4 min at 110 g to separate monocytes or granulocytes and bacteria (7).

Intracellular killing of S. aureus was determined as a decrease in the number of viable intracellular bacteria after a 3-min ingestion period (4-6). In short, 5 x 10^6 monocytes or granulocytes/ml and 5 x 10^6 pre-opsonized bacteria/ml were incubated for 3 min at 37°C under rotation (4 rpm), phagocytosis was then stopped by shaking the tube through crushed ice, and the non-ingested bacteria were removed by differential centrifugation (4 min at 110 g) and two washes with ice-cold HBSS.

Next, the cells containing ingested bacteria were re-incubated at 37°C under rotation (4 rpm) and at the appropriate time points a sample was taken and centrifuged for 4 min at 110 g, after which distilled water containing 0.01% albumin was added and the cells lysed. Lysis was performed in two ways: for granulocytes, by shaking the suspension for 1 min on a vortex mixer; and for monocytes, by freezing and thawing of the suspension three times in liquid nitrogen and a waterbath at 37°C. The number of viable micro-organisms in this suspension was determined microbiologically.

Receptors: The presence of Fc receptors was determined with IgG-coated sheep red blood cells as described elsewhere (4). The presence of receptors for lectin was determined similarly with lectin-coated sheep red blood cells.

RESULTS AND DISCUSSION

Before the effect of various lectins on intracellular killing was investigated, the binding of these lectins to phagocytes and their phagocytosis-promoting activity were studied. Incubation of mono-cytes and granulocytes with various concentrations of Con A; soybean, wheatgerm, peanut, or lentil lectin; or phytohemagglutinin, for 60 min at 37°C, led to agglutination of the phagocytes in the presence of all lectins except peanut. This indicated binding of the various lectins to granulocytes and monocytes, which was confirmed by rosette formation with lectin-coated sheep red blood cells. Although soybean lectin induced agglutination of phagocytes, rosette formation was not observed, presumably because binding of soybean lectin to sheep red blood cells leaves no lectin binding sites available for binding to phagocytes.

Although lectin-coated sheep red cells attached to monocytes and granulocytes, the red cells were not ingested. This absence of op-sonic activity was confirmed for the various lectins during incubation of monocytes and S. aureus in the presence of each of the lectins for 60 min at 37°C. The results showed no decrease in the number of viable extracellular bacteria, indicating that S. aureus was not ingested by the monocytes. These results were confirmed by light-microscopy.

That various lectins bound to phagocytes but failed to induce ingestion of micro-organisms makes lectins a valuable tool for inves-tigating intracellular killing. Under these conditions any decrease in the number of viable intracellular bacteria can only be due to intracellular killing and not to the ingestion of extracellular bacteria adherent to phagocytes.

Incubation of monocytes or granulocytes containing ingested S. aureus in the presence of the various lectins showed that lectins can stimulate intracellular killing by monocytes but not by granulo-cytes (Table 1). Incubation of monocytes containing ingested S.aureus with Con A for 30 min at 37°C, followed by two washes to remove free Con A and re-incubation in the presence of Con A or fresh serum, led to diminished intracellular killing. This indicated that interaction of Con A with monocytes prevented the functioning not only of Con A

Table 1.   Intracellular killing of S. aureus by Granulocytes
and Monocytes in the Presence of Various Lectins

		Intracellular Killing by	
Lectin	ug/ml	Granulocytes %	Monocytes %
Serum (10%)		88	75
HBSS		32	0
Con A	150	30	75
PHA	100	39	80
Lentil	100	23	72
Soybean	100	43	60
Wheatgerm	20	34	20
Peanut	100	31	0

Intracellular killing was measured during incubation of phagocytes containing ingested S. aureus in the presence of the indicated lectin.   HBSS = Hanks' balanced salt solution.

but also of Fc gamma and C3b receptors.   Subsequent incubation of
the Con A-treated monocytes with α-methyl mannoside, the sugar for
which Con A possesses specific binding sites, restored the activity
of the Fc gamma, C3b, and Con A receptors.   The inhibition of Fc
gamma and C3b receptor-mediated monocyte function by Con A was confirmed by the inability of Con A-treated monocytes to ingest either
IgG-coated sheep red blood cells or opsonized S. aureus.   Comparison
of the effect of tetravalent Con A and dimeric succinyl Con A on Fc
gamma and C3b receptor-mediated functions showed that the inhibition
of these functions by Con A was probably due to a sterical interference of Con A with the Fc gamma and C3b receptors.

To investigate the different responses of monocytes and granulocytes with respect to activation of intracellular killing by lectins,

phagocytes were incubated with FITC-labeled Con A. After incubation, monocytes displayed a cytoplasmic and membrane-bound fluorescence pattern, but granulocytes showed only membrane-bound fluorescence. This indicated that endocytosis of the lectin by monocytes but not by granulocytes had occurred. Therefore, it seems likely that the stimulation of intracellular killing by lectins required endocytosis. This hypothesis was supported by the absence of intracellular killing by monocytes in the presence of non-phagocytic Con A-Sepharose particles.

In conclusion, the present investigation shows that various lectins are able to stimulate intracellular killing by monocytes but not by granulocytes. This stimulation of the killing process probably requires internalization of the extracellular lectin.

Acknowledgement

This study was partially supported by the J. A. Cohen Institute of Radiopathology and Radiation Protection and by the Netherlands Organization for the Advancement of Pure Research (ZWO).

REFERENCES

1. Babior, B. M., New Eng. J. Med. 298:659 and 721, 1978.
2. Nathan, C. F., and Cohn, Z. A., J. Exp. Med. 152:198, 1980.
3. Sasada, M., and Johnston, R. B., J. Exp. Med. 152:85, 1980.
4. Leijh, P. C. J., Barselaar, M. Th. van den, Zwet, Th. L. van, Daha, M. R., and Furth, R. van, J. Clin. Invest. 63:772, 1979.
5. Leijh, P. C. J., Zwet, Th. L. van, and Furth, R. van, Inf. Immun. 30:421, 1980.
6. Leijh, P. C. J., Barselaar, M. Th. van den, Daha, M. R., and Furth, R. van, Inf. Immun. 33:714, 1981.
7. Furth, R. van, Zwet, Th. L. van, and Leijh, P. C.J., in "Handbook of Experimental Immunology" (D. M. Weir, ed.), Ch. 32, Blackwell Scientific Publications, Oxford, 1978.

DISCUSSION

PICK: Are you saying that Concanavalin A has to bind to the membrane at the same site as the pathogen destined to be killed, or are you implying an intracellular messenger is inducing hydrogen peroxide production in the phagolysosome, consequent to membrane stimulation by the non-interiorized lectin?

LEIJH: I think that the lectin must be internalized.

MAUEL: What is your evidence that the effect of serum is due to immunoglobulins?

LEIJH: We found that pure immunoglobulins were as effective in
stimulating intracellular killing as heat inactivated serum. We
also analyzed various subclasses of serum and found stimulation of
intracellular killing only by IgG1 and IgG3, which interacted with
the Fc receptor. Also, Fc fragments of immunoglobulin were able to
stimulate the process.

MAUEL: Must immunoglobulins be aggregated to obtain this effect?

LEIJH: No. Monomeric immunoglobulins are capable of producing the
effect.

MAUEL: Does either Con A or immunoglobulin induce a metabolic burst?

LEIJH: Con A evokes a metabolic burst as measured outside the cell,
but there is a discrepancy between the metabolic burst evoked in the
extracellular medium and stimulation of the intracellular killing.

GLUCAN IMMUNOMODULATION IN

EXPERIMENTAL E. COLI SEPSIS

David L. Williams[1], William Browder[2], Rose McNamee[1], and Nicholas R. Di Luzio[1]

[1]Department of Physiology, Tulane University School of Medicine, New Orleans, Louisiana   70112 USA
[2]Department of Surgery, Tulane University School of Medicine, New Orleans, Louisiana   70112 USA

INTRODUCTION

In spite of the increased efficacy of antibiotic therapy, sepsis, particularly with endotoxin containing gram-negative bacilli, is a serious complication in patients whose defense mechanisms have been compromised by such events as trauma, burns, hematological malignancies, x-irradiation or therapy (1,2). While a number of defects in homeostatic mechanisms have been identified in such patients, attempts to significantly alter host defense mechanisms in order to modify gram-negative sepsis have been limited, since all currently employed experimental immunotherapeutic agents profoundly increase the sensitivity of the host to endotoxins. These agents include Bacillus Calmette-Guérin (3), zymosan (4), C. parvum (5), glucan (6) and muramyl dipeptide (7).

Our laboratory has extensively employed glucan, a β-1,3 polyglucose derived from the inner cell wall of the yeast Saccharomyces cerevisiae, as a non-specific stimulant of the reticuloendothelial system (8). Administration of glucan causes proliferation and activation of macrophages resulting in increased phagocytic activity as well as cellular and humoral immunity (9,10). These properties have enabled glucan to be of benefit in a variety of experimental infection models including bacterial (11-13), viral (12-14), fungal (13-15), and parasitic (13-16).

In view of the inherent problem in the use of biological response modifiers during gram-negative infections where a significant endotoxin burden would be generated, the present study was designed to

701

evaluate the influence of the route of glucan administration in modifying Escherichia coli sepsis. Additionally, the influence of fulminant E. coli bacteremia on macrophage function and the effects of glucan thereupon were evaluated. Further delineation of the importance of macrophages to the protective effect of glucan was ascertained by studying methyl palmitate mediated reversal of glucan-induced macrophage hyperfunction on E. coli induced mortality. These composite studies denote that macrophage functional status is a determinant factor in the final outcome to E. coli bacteremia.

MATERIALS AND METHODS

ICR/Tex male mice weighing 20g were pretreated with glucan intraperitoneally (150 mg/kg) or intravenously (22.5 mg/kg) on days -5 and -3 prior to bacterial innoculation on day 0. Inoculation was accomplished by injecting 1 x 10^8 viable E. coli into the peritoneal cavity . Phagocytic function was evaluated by the vascular clearance of gelatinized ^{131}I-triolein labeled reticuloendothelial test lipid emulsion at nine hours following injection of E. coli (14). Additionally, glucan induced macrophage hyperactivity was restored to normal by the intravenous administration of an emulsion of methyl palmitate (6,17) at 35 mg/mouse 24 hours prior to E. coli injection. Control mice received either glucan alone, glucose, or 0.2% Tween 20 solution in 5% glucose, the suspending medium for methyl palmitate. Survival was ascertained for 72 hours.

RESULTS

The intravenous administration of glucan, which induces a profound state of macrophage activation and proliferation with resultant granuloma formation, was associated with a significant decrease in survival following E. coli administration. While control mice showed initial mortality at 9 hours post-injection and a 90% mortality by 13 hours, the mice pretreated with i.v. glucan showed initial mortality at 4 hours and 100% mortality at 5 hours post E. coli injection. In marked contrast, when glucan was administered intraperitoneally, a 90% survival was observed in contrast to a 90% mortality observed in glucose injected control mice.

Phagocytic function was measured by the intravascular clearance of RE test lipid emulsion and was significantly impaired in the control-E. coli infected mice where the intravascular half life was increased 200% (Table 1). The administration of glucan intraperitoneally produced a mean 23% enhancement in the half life of the RE test emulsion. Compared to the pronounced phagocytic impairment observed in the glucose-E. coli group, the glucan-E. coli infected group maintained a state of hyperphagocytic activity.

Table 1.   Impairment in the Intravascular Clearance of the
           [131]I Labeled Reticuloendothelial Test Lipid Emulsion
           Following E. coli Sepsis and its Modification by
           Glucan Administration[a]

Treatment Groups	E. coli	Half Life (minutes)
Glucose	-	7.9 ± 0.83
Glucose	+	23.7 ± 3.30^{b}
Glucan	-	6.1 ± 0.59^{c}
Glucan	+	4.3 ± 0.85^{d}

[a]Clearance was performed 9 hours post-challenge. Statistical comparisons were made using the glucose control, non-infected group
as the reference. Each group had 8 mice.
[b]$p < 0.001$
[c]$p < 0.05$
[d]$p < 0.01$

     The role of hyperfunctional macrophages in the protection of the
host to E. coli bacteremia was also evaluated. In this study, methyl
palmitate was utilized to reverse the glucan-induced macrophage hyperfunctional state while RE hypertrophy and hyperplasia was maintained
(6,17). In essential confirmation of our previous studies, a 100%
survival was noted in the i.p. glucan-treated mice. The administration of methyl palmitate to glucan treated mice completely abolished
the protective effect of glucan. The response of the methyl palmitate
treated group was comparable to the glucose control group as a 20%
and 0% survival was observed, respectively. Observation for an additional 48 hours revealed no additional mortality.

DISCUSSION

     It is apparent that glucan's ability to stimulate phagocytosis
may be of significant benefit in promoting survival in a murine E.
coli sepsis model. Infected mice which were pretreated with glucan
did not manifest impaired macrophage function as reflected in a decreased clearance of the RE test lipid emulsion. Indeed, glucan-

pretreated mice which were subsequently infected with E. coli displayed the characteristic enhanced phagocytosis induced by glucan. These findings are similar to our previous observations of a significant impairment in macrophage function following administration of murine hepatitis virus and its prevention by glucan (14). In agreement with the findings of Jennings et al. (18) the 8 hour interval following E. coli infection was found to be critical in the mediation of host response.

Previous studies have attempted to delineate the cellular dynamics that occur in the peritoneal cavity in the presence of gram-negative infections. Hau et al. (19) and Hau and Simmons (20) have stressed the importance of altered chemotactic and phagocytic activity of the polymorphonuclear leukocyte in a hemoglobin-E. coli peritonitis rat model. In addition, Hau et al. (21) reported that macrophage function is an important factor in protection from E. coli sepsis. Diminished bacterial clearance from the peritoneal cavity appears to be a significant cause of mortality in this model. Recently, Grover and Loegering (22) reported that reticuloendothelial depression is induced by injections of erythrocyte stroma in the peritonitis model. Since the intraperitoneal administration of glucan results in activated peritoneal macrophages (23) as well as an enhancement in granulocyte-monocyte progenitors (CFU-C) of bone marrow and spleen (23), the enhanced number and functional activity of activated macrophages in the peritoneal cavity as well as an increased systemic population and function of granulocytes and monocytes may also be a contributing factor in the beneficial effect of glucan in E. coli sepsis and in the reduction of bacteremia.

The importance of the hyperfunctional macrophage in mediation of host protection to E. coli is clearly denoted by the reversal of the protective effect of glucan when methyl palmitate was administered. This finding would suggest that macrophage functional status rather than macrophage cell number, which is unaltered following methyl palmitate administration (6,17), is the determining factor.

Our ability to materially modify the course of experimental E. coli infection by macrophage activation – if ultimately transferable to the clinical situation – portends the possibility of developing new therapeutic agents which could contribute to a reduction of morbidity and mortality in infectious episodes.

Acknowledgement

This investigation was supported, in part, by MECO Cancer Research Fund and National Cancer Institute grant #24326 and the American Cancer Society (IM-273).

REFERENCES

1. McCabe, W. R., and Jackson, G. G., Arch. Intern. Med. 127:120,
    1971.
2. Kreger, B. E., Craven, D. E., Carling, P. C., and McCabe, W. R.,
    Am. J. Med. 68:332, 1980.
3. Suter, E., J. Immunol. 92:49, 1964.
4. Benacerraf, B., Thorbecke, G. J., and Jacoby, D., Proc. Soc.
    Exp. Biol. Med. 100:796, 1959.
5. Ferluga, J., Kaplun, A., and Allison, A. C., Agents and Actions
    9:566, 1979.
6. Di Luzio, N. R., and Crafton, C. G., Adv. Exp. Med. Biol. 9:27,
    1970.
7. Ribi, E. E., Cantrell, J. L., Von Eschen, K. B., and Schwartzman,
    S. N., Cancer Research 39:4756, 1979.
8. Di Luzio, N. R., Adv. Exp. Med. Biol., 73A:412, 1976,
9. Wooles, W. R., and Di Luzio, N. R., Am. J. Physiol. 203:404, 1962.
10. Wooles, W. R., and Di Luzio, N. R., Science 142:1078, 1963.
11. Di Luzio, N. R., and Williams, D. L., Infec. Immun. 20:804, 1978.
12. Reynolds, J. A., Kastello, M. D., Harrington, D. G., Crabbs, C.
    L., Peters, C. J., Jemski, J. V., Scott, G. H., and Di Luzio,
    N. R., Infec. Immun. 30:51, 1980.
13. Song, M., and Di Luzio, N. R., Lysozymes in Biology Pathology
    6:533, 1979.
14. Williams, D. L., and Di Luzio, N. R., Science 208:67, 1980.
15. Williams, D. L., Cook, J. A., Hoffman, E. O., and Di Luzio, N.
    R., J. Reticuloendothelial Soc. 23:479, 1978.
16. Cook, J. A., Holbrook T. W., and Parker, B. W., J. Reticulo-
    endothelial Soc. 27:567, 1980.
17. Di Luzio, N. R., Lysosomes in Biology Pathology 6:447, 1979.
18. Jennings, M., Jennings, L., Robson, M., and Heggers, J., Can.
    J. Microbiol. 26:175, 1980.
19. Hau, T., Hoffman, R., and Simmons, R. L., Surgery 83:223, 1978.
20. Hau, T., and Simmons, R. L., Surgery 87:588, 1980.
21. Hau, T., Lee, J. T., and Simmons, R. L., Surgery 89:187, 1981.
22. Grover, G. J., and Loegering, D. J., Proc. Soc. Exp. Biol. Med.
    167:30, 1981.
23. Burgaleta, C., and Golde, D. W., in "Immune Modulation and
    Control of Neoplasia by Adjuvant Therapy" (M. A. Chirigos,
    ed.), p. 195, Raven Press, New York, 1978.

DISCUSSION

ADAMS: What happens to the E. coli in the peritoneal cavity?

DI LUZIO: As in the peripheral blood, there is a very significant reduction in number of E. coli in the peritoneal cavity. Glucan induces increased intracellular destruction of the bacteria.

UNKNOWN: Is this just an inflammatory response? Will non-active glucans protect?

DI LUZIO: The glucans that we used were all active in the sense that they are all macrophage activators.

UNKNOWN: What happens if you incubate glucan with the organisms?

DI LUZIO: Nothing at all; the reaction depends upon the host response.

UNKNOWN: Is there any LPS in the glucan?

DI LUZIO: No, none.

ACTIVATION OF MURINE MONONUCLEAR PHAGO-
CYTES FOR DESTROYING TUMOR CELLS: ANALYSIS
OF EFFECTOR MECHANISMS AND DEVELOPMENT

D.O. Adams and W. J. Johnson

Departments of Pathology and Microbiology-Immunology
Duke University Medical Center
Durham, North Carolina   27710 USA

INTRODUCTION

Over the past decade, the original concept of macrophage activation as development of the ability to destroy facultative or obligate
intracellular parasites has been broadened to include the destruction
of tumor cells (1). Macrophages from sites of infection with such
intracellular parasites, when cocultivated with tumor cells, efficiently destroy the tumor cells over several days (1). This form of
cytolysis, which has been termed macrophage-mediated tumor cytotoxicity, has generally been found to be selective for tumor cells, to
depend upon cell-to-cell contact, and to be independent of either
exogenous recognition factors or soluble lytic substances (2,3).
In the past two years, it has become apparent that macrophages can
also be activated for destroying tumor cells in another, quite distinct circumstance - the lysis of antibody-coated tumor targets (antibody-dependent cellular cytotoxicity or ADCC) (4,5). Considerable
evidence has been amassed that activation of macrophages is an important host defense against the development and spread of neoplasms in
vivo (1-3).

The mechanisms of target recognition and destruction in both
macrophage-mediated cytotoxicity and ADCC are currently the subject
of intense interest. The bases of target recognition and kill in
ADCC, a consideration of which is beyond the scope of this chapter
(for reviews, see 4,5) appear to be quite distinct from those in
macrophage-mediated cytolysis. Over the past 3 years, our laboratory
has been investigating macrophage-mediated tumor cytotoxicity. We
here briefly review the evidence indicating that such lysis involves,
at minimum, two properties of activated macrophages (a) the capacity
for augmented binding of tumor cells, and (b) the capacity for secret-

ing a novel cytolytic protease and the evidence that these capacities interact sequentially to complete tumor cytolysis. We then consider the development of these two capacities, in relation to the development of competence for completing tumor cytolysis.

TARGET BINDING

Experiments in many laboratories indicate macrophage-mediated cytotoxicity is target selective, since neoplastically transformed cells are generally lysed when their nontransformed counterparts are not (1-3,6). Most workers have found such cytolysis to be contact-dependent (1-3,6), and cine-microscopic studies have shown activated macrophages cluster about and attach to tumor cells that are to be lysed (7,8). After vigorous washes of such cultures, the tumor cells remain attached to the macrophages, so the two can be considered physically bound to one another. The extent of macrophage-target binding can be quantified by using labeled probe targets (9,10).

In our laboratory, murine macrophages activated for cytolysis demonstrate a high level of binding: 30 to 60% of added tumor cells become bound to activated macrophages (10). This augmented binding is highly selective for the combination of activated macrophages and neoplastic targets. By contrast, a low level binding (4-12% of targets bound) is observed between numerous cell pairs, such as activated macrophages plus lymphocytes; inflammatory macrophages plus tumor cells or lymphocytes; or adherent tumor cells or fibroblasts plus lymphocytes or nonadherent tumor cells. The two types of binding differ in several ways. First, the high level binding is saturable while the low-level is not (10). Second, the augmented binding, though not the low level binding, can be competitively inhibited by adding excess unlabeled targets (10). The augmented binding, once established, can be disrupted by adding a large excess of unlabeled targets. Third, augmented binding depends upon divalent cations while the low level binding does not (11). Fourth, augmented binding depends upon trypsin-sensitive structures on the surface of the activated macrophages while the low level binding does not (11). Fifth, tumor cells are bound in clusters of 3-5 to the central portion of activated macrophages, while single tumor cells are bound to the periphery of inflammatory macrophages (11). Sixth, the augmented binding of tumor cells or of plasma membranes from tumor cells to activated macrophages initiates secretion of a cytolytic protease (12). Seventh, the high level binding can be classified as mostly specific, while the low level binding can be classified as mostly non-specific, when binding in the presence of excess unlabeled targets is quantified (see 11 for details). Thus, the augmented binding of neoplastic cells to activated macrophages has several characteristics of the specific (high-affinity, low capacity) binding of ligands to receptors, while the binding of tumor cells to inflammatory macrophages has many characteristics of the nonspecific (low-

affinity, high capacity) binding observed between ligands and cells
or plastic (13).

The cell biology of augmented binding has been reviewed elsewhere (14,15) and will not be considered in this review.

The molecular basis of binding between neoplastic targets and
activated macrophages appears to depend upon structure(s) contained
within plasma membranes of neoplastic cells (16). Plasma membranes
from three neoplastic cells (a lymphoma, a leukemia, and a mastocytoma) have been found to bind to activated macrophages and inhibit
effectively the subsequent binding of either the homologous or heterologous neoplastic targets. Membranes from lymphocytes do not inhibit
augmented binding, and membranes from neoplastic cells do not inhibit
low level binding to inflammatory macrophages. These observations
suggest structures contained within and common to plasma membranes
of 3 neoplastic cells are responsible for the binding of these neoplastic cells to activated macrophages.

The augmented binding of tumor cells to activated macrophages
appears to be necessary for completion of macrophage-mediated tumor
cytotoxicity for several reasons. First, the selectivity of such
binding mimics the selectivity of cytolysis as to type of target and
type of macrophage (10). Second, increasing or decreasing binding
in various ways proportionately increases or decreases cytolysis
(10). Third, competitive inhibition of binding with unlabeled tumor
cells or with membrane preparations from tumor cells decreases cytolysis (10,16). Fourth, BCG-elicited macrophages from A/J mice
are not able to effect augmented binding and cannot complete macrophage-mediated cytotoxicity (17). Fifth, porous filters placed
between activated macrophages and tumor cells block binding and
cytolysis (18).

Augmented binding, however, is not sufficient for completion
of cytolysis. BCG-elicited macrophages from C3H/HeJ mice can effect
augmented binding but cannot mediate cytolysis (17). This finding
can also be demonstrated with macrophages which have been primed, in
that such macrophages bind neoplastic targets but do not lyse them
(18). Augmented binding thus appears to be an initial and necessary
but insufficient step toward completion of macrophage-mediated tumor
cytolysis.

TARGET CYTOLYSIS

The molecular basis of target injury in macrophage-mediated
tumor cytotoxicity remains to be established. Tumor cytolysis by
activated macrophages has generally been found to be contact dependent, and cytolytic substances have not generally been removed from
activated macrophages co-cultivated with tumor cells (see 6 and 19).

Macrophages are, however, known to secrete a wide variety of substances that can damage tumor cells, including hydrogen peroxide,
arginase, thymidine, prostaglandins, and various incompletely defined
factors (for review, see 6 and 19).

A novel serine protease secreted by activated macrophages, which
lyses neoplastic but not normal target cells, has been described in
this laboratory (19). Macrophages activated in a variety of ways
secrete this lytic substance, which we have termed cytolytic factor
(CF). The concentration of CF which can lyse one half of a population
of tumor target cells is about 1.0×10^{-9}M. Secretion of CF from
activated macrophages can be inhibited by treating the macrophages
with heat, sodium azide or cycloheximide. CF itself is inhibited
by fetal calf serum, heating the CF to 56°C for 30 minutes, or the
protease inhibitors - bovine pancreatic trypsin inhibitor (BPTI),
diisopropylfluorophosphate (DFP) and alpha-2-macroglobulin (12,19).
By contrast, arginine, catalase, supernatants of inflammatory macrophages, glucose, protein hydrolysate, or extensive dialysis against
fresh medium do not inhibit CF (19).

When supernatants of activated macrophages are separated by gel
filtration, lytic activity elutes in a fraction estimated to have a
molecular mass of about 40,000 daltons (19). The lytic fraction co-
chromatographs precisely with a novel neutral protease secreted only
by activated macrophages; DFP and BPTI block lytic activity of the
separated protease. Thus, CF appears to be a serine protease of
molecular weight about 40,000.

Cultures of macrophages, which are activated for tumor cytolysis,
secrete CF spontaneously (20). Macrophages elicited by various inflammatory stimuli do not secrete CF spontaneously but can do so when
pulsed with relatively large amounts of endotoxin (µg/ml). Resident
peritoneal macrophages do not secrete CF under either circumstance.
Thus, secretion of CF, like many other secretory products of macrophages, is regulated in a two step fashion: an initial priming signal
prepares the macrophage for secretion, while a second triggering
signal elicits actual release of CF (20).

Selective binding of tumor targets by activated macrophages induces increased secretion of CF, over basal levels (12). Only tumor
cells elicit enhanced secretion; and dead tumor cells or plasma membrane preparations are effective in this regard. Normal targets
cultured with activated macrophages or tumor targets cultured with
resident or inflammatory macrophages do not elicit augmented secretion
of CF. The secretion of CF triggered by target binding is rapid,
since it is detectable 30 minutes after addition of targets to activated macrophages.

Target effects of CF exhibit some degree of target selectivity
(19). A wide variety of tumor cells including leukemias, lymphomas,

sarcomas, carcinomas and a melanoma can be lysed by CF. On the other hand, the six normal target cells tested to date are not susceptible to CF. Cytolytic effects of CF are first observed after five to six hours and rise progressively until 16 to 20 hours, when maximal lysis is observed. The final degree of target destruction depends on the concentration of CF and can reach almost 100% kill. CF appears to interact directly with tumor targets, and this interaction has some of the characteristics of an enzyme-mediated reaction (21).

Several lines of evidence implicate CF in macrophage-mediated cytolysis. First, secretion of CF correlates closely with activation of macrophages for cytolysis in multiple circumstances (for review, see 21 and 14). Second, BCG-elicited macrophages from C3H/HeJ mice can bind tumor cells but cannot secrete CF or kill tumor cells (17). Third, BPTI and DFP, which inhibit CF, inhibit macrophage-mediated cytotoxicity (20). Furthermore, these protease inhibitors act only if present during the cocultivation of macrophages with tumor cells. Fourth, the course of target injury by CF (about 16-24 hours) is slow and quite similar to that of target injury produced by activated macrophages (21). Fifth, binding of neoplastic targets to activated macrophages induces release of CF (12).

These observations by no means imply CF is the sole mediator of macrophage-mediated cytolysis. It has been recently established that CF interacts synergistically with hydrogen peroxide to produce tumor cytolysis (22). Cooperation between various effector molecules may thus operate in macrophage-mediated tumor cytolysis, and, indeed, those molecules producing target injury may vary from target to target. Defining the precise effector substances which actually produce the lysis of neoplastic cells bound to macrophages is now the next step in unraveling the molecular basis of target injury.

A MODEL OF CYTOLYSIS AS A MULTISTEP EVENT

The initial evidence that macrophage-mediated tumor cytotoxicity is a multistep event was provided by cinemicroscopy (7,8). Activated macrophages cluster about tumor cells and generate prolonged periods of intimate contact, which are followed after several hours by lysis of the targets. More direct evidence on this point was obtained when activated macrophages were held in culture overnight (18). These macrophages could bind tumor cells effectively but could neither lyse them nor secrete CF. When such macrophages were pulsed with traces of endotoxin, both cytolysis and secretion of CF were restored. Of particular importance, the endotoxin was effective in inducing lysis only after completion of target binding. This endotoxin-responsive stage of the lytic cycle, though not the preceding stage of binding, is inhibitable by BPTI. These observations fit the data obtained in inbred mice indicating capacity for augmented binding and capacity for secretion of cytolytic factor are separate and independently

regulated capacities of activated macrophages, both of which are
necessary for completion of cytolysis.

Ultrastructural studies have shown the binding of tumor cells
to activated macrophages to create a narrow cleft (about 20 to 100
nm) between the two cells (10). Although CF is readily inhibited
by serum and by the large antiproteases (50,000-600,000 daltons)
contained within serum, recent evidence indicates binding of targets
to activated macrophages overcomes the inhibitory effects of serum
and permits cytolysis (18). We have speculated the cleft between
activated macrophages and targets is diffusion limited to macromolecules and blocks entry of serum antiproteases (14).

We have recently summarized these data into an integrated model
of macrophage-mediated tumor cytotoxicity (for review, see 14). In
this model, the initial step in macrophage-mediated tumor cytolysis
is the receptor-mediated recognition and binding of neoplastic targets. The augmented binding is followed by secretion of lytic substances, including CF, from the activated macrophage. In all likelihood, multiple lytic substances are involved in target destruction,
perhaps the effects of acting in combination with one another and
perhaps the combination depending on the particular target. The
cleft between activated macrophages and bound targets might represent
the principal site where lytic substances interacted with targets.
The relevance of this model has recently been shown by preliminary
studies on BCG-elicited macrophages from inbred mice (23). Cytolytic
competence can be induced in A/J macrophages by binding tumor targets
to the macrophage with multivalent concanavalin A, and in C3H/HeJ
macrophages by exposing the macrophages to an alternative second
signal that triggers release of CF.

INDUCTION OF ACTIVATION

Several laboratories have shown the activation of murine mononuclear phagocytes in vivo and in vitro proceeds by a defined sequence of stages (24,25). First, responsive macrophages (macrophages
from sites of sterile inflammation), though not resident peritoneal
macrophages, respond to lymphokines to become primed. Second, primed
macrophages, though not responsive macrophages, become activated for
cytolysis when exposed to ng of endotoxin. Third, activated macrophages mediate lysis of tumor cells. These stages of development,
which are thus operationally defined, are also pertinent to activation for tumor cytotoxicity in vivo and to activation for microbicidal function (26,27).

To characterize these stages of development, one could define
objective and quantitative markers, which distinguish macrophages
in the various stages. Cohn and colleagues have successfully used
this technique to distinguish macrophages resident in the peritoneal

cavity from macrophages taken from sites of sterile inflammation
(28). By analyzing macrophages in all of these operationally-defined
stages, we have found objective markers that characterize each of
these stages (Table 1) (29,30). To summarize the major findings,
responsive macrophages have several markers of the inflammatory phenotype. They also display two other such markers: decreased content
of 5' nucleotidase (<4.0 units/mg of macrophage protein) and extensive
phagocytosis via the C3 receptor (phagocytosis of >300 SRBC/100 macrophages) (compare with Ref. 28). Primed macrophages express these
inflammatory markers, have the capacity for augmented binding of
tumor cells, and have the capacity to secrete cytolytic factor when
pulsed with ng of endotoxin. Activated macrophages bear the markers
of inflammatory macrophages, effect augmented binding, and secrete
CF spontaneously. Other markers of activated macrophages are currently being described, including increased secretion of hydrogen
peroxide, increased content of leucine aminopeptidase, decreased
expression of Fc receptors, increased expression of Ia antigen, and
decreased expression of the mannose receptor (31-35).

   The essential features of our observations can be recapitulated
in vitro (36). Primed and fully activated macrophages can be induced
in vitro by pulsing responsive macrophages with lymphokines or with
lymphokines and then endotoxin respectively. Development of priming
for cytolysis coincides with development of capacity for augmented
binding (36). The kinetics, regulation, maintenance, and inductive
requirements for the lymphokine-mediated induction of augmented binding resemble closely those for the lymphokine-mediated induction of
priming. This suggests the lymphokine(s) which induces binding is
closely related to macrophage activating factor (MAF). Of note,
binding could readily be induced in C3H/HeJ mice which are resistant
to endotoxin, but binding could not be induced in A/J mice which are
resistant to lymphokine (36).

   It is of particular interest to relate these observations to
development of activation for cytolysis. Development of augmented
binding capacity is induced by exposure of responsive macrophages
to lymphokines; no other signal appears to be necessary (36). By
contrast, capacity for secretion of CF is regulated in two steps:
(a) an initial, priming step is necessary to prepare the macrophage
to secrete CF; and (b) a second triggering step is required to induce
actual release of the CF (29,30,38). Preparation for secretion
develops when macrophages are primed for cytolysis and is, thus, presumably lymphokine-mediated. Triggering actual release of CF requires a second signal, which can be supplied in a variety of ways
including traces of endotoxin or engagement of the receptor for
malelylated proteins (37). One would thus anticipate that macrophages could complete cytolysis when the capacity for binding and
the capacity for secreting CF were fully induced; i.e. when responsive macrophages were finally exposed to lymphokine and then a second
signal such as endotoxin. Since this is indeed the case, the de-

Table 1. Objective Markers of the Stages of Macrophage Activation

Property	Resident Macrophages	Responsive Macrophages	Primed Macrophages	Activated Macrophages
Spreading (% cells spread/hr)	5	72	68	74
Secretion of Plasminogen Activator (EU/106 macrophages/24 hr)	0	18	41	39
Phagocytosis of EIgG (CPM of 51Cr-SRBC/106 macrophages)	410	2600	3350	3200
Secretion of Cytolytic Factor (LU/106 macrophages/4 hr)	0	2	2	36
Secretion of Cytolytic Factor with LPS (10 ng/ml)	0	4	35	54
Binding of Tumor Cells (Targets bound/106 macrophages x 10^3)	4	11	42	43
Cytolysis of Tumor Cells (% net cytolysis)	0	0	4	56
Cytolysis of Tumor Cells with LPS (10 ng/ml)	0	0	48	69

Data are from Ref. 18,29,30,38. Table adapted from Ref. 29.

velopment of activation for cytolysis appears to consist of inducing
the capacity for augmented binding and the capacity for releasing CF.

CONCLUSIONS

We have analyzed the mechanisms operative in effecting macrophage-mediated tumor cytotoxicity, which appear to be divisible into
at least two steps: (a) an initial target recognition manifested
by augmented binding between activated macrophages and tumor cells;
and (b) secretion of lytic effector substances from the activated
macrophages, including secretion of a cytolytic proteinase. When
development of capacity for binding and of capacity for secreting
CF are examined, acquisition of these two capacities relates extremely closely to acquisition of competence of completing cytolysis.
Thus, those signals which induce activation for tumor cytotoxicity
appear to be the sum of those signals which induce capacity for binding and capacity for secreting CF.

This suggests a model of macrophage activation. If one considers
cytolysis of tumor cells in terms of systems analysis (for review,
see 6), a circuit can be drawn in which lysis of tumor cells is the
output (Figure 1A). Components of the circuit include capacity for
augmented binding and capacity for secreting CF. Lysis would be

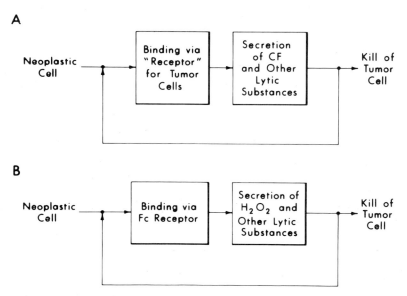

Fig. 1.   A systems model of macrophage function.   (A) Model of macrophage mediated cytolysis.   (B) Model of antibody dependent
cytotoxicity.

ccmpleted when all requisite components (capacities) were in place
(induced). Thus, the sum of signals leading to activation for cytolysis would equal the sum of all signals needed to induce the
requisite capacities. Systems analysis of cytolysis further distinguishes between complex functions (outputs of circuits) and less
complex capacities (components of circuits) (for a full discussion
of this point, see 6). Activation can thus be defined as competence
to complete a complex function. It should be emphasized that activation for one complex function (e.g., tumor cytolysis) will almost
surely be different from activation for another complex function
(e.g. ADCC), in reference to both requisite capacities and pathway
of induction (Figure 1B). Indeed, preliminary evidence indicates that
activation for ADCC requires different capacities and has a different
path of induction than activation for tumor cytotoxicity (23). This
model of activation for tumor cytotoxicity is being critically tested
in our laboratory. We anticipate the model can facilitate delineation
of the specific signals inducing various capacities and analysis of
the molecular bases of recognition and target lysis. Finally, it
will be interesting to see if this model will also pertain to activation for destruction of facultative and obligate intracellular parasites.

Acknowledgement

Supported in part by USPHS Grants CA 14236, CA 16784, and CA
29589.

REFERENCES

1. Hibbs, J. B., Remington, J. S., and Stewart, C. C., Pharmacol.
      Ther. 8:37, 1980.
2. Evans, R., and Alexander, P., in "Immunobiology of the Macro-
      phage", p. 536,(D. S. Nelson, ed.), Academic Press, New York,
      1976.
3. Keller, R., in "The Macrophage and Cancer" (K. James, B. McBride
      and A. Stewart, eds.), pp. 31-49, 1977.
4. Koren, H. S., and Adams, D. O.,in "Macrophage Mediated Antibody-
      Dependent Cellular Cytotoxicity" (H. S. Koren, ed.), Plenum
      Press, New York, in press.
5. Nathan, C. F., Fed. Proc., in press, April, 1982.
6. Adams, D. O., and Marino, P.,in "Contemporary Hematology-
      Oncology", vol. III, (A. S. Gordon, R. Silber and J. LoBue,
      eds.), Plenum Publishing Corp., New York, in preparation, 1982.
7. Meltzer, M. S., Tucker, R. W., and Breuer, A. C., Cell. Immunol.
      17:30, 1975.
8. Stewart, C., Adles, C., and Hibbs, J.B., in "The RES in Health
      and Disease", (H. Friedman, M. Escobar, and S. Reichard, eds.),
      Plenum Press, New York, 1975.
9. Piessens, W., Cell. Immunol. 35:303, 1978.

10. Marino, P. A., and Adams, D. O., Cell. Immunol. 54:11, 1980.
11. Somers, S. D., and Adams, D. O., manuscript in preparation, 1982.
12. Johnson, W. J., Whisnant, C. C., and Adams, D. O., J. Immunol. 127:1787, 1981.
13. Hollenberg, M. D., and Cuatrecasas, P., in "The Receptors", Vol. 1, General Principles and Procedures, Plenum Publishing Corp., New York, 1979.
14. Adams, D. O., Johnson, W. J., and Marino, P. A., Fed. Proc., in press, April, 1982.
15. Somers, S. D., and Adams, D. O., in "Natural Cell-Mediated Immunity. II" (R. Herberman, ed.), Academic Press, New York, in press, 1982.
16. Marino, P. A., Whisnant, C. C., and Adams, D. O., J. Exp. Med. 154:77, 1981.
17. Adams, D. O., Marino, P. A., and Meltzer, M. S., J. Immunol. 126:1843, 1981.
18. Adams, D. O., and Marino, P. A., J. Immunol. 126:981, 1981.
19. Adams, D. O., Kao, K. J., Farb, R., and Pizzo, S. V., J. Immunol. 124:293, 1980.
20. Adams, D. O., J. Immunol. 124:286, 1980.
21. Johnson, W. J., Weiel, J. E., and Adams, D. O., in "Natural Cell-Mediated Immunity. II." (R. Herberman, ed.), Academic Press, New York, in press, 1982.
22. Adams, D. O., Johnson, W. J., Fiorito, E., and Nathan, C. F., J. Immunol. 127:1973, 1981.
23. Johnson, W. J., and Adams, D. O., unpublished observations, 1982.
24. Hibbs, J. B., Taintor, R. R., Chapman, H. A., and Weinberg, J. B., Science 197:279, 1977.
25. Meltzer, M. S., Lymphokines 3:319, 1981.
26. Russell, S. W., Doe, W. F., and McIntosh, A. T., J. Exp. Med. 146:1151, 1977.
27. Buchmüller, Y., and Mauel, J., J. Exp. Med. 150:359, 1979.
28. Cohn, Z. A., J. Immunol. 121:813, 1978.
29. Adams, D. O., and Dean, J. H., in "Natural Cell-Mediated Immunity. II." (R. Herberman, ed.), Academic Press, New York, in press, 1982.
30. Adams, D. O., Johnson, W. J., and Marino, P. A., manuscript in preparation, 1982.
31, Nathan, C. F., and Root, R. K., J. Exp. Med. 146:1648, 1977.
32. Edelson, P., Lymphokines 3:57, 1981.
33. Ezekowitz, R. A. B., Austyn, J., Stahl, P. D., and Gomer, S., J. Exp. Med. 154:60, 1981.
34. Beller, D. I., Kiely, J., and Unkeless, E., J. Immunol. 124:1426, 1980.
35. Imber, M. J., Pizzo, S. V., Johnson, W. J., and Adams, D. O., J. Cell Biol., in press, 1982.
36. Marino, P., and Adams, D. O., J. Immunol., in press, 1982.
37. Johnson, W. J. Pizzo, S. V., Imber, M. J., and Adams, D. O., submitted for publication, 1982.

38.   Adams, D. O., Johnson, W. J., Marino, P., and Dean, J., sub-
      mitted for publication, 1982.

DISCUSSION

UNKNOWN: What is the nature of the molecules on the target cells
that are recognized and involved in the binding of macrophages and
targets?   Second, does binding of the macrophage to the target induce
cytostasis as opposed to cytolysis?

ADAMS: We do not know the recognition molecules, and we have not
yet examined cytostasis.

UNKNOWN: When you trigger cytolytic factor release by binding tumor
cells to activated macrophages, is there a concurrent release of
hydrogen peroxide?

ADAMS: We have attempted to measure hydrogen peroxide, but the tumor
cells and membrane preparations from the tumor cells both quench
the assay we use for hydrogen peroxide.

UNKNOWN: By using a glycolipid tehalose dimycolate extract from the
cell walls of Mycobacteria, we obtained cytostatic and even cytolytic macrophages which did not produce plasminogen activator.   Maybe
the pathway of response of macrophages exposed to lymphokine differs
from their response to purified bacterial products.

ADAMS: I suspect that this may be true.   In our laboratory at least,
we are using impure activating signals.   I suspect that specific
cellular functions will be affected differently as we use purer and
purer signals.

BELLER: There seems to be a number of similarities between activation
as assessed by tumoricidal capacity and as assessed by Ia induction.
One possible explanation may be the difficulty in physically separating macrophage activating factor (MAF) from macrophage Ia recruiting factor (MIRF).   Nevertheless, one difference between Ia
induction and tumoricidal activity seems to be that lymphokine alone
is sufficient to induce Ia expression whereas a second signal is
required in order to induce tumoricidal capacity.   One of the things
that second signals do in Ia expression, in concert with lymphokine,
is to accelerate dramatically the expression of Ia.   Since tumoricidal
experiments are normally done within 24 to 48 hours of isolating the
macrophage, is it possible that what you're really seeing is a difference in kinetics?   If you'd let the experiments go longer, you
might see that lymphokine actually did induce some detectable tumoricidal capacity.

ADAMS: We have examined macrophages for as much as 24 hours after

exposure to lymphokine and have seen no cytolytic activity. The dose
of lymphokine could be a problem as well.

BELLER: Let me point out that for resident peritoneal macrophages,
one needs to go out 6 or 7 days in order to see Ia induction with
lymphokine alone. With a second signal, you can see it in 3 to 4
days.

UNKNOWN: In regard to the specificity of binding, you showed several
targets that were cross competing. Do you find some heterogeneity
in target cell binding to macrophages?

ADAMS: We have now looked at approximately 10 non-adherent tumors,
and all bind to a high level. In contrast, we have examined 5 or 6
non-tumor targets and found that all of them bind to a low level.
These are not enough examples to make useful generalizations.

UNKNOWN: Do all the ones that bind cross compete with each other?

ADAMS: We don't know. We have cross-competed only 5 or 6 so we have
not made so large a checkerboard as one might wish.

RUCO: If you add endotoxin to BCG primed C3H/HeJ macrophages, do
they release cytotoxic factor? I would also like to know something
about the A/J defect and its relation to cytotoxic factor release?

ADAMS: With C3H/HeJ mice, we have had no success in inducing release
of lytic protease with endotoxin. We have gone as high as 10 micrograms of endotoxin per ml which is about 3 orders of magnitude greater
than is required in control mice. As for the defect in A/J mice,
they simply don't respond well to lymphokines. They do not have
augmented binding, and we cannot induce it with lymphokine in vitro.

RUCO: Since the BCG primed macrophages of C3H/HeJ mice can be induced
to kill with high doses of endotoxin, does this mean that there are
at least two different mechanisms of killing?

ADAMS: We didn't see killing, even at the doses we used. My analysis
is that these macrophages just do not see the second signal of endotoxin very well.

GORECKA-TISERA: You have suggested that fluidity of cell membranes
may be involved in cytolytic macrophage binding. I remember a paper
by Esser and Russell in which they reported no correlation between
membrane fluidity of cytolytic macrophages and their capacity to
kill neoplastic targets.

ADAMS: When Ralph Snyderman and I took macrophages and treated them
with concentrations of aliphatic alcohols sufficient to increase
membrane fluidity, we observed increases in target cell binding.

Therefore, we speculated that fluidity was necessary for binding.
I do not think the binding of tumor cells is at all the same thing
as the binding of a ligand. Some sort of physical interaction or
molding is involved in tumor cell binding. We have data indicating
that low level binding is quite weak, whereas high level binding
takes about 20 times more dynes of force to disrupt. We interpret
this data as indicating that simple attachment of a few ligands and
receptors may not be enough and that there must also be some membrane
molding, which would require membrane fluidity.

PRUZANSKI: Is it possible that neuraminidase, which is produced and
excreted by macrophages is chopping holes in the mucopolysaccharide
membrane of the tumor cells?

ADAMS: I am very open to many models of tumor cell killing.

LEONARD: One of the most interesting aspects of the activated macrophage is the fact that it's capacity to kill declines very rapidly
after about 12 hours in culture. Do you have any data yet on either
binding or protease production of activated macrophages subsequent
to 12 hours?

ADAMS: Yes. The basic observation is that binding capacity is maintained fairly well for at least 1 day, whereas production of pretease
declines very rapidly over 16 hours.

HUMAN MONOCYTE KILLING OF TUMOR CELLS:

CONTRIBUTION OF AN EXTRACELLULAR CYTOTOXIN

N. Matthews

Department of Medical Microbiology
Welsh National School of Medicine, Heath Park
Cardiff CF4 4XN, U.K.

INTRODUCTION

Cells of the macrophage series can be cytostatic or cytotoxic in vitro to certain tumor cell lines but they are usually without effect on non-transformed cells. For anti-tumor effects, macrophage-tumor cell contact is obligatory in certain systems but not in others where soluble mediators appear to be involved. Such mediators include arginase (1), the C3a complement component (2), cytolytic factor, a serine protease (3), hydrogen peroxide (4) and tumor necrosis factor (5,6,7). The contribution of each of these macromolecular mediators to the total anti-tumor effect of the macrophage is unclear, but could be resolved if neutralizing antisera to the mediators were available.

Most of the work on macrophage anti-tumor mediators has been confined to animal systems but recently we described the production of an anti-tumor cytotoxin by human monocytes. This cytotoxin is newly synthesized by stimulated monocytes, macrophages or myelo-monocytic leukemic cells. It is specific for certain tumor cell lines, has a molecular weight of 34,000 on gel-filtration and a slow electrophoretic mobility (8). Of the mediators described above the cytotoxin most closely resembles rabbit tumor necrosis factor. A number of agents can stimulate human monocytes to produce the cyto-toxin including endotoxin, zymosan, BCG, C. parvum, and pokeweed mitogen (9).

This paper describes further physicochemical properties of this human monocyte cytotoxin and in two experimental systems assesses the contribution of the cytotoxin to the total anti-tumor effect of the monocyte with the aid of an antiserum to the cytotoxin.

MATERIALS AND METHODS

   Cytotoxin Production: Monocytes were prepared from either
heparinized blood of laboratory staff or from fresh buffy coats from
blcod donors using Ficoll-Hypaque followed by plastic adherence (8).
After incubation overnight in medium containing 10% fetal calf serum
(FCS) and 10 µg/ml endotoxin, the cytotoxin-containing supernatant
was collected, centrifuged to remove cells and stored in small portions at $-70^\circ C$.

   Cytotoxin Assay: The mouse L929 tumor cell line was used as the
target cell. Seventy-five microliters of a target cell suspension
(3×10^5 cells/ml of MEM-FCS) were pipetted into 96-well microtiter
trays and incubated for at least 4 hours to allow the cells to adhere.
Dilutions (75 µl) of the monocyte supernatant in MEM-FCS with 2 µg/ml
actinomycin D were added to give 3 or 4 replicates/dilution. After
overnight incubation at $37^\circ C$, the supernatant containing the dead
cells was discarded and the adherent viable cells were fixed for
5 minutes with 5% formaldehyde and stained with crystal violet.
After drying, 100 µl of 33% acetic acid was added to each well to
dissolve and evenly distribute the dye. The amount of dye bound is
proportional to the number of viable cells and was quantitated photometrically using a Titertek Multiskan photometer. Reproducibility
was within the range 5-10%. The % cytotoxicity was calculated for
each supernatant dilution from the formula: 100 (a-b)/(a-c) where a,
b and c are the absorbance of wells with respectively L929 cells +
medium, L929 cells + monocyte supernatant and no cells. The titer
(defined as dilution causing 50% cytotoxicity) was then calculated
from the graph of cytotoxicity vs. $\log_{10}$ dilution using the least
squares method with the aid of a programmable calculator. The presence of actinomycin D in the medium increased the sensitivity of the
assay (8).

   Isoelectric Focusing: This was performed in polyacrylamide gel
with an LKB Ampholine PAG plate (pH range 3.5-9.5), Pharmacia stan-
dards and an LKB 2117 Multiphor tank. After focusing, the gel was
cut into 25 slices, each slice was disrupted with a rubber police-
man and the proteins were extracted by agitation at $4^\circ C$ overnight
in 1 ml PBS containing 0.025 M Hepes buffer.

   Antiserum Production: A female NZW rabbit was given 11 injections of partially purified cytotoxin over a period of 17 months.
The first 2 injections were given intradermally with Freund's complete
adjuvant, the next 3 injections were given intramuscularly with
Freund's incomplete adjuvant and subsequent injections were given
intravenously without adjuvant.

   Monocyte Anti-tumor Cytotoxicity: Monocyte enriched preparations
in 96 well microtiter trays (Sterilin M29 ARTL) were prepared by
incubation of 75 µl suspensions of mononuclear cells (prepared from

peripheral blood by centrifugation over Ficoll Hypaque) for 1½ hours at 37°C. The adherent monocytes were washed x3 with warm medium to remove non-adherent cells. Medium was completely aspirated from the wells and 150 µl of L929 or A549 tumor cell suspension (2.5×10^4/ ml) was added with 10 µl antiserum or medium as appropriate. Antiserum dilutions were expressed as the final dilution/160 µl culture volume and all tests were performed with 4 replicates. The plates were incubated for 3 days at 37°C and then processed as for the cytotoxin assay. The % cytotoxicity (for L929 cells) or % cytostasis (for A549 cells) was calculated from the formula: $100 (a-b)/(a-c)$ where a, b and c are the absorbances of wells with respectively target cells + medium (+ antiserum dilution as appropriate), target cells + monocyte suspension (+ antiserum dilution as appropriate) and no cells.

L929 is a mouse fibroblast line and A549 is a line derived from a human lung adenocarcinoma.

Enzyme Digestions: Supernatant containing the monocyte cytotoxin was diluted 1/10 with buffer containing 1 mg/ml trypsin (Sigma, bovine pancreas type I) or pronase (Sigma, Streptomyces caespitosus, type IV) and incubated overnight at 37°C. The positive control was supernatant diluted in buffer without enzyme and the negative control was the enzyme solution incubated with culture medium lacking cytotoxin activity. The buffers used were 46mM tris-HCl, 12mM $CaCl_2$ for trypsin and 80mM tris-HCl, 100mM $CaCl_2$ (pH 7.8) for pronase.

Effect of Enzyme Inhibitors and Other Agents: Monocyte supernatants were incubated with equal volumes of the enzyme inhibitors, Soybean trypsin inhibitor, 2 mg/ml (Sigma), Lima bean trypsin inhibitor, 2 mg/ml (Sigma), Trasylol, 10,000 KIU/ml (Bayer) or catalase, 4,400 U/ml (Sigma) for 1 hour at 37°C. The mixtures were then diluted in culture medium and tested for residual cytotoxin activity. The inhibitors and catalase were not toxic to the L929 target cells under the assay conditions.

RESULTS

Properties of the Monocyte Cytotoxin

Monocytes were incubated in medium containing 10% fetal calf serum and 10 µg/ml endotoxin and cytotoxin-containing supernatants were collected 20 hours later.

Previous work has shown that the monocyte cytotoxin has an apparent molecular weight of 34,000 on gel-filtration and a slow electrophoretic mobility at pH 8.6 in polyacrylamide gel (8). Consistent with its slow electrophoretic mobility, on ion-exchange chromatography using CM Sepharose, the cytotoxin was eluted at relatively high ionic strength along with the more basic serum proteins and on isoelectric focusing it had an isoelectric point in the range 6.0-6.5 (Figure 1). The heterogeneity apparent on isoelectric focusing has been noted repeatedly and merits further investigation.

Incubation of the cytotoxin for 24 hours with trypsin or pronase completely abolished activity as did heating for 20 minutes at 100° C (Table 1) suggesting that the cytotoxin is a protein. The cytotoxin was resistant to heating for 20 minutes at 56° C but lost 75% of its activity at 70° C (Table 1).

Soybean and lima bean trypsin inhibitors and Trasylol had minimal effect on the cytotoxin (Table 1) suggesting that it is not a serine protease. Cytotoxin activity was not abolished by catalase treatment, thus excluding a role for hydrogen peroxide.

Antiserum to the Monocyte Cytotoxin

A neutralizing antiserum to the cytotoxin was raised in a rabbit only after multiple injections of antigen over a period of 15 months. The capacity of the antiserum to neutralize the cytotoxin is shown in Figure 2.

It was equally effective at neutralizing monocyte cytotoxin induced by endotoxin, pokeweed mitogen or BCG (Figure 2) and in other

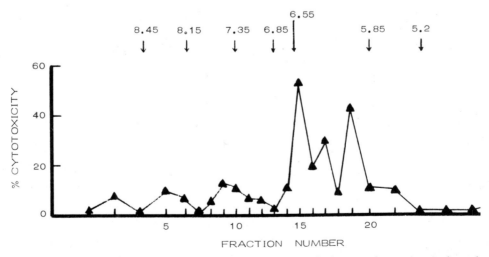

Fig. 1.   Fractionation of supernatant containing endotoxin-induced cytotoxin by isoelectric focusing.

Table 1. Effect of Various Treatments on Human Cytotoxin Activity

Treatment	Log_{10} Cytotoxin Titer ± Standard Deviation
Buffer alone	2.19±0.10
Trypsin	<0.7
Buffer alone	2.24±0.10
Pronase	<1.0
Medium alone	2.64±0.11
Soybean trypsin inhibitor	2.57±0.18
Lima Bean trypsin inhibitor	2.54±0.16
Trasylol	2.47±0.06
Catalase	2.55±0.15
Incubation 20 minutes at 20°C	2.36±0.07
56°C	2.34±0.10
70°C	1.76±0.15
100°C	<0.7

experiments C. parvum induced cytotoxin was inhibited to a comparable extent.

Effect of the Anti-Cytotoxin Serum on Monocyte Killing of Tumor Cells

The effect of the antiserum on human monocyte killing of tumor cells was investigated using two cell lines - L929 and A549. Human monocytes were cytotoxic to L929 cells with cytotoxicity being proportional to the effector:target ratio over the range 1:32 to 1:4 and then leveling off at ratios above 1:4. Figure 3 shows the inhibition of monocyte killing of L929 cells by the anti-cytotoxin serum. In this experiment the effector:target ratio was 1:1 but similar results have been noted at different ratios.

Human monocytes were less effective against A549 cells than L929 cells causing predominantly cytostasis which was proportional to the effector:target ratio over the range 1:1 to 8:1. To test the effect of the antiserum a monocyte:tumor cell ratio of 4:1 was used giving 41.5% cytostasis in the absence of antiserum. Cytostasis was not significantly different from this in the presence of antiserum dilutions ranging from 1/16 to 1/2048.

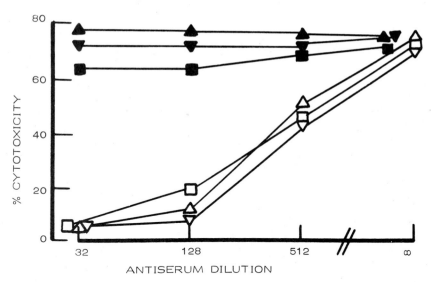

Fig. 2.   Neutralization of cytotoxin induced by endotoxin (∇), BCG
($\square$) or pokeweed mitogen ($\triangle$) by anti-cytotoxin serum (open
symbols) or control serum (filled symbols). Before testing
both antisera had been absorbed with mouse and human blood
cells and with insolubilized fetal calf and human sera.

DISCUSSION

     Human monocyte cytotoxin is a basic protein with an apparent
molecular weight of 34,000 and an isoelectric point in the range
6.0-6.5. It is probably not an arginase because of specificity differences (8). There is now some doubt about the ability of C3a to
damage cells (10) but in terms of molecular weight C3a is clearly
different from the monocyte cytotoxin. The cytolytic factor described
in murine systems has a similar molecular weight to the human monocyte
cytotoxin but differs in two other respects. Firstly, for production
of and assay of cytolytic factor a serum-free medium is required (3)
whereas production of monocyte cytotoxin is enhanced in the presence
of serum (8). Secondly, cytolytic factor is a serine protease and
can be inhibited by trypsin inhibitors of plant or animal origin
unlike the monocyte cytotoxin. The human monocyte cytotoxin is distinguishable from hydrogen peroxide on the basis of molecular weight
and insensitivity to catalase but does resemble rabbit tumor necrosis
factor in terms of specificity, molecular weight and mode of production (8). Although the isoelectric point of the human cytotoxin
differs appreciably from rabbit tumor necrosis factor (pH 5.15), there
are precedents for such large differences between species e.g. interleukin 1.

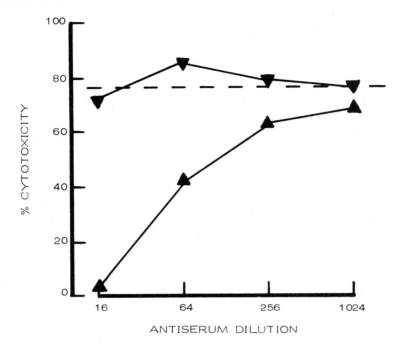

Fig. 3.   Effect of anti-cytotoxin serum (▲) or control serum (▼)
          on monocyte killing of L929 cells. The dashed line repre-
          sents the % cytotoxicity by monocytes in the absence of
          serum. Sera were absorbed as for Figure 2.

     The anti-cytotoxin serum raised against endotoxin-induced cyto-
toxin neutralized the cytotoxin induced by BCG, pokeweed mitogen and
C. parvum suggesting that these diverse agents induce the same cytotoxin.

     The antiserum has been used to prepare an immunoadsorbent for
purification of the cytotoxin (submitted for publication) and in
assessing the contribution of the cytotoxin to monocyte cytotoxicity
or cytostasis. The antiserum greatly reduced monocyte cytolysis of
L929 cells but was without effect when A549 cells were used as targets. This lack of inhibition in the L929 system was not simply due
to toxic effects on monocytes. Although the antiserum greatly reduced
the cytotoxicity against L929 targets we cannot be certain that the
antiserum did not react with other soluble mediators.

     Männel et al. (11) have shown that an antiserum against mouse tumor necrosis factor partially inhibits mouse macrophage cytotoxicity against the syngeneic 1023 tumor cell line. Comparable data from many more systems will be necessary before any generalizations can be made about the relative importance of the various mediators.

The monocyte cytotoxin is produced in maximal amounts when monocytes are incubated in the presence of serum and endotoxin or certain other inducers e.g. BCG, C. parvum, zymosan or pokeweed mitogen (9). Endotoxin is a contaminant of many batches of fetal calf serum but when care is taken to exclude endotoxin the cytotoxin is still produced but to a variable and lesser extent. In the human monocyte/L929 system, media with or without endotoxin support cytotoxicity to the same extent and it may be that the tumor cells themselves may act as cytotoxin inducers, as is the case with cytolytic factor (12). As well as differences in susceptibility to the various soluble mediators, tumor cells may also differ as inducers of the monocyte/macrophage mediators.

Acknowledgement

I thank the Cancer Research Campaign for financial support and Mrs. M. L. Neale for expert technical assistance.

REFERENCES

1. Currie, G. A. Nature 273:758, 1978.
2. Ferluga, J., Schorlemmer, H. U., Baptista, L. C., and Allison, A. C., Clin. Exp. Immunol. 31:512, 1978.
3. Adams, D. O., Kao, K. J., Farb, R., and Pizzo, S. V., J. Immunol. 124:293, 1980.
4. Nathan, C. F., Silverstein, J. C., Brukner, L. M., and Cohn, Z. A., J. Exp. Med. 149:84, 1979.
5. Männel, D. N., Moore, R. N., and Mergenhagen, S. E., Infect. Immun. 30:523, 1980.
6. Matthews, N., Br. J. Cancer 38:310, 1978,
7. Matthews, N., Br. J. Cancer 44:418, 1981.
8. Matthews, N., Immunology 44:135, 1981.
9. Matthews, N., Br. J. Cancer, in press.
10. Goodman, M. G., Weigle, W. O., and Hugli, T. E., Nature 283:78, 1980.
11. Männel, D. N., Falk, W., and Meltzer, M. S., Infect. Immun. 33: 156, 1981.
12. Johnson, W. J., Whisnant, C. C., and Adams, D. O., J. Immunol. 127:1788, 1981.

DISCUSSION

KIRCHNER: Would you reiterate why you feel that your cytotoxin is not interferon?

MATTHEWS: If it is interferon, it would have to be gamma interferon, because it's acid labile. But the charge on the cytotoxin is quite different from gamma interferon. So, I don't think it is gamma interferon.

KOLB: Is this cytotoxin which you obtain from human monocytes identical to the cytotoxin described for natural killer cells? The latter is a glycoprotein with terminal phosphorylated carbohydrates which bind to the mannose/fucosyl receptor on the target cell.

MATTHEWS: I tested one batch of our antiserum against the cytotoxin for inhibition of NK cytotoxicity, using K562 as the target cells. It inhibited about 100%. But when I looked closely at the preparation, the inhibition was due to contaminating immune complexes. So, I really don't know the answer yet.

KOLB: You have not tested mannose 6 phosphate as an inhibitor?

MATTHEWS: No.

MANTOVANI: Do all agents which augment the production of cytolytic factor such as C. parvum and Candida, increase the cytotoxicity of human monocytes?

MATTHEWS: I haven't looked at that. I've simply tested them to see if they will produce the cytotoxin.

MANTOVANI: Was it the unstimulated cytotoxicity of the blood monocytes which was blocked?

MATTHEWS: Yes.

THE EFFECT OF C. PARVUM THERAPY ON

INTRATUMORAL MACROPHAGE SUBPOPULATIONS

W. H. McBride and K. Moore

Department of Bacteriology
University of Edinburgh Medical School
Teviot Place, Edinburgh, Scotland

INTRODUCTION

Systemic administration of Corynebacterium parvum can cause partial or in a few instances complete (1,2) regression of subcutaneously growing transplanted tumors (3). It is reasonable to suggest that tumor regression of this kind is macrophage mediated in light of the knowledge that C. parvum injected by the same route stimulated the mononuclear phagocyte system into intense prolonged activity (3). In order to test this hypothesis, it seems best to examine the most relevant macrophage population which is within the tumor itself. Previous reports (4,5,6) had suggested that the intratumoral macrophages did not alter in number during C. parvum therapy. We therefore examined activation of Fc receptor expression. In addition to being a useful discriminating index, Fc receptors appear to equate with other indices of macrophage stimulation/activation (7,8,9).

We found that C. parvum therapy increases the level of activation of intratumoral macrophages as measured by an increase in Fc receptor expression and this increase is most associated with small macrophages.

MATERIALS AND METHODS

Most of the procedures have been described previously (1,2,9) and will be only mentioned briefly here.

C. parvum: 0.25 mg of C. parvum (Wellcome Foundation, Beckenham) was injected intravenously 4 days after tumor cell inoculation.

Mice and Tumors: The tumor used was a syngeneic C3Hf/Bu methyl-cholanthrene-induced fibrosarcoma (1) passaged 7-9 times.

Macrophages: Cell suspensions were prepared from tumors using 0.05% Dispase (Boehringer Corp., London) (9). Macrophages were adhered to plastic in the presence of 20% fetal calf serum (FCS) for 10 minutes (rapidly adherent) or for 30 minutes in the presence of Dispase and 20% FCS. The percentage of macrophages in cell suspensions was assessed by uptake of latex or sensitized calf red cells. The percentage of Fc receptor bearing cells was measured in suspension by rosetting using maximally sensitized calf red cells. Fc receptor avidity was measured by rosetting using calf red blood cells sensitized with different concentrations of rabbit anti calf red blood cell IgG. The end point, the EA_{50}, was the concentration of sensitizing IgG antibody that allowed 50% of the macrophages to form rosettes with 3 red cells (9).

RESULTS

C. parvum causes partial regression of our C3Hf/Bu fibrosarcoma when given intravenously 4 days after s.c. inoculation of tumor cells. The tumor contained approximately 20% macrophages. This percentage did not change with C. parvum therapy (data not shown; days 14-22 examined). However there was a subtle change following C. parvum treatment in that the Fc receptor avidity of the intratumoral macrophage populations was enhanced at all time periods examined in a repeatable fashion (Table 1). We had previously shown that high Fc

Table 1. The Effect of C. parvum on the Fc Receptor Avidity of Intratumoral Macrophage Subpopulations

Cells	EA_{50}
Control	2.15±0.09
C. parvum	2.4 ±0.06 (+84%)

The EA_{50} is the log mean concentration of IgG antibody/ml required to sensitize red blood cells so that 50% of rapidly adherent intratumoral macrophages form rosettes. Data are the mean of 5 separate experiments ± 1 standard deviation performed 14-22 days after tumor inoculation. C. parvum was administered on day +4. There were no significant alterations with time.

receptor avidity was the property of large stimulated macrophages
(1) and undertook studies to examine what subpopulation was affected
within the tumor. Tumor cell populations from control and C. parvum
treated mice were subjected to unit gravity velocity sedimentation.
The subpopulation of cells with the highest Fc receptors in the
control population sedimented at 6-8 mm/hr (Figure 1). The biggest

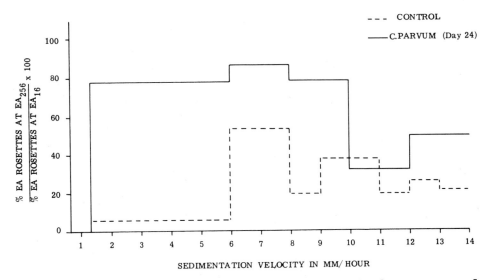

Fig. 1. Sedimentation of Fc receptor bearing cells from tumors of
control and C. parvum injected mice.

change following C. parvum administration was in the Fc receptor
avidity of slowly sedimenting (small) cells which increased dramatically. These experiments were performed with cells in suspension
but similar results were obtained with adherence macrophage monolayers (Table 2). Consistent increases in Fc receptor avidity were
only found in the slowly sedimenting population.

Table 2.   Fc Receptor Avidity of Tumoral Macrophage Subpopulations

Days After Tumor Injection		1–6mm/hour	6–9mm/hour
Day 14	Control	2.41	2.55
	C. parvum	2.50	2.47
Day 16	Control	2.11	2.33
	C. parvum	2.44	2.50
Day 21	Control	2.19	2.36
	C. parvum	2.33	2.52
Day 22	Control	2.05	2.09
	C. parvum	2.22	2.18

The EA_{50} value was determined as described in Table 1. C. parvum was given on day +4.

REFERENCES

1.   Milas, L., Hunter, N., Basic, I., and Withers, H. R., Cancer Res. 34:2470, 1974.
2.   Peters, L. J., McBride, W. H., Mason, K. A., and Milas, L., J. Reticuloendothelial Soc. 24:9, 1978.
3.   Milas, L., and Scott, M. T., Adv. Cancer Res. 27:257, 1977.
4.   Gebhardt, M. C., and Fisher, B., J. Natl. Cancer Inst. 62:1034, 1979.
5.   Moore, K., and McBride, W. H., submitted for publication.
6.   Thomson, A. W., Cruickshank, N., and Fowler, E. F., Br. J. Cancer 39:578, 1979.
7.   Rhodes, J., Bishop, M., and Benfield, J., Science 203:179, 1979.
8.   Rhodes, J., J. Immunol. 114:976, 1975.
9.   Moore, K., and McBride, W. H., Int. J. Cancer 26:609, 1980.

DISCUSSION

LAVIE: You used EA rosettes to determine the number of macrophages within the tumors. Now, several reports indicate many tumor cells also contain Fc receptors. Have you taken any precautions to insure that your tumor cells don't express an Fc receptor?

McBRIDE: We certainly don't pick them up in our assay if they do. We're dealing largely with adherent cells but we get the same kind of results in suspension. If you culture the tumor cells for any period of time, you get rid of the macrophages and the remaining tumor cells don't express any Fc receptors.

LAVIE: The expression of Fc receptors on tumor cells also varies if they grow <u>in</u> <u>vivo</u> versus <u>in</u> <u>vitro</u>, where they tend to lose this activity. So one really has to be careful with this kind of interpretation.

McBRIDE: I don't think we have that problem with these particular tumors. Certain tumors do appear to possess Fc receptors. I think it might depend on how you perform your assay.

UNKNOWN: Did you give <u>C. parvum</u> intratumorally or systemically?

McBRIDE: Systemically. We have not tried intratumoral injections.

UNKNOWN: When you inject <u>C. parvum</u> in the mice, do the numbers of macrophages in the tumor increase?

McBRIDE: The macrophage numbers do not change.

UNKNOWN: That's quite striking, for it seems that your system reproduces the conditions described by Normann or Synderman. There you have a growing tumor, and, upon inflammatory challenge, a reduced influx of macrophages, at least in the area of induced inflammation. Have you ever noticed anything like this?

McBRIDE: There are about the same number of macrophages within the tumors following <u>C. parvum</u> injection. However, we do see an increase in small macrophages with high Fc receptor avidity. If you look at the blood, you see no change in blood monocytes at day 14. But if you look earlier, up to about day 7, then you find in the monocyte population changes associated with Fc receptor avidity. You get a subpopulation of monocytes which are very small with high Fc receptor avidity.

UNKNOWN: Have you ever checked the number of macrophages in the peritoneal cavity?

McBRIDE: We give the <u>C. parvum</u> intravenously rather than intraperitoneally. If you give the <u>C. parvum</u> intraperitoneally, you get a big influx of cells with increased Fc receptor avidity. The interesting thing in that situation is that it's totally different from the tumor. With <u>C. parvum</u>, it is the large cells which are the ones which express enhanced Fc receptor avidity, not the small cells.

WEINER: Have you had an opportunity to look at cell cycle kinetics of your infiltrating monocytes and the effect of C. parvum on the cell cycle?

McBRIDE: We would like to do that. It's not very easy to do in the tumor system, but we probably could using the cell separation techniques that we have available.

TUMORICIDAL AND IMMUNOREGULATORY ACTIVITY

OF MACROPHAGES FROM HUMAN OVARIAN CARCINOMAS

Giuseppe Peri, Andrea Biondi, Barbara Bottazzi, Nadia Polentarutti, Claudio Bordignon and Alberto Mantovani

Istituto di Ricerche Farmacologiche "Mario Negri"
Via Eritrea 62
20157 Milan, Italy

INTRODUCTION

Macrophages are a major component of the lymphoreticular infiltrate of experimental and human tumors but their in vivo significance remains to be elucidated (1). Macrophages have diverse functions, including the capacity to interact directly with tumor cells and with lymphoid cells. The results summarized here were aimed at elucidating the tumoricidal and immunomodulatory capacity of diverse human macrophage populations including tumor-associated macrophages.

ANTIBODY DEPENDENT (ADCC) AND INDEPENDENT CYTOTOXICITY OF HUMAN MACROPHAGES

Human peripheral blood monocytes have appreciable natural cytotoxicity unrelated to endotoxin contamination of reagents (unpublished data), against susceptible targets such as TU5 (2). Direct cytotoxicity of blood monocytes can be modulated by in vitro exposure to interferons (IFN) and lymphokines (2). Using mononuclear phagocytes from diverse anatomical sites, we found considerable heterogeneity in the capacity of these cells to mediate direct (TU5) or antibody-dependent (TLX9) cytotoxicity (Table 1 and 2). For instance, milk macrophages were totally unresponsive to IFN and lymphokines but mediated ADCC as efficiently as other mononuclear phagocyte populations (Table 2).

Table 1.   Expression of Direct Tumoricidal Activity and ADCC
           by Different Human Mononuclear Phagocyte Populations

| Effector Cells | ADCC | Direct Cytotoxicity | | |
		Natural	IFN Boosted	Lymphokine Boosted
Monocytes	+	+	+	+
Peritoneal macrophages	+	+	+	+
Alveolar macrophages	+	-	-	+
Milk macrophages	+	±	-	-

Table 2.   Tumoricidal Activity of Human Milk Macrophages

Target Cells	Effector Cells	IFN (Units/ml)	Anti-TLX9 Antibody	Specific Lysis (%)
TU5	monocytes	-	-	14
		1000	-	57
	milk macrophages	-	-	17
		1000	-	15
TLX9	monocytes	-	-	2
		-	1/10,000	60
	milk macrophages	-	-	4
		-	1/10,000	58

MODULATION OF NATURAL KILLER (NK) ACTIVITY BY VARIOUS MACROPHAGE
POPULATIONS

When we examined the effect of various human macrophage populations on NK activity, we found considerable heterogeneity. As shown
in Figure 1, unlike blood monocytes, bronchoalveolar macrophages
caused a profound, dose-dependent inhibition of NK activity, which
was observed at suppressor/effector (S/E) ratios as low as 0.125:1.
In contrast, blood monocytes and macrophages from peritoneal exudates,
milk, or ascites ovarian tumors lacked consistent suppressive activity
up to ratios of 2:1.

Low levels of NK activity are associated with the human pulmonary
tissue (3). Depletion of macrophages by adherence significantly enhanced NK activity and, to an even greater extent, augmented responsiveness of lung cells to IFN (3). Therefore, the suppressive capacity of alveolar macrophages could play a role in determining the
defective NK cytotoxic capacity in human lungs.

Pulmonary alveolar macrophages share many properties with mononuclear phagocytes from other tissues, but have distinctive struc-

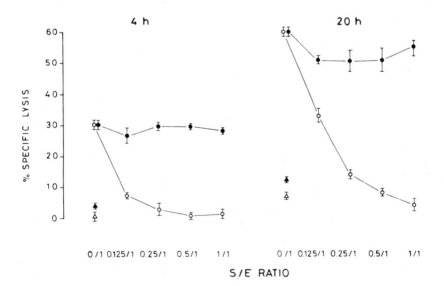

Fig. 1. Effect of broncho-alveolar macrophages and blood monocytes
         on NK activity. The E/T ratio was 25/1. Peripheral blood
         lymphocytes containing NK activity were cultured with bron-
         cho-alveolar macrophages (open circles) or with blood mono-
         cytes (closed circles). S/E = suppressor to effector ratio.

tural, metabolic and functional characteristics. Alveolar macrophages were unique, among the human mononuclear phagocyte populations
tested, in their capacity to inhibit NK activity effectively at low
concentrations. This finding emphasizes the heterogeneity in this
respect within the human monocyte-macrophage lineage and suggests
the existence of organ related differences in the mechanisms of
regulation of NK cytotoxicity. Alveolar macrophages should provide
a useful tool to analyze macrophage-mediated regulation of NK activity
in humans.

HUMAN OVARIAN CANCER-ASSOCIATED MACROPHAGES

     There are indications that mononuclear phagocytes may enter
tumors as a consequence of specific immune responses (1), but data
suggest that this may not be the sole or most important mechanism of
regulation of macrophage levels in neoplasms (4). We are currently
examining the possibility that tumor-derived chemoattractants play
a role in determining the entry of mononuclear phagocytes into tumors.
As illustrated by the experiment shown in Table 3, supernatants from
a human sarcoma line (8387) have chemotactic activity on human monocytes. Chemotactic activity was also found in supernatants from
human melanomas (1080/80 and SKBR-3), whereas the K562 line and the

Table 3. Chemotactic Activity of Supernatants from a Human
          Sarcoma Line (8387)

Chemoattractant	Number of Macrophages in 10 Fields
Medium	202±9
Supernatant undil.	438±7
1/3	351±27
1/9	310±29
Lymphokine	436±43

Chemotaxis of human monocytes was measured in modified Boyden
chambers. A PHA-elicited lymphokine was used as positive control.

Raji and CEM lymphoma produced little chemotactic activity. Various murine sarcomas, transplanted or primary, also had chemotactic activity, as measured either in culture supernatants or in tissue homogenates. Preliminary observations suggest that there may be some gross relationship between production of chemoattractants and macrophage concentrations in murine sarcomas, measured as previously described (5). Hence we suggest that production of chemotactic factors may be one of the mechanisms determining macrophage levels in tumors.

We have investigated the functional status, in terms of tumoricidal and immunomodulatory activity, of macrophages from ascites or solid ovarian carcinomas. Tumor associated macrophages did not appreciably inhibit the expression of NK activity when this parameter was tested immediately after mixture of lymphoid cells and macrophages (6). In contrast, in 5/9 patients tumor associated macrophages cultured overnight with blood lymphocytes caused some inhibition of NK lysis. However, in this regard, it must be stressed that the major mechanism responsible for depressed NK activity in ovarian carcinomas is probably a low frequency of large granular lymphocytes, the effectors of this reactivity (7). The suppressive activity of tumor associated macrophages on lymphoproliferative responses is being currently investigated. Preliminary data (Table 4) suggest that tumor associated macrophages may have enhanced suppressive activity compared to blood monocytes from the same subjects.

The tumoricidal activity and responsiveness to IFN and lymphokines of macrophages from ascitic carcinomas was similar to that of control peritoneal exudate macrophages (7,8). In contrast, as shown in Figures 2 and 3, preliminary observations suggests that tumor associated macrophages of solid tumors may have defective responsiveness to stimuli.

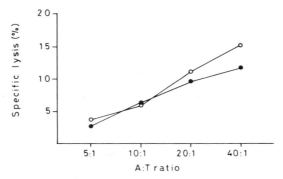

Fig. 2.   Unstimulated cytotoxicity of tumor associated macrophages from ascites (open circles) and solid tumor (closed circles). Patient TU, target cells were TU5.

Table 4.   Suppressive Activity of Blood Monocytes and Tumor
           Associated Macrophages from an Ovarian Cancer Patient
           on Proliferative Response of PHA Stimulated Lymphocytes

Number Suppressor Cells Added	Suppressor Cells	
	Monocytes	Tumor Associated Macrophages
None	59,000	72,000
1×10^4	68,000	52,000
2×10^4	68,000	3,000
5×10^4	68,000	2,000

Data are reported in counts per minute. Blood lymphocytes (1×10^5/
well) were stimulated by 0.2% PHA for 72 hours.

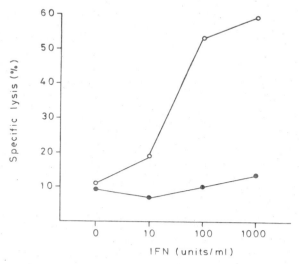

Fig. 3.   Responsiveness to interferon of tumor associated macrophages
          from ascites (open circles) or solid ovarian tumor (closed
          circles).

CONCLUDING REMARKS

The ongoing studies briefly summarized here suggest that human mononuclear phagocytes may be considerably heterogeneous in the expression of tumoricidal activity and in the capacity to modulate lymphoid cell functions. Characterization of the biological properties of the macrophage population(s) present at the tumor site may provide information concerning the role of these cells in the regulation of tumor growth and metastasis.

Acknowledgement

This work was supported by Grant R01 CA 26824 from the National Cancer Institute, USA, by Contract 80.01579.96 from CNR, Rome, Italy, and by a generous contribution of the Italian Association against Cancer.

REFERENCES

1. Evans, R., J. Reticuloendothelial Soc. 26:427, 1979.
2. Mantovani, A., Peri, G., Polentarutti, N., Allavena, P., Bordignon, C., Sessa, C., and Mangioni, C., in "Natural Cell-Mediated Immunity Against Tumors" (R. N. Herberman, ed.), pp. 1271-1293, Academic Press, New York, 1980.
3. Bordignon, C., Villa, F., Vecchi, A., Giavazzi, R., Introna, M., Avallone, R., and Mantovani, A., Clin. Exp. Immunol., in press.
4. Evans, R., and Lawler, E. M., Int. J. Cancer 26:831, 1980.
5. Mantovani, A., Int. J. Cancer 27:221, 1981.
6. Allavena, P., Introna, M., Mangioni, C., and Mantovani, A., J. Natl. Cancer Inst. 67:319, 1981.
7. Introna, M., Allavena, P., Acaro, R., Colombo, M., Molina, P., and Mantovani, A., in "Natural Cell-Mediated Cytotoxicity" (R. B. Herberman, ed.), Academic Press, New York, in press.
8. Peri, G., Polentarutti, N., Sessa, C., Mangioni, C., and Mantovani, A., Int. J. Cancer 28:143, 1981.

DISCUSSION

SWINBURNE: Have you tested human macrophages, obtained from ovarian ascitic and solid tumors, against human ovarian tumor cells?

MANTOVANI: In these studies and our previous studies, we were concerned only with baseline cytotoxicity, and we used human tumor targets. We have also used TU5 murine targets as a standard way to quantitate cytotoxicity. As for human targets, we have been using primary ovarian carcinoma cultures. The results can be summarized as follows: a) there is killing of ovarian carcinoma targets by autologous normal monocytes or activated macrophages as well as by allogeneic normal or activated macrophages; b) there is great variability in fresh primary cultures of ovarian carcinomas in terms of their susceptibility and we even have a few primary cultures which are apparently resistant to macrophage killing.

SECTION 11

NATURAL RESISTANCE MECHANISMS

This symposium emphasized free and prolonged discussion with each of 5 segments or topics introduced by an invited discussion introducer. The introducer's presentations are reproduced here either as they were delivered or in the form of a prepared paper. These presentations served as the stimulus for a general discussion which was recorded and now transcribed as part of this volume.

IN MEMORIUM

Dr. Cudkowicz organized this symposium on Natural Resistance Mechanisms. Shortly after the symposium concluded, Dr. Cudkowicz underwent treatment for cancer from which he did not recover. This symposium is dedicated to him by his friends and colleagues throughout the scientific community.

Dr. Cudkowicz was well known and respected for his work on natural resistance to foreign hemopoietic stem cell grafts. He demonstrated that transplantation of non-syngeneic marrow to irradiated recipients was often rejected and suggested that the genetics and immunobiology of hemopoietic transplant rejection was different from that of kidney, skin, or carcinoma. Natural resistance implies immunity in the absence of overt stimulation and the study of the phenomena with marrow grafts led to the awareness that a fundamental system of natural resistance exists to microorganisms and tumors as well. Dr. Cudkowicz's work was his life. His research was imaginative and his contributions helped establish the field of natural resistance as a significant new development in immunobiology.

NATURAL RESISTANCE MECHANISMS:

INTRODUCTION

Gustavo Cudkowicz

Department of Pathology
State University of New York
Buffalo, New York USA

This symposium entitled "Natural Resistance Mechanisms" is meant to focus on immunity conferred by cells that don't require antigen specific induction for acquisition of their function. Natural resistance also implies a degree of immunity demonstrable in the absence of overt stimulation, although modification of activity can be induced either by the host in response to challenge or by iatrogenic intervention. One class of cells possessing these properties are natural killer (NK) cells and there has been a lot of interest lately in their origin, function, and regulation. While we will discuss these cells at length, our focus should not be exclusively upon NK cells. Other cell types also are involved in natural resistance of which the monocyte/macrophage is a prime example.

We define the cells of natural resistance in an empirical manner. Most of us use tumor targets and define the effector cells by the damage they render to the membranes of the tumors. Some of us use virus infected cells or other microbial systems but again we define the effector cells operationally by the kind of injury or damage that is inflicted upon the target cells. Unavoidably, we tend to think that NK cells possess the function of killing tumors simply because that is the way we detect them. This may well be true, but I remind you that it is far from being proven.

When we work with virus infected cells and observe that NK cells contribute to controlling the spread of infection, again we are tempted to view the effector cells in terms of a very specific objective: namely, controlling infection. All of us would like this to be the case, but my personal feeling is that we are very much persuaded to think this way simply because of the assays we are using.

I would remind you also that we should consider this family of effector cells as being auto-reactive cells. Cells conferring resistance are capable of interacting with other cells within the organism and such interactions need not necessarily lead to cell death or killing. Accordingly, the major function of these cells in vivo may not be cell killing but rather the regulation of the lymphoid-myeloid system by cell-cell interactions or secretion of factors. This concept implies an important function apart from resistance to cancer or infection and underscores a potential homeostatic role for natural resistance systems in the regulation of hemopoietic proliferation and differentiation.

NATURALLY OCCURRING CYTOTOXIC CELLS

Gert Riethmüller, J. Lohmeyer and E. P. Rieber

Institut fur Immunologie der Universitat
München, Germany

SUMMARY

From its beginning, the field of natural cytotoxicity has been
beset by two major problems which have a direct bearing on the continuing quest for the in vivo relevance of spontaneous cytocidal
mechanisms. First, the remarkable phenotypic heterogeneity of NK
cells has remained until recently an intriguing open question. The
second problem concerns the as yet unknown recognition mechanisms
and the undefined membrane structures on the target cells. Many
studies have been undertaken to define the relationship of NK cells
to the three main classes of lymphoid cells: macrophages, T cells,
and B cells. This problem has become amenable recently with the
help of monoclonal antibodies and by cloning of effector cells. The
designation of NK cells as null cells has been revised for several
species. Thus, in the mouse promonocytes have been characterized
as NK cells with a defined spectrum of susceptible target cells.
However, other findings point to a considerable heterogeneity which
shows that NK cells express Thy-1 antigens as well as Ly related
antigens characteristic for the T cell lineage. Expression of the
Fc receptor remains controversal since Fc receptors may be absent
on some NK cells depending on their organ origin. In man, large
granular lymphocytes (LGL), as a characteristic minor subpopulation,
seem to account for all of the NK activity. Analysis with monoclonal
antibodies revealed that malignant cells as well as normal cells
which resemble LGL may express a hybrid phenotype; that is, they
may show a simultaneous expression of thymus-associated and monomyelo-
cytic-associated antigens. This finding on human cells may shed light
on the unusual heterogeneity observed in mice where NK cells with
monocytic and thymic phenotypes have been discerned.

No conclusive evidence has so far been provided for the in vivo operation of these cells. Cloning of effector cells, though fashion-able and widely applied, has not yet resolved the Chamaeleon-like phenotype. No conclusion can be reached on the basis of cloned cells as to the problem of clonal distribution of NK-like specificity. Phenotypic analysis by monoclonal antibodies as well as the examina-tion of the specificity pattern of cloned cells is expected to yield some insight into these problems.

REFERENCES

1. Schlimok, G., Thiel, E., Rieber, E. P., Huhn, D., Feucht, H., Lohmeyer, J., Riethmüller, G., Chronic leukemia with a hybrid surface phenotype (T lymphocytic/myelomonocytic): leukemic cells displaying natural killer activity and antibody dependent cellular cytotoxicity, Blood, 1982, in press.
2. Lohmeyer, J., Rieber, P., Feucht, H., Johnson, J., Hadam, M., and Riethmüller, G., A subset of human natural killer cells iso-lated and characterized by monoclonal antibodies, Eur. J. Immunol. 11:997-1001, 1981.

DISCUSSION INTRODUCER GERT RIETHMÜLLER: It comes as no surprise that NK cells were defined originally as non-induced, non memorizing tumoricidal cells and were looked upon as a special type of phagocyte. However, NK cells are non-adherent and non-phagocytic. Therefore the question arises as to the origin and nature of this cell. The quest for genealogy of NK cells is not trivial, particularly if one assumes that profound genomic rearrangements may take place as is the case in B and T cell development.

From the very beginning, a number of characteristics spoke in favor of a close relationship between NK cells and T cells. In the mouse, it is evident that NK cells share a number of surface antigens with T cells including Thy-1, Qa-4 and Qa-5. In addition, the new marker NK-1 does not overlap with monocytes although it is also not present on mature T or B cells. Since most of these surface antigens are not directly associated with classical T cell function, their presence did not seem to be stringent proof for T or a pre-T cell origin of NK cells.

In man, we in fact have at least one identifiable NK entity: namely, the large granular lymphocyte (LGL). It is astonishing that NK cells have been known for about 8 years yet little is known about spontaneous tumors arising from these cells. We have looked for tumors in man with NK activity and I would like to discuss briefly our findings on one such human NK type tumor.

After having screened several leukemias, we came across a patient with chronic lymphatic leukemia of T cell type whose tumor had a remarkable morphologic resemblance to LGL. LGL are believed responsible for a major part of NK activity in man. In our patient, we found about 470,000 cells/mm^3 and the majority were medium to large lymphocytes. We determined early that these cells were positive for sheep red blood cell receptor (ER) and positive for Fc gamma but not for Fc mu receptor. The tumor cells were active in antibody dependent cytotoxicity and killed directly three different tumor targets, including MOLT-4 and adherent melanoma cells. The tumor cells had no effect, however, on B cell differentiation. We characterized the cells using either commercially available monoclonal antibody OKT4 or our own monoclonals designated T811 and M522 (Table 1). We detected three antigens on tumor cells: a pan T antigen that is also expressed on most ER positive T lymphocytes and 20% of thymocytes; a second antigen (T811) which is found on 20% of peripheral blood lymphocytes and 90% of the thymus cells; and a third antigen (M522) which also is expressed on monocytes and granulocytes. Antigen T811 is similar in structure to OKT8, being a 70,000 molecular weight entity which can be split into two subunits of 32,000 and 34,000 daltons. Antigen M522 has a MW around 200,000. We next phenotyped other chronic lymphatic leukemias and T cell lines using the monoclonal antibodies for antigen T811 and M522. Only the tumor cells used for immunization reacted with the T811 (OKT8 type) and the M522 reagents, all the other tumor cells being negative. So we detected on the tumor one antigen which characterizes immature T cells or thymocytes and presumably cytotoxic suppressor T cells and another (M522) which characterizes mostly monocytes-granulocytes.

The next question was whether or not such double marked cells exist in the normal lymphocyte population. This question was approached in two different ways: double immunofluorescence with monoclonals M522 and T811 and by autoradiography using I-125 labeled T811, combined with rosetting of M522 coated erythrocytes (Table 2). By this approach, we found that about 4% of the peripheral leukocyte population carries both markers and the major part of these cells (70%) also bear the ER or the Fc gamma markers. On the other hand, 28% of the M522 positive cells carry T811 antigen whereas 15% of T811 positive cells, presumably T cells, carry the monocytic related antigen M522. The simultaneous expression of T and M antigens has been suggested by others using the OKM and OKT monoclonal antibodies.

We next asked whether or not these double marker cells possessed NK like functions. We purified the cells using rosette fractionation. When we removed M522 positive cells from non-adherent lymphocytes, we completely removed NK activity (Table 3). When we tested the positively enriched cells, lytic activity was increased against K562. This is also true if we tested M522 positive cells which are T811 negative (about 70%). Although the latter cells are competent, the cells which are positive for both markers had the highest NK activity.

Table 1. Membrane Phenotype of the Cytotoxic T Cells of
         Patient Si

Marker	Cytotoxic T Cells %	Normal Range of Pheripheral Blood Lymphocytes[a] %
OKT3	84	76±7
OKT4	1	45±6
OKT8	95	22±7
OKM1	96	21±6
T411	96	62±8
T811	98	20±10
M522	95	35±8
T811/M522[b]	96	4±2
Anti-Ia	<1	12±7
Anti-Ig Poly[c]	<1	12±4
Rosette Formation		
E$_{AET}$	98	74±12
Fc-IgG Receptor	75	24±7
Fc-IgM Receptor	<1	52±7
C$_{3b}$ Receptor	<1	10±4
C$_{3d}$ Receptor	1	14±5

[a]Normal peripheral blood lymphocyte values were determined on
 healthy donors and are reported as mean ± SD.
[b]Simultaneous detection of T811 and M522 antigen by use of FITC
 conjugated M522 and TRITC labeled T811 antibody.
[c]Besides polyvalent anti-Ig reagent, light (K and L) and heavy
 (G, M, D) chain specific antisera were used; all antisera gave
 negative results.

Table 2. Simultaneous Detection of Different Cell Surface Markers on Normal Human Non-adherent Lymphocytes

Cells Analyzed	Antigen T811	Antigen M522	Antigens T811 + M522	Sheep Red Blood Cell Receptor (ER)	Fc Gamma Receptor
Non-adherent lymphocytes	24±3	11±2	4±2	70±5	32±5
Stained by T811	-	15±2[a]	-	98±2	19±3
Stained by M522	28±5[a]	-	-	30±8	60±10
Stained by T811 and M522	-	-	-	85±5	91±2

[a] Cells bearing the additional marker as percentage of cells analyzed (mean ± SD of 4 experiments; 2000 cells evaluated per experiment).

Table 3. Natural Cytotoxicity Against K562 of Human Lymphocyte Subpopulations Isolated by Monoclonal Antibodies

Population	Recovery %	Surface Markers		Cytotoxicity Against K562 Lu33/10^7 Cells
		M522	T811	
Non-adherent lymphocytes	100	15	20	49
Separation control	68	12	22	34
M522 negative cells	50	<1	18	<1
M522 positive cells	11	97	35	268
M522 positive, T811 negative	5	98	1	200
M522 positive, T811 positive	2	95	95	282

Surface markers are reported in % of positive cells by immunofluorescence.

An important question is whether the double marker phenotype reflects an independent differentiation pathway or is a transitional phenomenon. The question of delineation of this cell lineage and its relationship to monocytes will be discussed by Dr. Rumpold from Vienna.

RUMPOLD: We studied NK cells in man in two ways. Firstly, we attempted to define surface antigens using monoclonal antibodies directed against lymphocytic or myelo-monocytic antigens and, secondly, we tried to raise monoclonal antibodies by immunizing mice with LGL. The monoclonal antibodies VEP8 raised against granulocytes and bone marrow precursors of granulocytes reacted with granulocytes and bone marrow precursor cells and VEP9 raised with monocytes reacted with monocytes, but not with NK cells. If we tested the antilymphocyte antibodies, (Leu 2a, OKT8, Lyt 3) a reactivity with LGL was observed. We confirmed the expression of OKM1 on NK cells. If cell loss is taken into account, however, the total lytic capacity was reduced. We obtained two monoclonals by immunizing mice with LGL. Both antibodies reacted with LGL and in addition with other cell types. The VEP10 antibody reacted with thymocytes presumably with an antigenic determinant similar to that defined by OKT10. The VEP13 antibody reacted with LGL and also with granulocytes but not with an immature myelo-monocytic cell nor with lymphocytes other than LGL. It is strange that many people have tried to immunize with LGL or LGL-type leukemias yet nobody has obtained a monoclonal antibody reacting selectively with LGL. The only one described is HNK-1 obtained by immunization with a cell line.

RIETHMÜLLER: Would Hillel Koren give his views on this phenotype in man?

KOREN: There are 3 questions which we addressed. (a) Can we identify monoclonal antibodies which support a monocytic or lymphocytic origin of NK cells? (b) Does treatment with interferon evoke phenotypic changes in LGL? (c) Do monoclonals reacting with LGL interfere with the binding or killing steps of the effector-target interaction? By fluorescence analysis, a variety of T cell monoclonal antibodies, especially the 3A1 which is a pan T antibody, reacted with LGL purified on Percoll gradients. In addition, the 4F2 which is Bart Haynes' antimonocyte antibody and the OKM1 reacted with a high percentage of LGL. L243 and OK11 both recognizing Ia antigens had almost no activity against LGL. Using the 5E9 monoclonal antibody (anti-transferin receptor) no positively stained cells were detected. Interestingly, LGL stimulated for 18 hours with alpha interferon sufficient to detect increasing NK activity did not change the phenotypic expression for any of the monoclonal antibodies studied. The transferin receptor often associated with activation was also absent on the interferon activated cells. FACS analysis showed that the staining intensity was also unchanged.

To answer the last question, some of the monoclonal antibodies

were added to the conjugate assay using K562 as target cells. OKT10
present on the majority of LGL was the only antibody which inhibited
the killing (50 to 90%) without affecting binding at the single cell
level. 4F2 which detects nearly 100% of the LGL caused neither inhibition of binding nor killing. The results suggest that the OKT10
interferes with a lytic step possibly with a lytic molecule involved
in the killing of K562 targets by LGL.

RIETHMÜLLER: Does the negativity for the transferrin receptor mean
that it is a non proliferating cell? Most proliferating cells express
the transferrin receptor.

KOREN: I'm not sure that conclusion is right because we know you
can induce LGL to proliferate. However, we have not actually looked
for LGL proliferation. It could well be that interferon stimulates
the cells without turning on DNA synthesis. While interferon increases the expression of H-2 and HLA antigens, it does not seem to
increase the surface antigens detected by our monoclonal antibodies.

RIETHMÜLLER: Vose has cloned T cells and LGL and could he tell us
about their proliferating properties?

VOSE: We attempted to induce LGL to proliferate by culturing them
in the presence of lectin free interleukin 2. We isolated LGL and
removed the subpopulation which forms SRBC rosettes at $29^{\circ}C$.
By morphology, the resulting population was more than 98% LGL. We
then plated out the cells under limiting dilution conditions. We
measured proliferation at day 10 by incorporation of tritiated thymidine. Under mitogen free conditions, something like 1% of the
LGL proliferate. By adding PHA at 1 μg/ml, the proliferative frequency of T cells was up to 10%. We measured also the cytotoxic
cell frequency by adding chromium labeled K562 cells to the limiting
dilution assay and found no significant difference between the number
of cells which proliferate and the number of cells which kill K562.
So all of the proliferative units seem to be cytotoxic precursors.
The cytotoxic frequency of T cells against K562 is less than one in
10,000. Proliferating LGL and T cells from positive wells were
expanded and tested against a variety of other targets. The cells
which were cytotoxic for K562 also killed Raji, some explanted human
tumor cells but not L 1210, RL ♂1 mouse leukemias and alloblasts.
The phenotype of the killer cells was T 28 positive (pan T antigen)
and OKT8 negative, some were OKT4 positive and OKT10 and OKM1 negative.

RIETHMÜLLER: Is there a change of phenotype during culture?

VOSE: According to Ortaldo, they lose OKM1, OKT10 and gain Ia and
OKT3 over 14 days in culture.

RIETHMÜLLER: Do you know anything about the karyotype?

VOSE: There is no chromosomal abnormality at the time points examined.

KIRCHNER: Is the expression of surface molecules defined by monoclonal antibodies dependent on LGL proliferation?

VOSE: LGL do have a higher rate of spontaneous thymidine incorporation
than small T cells (2,000 cpm for 1×10^5 cells compared to 200 to
300 cpm respectively).

RIETHMÜLLER: Do all NK cells have the interleukin 2 receptor?

VOSE: Apparently not, since we can only grow 1%.

RIETHMÜLLER: I would like to ask Dr. Leibold to comment on the antigenic pattern of LGL that makes them become more T-like during culture.

LEIBOLD: The patient, Dr. Riethmüller has described, provided Dr.
Stünkel, Dr. Thiel and myself with peripheral blood cells which were
subsequently OKM1 purified.   Treatment of the cells with OKM1 antibody
reduced cytotoxicity to 70% of the original.   FACS purified OKM1
positive and OKM1 negative cells segregated into cytotoxic and non-
cytotoxic cells respectively when the target cell K562 was used.
The same phenotype appears on both effector cells from normal human
donors and from the patient described by Riethmüller.   Cytotoxic effector cells were found among OKT3 positive and OKT3 negative populations typed by a pan-T reagent (OKT3).   We confirm that there are
at least two functional phenotypically distinct subpopulations of
effector cells: OKT3 positive and OKT3 negative that also bear the
OKM1 marker.   Using the K562 target, only two effector subtypes could
be detected: OKM1 positive, OKT3 positive and OKM1 positive, OKT3
negative.   However, when from the same effector cells a subpopulation,
being OKM1 negative but sheep red blood cell rosetting positive (ER
positive) on the target cell PDe-B-1 (human-B cell line transformed
by EB virus), a significant spontaneous cytotoxic effect was observed.
Thus if one phenotypes NK subpopulations, it is not sufficient to
employ one target.   By using different target cells, we were able
to define three different subsets of NK effector cells in the peripheral blood of normal human donors without cultivation:

        OKM1 positive and OKT3 positive
        OKM1 positive and OKT3 negative
        OKM1 negative and ER positive

Thus, if Dr. Vose shows that during cultivation LGL lose the OKT10
and OKM1 markers, the possibility arises that the OKM1 positive subpopulation which is present in peripheral blood of normal individuals
is lost during cultivation in vitro.

HERBERMAN: I am not entirely sure I agree that growth in culture

selects for or against a particular subpopulation of cells. When LGL
are purified and positively or negatively selected for OKT10 positive,
OKM1 positive or OKT3 negative cells, the phenotype of the cells in
culture are nearly identical after 7 days. Thus, it appears from the
work of Ortaldo that the phenotype of the cells does change in culture, even if you start with a preselected population. Another difficulty with phenotyping is that even if you carefully purify LGL
to remove any contaminating T cells by E rosetting at 29 degrees and
remove any OKT3 positive cells, the remaining cell population will
proliferate very strongly in response to OKT3. This suggests that
even if the cells appear to be entirely negative for OKT3 antigens
by FACS, they still have some of these molecules on their surface
and can proliferate at least as well in response to OKT3 as a typical
T cell preparation.

LEIBOLD: Does the phenotypic change only occur in one direction
or does the expression of markers fluctuate? Vose indicated that
markers were always lost from LGL.

HERBERMAN: The cells become OKT3 and Ia positive within a few days,
so that there is a gain of certain markers and a loss of others.

RIETHMÜLLER: The point should be made as to whether there is a clear
distinction between monocytic and NK cell mediated cytotoxicity.

MANTOVANI: Several laboratories have shown that human monocytes are
spontaneously cytotoxic and a number of papers were presented on this
topic. It happens that NK-oriented people ask me whether I am sure
that monocytic cytotoxicity may be due to contamination with adherent
NK cells. On the other hand, macrophage-oriented people usually
ask whether NK activity is in fact mediated by monocytes. Therefore, it may be worthwhile to list the reasons why I feel that these
are two distinct mechanisms of cytotoxicity. The first criterion
is adherence, NK cells being regarded as nonadherent. However, it
can be shown that some NK cells adhere and that the property of adherence increases upon treatment with interferon. The morphology of
adherent cytotoxic cells may resolve the issue. Monocytes of some
donors are contaminated (up to 1%) with LGL. The second criterion
is target cell susceptibility. Mouse target cells are usually susceptible to human monocytes but not to human LGL. Other criteria
include susceptibility of the cells to silica and their tissue distribution. Cytotoxic mononuclear phagocytes have been isolated from
anatomical sites from which we have not been able to recover NK
activity, such as the peritoneal cavity, milk, and the alveolar
spaces. Further, the activity of blood monocytes is stable in culture
with human serum unlike that of NK cells. So I feel that these are
two distinct mechanisms. Nonetheless, one should be aware that minor
contaminations with either monocytes or LGL can complicate the interpretation of in vitro assays. As to why monocytes have spontaneous
cytotoxicity, possible causes include endotoxin activation, mycoplasma
activation and microbial status.

RIETHMÜLLER: Dr. Habu has used another marker for studying NK cells, asialo GM1, but this marker is also present on some T cells and brain cells.

HABU: Antibody raised to purified asialo GM1 preferentially reacts with NK cells and removes NK activity. The very same antiserum reacts with immature thymocytes of the embryo which do not yet express the Thy-1 antigen. At 12 days of gestation, 99% of the thymocytes express asialo GM1 antigen but not Thy-1. The proportion of asialo GM1 positive thymocytes decreases rapidly in late gestation, while Thy-1 positive, asialo GM1 negative cells increase. This reciprocal relationship suggests that the asialo GM1 antigen is expressed by immature thymocytes that in the thymic environment differentiate and express the Thy-1 antigen. About 20 to 30% of fetal liver cells are also asialo GM1 positive. We performed transfer experiments in which fetal liver cells from Thy-1.2 positive CBA mice were inoculated into irradiated Thy-1.1 AKR mice. The fetal liver cells were treated either with normal rabbit serum plus complement or with anti-asialo GM1 antibody and complement. Removal of asialo GM1 positive cells resulted in decreased host thymus repopulation with donor-type Thy-1.2 positive cells. Taken together, these data support our contention that NK cells and T cells are derived from the same lineage.

CUDKOWICZ: Phenotyping NK cells is important in its own right, but what is it really telling us about the origin and nature of NK cells? What are NK cells? Are they really different from other cells that we already know about? Do they represent a separate cell line? Let me remind you that a few years ago phenotyping of lymphocytes yielded clear cut results: B cells possessing markers distinct from those of T cells. We know now that T and B cells share a number of markers and yet we don't doubt that T and B cells belong to distinct lineages.

RIETHMÜLLER: B and T cells have very stable markers that make them different and the markers are related to function. There are other markers which I think are associated with different differentiation stages and these may be the shared ones. We have not spoken about receptors, receptor mechanisms or NK cell functions in vivo. The 200,000 molecular weight OKM1 or M522 antigens may be related either to differentiation along a common pathway or to the spontaneous cytotoxic activity of these cells. These antigens may even be an element that obviates the so called MHC restriction characteristic for T cells. In conclusion, I think that the examination of markers is only worthwhile if we know their nature, their occurrence during differentiation, and their probable involvement in function. I vote for NK cells as a distinct subpopulation of effectors, probably branching off from immature T cells.

REGULATION OF NATURAL CYTOTOXIC ACTIVITY

Gustavo Cudkowicz

Department of Pathology
State University of New York
Buffalo, New York   USA

DISCUSSION INTRODUCER GUSTAVO CUDKOWICZ: If naturally occurring
cytotoxic cells are to fulfill an important biological function,
they will be integrated and they will be regulated. Only in the
last few years have studies been conducted on NK cell regulation.
In trying to present an overview of these studies, I concluded that
the best approach was to follow the chronology of these studies.

The first evidence of regulation came with the observation made
in 1977-1978 in a number of laboratories that interferon was a natural
mediator with an exquisite effect on natural killer cell activity.
At the time, the assay was not quantitative and all one could say
was that activity increased. Subsequent studies at the cellular
level have revealed that interferon does a number of things to natural
killer cells. Among others, it promotes differentiation of a pre-
NK cell into a non-lytic NK cell into an active lytic NK cell.
Concurrently, the effect of prostaglandin on NK cells was recognized
and a few years later that of interleukin. I would group these
regulators under the heading of defined soluble mediators or defined
growth factors. These substances have a pronounced effect on NK
cell activity and their mechanism of action is being investigated,
not only individually but also collectively as a regulatory series
of events.

I would like to talk now about regulatory interactions between
cells. Do NK cells interact with other cells at various stages,
either during differentiation or during the actual stage of target
cell lysis? This question was first addressed using cell mixtures.
The experimental approach consisted in taking a cell preparation
devoid of NK activity, mixing it with NK cells, and then seeing if
the mixture resulted in inhibition of the effector cell activity.
The important variables in these cell mixing experiments were (a)

761

the purity of the cells that inhibit or promote NK activity and
(b) the step in NK cell activity that was influenced. Specifically,
was binding of the effector cell with the target inhibited? Was
the lytic event inhibited or promoted? Was the maintenance of the
NK cell altered? While it has not been possible to clearly demonstrate a cellular interaction which promoted NK activity, it was
rather easy to demonstrate the existence of cells that inhibited NK
activity either at the lytic phase or at the phase concerned with
maintenance of NK cells. In the latter instance, the inhibitor cell
and the NK cell interact for a long period of time prior to the lytic
event. During this prolonged interval, NK cells that otherwise would
be functional are inhibited by the extraneous cell being added.

In 1980-81, another approach evolved for studying regulation:
namely the use of inbred strains of animals in which NK activity is
either extremely low or extremely high. The investigator tries to
learn from this unusual situation what underlies the level of effector
cell activity. A particularly informative strain of mouse has been
the SJL strain, because this strain is characterized by low NK activity. Like the SJL strain, there are a number of other mouse strains
(like I, PL, or A/Jax) that more or less phenotypically resemble
each other in having low NK activity and also in not being amenable
to stimulation of NK activity even with interferon or interferon
inducers. One could consider a spectrum of NK activity with some
strains towards the high end of the spectrum opposite to those strains
listed here. This approach is just beginning and to my knowledge
studies have not yet been reported on the mechanism of low NK activity in I, PL or A/Jax strains. Studies performed with SJL have
indicated a role for the thymus in this particular strain. Removal
of the thymus increases the animal's own NK activity as well as the
response of the NK cells to interferon.

The latest development in the area of regulation addresses the
fundamental question of whether or not regulation by normal cells
of the lymphoid/myeloid system occurs all the time. When we assay
NK cell activity, are we really assaying the sum of the promoting
and inhibiting influences that are being exerted all the time? This
approach has led to the identification of factors which regulate NK
activity in culture supernatants of normal cells such as non-
stimulated lymphoid cells. I stress the word lymphoid to indicate
that the preparations weren't purified enough to identify the source
of the inhibitory factor. The inhibitory factors also haven't been
characterized satisfactorily, although they seem to be of the order
of 20,000 molecular weight. Interestingly, the factors seem to
inhibit different types of effector cells with different efficiency.
The factors seem to be most efficient in inhibiting NK cells, somewhat
less efficient in inhibiting K cell activity, and ineffective in
inhibiting cytotoxic T cell activity or the cytotoxic activity of
macrophages. Such data suggest that normal cells do as a matter of

course release defined factors in culture that selectively inhibit
NK cells.

    In our own laboratory, we have demonstrated regulatory cells
that interact with natural effectors using co-culturing techniques.
When carrageenan was added to cultures of mouse spleen cells, it
induced cells that suppressed NK activity. Spleen cell suspensions
were cultured for five days with or without the addition of carrageenan. The addition of carrageenan to splenocytes resulted in
absolutely no NK activity after 5 days. These carrageenan treated
spleen cells were then used as the source of suppressor cells. In
a cell mixing assay, the cells not exposed to carrageenan failed to
inhibit NK activity whereas the cells exposed to carrageenan markedly
inhibited NK activity. This is a rather simple way of inducing suppressor cells in a short time. Two days is sufficient. This experimental approach provides a starting preparation for the purification
and identification of the cell type involved and eventually for
studies on the mechanism of inhibition. It turns out that the cells
responsible for the inhibition are adherent cells belonging to the
monocyte/macrophage lineage and that the inhibition is reversible.
If the suppressor cells are removed from the mixture via their adherence property, the NK activity is restored. Variations of this model
have been produced in different laboratories. In some experiments,
adherent cells were obtained from pleural effusions of lung cancer
patients which were inhibitory to NK cells from human peripheral
blood. The inhibition was reversed once adherent cells were removed
and the NK cells allowed to regenerate.

    The question naturally arises as to whether or not these sup-
pressor cells act by releasing factors. Our approach was to examine
supernatants of spleen cells after 4 hours in culture for their
capacity to depress NK activity. Supernatants of normal spleen cells
showed a marginal inhibition in contrast to supernatants of spleen
cells exposed to carrageenan. The latter produced a profound depression. Although one can recover from the supernatant an inhibitory
factor, that fact alone does not prove that the mechanism of inhibition when cells are introduced into the mixture is due to an inhibitory factor. But it opens the possibility.

    NK activity is reduced in mice treated with estrogens or hydro-
cortisone. BCG and C. parvum given by a certain route and dose will
inhibit NK activity but when given at a different dose or route
will increase NK activity. Further, the infant mouse is characterized by either low (from 10-15 days) or essentially no NK activity.
In all these instances, suppressor cells have been identified by
cell mixing experiments. In some cases, the suppressive effect was
reversible. This can not be tested in every instance since the suppressor cells are not always adherent cells that can be removed at
will and then added again.

Do suppressor cells inhibit only the lytic phase or do they inhibit also some process of maturation or differentiation? In most cases of low NK activity, it can be enhanced by the interferon inducer poly I:C. In these cases, the suppressor cells have not affected the generation of cells capable of responding to interferon. However, this does not appear to be the case following estrogen treatment, irradiation of the bone marrow by strontium 89, or total body irradiation. In these cases, the interaction of suppressor cells with effector cells results in some inhibition of differentiation or maturation. We really don't know at which stage the inhibition occurs.

Let me conclude by summarizing some of our knowledge about the various types of suppressor cells. First, differentiation of NK progenitor cells is effected by the thymus as evident from studies with the SJL strain. Whether or not other mouse strains present this abnormality remains to be determined. Second, a large category or non-T, non-adherent splenocytes, act as suppressor cells which are negative for most markers: they are Ia negative, Thy-1 negative and Ly antigen negative. They occur in irradiated mice and in irradiated mice treated with estradiol. These cells are quite effective in inhibiting differentiation of progenitor cells. When we look at lytic activity, we find that the same population of non-adherent suppressor splenocytes are also capable of interfering with the effector target interaction. In addition, we find that non-T but adherent splenocytes are quite effective in reversibly inhibiting the NK activity.

I would now like to ask Dr. Riccardi to discuss his in vitro method of growing NK cells. This method allows him to influence differentiation or maturation and lends itself to the study of regulatory interaction.

RICCARDI: A number of factors regulate NK activity and influence the actual levels of natural cytotoxicity. The majority of studies have concerned the regulation of the lytic phase of natural cytotoxicity. However, it would be valuable to have available a means to study the interaction between factors and cells in the differentiation of the NK cell lineage. It has been shown recently that mouse natural killer cells will grow in a medium containing interleukin 2. Using this system, we studied the growth of natural killer cells using a limiting dilution assay for cytotoxic cells against chromium labeled YAC-1 target cells. We cultured the cells for 8 days and then measured cytotoxicity. If we cultured the mouse spleen cells in the presence of irradiated, feeder spleen cells and interleukin 2, a different frequency of cytotoxic cells was obtained depending upon the target cells used (Figure 1). We also found a difference if you cultured spleen cells from high reactive mouse strains such as CBA/J versus low reactive strains such as SJL. In addition, we found that the frequency of cytotoxic cells from spleens of conventional

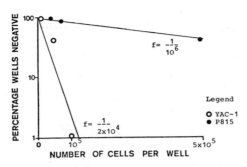

Fig. 1.   Frequency of CBA/J cytotoxic cells against YAC-1 and P-815
          targets. Different numbers of spleen cells were dispersed
          into 0.2 ml round bottomed microtiter plates (12-24 wells/
          responder concentration), together with irradiated (3,000R)
          feeder cells, and conditioned medium containing interleukin
          2. Cytotoxicity was measured against chromium labeled tar-
          gets (5,000 cells/well) in a 4 hour cytotoxicity assay.

BALB/c mice was about 1 in 20,000 cells plated compared to that of
BALB/c nu/nu mice of one cell in 1,000,000 spleen cells plated (Figure
2). These results suggest that the presence of thymic cells is important to the growth of NK cells.

     In another experiment, we cultured responder BALB/c nu/nu spleen
cells in the presence of BALB/c nu/nu feeder cells. The frequency
of natural killer cells was again about 1 cell per 1,000,000 plated
cells. However, if we used nude heterozygous feeder cells, the
frequency of natural killer cells goes up to 1 cell per 100,000 spleen
cells plated. This helper activity is destroyed if we pretreated
the nude feeder cells with anti-theta and complement (Figure 3).

     Another way to regulate and to augment the frequency of BALB/c
nu/nu spleen NK cells growing in vitro is to pretreat the cells with
interferon. Pretreatment with interferon increases the capacity of
mouse natural killer cells to grow in the presence of interleukin 2.
We feel that this is a physiological means to increase the number of
natural killer cells growing, both in conventional mice and in homozygous nude mice. So using homozygous nude feeder cells the number
of NK cells growing from homozygous nude mice can be augmented by
addition of T cells or by treatment with interferon.

     While interferon in the culture increases the frequency of NK

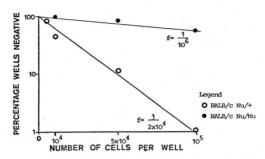

Fig. 2.   Frequency of cytotoxic cells reactive against YAC-1 of
          BALB/c nu/+ and BALB/c nu/nu responder spleen cells cultured
          in the presence of syngeneic feeder spleen cells.

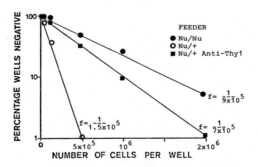

Fig. 3.   Frequency of cytotoxic cells reactive against YAC-1 of
          BALB/c nu/nu spleen cells cultured in the presence of splenic
          feeder cells of BALB/c nu/nu or BALB/c nu/+ mice.

cells developing from the homozygous nude spleen cells, the converse occurs when spleen cells from conventional mice are cultured in the presence of interferon. Under these latter conditions, there is a significant inhibition in the number of natural killer cells growing. These results suggest that this inhibition of proliferation or differentiation of NK cells by interferon is mediated by T cells.

Figure 4 summarizes information about the regulation exerted by interferon and interleukin 2 on the differentiation of natural killer cells. We know that both interleukin 2 and interferon enhance NK activity. Interleukin 2 can induce proliferation and probably differentiation of natural killer cells in the mouse. Parenthetically interferon can induce high expression of T cell growth factor possibly by increasing the number of T cells producing it.

We also found that T cells help the growth of mouse natural killer cells in vitro in the presence of a lectin free interleukin 2 preparation. However, if interferon is present in the culture, it inhibits such T cell dependent proliferation. While such information doesn't assist us in deciding whether natural killer cells derive from a macrophage or T cell lineage, it does tell us that the mouse natural killer cell system is a thymus dependent system.

CUDKOWICZ: These results provide evidence for an effect of thymic cells on NK production or differentiation in vitro. I now ask Dr. Kaminsky to discuss the effect of the thymus in vivo.

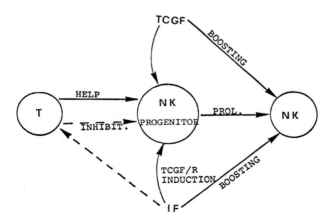

Fig. 4.   Schematic representation of interleukin 2 (TCGF), interferon (IF), and T cell regulatory mechanisms influencing NK cell activity.

KAMINSKY: SJL mice have low NK activity and there is minimal NK cell
induction with interferon or interferon inducers. We were very
fortunate to acquire SJL mice that possessed the nude gene. The
NK activity of euthymic heterozygous mice is low and it is not very
inducible with poly I:C. In the athymic SJL mice, the endogenous NK
activity is still low but an injection of poly I:C now greatly increases NK activity. If the athymic mice are treated with anti-
interferon antibody 6 hours before poly I:C,no augmentation is
observed. Thus, interferon appears to be the agent causing the induction of NK activity in this strain of mouse. K cell activity
in SJL mice, whether thymic or athymic, follows that of NK activity.

We next wanted to know if the nude gene was controlling NK
activity or whether it was the thymus itself. Accordingly, we
examined neonatal and adult thymectomized SJL mice. The results
showed that it is the thymus itself which affects NK activity in SJL
mice. The neonatal thymectomized SJL mice still have low endogenous
NK activity, but an injection of poly I:C now greatly augments NK
activity similar to that seen in the nude SJL mice. Sham or hemi-
thymectomized neonatal mice behave like the wild type or the SJL
mice with a thymus. Adult thymectomy does not have an effect on the
NK activity.

In summary, neonatal but not adult thymectomy in SJL mice removes
a regulating or suppressing effect on NK activity.

CUDKOWICZ: I now ask Dr. Hillel Koren to discuss his work on the
effect of prostaglandins and interferon.

KOREN: I think it's clear that the regulation of NK cells is very
complex. I will focus on one phenomenon, namely the effect of interferon on the activation of NK cells and its suppression by PGE_2. I
believe it is well documented that the natural killer cells of both
mouse and human are sensitive to suppression by PGE_2 and in this
respect resemble macrophages and cytotoxic T cells. Macrophages
are good producers of PGE_2 especially in humans and so are some tumor
cells. Thus the effect of PGE_2 on NK cells is an important issue.
We addressed this issue by incubating NK cells in the presence of
different amounts of poly I:C or interferon. After washing, the cells
were assayed for cytotoxicity. In the absence of PGE_2, a reasonable
amount of killing was observed with unstimulated NK cells and this
activity was markedly increased by both poly I:C and interferon. If
we now added PGE_2, the cytotoxicity of cultures not receiving poly
I:C or interferon was markedly suppressed. With increasing amounts
of either poly I:C or interferon, however, a partial resistance to the
suppressive effects of PGE_2 was observed reaching almost total resistance at 100 µg/ml of poly I:C or 1000 units of interferon per
ml. The mechanism of this resistance is being investigated exploring
several possibilities such as desensitization or loss of PGE_2 receptors. Whatever the mechanism, it is important to remember that

inactive NK cells could be NK cells suppressed by PGE$_2$ elaborated by tumor cells, macrophages or other sources. Our concept of the regulation of NK activity by interferon and PGE$_2$ is presented in Figure 5.

Human monocytes produce large amounts of PGE$_2$ especially in the presence of poly I:C, certain viruses or immune complexes. If large granular lymphocytes are activated by interferon (which is produced by NK cells themselves) or exposed to poly I:C or interleukin 2, they seem to be less sensitive to the suppressor effect of PGE$_2$. This could be an interesting mechanism whereby the natural killer cells avoid the suppressive effects of PGE$_2$.

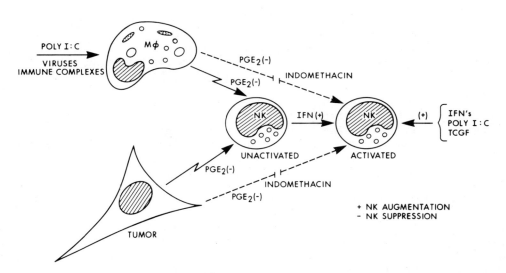

Fig. 5.   Involvement of interferon (IFN) and PGE$_2$ in the regulation of NK activity. Major sources of PGE$_2$ are macrophages and certain tumor cells. Human macrophages produce PGE$_2$ in response to poly I:C, certain viruses, and immune complexes. PGE$_2$ suppresses activity of NK cells. However, if NK cells are activated by IFN, poly I:C or T cell growth factor (TCGF), they become partially resistant to suppression by PGE$_2$. In a system where mononuclear cells are co-cultured and stimulated by PGE$_2$ inducing agents (e.g.: poly I:C), the synthesis of PGE$_2$ can be inhibited by the addition of indomethacin (10^{-6}M).

VARESIO: I wish to briefly summarize our results on the role of macrophages in the maintenance of NK activity in vitro. In our system we used CBA spleen cells passed over nylon wool columns and then cultured overnight in the presence of different preparations of macrophages. We already knew that resident macrophages didn't affect lymphokine production by T cells while Corynebacterium parvum elicited macrophages strongly suppressed lymphokine production. We used these two populations of macrophages at a ratio of 1 macrophage to 10 spleen cells with the following results. With respect to lymphokine production, resident macrophages had no effect while C. parvum elicited macrophages were inhibitory. With respect to NK activity, the resident macrophages were strongly inhibitory while C. parvum induced macrophages were much less inhibitory. We have repeated this experiment many times with the same results. We concluded that resident macrophages which are unable to suppress T cell lymphokine production do inhibit NK activity in vitro. This is also true when interferon is added to the system. We then looked at the splenic macrophages of normal and C. parvum injected mice. They did not affect NK activity in vitro either in the presence or absence of interferon.

Although resident peritoneal macrophages are highly suppressive for NK activity, normal macrophages from the spleen are not. That might be a reasonable explanation for why in the peritoneal cavity NK activity is generally quite low, while in the spleen you find the most NK activity.

CUDKOWICZ: The correlation between NK activity in the peritoneal cavity or spleen to the presence and functional state of the macrophage is interesting. Let me add that the elaboration of inhibitory factors in culture by peritoneal cells as compared to spleen cells is much higher.

SILVERSTEIN: How was the C. parvum given?

VARESIO: The C. parvum was given intraperitoneally in a concentration of 0.7 mg/mouse. We did not test the affect of indomethacin.

KIRCHNER: Why do newborns have very low NK activity and 8 weeks later very high activity? This is a central issue because newborns are highly susceptible to infection by both viruses and bacteria. Exactly what is defective in newborns? It appears that they have an intact NK cell system but do not have an intact interferon system. If you give interferon directly or certain interferon inducers, you can activate the NK cells. As regards the second point: What really causes the increase in NK activity at 8 weeks? I think this has never been settled. I feel that it is due to a viral infection which occurs during those weeks. Has anyone had experience with germ free animals?

LEIBOLD: We did some experiments with gnotobiotic pigs and they did have NK cell activity upon delivery.

CUDKOWICZ: Dr. Kirchner has raised an interesting point. His statements are correct for the mouse. Newborn mice have very little NK activity and this situation might arise if the NK system were underdeveloped or heavily suppressed. But let me point out that this is not the case in all species. Some newborns have a fully developed NK system. For instance, in humans you can find functional NK cells in the cord blood. We should keep in mind that the mouse is an unusual situation in terms of neonatal competence for NK activity.

THE ROLE OF NK CELLS IN

RESISTANCE OF IN VIVO TUMORS

Ko Okumura,[1] Sonoko Habu,[2] and Masataka Kasai[3]

[1]Department of Immunology, Faculty of Medicine
University of Tokyo, Bunkyo-ku, Tokyo, Japan
[2]Department of Pathology, School of Medicine
Tokai University, Isehara, Kanagawa, Japan
[3]Department of Tuberculosis, Japanese National
Institute of Health, Meguro-ku, Tokyo, Japan

INTRODUCTION

The significance of natural killing of tumor cells by lymphocytes, previously investigated as an in vitro phenomena, has now been
broadened by examining mechanisms of in vivo tumor resistance.
Although the origin and differentiation pathway of murine natural
killer (NK) cells is not clearly defined, NK cells express the glycolipid marker asialo GM1 which is not expressed on mature T or B
cells (1,2). Taking advantage of the fact that microliter amounts
of intravenously injected anti-GM1 antibody could abolish NK activity
in vivo, we studied the role of NK cells in the incidence and growth
of various tumors.

EFFECT OF DELETION OF ASIALO GM1 POSITIVE CELLS ON TRANSPLANTED TUMOR GROWTH IN NUDE MICE

Repeated injections of anti-GM1 (20 μl) every three days into
mice suppressed NK activity to less than 5% of normal and abrogated
GM1 positive cells from the spleen without eliminating T and B cells.
Thus, this treatment was judged adequate for studying the biological
role of NK cells in vivo. Two million syngeneic tumor cells (RLδ-1)
or two million allogeneic tumor cells, (YAC-1), were subcutaneously
injected into BALB/c nude mice. On the same day, each mouse also
received intravenously(i.v.) an injection of rabbit anti-GM1 or normal
rabbit serum (NRS). For a duration of 24 days, each mouse received
repeated injections of the antiserum every three days and the size

773

of palpable tumors was recorded. The mice which received anti-GM1
had a markedly higher incidence of RL♂-1 tumor takes and the size
of their tumors was larger than those which received the tumor cells
and NRS. As seen in Table 1, palpable RL♂-1 tumors appeared in 100%
of the mice injected with anti-GM1, whereas the tumors were detectable
in less than 25% of mice injected with NRS. The difference was even
more striking in mice transplanted with the YAC-1 tumors where no
palpable tumors appeared in NRS injected mice within 3 weeks. This
correlation between elimination of NK activity and enhancement of
tumor growth was consistent with the observation that beige mice
which have low NK cell activity have a low natural resistance to
transplanted leukemia (3). The same effect of the antiserum was
observed when heterogeneic tumors, such as human malignant lymphoma
and gastric carcinoma, were transplanted into the nude mice (4).
These observations suggest that surveillance by NK cells extends
to transplanted syngeneic, allogeneic and heterogeneic tumors.

ROLE OF ASIALO GM1 POSITIVE CELLS IN AUTOCHTHONOUS TUMOR GROWTH AND
METASTASIS

    That NK cells participate in the suppression of transplanted
tumor growth in vivo raises the question whether or not NK cells
function in surveillance of primary tumors or of metastasis. There
is as yet no direct evidence that NK cells are active against
autochthonous tumor growth. Indeed, nude mice develop similar numbers of tumors as their normal litter mates (5). Since cells other
than NK cells may be involved in the regulation of autochthonous
tumor growth and metastasis, it is difficult to define the role of
one particular cell population. However, such a study should be
possible by injecting nude mice with anti-GM1. As shown in Table 2,
tumors induced by methylcholanthrene in nude mice receiving anti-
GM1 were detectable two weeks earlier than in control mice. By the
10th week following injection of the carcinogen, the first tumors
were detectable in 6% of the mice injected with anti-GM1 and by the
16th week all mice injected with anti-GM1 had tumors. In the control
mice, the tumor first appeared 12 weeks after carcinogen application
and 20% of mice still had no palpable tumors after 36 weeks. This
is the first evidence which suggests that NK cells act in autochthonous tumor growth. The frequency of tumor appearance in the control
mice was low but not zero. The reason for this may relate to the low
sensitivity of the methylcholanthrene induced tumor to NK cell activity. This explanation is consistent with the finding by Talmage et
al. that the growth rate of transplanted tumors insensitive to NK
cell lysis in beige mice (low NK cell activity) was the same as that
in C57BL/6 mice (moderate NK cell activity) (6). They also reported
that the frequency of metastasis of syngeneic NK cell sensitive
transplanted tumors was markedly greater in NK deficient beige mice
than in normal C57BL/6 mice. A role of NK cells in metastasis was
initially suggested by the observation that transplanted tumors

Table 1.   Tumor Growth in Nude Mice Inoculated Subcutaneously
with 3×10^6 YAC-1 or RLδ-1

Days after Inoculation	Injection of Anti-asialo GM1	Tumor Growth	
		YAC-1	RLδ-1
4	+	0/20	0/20
	−	0/20	0/20
8	+	16/20	16/20
	−	0/20	0/20
16	+	20/20	20/20
	−	0/20	3/20
20	+	20/20	20/20
	−	0/20	4/20

rarely metastasized in nude mice (7).   We obtained direct evidence
for this contention using nude mice injected with anti-GM1.   Metastasis of the transplanted human gastric carcinoma was observed in nude
mice injected with anti-GM1, whereas no metastasis was found in the
control mice.   The above observations suggested that NK cells have
an important function in the host's control of metastasis as well
as growth of tumors even in mice having normal T cell function.

EFFECT OF ANTIBODY AGAINST ASIALO GM1 ON NK ACTIVITY IN VIVO

Considering the various biological roles of glycolipids such as
receptor or recognition molecules in cellular interaction, we set
up an experiment to determine the in vivo effect of anti-GM1 on NK
activity.   The amount of anti-GM1 required for elimination of NK
activity was much smaller than we expected.   The dose response curve
clearly indicated that even a few microliters of anti-GM1 could deplete virtually all NK activity.   This depletion lasted from several
days to 1 week after one injection of the antiserum (4).   This strong
correlation between the presence of NK function and resistance to
tumor growth agrees with previous experiments indicating that positively selected NK cells can retard tumor growth (8).

As the reduction in NK activity was observed within 30 minutes

Table 2. Effect of anti-Asialo GM1 on Methylcholanthrene
          Induced Tumor Growth

Weeks after Methylcholanthrene	Injection of Anti-asialo GM1	Incidence of Tumor Growth (%)
9	+	6
	–	0
10	+	13
	–	0
11	+	26
	–	6
12	+	40
	–	20
13	+	60
	–	33
15	+	100
	–	66

after injection of anti-GM1, we set up an experiment to determine
the mode of action of anti-GM1 on NK activity. We incubated BALB/c
nu/nu spleen cells with anti-GM1 for 30 minutes at 37°C. A strong
reduction of NK activity occurred only when normal mouse serum was
present. The effect of the mouse serum was abolished by heating
it for 30 minutes at 56°C. As was the case in vivo, NK activity was
inhibited even with minute amounts of anti-GM1, whereas normal mouse
serum alone was not effective. Because of the weak activity of mouse
complement, it seemed unlikely that abolition of NK activity was due
to a cytolytic effect caused by microliter amounts of anti-GM1. NK
activity was reduced equally in the presence of anti-GM1 and serum
from AKR mice with a genetic deficiency of C5 as with serum from a
mouse strain possessing the complete spectrum of complement components. Serum from C4-deficient guinea pigs was also effective in
removing NK activity in the presence of anti-GM1. These results
indicated that a complement dependent cytolytic mechanism was not
required for this effect. Rather, the results suggested that the
function of NK cells was seriously damaged by anti-GM1 antibodies
and an unknown serum factor(s).

As interferon seems to influence NK function both in vivo and in vitro, we thought it appropriate to determine whether anti–GM1 would also alter the effect of interferon on NK cell activity. Table 3 shows that intraperitoneal injection of interferon markedly augmented the cytotoxic activity of NK cells in normal nude mice. However, interferon could not restore NK activity that was depressed by injection of anti–GM1. Similar results were also obtained in vitro. Based on these experiments, we postulated that the receptor able to bind interferon and increase NK activity is expressed on NK cells at a differentiation stage when these cells are also reactive to anti–GM1. Other NK cell surface antigens, such as Ly–5 and Qa–5, have been determined serologically and it was recently reported that interferon triggers the differentiation of NK precursor cells (Qa–5 positive, Ly–5 negative) into NK effector cells (Qa–5 positive, Ly–5 positive) (9). If so, the present results indicate that the phenotype of NK precursor cells is asialo GM1 positive, Qa–5 positive, Ly–5 negative. An attractive possibility is that asialo GM1, the cell surface structure responsible for killing, and the receptor for interferon are closely inter–connected in the NK cell membrane. The results presented here might aid in understanding the cytolytic mechanism of NK cells and the mode of action of interferon.

CHARACTERIZATION OF ASIALO GM1 BEARING CELLS

The preceding experiments show that the in vivo administration of anti–GM1 depresses NK activity as well as the number of GM1 positive cells in the spleen. The next question concerned the kinds of cells that express asialo GM1. We have already reported that anti–GM1 does not influence T cell functions such as T killer or T helper activities (1,10). Furthermore, concanavalin A (Con A) responding T cells as well as Ia–bearing macrophage like cells which are essential for Con A responses are unaffected by anti–GM1 (4). Asialo GM1 positive cells are radiation resistant, so GM1 positive cells and NK activities are enriched in spleens of lethally irradiated mice. In these GM1 positive enriched cell populations, we could morphologically distinguish mainly two kinds of cells: large and small mononuclear cells. Immunoelectromicroscopy of NK enriched cell populations from nude mice indicated that the cells with cytolytic activity against YAC–1 targets resembled small lymphocytes. The larger GM1 positive cells were predominately monocytes which did not participate in cytolysis during a short 4 to 6 hour incubation period (11). Since the in vivo administration of anti–GM1 caused a decrease in both large and small GM1 positive cells, it is difficult to know whether the participation of asialo GM1 positive monocytes is essential to the NK system. We have also observed by cell membrane staining that the asialo GM1 positive macrophages which represent less than 30% of total splenic macrophages do not bear detectable amounts of Ia antigen (unpublished data). Thus it may be concluded that Ia positive macrophages are not involved in NK systems.

Table 3. Effect of Interferon after Treatment with Anti-
          asialo GM1 In Vivo

Injection of Anti-asialo GM1	Interferon	% Lysis of YAC-1
-	-	18.5
-	+	41.7
+	-	0.5
+	+	1.3

BALB/c nu/nu mice were injected i.v. with 5 µl of anti-GM1 or
with 500 U of interferon. Interferon was also injected 30 min
after injection of anti-GM1. NK activity of spleen cells was
determined 24 hours after administration of each reagent. Ef-
fector/target ratio was 50:1.

As reported elsewhere, anti-GM1 reacts with fetal thymocytes
as well as some immature cells in fetal liver which migrate into the
thymus (2,12). We postulate that the asialo GM1 positive small lymphocytes of adult spleens with NK activity belong to the T cell
lineage. Since asialo GM1 is a marker common to NK cells, to the
effector cells of natural resistance (13) and to pre-T cells (12),
it may provide a useful tool to further understanding of the natural
immune systems.

Igh-1 LINKED MARKER ON NK CELLS DETECTED WITH MONOCLONAL ANTIBODY

If NK cells belong to a branch of the T cell differentiation
lineage, one can postulate that both cell types utilize similar receptor molecules. Recent reports from F. Owen's group demonstrate that
T cells bear cell surface determinants which are closely linked to
the Igh-1 allotype locus (14) and which are thought to be the constant
portion of an antigen-binding receptor. With this in mind, we produced a monoclonal antibody that would react with NK cells in an
allotype specific fashion. B cell hybridomas were made from SJL
(Igh[b])or SJA (Igh[a]) mice by repeatedly immunizing with SJA or SJL

spleen cells, respectively. In the presence of complement, culture supernatants from three hybridoma clones derived from immunized SJL mice (anti-"Igh^a") eliminated the NK activity of BALB/c (Igh^a) but not CB-20 (Igh^b), an Igh allotype congenic strain possessing the background genes of BALB/c. Conversely, three clones (anti-"Igh^b") were also established whose products could deplete NK activity of CB-20 but not BALB/c. In addition, some monoclonal antibodies inhibited NK activity in an allotype-specific manner when they were added to in vitro cultures (manuscript in preparation).

These results indicate that analogous to T killer or T suppressor cells, NK cells also express products that are closely linked to Igh-1. These products may correspond to a differentiation marker or may be a constant portion of the NK cell receptor. In any case, these monoclonal antibodies may be useful in studying NK cell receptors as well as the differentiation process.

REFERENCES

1. Kasai, M., Iwamori, M., Nagai, Y., Okumura, K., and Tada, T., Eur. J. Immunol. 10:175, 1980.
2. Habu, S., Kasai, M., Nagai, Y., Tamaoki, T., Tada, T., Herzenberg, L. A., and Okumura, K., J. Immunol. 125:2285, 1980.
3. Karre, K., Klein, G. O., Kiessling, R., Klein, G., and Roder, J. C., Nature 284:624, 1980.
4. Habu, S., Fukui, H., Shimamura, K., Kasai, M., Nagai, Y., Okumura, K., and Tamaoki, N., J. Immunol. 127:34, 1981.
5. Rygaard, J., and Povlsen, C. O., Transplantation 17:135, 1976.
6. Talmage, J. E., Meyers, K. M., Prieur, D. J., and Starkey, J. R., Nature 284:622, 1980.
7. Herberman, R. B., in "Nude Mouse in Experimental and Clinical Research", p. 135, Academic Press, New York, 1978.
8. Kasai, M., Leclerc, J. C., McVay-Boudreau, L., Shen, F. W., and Cantor, H., J. Exp. Med. 149:1260, 1979.
9. Minato, N., Reid, L., Cantor, H., Lengel, P., and Bloom, B. R., J. Exp. Med. 152:124, 1980.
10. Habu, S., Hayakawa, K., Okumura, K., and Tada, T., Eur. J. Immunol. 9:938, 1979.
11. Shimamura, K., Habu, S., Fukui, H., Akatsuka, A., Okumura, K., and Tamaoki, N., J. Natl. Cancer Inst., in press.
12. Habu, S., and Okumura, K., in "Natural Cell Mediated Immunity Vol. II" (R. B. Herberman, ed.), Academic Press, New York, in press.
13. Okumura, K., and Habu, S., in "Natural Cell Mediated Immunity Vol. II" (R. B. Herberman, ed.), Academic Press, New York, in press.
14. Owen, F. L., Finnegan, A., Gates, E. R., and Gottlieb, P. D., Eur. J. Immunol. 9:948, 1979.

DISCUSSION

KOREN: Could you summarize again the evidence that natural killer
cells were responsible for the suppression of tumor growth? I ask
the question for two reasons. First, anti-asialo GM1 has been shown
to react with macrophages. Second, since you injected whole antibody,
how do you know that you didn't get blocking of Fc receptors on some
of the effector cells?

OKUMURA: We cannot say that asialo GM1 positive macrophages are not
involved in the rejection of these tumors. However, cell sorter
studies showed that after the anti-GM1 injections, the remaining
GM1 positive cells were almost all monocytes. So the injection of
anti-GM1 deleted the small lymphocytes but not the monocyte like
cells. Since the treatment accelerated tumor growth, we feel that
the natural resistance of untreated mice was due to NK cells.

     Concerning the second question about Fc receptors, we have in-
jected the Fab monomer of anti-GM1 into mice. This antibody also
eliminated the small lymphocyte as well as NK functions. This information strongly suggests that blocking of Fc receptors is not the
mechanism of action of our antibody and further that complement is
not very important to the deletion of the asialo GM1 positive cells.

RIETHMÜLLER: Everybody is fascinated by your experiments with the
anti Igh linked allotype. I understand that the antibody was induced
with spleen cells from the other strain: SJA into SJL and vice
versa. Since you did not use thymocytes, how do you rule out that
the antibody actually is directed against the allotype immunoglobulin
bearing B cells?

OKUMURA: We confirmed that the clone which we used to delete the NK
cells was not reacting with immunoglobulin.

RIETHMÜLLER: So that is ruled out. Francis Owen's group also induced
with thymocytes.

OKUMURA: In our case, we just used spleen cells from SJL or SJA.

HERBERMAN: Do you have any evidence as to whether this antibody in
the absence of complement will inhibit NK cell binding to target
cells? Is the inhibition in fact at the level of the recognition
site on the NK cell? You showed that it inhibited cytotoxicity but
do you know if it is inhibiting at the binding (recognition) step
or by some other mechanism?

OKUMURA: I don't know.

CUDKOWICZ: What is the frequency of asialo GM1 positive cells after
using the anti-GM1 antibody in vivo?

HABU: The percentage of small lymphocytes that are asialo GM1 positive in the spleen is less than 1% after injection of anti-GM1.
Without treatment, we generally find about 16-20%.

CUDKOWICZ: So the percentage drops from 15 to 1.   That information
may help answer Dr. Herberman's question about how the anti-allotypic
antibody is working.

HERBERMAN: But that change in percentage is with anti-asialo GM1
not the anti-Igh[a].

CUDKOWICZ: Did you give the allotypic antibody in vivo?

OKUMURA: No, we used it only in vitro.   I would now like to change
the subject and ask Dr. Habu, my collaborator, to discuss NK cells
in resistance to bone marrow grafts.

HABU: In addition to the role of natural killer cells in controlling tumor growth in vivo, I wish to comment on their role in natural
resistance to allogeneic or semi-allogeneic bone marrow transplantation.   We used the basic experimental system originally described
by Dr. Cudkowicz, except that we treated some of our recipient mice
with asialo GM1 antisera.   It is well known that NK cells are resistant to irradiation and that is why we used the antisera.   In our
system, recipient C57BL/6 x C3H/J F1 mice were irradiated with 650
rads and injected with either anti-asialo GM1 antisera or normal
rabbit serum as a control. Then we transferred the parental semi-
syngeneic B6 or syngeneic F1 bone marrow cells into the irradiated
recipient mice.   After 5 or 7 days, we sacrificed the mice and looked
at colony formation in the spleen (Table 4).   In the group of mice
injected with anti-asialo GM1 antisera and transplanted with semi-
syngeneic B6 bone marrow cells, we found abundant colony formation.
In contrast, very few colonies were observed in the spleens of mice
injected only with normal rabbit serum.   We interpreted these results
to mean that NK cells suppressed colony formation with semi-syngeneic
bone marrow transplantation and that the injection of anti-asialo
GM1 antisera removed this resistance.   In order to rule out the effect
of this antisera on T cells, we used nude mice instead of the B6 x
C3H F1 mice.   The BALB/c nu/nu mice received asialo GM1 antisera and
the semi-syngeneic B6 x BALB/c bone marrow.   The results were the
same as in the B6 x C3H F1 mice that received semi-syngeneic bone
marrow and anti-asialo GM1 antisera. · In both recipients, there was
abundant colony formation in the spleens.   We concluded that the
asialo GM1 positive cells had NK like function and that the resistance to semi-syngeneic bone marrow transplantation in vivo reflects
NK activity.

OKUMURA: Dr. Cudkowicz has some interesting data concerning the
genetically determined low NK activity beige mouse.

Table 4. Effect of Anti-asialo GM1 on Hybrid Resistance

Recipient Mice		Strain of Bone Marrow Donors	Number of Colonies[a]	Growth Index[b] %
Strain	Injection of Anti-asialo GM1			
C57B1/6xC3H/J F1	+	C57B1/6	++	24
C57B1/6xC3H/J F1	-	C57B1/6	+/-	3
C57B1/6xC3H/J F1	+	B6xC3H F1	+++	
C57B1/6xC3H/J F1	-	B6xC3H F1	+++	
C57B1/6xBALB/c F1 nude	+	C57B1/6	++	ND
C57B1/6xBALB/c F1 nude	-	C57B1/6	-	ND
C57B1/6xBALB/c F1 nude	+	B6xC3H F1	++	
C57B1/6xBALB/c F1 nude	-	B6xC3H F1	++	

Antibody was injected intraperitoneally one day before bone marrow cells (2×10^6) were transferred into either non-irradiated or irradiated (650R) recipient mice. Recipient mice were sacrificed 5 days after bone marrow cell transfer and their spleens fixed in 10% formalin for colony counting.

[a] The number of colonies in median sections of spleens were scored as +++:>50, ++:50 to 30, +:30 to 10, -:<10. Four recipients for one group were examined.

[b] ($5-^{125}$I)-2-deoxy-uridine (100 μci) was injected into recipient mice 6 hours before sacrifice and the spleens counted. Growth index was calculated as follows:

$$\frac{\text{CPM of spleens receiving C57B1/6 bone marrow cells}}{\text{CPM spleens receiving F1 bone marrow cells}} \times 100$$

CUDKOWICZ: The issue I think Dr. Habu and Dr. Okumura and myself
are addressing is whether we can gather evidence that hybrid resistance to parental marrow graft in vivo is mediated by NK cells.   If
this were the case and there's evidence towards that conclusion,
it's an unusual NK cell because the effector cell is an antigen specific cell that only interacts with target cells that bear certain
major histocompatibility complex structures on their surfaces.   I
wish to describe an experiment performed in beige mice in which we
transplanted parental bone marrow cells into syngeneic and F1
recipients.   In untreated F1 hybrid mice, there was very little cell
growth, as measured by splenic uptake of IUDR, regardless of whether
the animal was +/+ or heterozygous at the beige locus.   When the host
was homozygous, there was substantial growth but not as much as in
syngeneic recipients.   The conclusion was made that a major problem
in the beige mouse was one affecting NK cells.   On the other hand,
if these animals were stimulated with the interferon inducer poly
I:C,they acquired some resistance although not full resistance.   In
a similar experiment involving allogeneic and xenogeneic donors, we
transplanted WB bone marrow cells into C3H, the choice of strains
being dictated by the availability of beige mutant alleles.   Again,
the heterozygous litter mate was strongly resistant to the graft,
but the beige litter mate had lost most of its resistance.   This
observation extends to the xenogeneic system as well, where the
heterozygous C3H was resistant to a large number of hamster bone
marrow cells, but the C3H beige homozygote was not.   We don't have
syngeneic hamster recipients but as a positive control we used the
SJL, another NK deficient mouse strain.   The SJL was susceptible to
this graft.

     This data suggests than an animal with a pronounced NK cell de-
fect also has a defect in resistance to bone marrow transplantation.
Dr. Habu and Dr. Okumura have shown that the asialo GM1 positive
cells in the spleen or marrow of the beige mouse is reduced.   The
reduction seems to be mostly for lymphocytic GM1 positive cells as
opposed to the monocytic series.   This is additional evidence that
this in vivo phenomenon involves an interaction between NK like cells
and hemopoietic stem cells.

DORIA: Has Dr. Okumura tried to immunize nude mice pretreated with
asialo GM1 antibody with a T independent antigen? How would such
pretreatment affect the antibody response? What is the regulatory
role of NK cells on antibody responses?

OKUMURA: We haven't used T independent antigens, only T dependent
antigens.   In general, if you delete NK cells during in vitro antibody formation, a great enhancement of antibody production is observed.   One can postulate that the NK cells attack some of the cells
involved in antibody production.   Alternatively, the asialo GM1
antisera may be affecting suppressor T cells as well as NK cells.
There are two possibilities here.

KIRCHNER: Using fluorescent staining and anti-asialo GM1 antibody, how does the percentage of positive cells in the peritoneal exudate population compare to that of the spleen?

HABU: We don't have data on peritoneal cells because peritoneal cells non-specifically bind the antibody.

FRIEDLANDER: Could you reiterate which cells are positive for asialo GM1 antibody besides NK cells?

OKUMURA: Generally, about 70% to 80% of the small lymphocytes react, as do about 30% of the monocytes.

TOKUNAGA: If we induce macrophages with proteose peptone, the percentage of asialo GM1 positive macrophages is very low. If we inject BCG or C. parvum and harvest after 4-5 days, the percentage of positive macrophages is 35%. In mice, activated macrophages are stained with asialo GM1 antisera.

KIRCHNER: After injecting C. parvum, there is very high NK cell activity in the peritoneal cavity. Are you really sure that these cells are macrophages?

TOKUNAGA: We separated the cells by adherence and we tested cytotoxicity for NK cells using 4 hour chromium release from labeled YAC-1 cells and a 24 hour chromium release from EL-4 cells for macrophages. Asialo GM1 antisera abrogated both cytolytic activities.

ROLE OF NATURAL KILLER CELLS AND OF INTERFERON

IN NATURAL RESISTANCE AGAINST VIRUS INFECTIONS

H. Kirchner, H. Engler, R. Zawatzky and L. Schindler

Institute of Virus Research
German Cancer Research Center
Heidelberg, F.R.G.

INTRODUCTION

There are a number of experimental findings that are fairly well established:

(a) Interferons cause an antiviral effect in tissue culture (1).

(b) Injection of interferon causes antiviral protection in vivo (for a review see Ref. 2).

(c) Injection of a variety of viruses causes rapid interferon production in vivo, both at the infection site (for example in the peritoneal cavity) and in the serum (3).

(d) Injection of interferon or interferon inducers (including viruses) causes activation of Natural Killer (NK) cells in vivo (4).

The following points, however, need to be established:

(a) Is the in vivo activation of NK cells exclusively due to interferon; or, if not, are there additional mechanisms by which viruses or perhaps virus-induced cellular neoantigens activate NK cells?

(b) Do interferon and/or NK cells as endogenously produced during the course of an infection play a role in the primary resistance against this infection.

In the following we want to first summarize some of the pertinent
literature and subsequently give a brief account on the data of our
laboratory in one experimental virus infection in mice.

WHAT IS THE EVIDENCE FOR A ROLE OF NK CELLS IN ANTIVIRAL RESISTANCE

The recent literature has been summarized by Welsh (5). The
hypothesis that NK cells play a role in antiviral resistance is supported by several lines of evidence, some of which are represented
by in vitro data. For example, it has been found that virus-infected
cells may represent targets that are more readily lysed by NK cells
than their uninfected counterparts (6). However, the situation
appears to be complex and to differ very much among the different
virus systems studied. It is our opinion that thus far none of the
claims based on in vitro studies sufficiently supports a role for
NK cells in antiviral defense.

Basically, there are three in vivo approaches to test if NK
cells have a role in antiviral defense:

First, adoptive transfer experiments may be performed. By transfer of cells from adult, resistant mice to newborn, susceptible mice
Tardieu et al. have obtained data suggesting a role of NK cells in
the resistance against infection with Mouse Hepatitis Virus type 3
(MHV3) (7). However, adoptive transfer protocols in NK cell work
have been plagued by serious problems since it is virtually impossible
to obtain populations sufficiently enriched for NK cells and sufficiently depleted of other cell types.

A second approach to prove the relevance of certain effector
cells is to inject immunosuppressive substances. For example, our
studies (see below) have shown that injection of cyclophosphamide
(or other substances) abolishes NK cell activation and breaks resistance of C57BL/6 mice to Herpes simplex virus (HSV). However, substances such as cyclophosphamide affect other cell types besides NK
cells.

Third, to prove the relevance of NK cells in antiviral resistance
one has searched for genetic associations between the magnitude of
the NK cell response after injection of certain viruses and the
degree of resistance against this virus.

In at least two different virus systems, a correlation was found
between the magnitude of NK cell activation and genetically determined
resistance. One is our model of experimental infection of mice with
HSV (see below) and the second is a model of murine cytomegalovirus
(MCMV). A correlation existed between resistance to the lethal effects of MCMV and the magnitude of the virus-induced NK cell activity
in 10 of 11 mouse strains tested (8).

In a Sindbis virus system it was found that susceptible newborn mice did not differ in the degree of NK cell activation from adult resistant mice after virus injection (9). However, in the HSV system different results were obtained as will be shown below.

Finally, homozygous beige mice (bg/bg) that are deficient in NK cell activity were compared in two viral models (10). There was no evidence of increased susceptibility to either virus infection in NK cell deficient bg/bg mice.

WHAT IS THE EVIDENCE FOR A ROLE OF INTERFERON IN ANTIVIRAL RESISTANCE

Based on numerous experiments, there is no doubt that interferon protects cells in vitro from a viral infection if added at a sufficient time before virus infection. It is also well established that injection of interferon or interferon inducers protects experimental animals from virus infection in vivo. Under favorable experimental conditions, interferon may even influence the outcome of a viral infection when administered after virus infection. However, the crucial issue remains if the interferon that is produced during the course of an infection is able to influence the outcome of this very infection. Often one observes a coincidence between high virus titers and high interferon titers occurring at the same time. However, by the use of anti-interferon serum, in several virus systems one has been able to break resistance against the infection (11). Although the anti-interferon serum was not monospecifically directed against interferon, these data have strongly suggested a role of interferon in primary antiviral resistance. The use of antisera certainly is a much finer tool than using cytotoxic drugs that more or less affect all cells of the defense system. However, there are few data that show a correlation between the magnitude of the primary interferon response and resistance. One system where such a correlation was found will be described below.

ACTIVATION OF NK CELLS BY HSV IN VIVO

Peritoneal exudate cells of untreated C57BL/6 mice in our laboratory show low if any NK cell activity. Injection of HSV caused a marked activation of NK cells in the peritoneal exudate cell population. This activation was fast (12-16 hours) and was caused by the virions themselves and not by any contaminant of the virus pool as appropriate controls have shown(12). NK cells activated by HSV lysed a variety of different targets including YAC-1 lymphoma cells, the targets most commonly used in NK cell studies. As previously decribed, the NK cells found in the peritoneal exudate of HSV-injected mice shared all typical characteristics of NK cells and there was no need to assume that HSV-induced NK cells differ from conventional "uninduced" NK cells (12,13).

HSV virions were inactivated by a variety of techniques including heat and UV light, and the complete absence of infectious particles was confirmed by a sensitive plaque assay. Non-infectious virions were fully capable of activating NK cells in the peritoneal exudate of C57BL/6 mice.

Different strains of HSV, including HSV-1 (ANG) which is apathogenic for mice after intraperitoneal injection (14), and isolates of HSV-2 were found to be equal to our standard strain HSV-1 (WAL) in their capacity to induce NK cell activity in the peritoneal exudate of C57BL/6 mice.

INDUCTION OF INTERFERON IN THE PERITONEAL CAVITY BY INJECTION OF HSV

High titers of interferon were observed in the peritoneal exudate of HSV-injected adult C57BL/6 mice as early as 2-4 hours after infection. This interferon was found several hours before any newly formed infectious particles could be detected (15) and therefore must have been induced by the input virions or by early virus-induced cellular functions (see below).

The interferon induced in the peritoneal exudate of HSV-injected mice represented interferon alpha/beta and not interferon gamma because of its physicochemical properties and as determined by the use of appropriate antisera.

Virions inactivated by heat or by UV light did not induce detectable amounts of interferon in the peritoneal exudate fluid. A strain of HSV-2 that was a potent inducer of NK cell activity was a poor inducer of interferon. These discrepancies between the NK cell data and the interferon data will be discussed below. Interestingly a temperature-sensitive mutant of HSV-1 ANG was equally capable of in vivo interferon induction as the wild type indicating that only early virus-coded functions are required for induction of interferon production.

Obviously, the cells that produce the interferon that is recovered from the peritoneal cavity cannot be characterized. We have therefore used a modified technique in which peritoneal exudate cells were recovered after in vivo injection of HSV and subsequently cultured for short periods in vitro (12). These cells produced interferon alpha/beta and by several techniques we have approached the question as to the producer cell of HSV-induced interferon. Cells were treated with several monoclonal antibodies including anti-theta, anti-Qa 4 and anti-Qa 5. These reagents, under controlled conditions, did not affect HSV-induced interferon production, indicating that the interferon producing cells were not mature T lymphocytes nor NK cells.

A variety of additional data were not fully conclusive. In

spleen cell cultures, in contrast to peritoneal exudate cells, we have
found interferon induction by non-infectious HSV and the producer
cells appeared to be B lymphocytes (16). However, peritoneal exudate
cell cultures responded only to infectious HSV and the producer cells
may be macrophages. These conclusions are supported by data indicating that cultures of pure bone marrow-derived macrophages responded
to infectious HSV with production of interferon. However, they did
not respond to inactivated HSV (unpublished data by H. Kirchner and
E. Storch).

EFFECT OF IMMUNOSUPPRESSANTS ON NK CELL ACTIVATION AND INTERFERON
PRODUCTION

     Three compounds which were found to increase the lethality of
C57BL/6 mice after intraperitoneal infection with HSV were tested
for their effect on NK cell activation and on interferon production.
None of these, including silica, hydrocortisone, and cyclophosphamide,
affected early in vivo interferon production, suggesting that the
producer cell of interferon was not sensitive to any of these drugs.
However, HSV-induced NK cell activity in peritoneal cell populations
of mice injected with any of the drugs was completely abolished.
Although the cytotoxic drugs affect a variety of other cells besides
NK cells, the data support a role of NK cells in antiviral resistance
(see below for further discussion of this aspect).

     Administration of one injection of cyclosporine A also abolished
HSV-induced NK cell activation. However, in contrast to other drugs,
cyclosporine A increased survival of C57BL/6 mice and increased the
titer of interferon in the peritoneal exudate fluid after injection
of HSV.

HSV-INFECTION OF NEWBORN C57BL/6 MICE

     Newborn C57BL/6 mice, in contrast to adult mice, are highly
susceptible to intraperitoneal infection with HSV (15). Resistance
of C57BL/6 mice develops at about 20 days of age (see below). Newborns do not show early interferon production at 4 hours in the
peritoneal cavity and have no detectable NK cell response at 24 hours
(15).

     However, high interferon titers are observed 24 hours after
infection. These titers are observed concomitantly with high titers
of newly formed HSV, since HSV is replicated readily in the peritoneal
cavity of newborn C57BL/6 mice, in contrast to the situation observed
in adult C57BL/6 mice (15). After 2-3 days NK cell activation is
also observed in the peritoneal exudate cell population of newborn
mice. This late activation of NK cells is most probably due to the
late interferon.

The capacity of early interferon production and subsequent early NK cell activation develops in newborn mice at about 20 days of age in striking parallel with the development of resistance against HSV in C57BL/6 mice.

Incidentally, our data in newborn mice show that pre-NK cells are present in newborn mice, which after appropriate stimulation (i. e. by the late interferon) may be activated to kill YAC-1 target cells.

TESTING OF ADULT MICE OF DIFFERENT INBRED STRAINS

As previously described, adult mice of different inbred strains differ in their relative resistance to intraperitoneal infection with HSV-1 or HSV-2 (17,18). The data which we have summarized in the pregoing paragraphs have all been obtained with C57BL/6 mice that are resistant. Additionally, we have studied a variety of inbred strains of mice. The situation which we have observed in most resistant mice was similar to the situation in C57BL/6 mice. In susceptible mice data were obtained that closely resembled those of newborn C57BL/6 mice. Thus, there was a lack of early HSV-induced interferon at the infection site and subsequently a lack of early NK cell activation in mice that are genetically susceptible to HSV infection.

Among the resistant mice, SJL mice were of interest since they showed an interferon response that was comparable to the response of C57BL/6 mice but they did not show activation of NK cells. We will discuss this finding further below.

Homozygous nude mice of the C57BL/6 background were as resistant to HSV infection as their heterologous littermates (19) and showed high early interferon titers and high NK cell activity. This finding reiterates that neither interferon production nor natural cytotoxicity are functions of mature T cells and that in our model of primary antiviral resistance T cells play no major role.

DISCUSSION

Collectively, our data in newborn C57BL/6 mice and in adult mice of different strains seem to suggest that both NK cells and interferon play a role in the genetically controlled resistance of mice to HSV.

That viruses induce interferon and that interferon in turn activates NK cells is experimentally well established. However, the relationship between NK cells and interferon and the respective contribution of NK cells and interferons to host resistance are not fully understood. It is not known if virus-induced interferon is the only mechanism by which viruses activate NK cells.

Our data using inactivated preparations of HSV (or certain strains of infectious HSV) might indicate that NK cell activation can occur in the absence of interferon induction. Similarly, in experiments not shown here we have observed that low doses of HSV-1 (WAL) did not induce interferon but were able to activate NK cells in the peritoneal exudate (Kirchner and Engler, unpublished observations). However, it cannot be excluded that locally produced interferons, at concentrations too low to be detected in the wash-out fluid, are sufficient for activation of NK cells.

The data in SJL mice show a correlation between resistance and high titers of early interferon. However, in confirmation of the data of Cudkowicz, we have found that SJL mice have a defective NK cell system, even after induction of high titers of interferon. From these data one might conclude that interferon plays a major role in primary antiviral resistance in our model, but only by its direct effect on the targets of viral infection. However, SJL mice are less resistant than C57BL/6 mice that show a high early interferon response and a high NK cell response.

The data from the experiments in which the effects of immuno-suppressants were tested showed that several agents were without effect on the interferon producing cell but abolished NK cell activity and increased lethality. These data seem to suggest that NK cells do contribute to resistance in our model, although the compounds tested (e. g. cyclophosphamide) certainly affect other cells besides NK cells.

Treatment with cyclosporine A, in contrast, decreased mortality although it severely diminished the NK cell response. Cyclosporine A increased the amount of interferon locally produced which may be responsible for the observed increase in survival. These data seem to suggest again that NK cells are of less relevance than interferon in our model.

Thus, obviously the situation is complex in regard to the relative contributions of interferon and NK cells in the primary defense of mice against HSV and one might conclude that both the interferon producing cell as well as the NK cell play a role in resistance and that the activation of NK cells is of greater significance than simply representing an indicator system of previous interferon production.

Herpesviruses are infectious agents that occur in most if not all species. Usually, the host range of a given herpesvirus is restricted to one species. Herpesviruses have coevolved with the species in which they are presently found, for example MCMV with mice and EBV with the human species. They live in good harmony with the host and probably the individuals genetically susceptible to the virus have been eliminated by evolution. Thus, herpesviruses are ubiquitous and usually are not pathogenic for the species in which they occur. Exceptions have been found lately when iatrogenic procedures have been developed that massively interfere with the defense system of the body, such as the use of immunosuppressive drugs and the transplantation of organs, particularly bone marrow, the latter leading to graft vs host reactions.

HSV, in contrast to MCMV and EBV, is pathogenic for mice besides humans as their natural hosts. HSV, after i.p. infection, kills newborn mice of all strains and adult mice of genetically susceptible strains. In this experimental system, Lopez (20) has documented a role of a T cell-independent, bone marrow-dependent defense system. We have presented evidence that an early interferon response at the infection site is relevant for defense. This early interferon response is caused by the infecting virus and formed well in advance before new virus is replicated. Interferon is produced by non-T cells.

It remains to be determined if these findings in one experimental system apply to other viral infections. It is of interest that in the MCMV system, a genuine mouse model of herpesvirus infection, in contrast to the HSV system, a role of the H-2 locus in resistance has been documented (8) which is not the case in our system.

REFERENCES

1. Isaacs, A., and Lindenmann, J., Proc. Royal Soc. B 147:258, 1957.
2. Stewart, W. E., "The Interferon System", Springer Verlag Wien, New York, 1979.
3. Baron, S., and Buckler, C. E., Science 141:1061, 1963.
4. Gidlund, M., Örn, A., Wigzell, H., Senik, A., and Gresser, I., Nature 273:759, 1978.
5. Welsh, R. M., Antiviral Res. 1:5, 1981.
6. Welsh, R. M., and Hallenbeck, L. A., J. Immunol. 124:2491, 1980.
7. Tardieu, M., Héry, C., and Dupuy, J. M., J. Immunol. 124:418, 1980.
8. Bancroft, G. J., Shellam, G. R., and Chalmer, J. E., J. Immunol. 126:988, 1981.
9. Hirsch, R. L., Immunology 43:81, 1981.
10. Welsh, R. M., and Kiessling, R. W., Scand. J. Immunol. 11:363, 1980.
11. Gresser, I., Tovey, M. G., Maury, C., and Bandu, M. T., J. Exp. Med. 144:1316, 1976.
12. Engler, H., Zawatzky, R., Goldbach, A., Schröder, C. H., Weyand, C., Hämmerling, G. J., and Kirchner, H., J. Gen. Virol. 55: 25, 1981.
13. Armerding, D., and Rossiter, H., Immunobiol. 158:369, 1981.
14. Schröder, C. H., Engler, H., and Kirchner, H., J. Gen. Virol. 52:159, 1981.
15. Zawatzky, R., Engler, H., and Kirchner, H., J. Gen. Virol., in press, 1982.

16. Kirchner, H., Keyssner, K., Zawatsky, R., and Hilfenhaus, J.,
       Immunobiol. 157:401, 1980.
17. Lopez, C., Nature 258:152, 1975.
18. Kirchner, H., Kochen, M., Hirt, H. M., and Munk, K., Z. Immun-
       forsch. 154:147, 1978.
19. Zawatzky, R., Hilfenhaus, J., and Kirchner, H., Cell. Immunol.
       47:424, 1979.
20. Lopez, C., Ryshke, R., and Bennett, M., Infect. Immun. 28:1028,
       1980.

DISCUSSION INTRODUCER HOLGER KIRCHNER: I would like to call on Dr.
Habu to discuss her results with Herpes simplex virus infection in
mice treated with anti-asialo GM1.

HABU: NK cells contribute to primary resistance to viral infection.
We obtained evidence on this point by demonstrating that mice injected
with asialo GM1 antisera are more susceptible to viral infection.
In our experiment, we injected Herpes simplex type 1 virus into
nude mice. The group of infected mice that had received anti-asialo
GM1 died faster than the mice treated with normal rabbit serum (Figure
1). We obtained the same results using euthymic C57BL/6 mice injected
with the same amount of antiserum and Herpes simplex type 1 virus
(Figure 2). We interpreted these results to mean that NK activity
and survival after viral infection are correlated and that NK cells
mediate resistance against the virus.

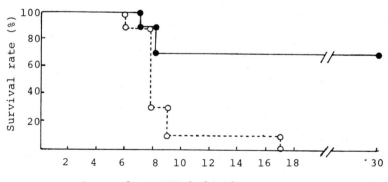

Fig. 1. Herpes-simplex virus type 1 (HSV) was inoculated intraperi-
         toneally in C57BL/6 nude mice receiving 0.01 ml anti-asialo
         GM1 (o----o) or normal rabbit serum (•——•) on day-1.

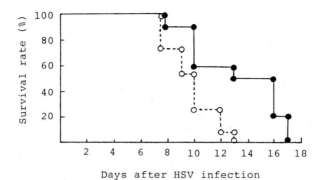

Fig. 2. Herpes-simplex virus type 1 (HSV) was inoculated intraperi-
         toneally in C57BL/6 mice receiving 0.01 ml anti-asialo GMl
         (o----o) or normal rabbit serum (●——●) on day-1.

        As suggested by Dr. Kirchner, resistance against viral infection
is related to interferon production. Interferon production induced
by poly I:C was decreased in mice receiving anti-asialo GMl versus
those receiving normal rabbit serum (Table 1). These results suggest
that interferon may contribute to the resistance against the virus
and be diminished in response to asialo-GMl antisera.

KIRCHNER: Several groups have shown that a human NK like cell is
a producer of interferon. In my opinion, this is a very important
issue not only to resistance to viral infection but also to tumor
resistance. Does Dr. Herberman have a comment on human large granular
lymphocytes or on murine NK cells being interferon producers?

HERBERMAN: In the human, it appears that large granular lymphocytes
respond to a variety of interferon inducers including Herpes simplex
virus. At least in vitro, the large granular lymphocytes are the
main cells making interferon. We don't have comparable data in the
mouse.

KIRCHNER: We know that bone marrow cultures containing pure macrophages respond to Herpes simplex virus with interferon production.
It would be important to see now if the anti-asialo GMl antibody also
acts on bone marrow cells. Aside from viruses, bacteria are important
interferon inducers. One can speculate on how resistance to bacteria
is also influenced by interferon. Leibold has some data about the
effect of bacteria on NK cells.

LEIBOLD: We induced rheumatoid arthritis in specific pathogen free

Table 1. In vivo Effect of Anti-asialo GM1 on NK Activity and
          Interferon Production

Strain	Anti-asialo GM1	% Lysis YAC-1 Targets (E/T ratio 100/1)		Units of Interferon	
		poly I:C	PBS	poly I:C	PBS
C57BL/6	+	2	1	60	<20
	−	41	27	1,280	<20
C57BL/6 nude	+	3	2	620	<80
	−	55	42	1,520	<20

All mice were injected (intravenously) with either anti-asialo
GM1 or normal rabbit serum on day 0. On day 1, they received
either 100 µg poly I:C or phosphate buffered saline (PBS). Four
hours later, their sera were pooled to examine for circulating
interferon (IF) level.

pigs. Among other immunological functions, we investigated natural
killer activity. Our biochemists have obtained from the causative
bacteria several components with varying effects on the immune system.
One component appears to modulate natural killer activity.

KIRCHNER: When tumor cells are co-cultivated with lymphocytes or
NK cells, the data indicate that interferon is produced. We have
been interested in interferon produced in response to a mycoplasma
infection of the tumor cells. Are there any systems of co-cultivation
of tumor cells and leukocytes where it has been totally ruled out
that the interferon has not been induced in response to a contaminating microorganism?

GALLILY: When we co-cultivated tumor cells with non-activated macrophages, we found very active killing of the tumor cells. We finally
looked for mycoplasma contamination and discovered that we could
attribute the cytotoxicity to the presence of mycoplasm. When we
eliminated the mycoplasma, the macrophages did not kill the tumor
cells. After adding various amounts of mycoplasma to the assay
system, we could achieve cytotoxicity as observed before. Cytotox-

icity correlated with the amount of mycoplasma added. We don't understand as yet the mechanism of the cytotoxicity but we assume that it
involves activation of the macrophages.

HERBERMAN: The studies that Dr. Julie Djeu did in collaboration
with my laboratory were done using human large granular lymphocytes.
She saw that a variety of cell lines were inducing interferon but in
fact most of these were contaminated with mycoplasma. When the same
cell lines were studied in the absence of mycoplasma contamination,
almost all of this interferon production went away. We have discussed
the same issue with Giorgio Trinchieri who was the first to show that
NK cells could make interferon in response to various cell lines.
He has only one cell line which he feels makes interferon in the
absence of mycoplasma contamination.

In the mouse, Dr. Julie Djeu has data indicating that inoculation
of these cell lines can induce interferon production. Here the
situation seems to be a bit different as some of the cell lines can
induce in vivo production of interferon even if they are not contaminated with mycoplasma. But I would agree that in vitro one would
be in a rather difficult position right now to argue that it is anything more than mycoplasma contamination.

LIEBOLD: In the laboratory of Hartmut Peter, we have investigated
this topic. Using a new method, we were able to clean certain melanoma cell lines and the K562 cell line free of mycoplasma. Using
these cell lines now freed of mycoplasma, we found the following.
The K562 cell line still acted as an inducer of interferon and still
acted as a natural killer cell target. The level of killing was
reduced but there was still significant killing. In contrast, the
melanoma cell lines which were easily killed in the presence of mycoplasma were not killed in the absence of mycoplasma. This result
was true if we used the effector cell on day 0 after harvest. If
we cultivated the effector cells for 2 or 3 days in normal medium,
however, then substantial natural killer activity was observed despite
the fact that interferon was barely detectable. Lastly, I would like
to mention work done in collaboration with Hartmut Peter and Holger
Kirchner. We co-cultivated infected tumor cells with normal peripheral blood lymphoid cells and assayed in parallel the interferon
induced and the natural killer activity. We found no correlation
between the amount of interferon induced and the degree of cytotoxicity.

KIRCHNER: We need to develop single cell assays for interferon
production in order to approach the question of the cell type producing interferon. I would like to ask Dr. Alm to discuss his system.

ALM: As you have heard, certain cell lines can induce interferon in
peripheral blood leukocytes in the human system. Why they are inducers of interferon is not clear, but mycoplasma contamination is

obviously one possibility. What we did was to establish monolayers
of amnion cells in microtiter plates to which we added human peripheral blood leukocytes. We found interferon production. We then
checked the cells for resistance against the common vesicular stoma-
titis virus. Since they produced interferon, they were resistant
to the virus. We then diluted the peripheral blood leukocytes and
found that some of these cultures were protected against the virus
while other cultures were not. It was an all or none phenomenon.
Using a limiting dilution type assay, we calculated the frequency of
a protecting unit in the system. This protecting unit had a frequency
of about 1 in 1,000 blood leukocytes. The protesting unit was a cell
and that cell was producing alpha type interferon. Further, the
producing cell could be activated. The frequency of the protecting
unit could be significantly increased 2 or 4 times by interferon preincubation. When we examined the nature of the cell, we found that
it was an NK like cell. It was a non B, non T, non-adherent, Fc
receptor bearing cell. This cell was non-specific. Its frequency
could not be increased by adding other types of inducers to these
co-cultures. It always stayed the same about 1 in 1,000. It was
remarkably constant if you assayed the same individual several times.

OVERVIEW AND PERSPECTIVES:

NATURAL RESISTANCE MECHANISMS

Ronald B. Herberman

National Cancer Institute
Division of Cancer Treatment
Biological Response Modifiers Program
Frederick Cancer Research Center
Frederick, Maryland   21701 USA

I've been asked to summarize today's session and to give an
overview of the field of natural resistance. It would be worthwhile
first to go back to the issue raised by Dr. Cudkowicz at the beginning of the session: that there is a heterogeneity of natural effector cells. To think broadly about the natural resistance system,
one needs to consider effector cells other than NK cells. This is
an important point to reiterate, particularly since we heard in this
symposium very little about cells other than NK cells. In order to
correct that imbalance, I'll make some remarks not only about NK cells
but also about the so-called natural cytotoxic (NC) cells and about
naturally cytotoxic macrophages. For completeness, it should be
noted that granulocytes from normal individuals can have at least
cytostatic effects against tumor cells as well as their known effects
on various microorganisms.

One of the major issues discussed during the early part of the
symposium addressed the question, "What is the nature of the natural
killer cell?" As was rightly pointed out by Dr. Cudkowicz, most of
the definition of the NK cell has been a functional and empirical
one; namely, in the mouse, "the cell that kills YAC-1 targets primarily" and in the human, "the cell that kills K562 targets". If one
had to remain entirely at that functional level, there clearly would
be some difficulties. However, it is becoming possible to get away
from a purely functional definition. Probably the best alternative
has been morphology for it is now established that large granular
lymphocytes are intimately associated with natural killer cell activity. Indeed, one can state quite categorically that virtually all
of the human and rat NK activity against typical NK target cells is

mediated exclusively by large granular lymphocytes and that a high
proportion of cells with that morphology have the capability to
mediate NK activity. It's been very difficult to make the same association in the mouse for various technical reasons: For one thing,
the cells in the mouse that have azurophilic granules are somewhat
smaller, so they aren't really <u>large</u> granular lymphocytes--they're
sort of medium-sized ones. However, they are a bit larger and less
dense than a typical lymphocyte, and by varying the Percoll gradient
conditions, Dr. Tagliabue in Italy and also Dr. Kumagai in Japan have
been able to get reasonably good separation of these cells from other
lymphoid cell types of the spleen or blood. This has been repeated
in my laboratory and the cell fractions enriched for these granulated
cells in the mouse, as well as in the other species, account for
virtually all of the NK activity against YAC-1 targets.

"What is the NC cell?" This is a cell that Stutman and co-
workers have described as reacting predominantly with certain monolayer target cells, like the WEHI-164 cell and certain methylcholanthrene-induced sarcomas (1). The definition of the NC cell in regard
to antigenic markers has largely been a negative one, since these
cells do not seem to have easily detectable cell surface antigens.
To the extent that we've looked (and we're just at the beginning),
it appears that the effector cell that is cytotoxic against the
WEHI-164 target cell line, and thus is a prototype for the NC cell,
is enriched in the Percoll fraction that contains the granular lymphocytes and is depleted from the populations devoid of these cells.
Thus, it would appear that the heterogeneity that may exist in NC,
NK, or other types of natural nonadherent effector cells is probably
heterogeneity within the large granular lymphocyte population and
not heterogeneity due to entirely distinct cell populations. Considerably more work needs to be done on this point but this is the
way the data are going right now.

A further issue about the nature of the NK cell is its cell
surface phenotype. There was a lot of discussion about the various
markers and how reliable they really are. In regard to the T-cell
associated markers or those associated with macrophages or granulocytes on NK cells, I've been impressed that these markers tend to be
unstable. For example, if NK cells are placed in culture with IL-2,
within a few days the phenotype changes rather dramatically. I feel
that this is not unique to NK cells but rather is a reflection of
the markers themselves. My impression is that most of these markers
are differentiation markers rather than stable lineage markers.
If one changes the state of differentiation of the cells, there can
be up- or down-regulation of the expression of most of these markers.
There are a number of instances that support this contention independent of studies on NK cells. What one would like to have most would
be markers that are really selective for NK cells and that are not
found on other types of cells. At the moment, I'm not convinced that
we have any such marker in any of the species that have been studied.

We have some markers that are relatively selective, but none of them (to my mind) can be considered completely documented as entirely selective for the NK population. I might take the asialo GM1 antigen as an example. Asialo GM1 seems to be a good marker for the NK population, but in studies with the fluorescence-activated cell sorter with separated cell populations in both the rat and the human, it has been seen in my laboratory that not only do most of the large granular lymphocytes have detectable amounts of asialo GM1 but also most of the peripheral blood monocytes in both the rat and the human have detectable (although lower) amounts of the asialo GM1. In addition, many granulocytes express this antigen. So one needs to be cautious as to just how reliable one of these markers really is.

In regard to the asialo GM1, we've been able to confirm the reports of Okumura and Habu that the in vivo administration of this antibody very effectively depletes NK activity in both the mouse and the rat. In the rat, most large granular lymphocytes (LGL) disappear and stay away for at least 4 to 5 days after administration of the antibody. It's interesting that although most of the peripheral blood monocytes express this marker, administration in vivo of the antibody has had no effect on the number of circulating monocytes. In preliminary studies, we have not seen a decrease in monocyte-mediated cytotoxic activity in rats treated with this antibody. Therefore, although the marker is not entirely selective, it appears as if the functional effects are rather selective. However, I think this needs to be looked at more carefully.

Several markers have been claimed to be rather specific for NK cells and include the alloantibodies in the mouse, anti-NK-1 described by Glimcher, Shen and Cantor (2) and anti-NK-2 described by Burton. Although the current data indicate high selectivity, they've not been studied extensively enough to really be sure. The only thing that can be said is that these markers are on a small percentage of cells and they don't seem to be on cytotoxic T cells. I think that more extensive studies are needed before we can be satisfied that these markers are really restricted to NK cells. Similarly, for the HNK-1 that has been described by Abo and Balch (3), it would appear that more cells are positive for HNK-1 than one would expect for the percentage of LGL in human peripheral blood; in fact, they have some very recent indications that there is a subset of HNK-1 positive cells that express OKT3 which have very little NK activity associated with them. So again, this marker probably is not quite as selective as we would like.

In regard to the overall question of the lineage of NK cells, the question is still open. I personally favor the interpretation that NK cells are in some way related to a T-cell lineage, but the evidence is not conclusive by any means and one can not discount completely the sharing of antigens between NK cells, macrophages, and granulocytes. One also needs to entertain the possibility that

the NK cells may be an entirely separate lineage, coming from pleuri-
potential stem cells and simply sharing attributes between the T-cell
lineage and the myelomonocytic lineage. Perhaps it would be worthwhile to summarize briefly some of the evidence for the association
of the NK cells with the T-cell lineage. The sharing of certain
antigens and the presence of E-receptors on human NK cells is fairly
clear. Also the growth of the NK cells with IL-2 is fairly striking
and stands in contrast to the inability of these cells to grow with
colony stimulating factor. Another factor that distinguishes the NK
population from the myelomonocytic series is that a purified human
LGL population, in response to a range of different stimuli, was not
able to undergo an oxidative burst. These latter studies were performed in collaboration with Drs. Ronald Goldfarb and Edgar Pick.

     It originally was thought that NK cells were early or pre-thymic
T cells, largely based on the observations of high NK activity in
nude or neonatally thymectomized mice or rats. However, a more
plausible argument may be that the NK cells represent a side pathway
of differentiation from the precursors for T-cells; perhaps cells in
the T-cell lineage may have two choices when a thymus is present:
they may either go through the thymus and become typical T-cells with
typical T-cell functions, or they may develop through a thymic-
independent pathway and become NK cells. There is also the possibility of an interchange between the NK compartment and the typical
T-cell compartment. The strongest argument for that comes from clones
prepared from cells growing in culture with IL-2. These apparent
clones were selected first for having specific T-cell cytotoxic
activity but, simultaneously or more frequently after losing their
T-cell function, gained the ability to kill a typical NK target.
I don't think any of these claims are well-substantiated but since
they've come from several different laboratories, they raise the
possibility of a transit from cytotoxic T cells to an NK cell and
possibly also in the other direction. The type of evidence that
Dr. Okumura presented about the allotypes is quite intriguing and
would support such an argument; perhaps the difference between an
NK and a typical cytotoxic T cell may largely be what its previous
history has been and what its state of differentiation might be.
Although I favor this possibility, it is important to reiterate that
one could make an equally strong argument for the NK cells being
in a separate lineage and simply sharing some attributes with the
T-cells. For example, the argument about growth with IL-2 becomes
a bit circular. You could say that IL-2 is specific for T-cells and
therefore NK cells that grow in response to IL-2 are in the T-cell
lineage. On the other hand, you could say that IL-2 simply promotes
growth of both T-cells and cells in the NK lineage. It's difficult
to distinguish between these alternatives. The definitive experiments
would be to either document completely the above-mentioned in vitro
observations and show that there is a true clone of cytotoxic T cells
that either simultaneously or subsequently develops NK activity or
to show that administration of a purified population of NK cells in

vivo can reconstitute or develop some T-cell function in the recipients.

I might mention some intriguing recent data that Dr. Julie Djeu
has obtained in collaboration with Dr. James Ihle and Andrew Happel,
with cells grown with another interleukin, IL-3. This factor has
been described as causing the induction, particularly in nude spleen
cells, of a progesterone metabolic enzyme, 20-alpha-corticosteroid
dehydrogenase. This enzyme seems to cause differentiation of T cells
and growth of some cells in the T-cell lineage as well as growth of
cells with the general appearance of mast cells. This is another
point of confusion, with T cells and mast cells having some unexpected
connections. Dr. Djeu has recently tested several IL-3 dependent
cell lines for cytotoxic activity in vitro. None of these had any
cytotoxic activity against YAC-1 targets, even in an 18-hour assay.
The striking thing has been that most of these IL-3 dependent cell
lines killed the WEHI-164 cell line very well. So in the fresh state,
mouse LGL seem to account for both the YAC-1 and the WEHI-164 killer
activity. Cells cultured with IL-2 have more typical NK activity
whereas cells cultured with IL-3 seem to lack the anti-YAC-1 activity
but have quite strong anti-WEHI-164 activity. Observations along
this line may help to sort out some of the interrelations between
those two types of natural effector cells.

I'd like to turn now to a few comments about regulation, particularly about how interferon might affect NK cell activity. It has
turned out to be a very complex story. Briefly, there appear to be
several levels at which interferon can affect the amount of NK activity. By doing conjugate assays with purified LGL and single-cell
cytotoxicity assays in both the rat and the human, it has been possible to show that interferon can act at several different levels,
depending on the target cells used in the analysis. First, one can
obtain a very rapid induction in some cells with LGL morphology from
cells which lack the ability to recognize a given target into target-
binding cells. Second, one can show that some cells that already
bind targets but which are lytically inactive become lytically active.
Third, interferon can accelerate the rate of killing by cells that
already have the ability to kill spontaneously. Fourth, as reported
by some indirect calculations from Ull and Jondal and measured more
directly in my laboratory by Tuomo Timonen, it appears that interferon can increase the amount of recycling, i.e. cause the NK cells
to interact with more than one target over the period of a 4-hour
assay. In addition, as we heard earlier from Carlo Riccardi, it
appears that interferon also can stimulate the differentiation of
NK cells and their ability to grow in response to IL-2. Finally,
there are some interesting recent observations made by Diana Buraski
and Angela Santoni that exposure of the macrophage-type of adherent
suppressors, as Dr. Cudkowicz has described for the mouse NK system,
to poly I:C or to interferon resulted in their becoming less suppressive. Thus, this is another way in which NK activity can be in-

creased: namely, by interfering with the inhibitory activity of
adherent suppressor cells. It is important to also mention some
of the effects of interferon which can negatively affect NK activity.
One has been the demonstration by Riccardi that interferon can inhibit
the differentiation of NK cells, which seems to be an indirect effect,
dependent on the presence of T-cells in the culture. Another observation, originally made by Giorgio Trinchieri, was that interferon
treatment of a number of target cells makes them more resistant to
being killed by NK cells. Clearly, there are quite a lot of interesting ways in which interferon can regulate NK activity.

    One point which has not been brought up concerns positive regula-
tion of NK activity by IL-2 itself. In addition to its growth effect
on NK cells, IL-2 seems also to stimulate their function. It was
originally reported by Kurabayashi, Gillis and Henney that in the
mouse IL-2 stimulated NK activity. This has been repeated in my
laboratory with human NK cells. This effect can be fairly rapid,
raising the possibility that IL-2 may not only be a proliferative
signal but also a signal for stimulating the function of these cells.

    With these types of observations, we might come back to the
question that Dr. Kirchner raised earlier: "What is the basis for the
spontaneous NK activity in either the mouse, the rat or the human?"
It has been suggested by various people, including Dr. Kirchner,
that perhaps endogenous production of interferon, for example by
virus infection, may be to some extent responsible for this. However,
the finding of some NK activity in the newborn human tends to speak
against this hypothesis. Our experience also has been that germ-
free or pathogen-free mice have fairly good levels of NK activity.
Thus, the requirement for environmental stimulation is not so clear.

    In regard to the potential role of IL-2 as a signal for sponta-
neous NK activity, I might briefly mention some interesting experiments that Wolfgang Domzig has performed in my laboratory using a
monoclonal anti-IL-2 antibody (this is in a human system). If he
incubates human LGL with this monoclonal anti-IL-2 overnight, most
of the spontaneous NK activity disappears, suggesting that, at least
in the human, it is important to have some endogenous production of
IL-2 in order to maintain the NK activity. The relative importance
of interferon versus IL-2 production is something which will need
to be sorted out further.

    Coming back to NC cells, I might mention that it has now been
clearly shown by Osias Stutman that for target cells only susceptible
to lysis by NC cells (e.g., meth-A target cells), interferon is not
an activator of that activity. So NC effector cells seem to be
regulated differently from typical NK cells. I might also mention
parenthetically that cytotoxic T cell activity has been reported by
Lindahl and Gresser to be positively regulated at the effector phase
by interferon. However, we've looked at that issue more carefully

and have not been convinced that either mouse or human cytotoxic T
cell activity at the effector phase can in fact be increased by interferon. My interpretation of those early experiments is that probably
the observed increase was related to NK activity in the population,
which would be supported by the fact that the experiments were done
with tumor targets in which there may have been some role for NK
cells as well as cytotoxic T cells. It's also worthwhile to stress
that macrophage cytotoxicity can be increased not only by lymphokines
and LPS but also by interferon in both mouse and human systems.
Gamma interferon is a particularly potent augmenter of macrophage
or monocyte cytotoxic activity, to the extent that some people are
raising the question as to whether some of the MAF preparations are
largely gamma interferon. I think that's unlikely: there presumably
is a true MAF separate from interferon. But one needs to be very
careful and rule out a role for gamma interferon before one speaks
about a true, separate macrophage activating factor.

Let me now turn briefly to the issue of the in vivo relevance
of natural effector cells. First, I think it's important to point
out that there is substantial evidence for natural cell-mediated
resistance mechanisms against the growth of various transplantable
and primary tumors. There is also evidence for a role of natural
antibodies in some of these systems. Just what form of natural
cellular immunity underlies this resistance is not so easy to sort
out. Most of the evidence presented in this symposium has been related to NK cells or NK-like phenomena, but none of this evidence
is entirely definitive. However, the evidence for other effector
cell types being important in resistance to tumor growth is not as
compelling as that for NK cells. For example, since this whole
meeting has been predominantly oriented around the macrophage, it
seems important to at least make a provocative statement: namely,
I'm not convinced of the documentation or evidence in the literature
that macrophages are important natural defense cells against tumors.
This has been taken largely as an assumption in the literature and
there have been very few experiments which attempt to document or
prove that it's a macrophage and not another cell type. If investigators working with macrophages really want to sustain the argument
that macrophages are important natural effector cells against tumor
growth, I think it's incumbent upon them to do more definitive experiments than are in the literature at the moment.

Turning to the evidence for NK cells in tumor resistance, much
of this evidence is largely of the correlative type showing that
resistance to various transplantable tumors, mainly lymphomas, correlates with the levels of NK activity in the individuals. In addition, I've been particularly impressed by the studies performed in
my laboratory by Elieser Gorelik which showed that resistance to
metastases, both artificial and spontaneous metastases from locally
growing tumors, may be mediated in a very potent way by NK cells.
This has been observed not only with lymphomas but also with the B16

melanoma and with two different lung adenocarcinomas in the mouse.
The data of Habu and Okumura with the asialo GM1 antibody also are
quite impressive. Nevertheless, we cannot be absolutely sure that
NK cells are the effector cells in these different systems. Both
with transplantable tumors and with primary tumors, we have to define
as stringently as possible that only NK cells are being affected by
the in vivo treatment and that no other effector cells are being
influenced. As an example of the complexity of this situation, we
have seen that mice treated in vivo with anti-asialo GM1 have impaired
clearance from the lungs of the WEHI-164 tumor as well as typical NK
targets, suggesting that the NC cell which does not seem to express
asialo GM1 in vitro may do so in vivo.

In the last part of the symposium, we discussed other in vivo
functions of NK cells, particularly resistance to hybrid bone marrow
transplantation and against infection. One aspect of the former
that was not mentioned by Dr. Cudkowicz is that autologous bone marrow
cells also can be seen by NK cells in vitro although to a lesser
extent. In addition to earlier studies from my laboratory and by
Carlo Riccardi in the mouse, Rolf Kiessling has recent data indicating
that both autologous and allogeneic human LGL can cause an inhibition
of colony forming units. It appears that NK cells recognize colony
forming cells even in the autologous individual, raising the possibility that they may be involved in the regulation of the host's own
hematopoietic differentiation. In regard to natural resistance to
microorganisms, it's worthwhile to point out that there is some
evidence for a role of NK cells not only with viruses but also with
certain parasites, particularly malaria or malaria-like parasites.
The best evidence that I have seen has been with Babesia, in which
there is some increased susceptibility to Babesia during the early
part of the infection in beige mice as compared to normal C57BL/6
mice. Also, Dr. June Ann Murphy has data that natural resistance
to Cryptococcus seems to be mediated to some extent by NK cells, and
she has been able to show in vitro that mouse cells with all the
features of NK cells are able to have at least a cytostatic effect
on the Cryptococcus.

One point about resistance to viral infections by NK cells is
that almost all the evidence has been acquired with herpes-type
viruses. Apart from the herpesviruses, the evidence becomes either
much weaker or nonexistent.

To follow up on the point made by Dr. Cudkowicz at the beginning
of the session, we have almost all focused on cytotoxic activity
but there may be other important functions for these cells. An important separate function may be their ability to secrete interferon.
Dr. Julie Djeu studied separated cell populations in the human (i.e.
purified LGL, monocytes, and T-cells) in overnight cultures with
6 or 8 different stimuli. The highly purified LGL population, which
was depleted of monocytes and typical small T-cells, produced inter-

feron in response to almost the entire range of stimuli used, including herpesvirus, influenza virus, some typical T-cell mitogens
like Con A and staphylococcal enterotoxin, and also poly I:C, BCG and
C. parvum. This raises the possibility that in certain situations
LGL may be the predominant and most rapid interferon-producer. A
further intriguing aspect of her study has been that different types
of interferons can be detected depending on the stimulus used. With
some of the bacterial or viral stimuli, mainly alpha type interferon
was made. In response to the T-cell mitogens, however, it was mainly
gamma interferon. With staphylococcal enterotoxin, it was predominantly gamma but also a small amount of beta interferon. It would
appear the LGL have a considerable amount of versatility in their
production of various species of interferon. It is not settled yet
whether a given LGL makes more than one kind of interferon, but this
should be an answerable question with the clones of NK cells that are
now being established. Dr. Wolfgang Domzig also has some suggestive
evidence that the LGL themselves make at least low levels of IL-2
and that a highly purified population of LGL, depleted of OKT3-
positive cells, proliferate strongly in response to OKT3. The easiest
interpretation is that the LGL make enough IL-2 in order to promote
their proliferation. Perhaps one of the reasons why only small
amounts of IL-2 have been detected in the supernatants from these
cells is that a subset of LGL have receptors for IL-2 spontaneously
and that a fair amount of the IL-2 produced is absorbed out immediately on the cells with the receptors. Rather than seeing an excess
in the supernatant, one gets proliferation of these cells in response
to that which is produced.

The final point that I might mention concerns the question
raised by Dr. Reithmüller, "Are there tumors of the NK cell population which would provide an opportunity for more detailed studies?"
In addition to the one case of T-cell chronic lymphatic leukemia
described by Reithmüller, Dr. Franco Pendolfi in Italy has described
several others and Manlio Ferarini in Italy independently has described some additional leukemias with NK activity. In the human at
least, I am only convinced by the reports of CLL leukemic cells
having NK activity. There has been one report by Peter Hokland of a
monocytic leukemia having NK activity, but if one looks carefully
at that report only about 60% of the cells had the characteristics
of monocytes. One could at least raise the question that it was the
other, non-leukemic cells in the population that had the NK activity.
There is a recent observation by Craig Reynolds in my laboratory
in collaboration with Jerry Ward, of spontaneous NK tumors in rats.
Our Fischer rats develop a rather high incidence of leukemias and
lymphomas after about 2 years of age which largely involve the spleen.
The morphology of these cells is almost exclusively that of a large
granular lymphocyte and about one-third to one-half of these LGL-
type tumors in the Fischer rats have quite high NK activity. To the
extent studied so far, it appears that the NK activity resides in
the leukemic cell itself rather than the host infiltrating cells

inasmuch as the phenotype of these cells seems to be somewhat different than what one sees with a typical normal NK cell or LGL in the rat. This may serve as a good model for the detailed studies of large numbers of NK cells in an animal system.

REFERENCES

1. Stutman, O., Paige, C. J., and Figurella, E. F., Natural cytotoxic cells against solid tumors in mice. I. Strain and age distribution of target cell susceptibility, J. Immunology 121:1819, 1978.
2. Glimcher, L. F., Shen, F. W., and Cantor, H., Identification of a cell surface antigen expressed on the natural killer cell, J. Exp. Med. 145:1, 1977.
3. Abo, T., and Balch, C. M., A differentiation antigen on human NK and K cells identified by a monoclonal antibody (HNK-1), J. Immunology 127:1024, 1981.

MEETING REPORT

9TH INTERNATIONAL RES CONGRESS

DAVOS, SWITZERLAND

FEBRUARY 7TH-12TH, 1982

A PESSIMIST AT THE CONGRESS

Nothing to hear but science
Nothing to do but listen
Nothing to wear but ski clothes

Nothing to pay but registration fee
Nowhere to sleep but in first class hotel
Nobody to talk to but scientists

Nothing to breathe but alpine air
Nowhere to fall but on snow
Nowhere to stand but on feet

Nothing to show but results
Nothing to look at but posters
Nothing to eat but Swiss food

Nothing to enjoy but beauty
Nothing to complain about but self
Nowhere to go but to Congress.

E. Sorkin

11, February 1982

THE MACROPHAGE AND CANCER PAS DE DEUX

With due respect to Elie Metchnikoff, George Bernard Shaw and Sir Almroth Wright.

<u>Ignorance is on our side</u>

<u>Prologue</u>

To solve the riddle of malignant cancer
some doctors think the macrophage must be the answer;

So they met, of course, in a valley of beauty
for five days to do what they felt their duty.

<u>Les Danseurs</u>

Out of the marrow into the blood
the promonocytes do move in flood,

Finally landing in the tissues
becoming involved in many issues

As macrophages enjoy warm cooperation
with other cells in pleasing animation.

Yet their major task is to survey
poor chaps, every single hour per day;

To clear away the infection mess
which as everyone will guess

Is a full-time job per se
(and this without a cup of tea).

Beyond this too enormous task
evolution also asked

That they look after the cancer cell
which the macrophages think is Hell.

Having for this no inclination
they amalgamate in consolation;

Thus not tonight will fall the blow
for they must first be taught to grow;

Getting stimulated, recognize and split (what not?)
this is the macrophages lot.

Entreacte

About these scavengers of ours
we spoke for many, many hours

On how they move and where they rest
and what they like for dinner best,

And what they drink and how they grow;
why in inflamed tissue they are slow.

How when activated they can kill
different cancer cells at will;

What they secrete on stimulation
depends also here on titillation.

Monocyte precursors are no good in killing
even if they were quite willing;

They know how to be primed and loaded
and close to cancer cells exploded;

But these beastly cells release
what blocks the macrophage with ease

So that they perform no longer
the malignant cell is ever stronger.

Diddle, duddle, dumble dee
a macrophage which can not see;

The devillish cells within the tumour
which we hear, so goes the rumour

Must be in a special state
yet these defenses come too late.

Les Spectateurs et le Spectacle

Some results did seem to vary;
other stories sounded fairy

The skiers had of course, great hopes
that soon they would be on the slopes.

Next slide, please, full thirty shots
evoked in minutes thirty nods.

Although one heard much published stuff
nobody played it really rough

With the speakers all were fair;
there was no reason for despair.

Finale

To sum up such a meeting hot
one must know what one has got;

That cancer cells and bugs we do not crave
that they still put us in the grave.

The scientists homeward bound
wondered if anything they'd found

Would take away their tired look
Yes! To see their names printed in this book.

                    E. Sorkin

                        11, February 1982

CONTRIBUTORS

ADAMS, D.O.
Department of Pathology
Duke University Medical Center
Durham, North Carolina 27710

AHMANN, G.B.
Department of Internal Medicine
University of Iowa
Iowa City, Iowa 52240

AKAGAWA, K.S.
National Institute of Health
Shinagawa-ku
Tokyo 141, Japan

ALM, G.
Uppsala, Sweden

ALTMAN, A.
Dept. of Molecular Immunology
Research Inst. of Scripps Clinic
La Jolla, California

ANDREW, P.W.
M.R.C. Unit
Lab. Studies of Tuberculosis
Royal Postgraduate Med. School
London W12 OHS, U.K.

BAGGIOLINI, M.
Wander Research Institute
Wander Ltd., P.O. Box 2747
CH-3001 Berne, Switzerland

BAGLEY, M.B.
Laboratory of Immunodiagnosis
National Cancer Institute
Bethesda, Maryland

BAR-SHAVIT, Z.
Department of Membrane Research
Weizmann Institute of Science
76100 Rehovot, Israel

BECKER, S.
Dept. of Obstetrics/Gynecology
University of North Carolina
Chapel Hill,
North Carolina 27514

BELLER, D.I.
Department of Pathology
Harvard Medical School
Boston, Massachusetts 02115

BERLIN, R.D.
Department of Physiology
Univ. Connecticut Health Center
Farmington, Connecticut 06032

BEYER, J.-H.
Division of Oncology
Department of Medicine
University Clinics
Robert-Koch-Strasse 40
D-3400 Goettingen, Germany

BIONDI, A.
Istituto di Ricerche
Farmacologiche "Mario Negri"
Via Eritrea 62
20157 Milan, Italy

BITTER-SUERMANN, D.
Inst. of Med. Microbiology
University of Mainz
Hochhaus Augustuspl.
6500 Mainz, Germany

BLUMBERG, S.
Department of Biophysics
Weizmann Institute of Science
76100 Rehovot, Israel

BLUMENSTOCK, F.A.
Department of Physiology
Albany Medical College
Albany, New York 12208

BODOLAI, E.
Department of Medicine
University Medical School
Debrecen, Hungary

BORASCHI, D.
Sclavo Research Center
Siena, Italy

BORDIGNON, C.
Istituto di Ricerche
Farmacologiche "Mario Negri"
Via Eritrea 62
20157 Milan, Italy

BORTH, W.
Institute of Immunology
University of Vienna
Borschkegasse 8A
1090 Vienna, Austria

BOTTAZZI, B.
Istituto di Ricerche
Farmacologiche "Mario Negri"
Via Eritrea 62
20157 Milan, Italy

BRADFIELD, J.W.B.
Department of Pathology
Medical School
University Walk
Bristol, U.K.

BRANDSLUND, I.
Department of Clinical Chemistry
Odense University Hospital
DK-5000 Odense C, Denmark

BRAY, M.
Basel, Switzerland

BRAY, R.
Sunninghill, U.K.

BROMBERG, Y.
Department of Human Microbiology
Sackler School of Medicine
Tel-Aviv University
Tel-Aviv 69978, Israel

BROOKS, A.
National Institute Allergy
 and Infectious Diseases
Bethesda, Maryland

BROWDER, W.
Department of Surgery
Tulane Univ. School of Medicine
New Orleans, Louisiana 70112

BRUNNER, K.T.
Department of Immunology
Swiss Inst. for Exp. Cancer Res.
Epalinges, Switzerland

BUCHMULLER, Y.
University of Lausanne
Institute of Biochemistry
1066 Epalinges, Switzerland

BURMESTER, G.R.
Hospital for Joint Diseases
Mount Sinai School of Medicine
New York, New York 10003

BUYSSENS, N.
Department of Pathology
University of Antwerp
Antwerp, Belgium

CAMPBELL, D.A.
Division of Surgical Oncology
Univ. Michigan Medical Center
Ann Arbor, Michigan 48109

CECKA, M.J.
Institute for Microbiology
University of Basel
Petersplatz 10
CH-4003 Basel, Switzerland

CEROTTINI, J.-C.
Department of Immunology
Ludwig Inst. for Cancer Research
Epalinges, Switzerland

CHAINTREUIL, J.
Laboratorie de Biochemie
Faculte de Medecine
34060 Montpellier Cedex, France

CHARLESWORTH, J.A.
Department of Medicine
Royal Postgraduate Med. School
Hammersmith Hospital
London, U.K.

CIANCIOLO, G.
Dept. of Microbiol. and Immunol.
Duke University Medical Center
Durham, North Carolina 27710

COHEN, D.A.
Department of Surgery
Virginia Commonwealth University
Medical College of Virginia
Richmond, Virginia 23298

CORNELIUS, J.
Department of Pathology
University of Florida
Gainesville, Florida 32610

CORRADIN, G.
University of Lausanne
Institute of Biochemistry
1066 Epalinges, Switzerland

CRUCHAUD, A.
Division Immunol. & Allergology
Department of Medicine
Hopital Cantonal Universitaire
Geneve, Switzerland

CUDKOWICZ, G.
Department of Pathology
State University of New York
Buffalo, New York

CZARNETSKI, B.M.
Universitats-Hautklinik
Institut fur Biochemie
  und Technologie
Munster, Germany

CZITROM, A.A.
Tumour Immunology Unit
Department of Zoology
University College London
Gower Street, London,
WCIE 6BT, U.K.

DAHLGREN, C.
Dept. of Medical Microbiology
Linkoping Univ. Medical School
S-581 85 Linkoping, Sweden

DAMON, M.
Laboratoire de Biochemie
Faculte de Medecine
34060 Montpellier Cedex, France

DAVID, C.S.
Department of Immunology
Mayo Clinic and Medical School
Rochester, Minnesota 55905

DAVIES, W.A.
Kolling Inst. of Med. Research
Royal North Shore Hospital Sydney
St. Leonards, NSW 2065, Australia

DE BAETSELIER, P.
Inst. voor Moleculaire Biologie
VUB, Paardenstraat 65
Sint-Genesius-Rode, Belgium

DE PAULET, A.C.
Laboratoire de Biochemie
Faculte de Medecine
34060 Montpellier Cedex, France

DE VRIES, E.
University Hospital
Leiden, The Netherlands

DE VRIES, J.E.
Netherlands Cancer Institute
Plesmanlaan 121
1066 CX, Amsterdam, Netherlands

DEWALD, B.
Wander Research Institute
Wander Ltd., P.O. Box 2747
CH-3001 Berne, Switzerland

DIERICH, M.P.
Institute for Med. Microbiology
Johannes Gutenberg University
6500 Mainz, Germany

DIESSELHOFF-den DULK, M.M.C.
Dept. of Infectious Diseases
University Hospital
Rijnsburgerweg 10
2333 AA Leiden, Netherlands

Di LUZIO, N.R.
Department of Physiology
Tulane Univ. School of Medicine
New Orleans, Louisiana 70112

DIMITRIU-BONA, A.
Hospital for Joint Diseases
Mount Sinai School of Medicine
New York, New York 10003

DORIA, G.
CNEN-Euratom Immunogenetics Gr.
Laboratory of Radiopathology
C.S.N. Casaccia, Rome, Italy

EIDLEN, L.G.
The Jackson Laboratory
Bar Harbor, Maine 04609

ELZENGA-CLAASEN, I.
Dept. of Infectious Diseases
University Hospital
Leiden, The Netherlands

ENGERS, H.D.
Department of Immunology
Swiss Inst. for Exp. Cancer Res.
Epalinges, Switzerland

ENGLER, H.
Institute of Virus Research
German Cancer Research Center
Heidelberg, Germany

ERB, P.
Institute for Microbiology
University of Basel
Basel, Switzerland

EVANS, R.
The Jackson Laboratory
Bar Harbor, Maine   ·

EZEKOWITZ, R.A.B.
Sir William Dunn
 School of Pathology
University of Oxford
South Parks Road, Oxford U.K.

FALK, W.
National Cancer Institute
Bethesda, Maryland 20205

FELDMAN, M.
Department of Cell Biology
Weizmann Institute of Science
Rehovot, Israel

FELDMANN, M.
Department of Zoology
University College London
Gower Street, London,
WC1E 6BT, U.K.

FERNE, M.
Streptococcus Ref. Laboratory
Ministry of Health
Jerusalem, Israel

FIDLER, I.J.
Cancer Metastasis and
 Treatment Laboratory
Frederick Cancer Research Center
Frederick, Maryland 21701

CONTRIBUTORS

FLAD, H.-D.
Forschungsinstitut Borstel
D-2061 Borstel, Germany

FLANDRE, O.
Laboratoire de Biochemie
Faculte de Medecine
34060 Montpellier Cedex, France

FLORY, E.
Department of Medicine
University of Melbourne
Melbourne, Australia

FOERSTER, O.
Inst. of Experimental Pathology
University of Vienna
Vienna, Austria

FREUND, M.
Department of Human Microbiology
Sackler School of Medicine
Tel-Aviv University
Tel-Aviv 69978, Israel

FRIDKIN, M.
Department of Organic Chemistry
Weizmann Institute of Science
Rehovot 76100, Israel

FRIEDLANDER, A.M.
U.S. Army Medical Research
Institute for Infectious Diseases
Frederick, Maryland

FRITZ, P.
Robert Bosch Hospital
Stuttgart, Germany

FUKS, Z.
Dept. Radiation & Clin. Oncology
Hadassah Univ. Medical Center
Jerusalem 91120 Israel

GALLILY, R.
Lautenberg Center
General and Tumor Immunology
Hebrew University
Hadassah Medical School
Jerusalem, Israel

GALSWORTHY, S.
Dept. Microbiol. and Immunology
University of Western Ontario
London, Ontario Canada

GEMSA, D.
Institute of Immunology
University of Heidelberg
Heidelberg, Germany

GHEZZI, P.
Instituto di Ricerche
Farmacologiche "Mario Negri"
Via Eritrea 62
20157 Milan, Italy

GINSBURG, I.
Department of Oral Biology
Hadassah Sch. Dental Medicine
Hebrew University
Jerusalem, Israel

GLASEBROOK, A.L.
Department of Immunology
Swiss Inst. for Exp. Cancer Res.
1066 Epalinges, Switzerland

GODARD, P.
Clinique Maladies Respiratoires
Hospital l'Aiguelongue
Avenue du Major Flandre
34059 Montpellier Cedex, France

GOLDMAN, R.
Department of Membrane Research
Weizmann Institute of Science
76100 Rehovot, Israel

GORDON, S.
Sir William Dunn
 School of Pathology
South Parks Road
Oxford OX1 3RE, U.K.

GORECKA-TISERA, A.
Allegheny-Singer Research Corp.
Pittsburgh, Pennsylvania

GOUD, Th.J.L.M.
Dept. of Infectious Diseases
University Hospital
Rijnsburgerweg 10
2333 AA Leiden, Netherlands

HABU, S.
Department of Pathology
School of Medicine
Tokai University
Isehara, Kanagawa, Japan

HADAM, M.
Immunology Laboratories
University of Tubingen
Tubingen, Germany

HADDING, U.
Inst. of Medical Microbiology
University of Mainz
Hochhaus Augustusplatz
6500 Mainz, Germany

HAIMOVITZ, A.
Unit of Gastroenterology
Hadassah Univ. Medical Center
Jerusalem 91120 Israel

HALME, J.
Dept. Obstetrics/Gynecology
University of North Carolina
Chapel Hill,
North Carolina 27514

HARTUNG. H.-P.
Inst. of Medical Microbiology
University of Mainz
Hochhaus Augustusplatz
6500 Mainz, Germany

HARVATH, L.
National Cancer Institute
Bethesda, Maryland 20205

HASKILL, S.
Dept. of Obstetrics/Gynecology
University of North Carolina
Chapel Hill,
North Carolina 27514

HERBERMAN, R.B.
National Cancer Institute
Division of Cancer Treatment
Biol. Response Modifiers Program
Frederick Cancer Research Center
Frederick, Maryland 21701

HIRSCH, S.
Sir William Dunn
  School of Pathology
University of Oxford
South Parks Road
Oxford OX1 3RE, U.K.

HOLDEN, H.T.
Laboratory of Immunodiagnosis
National Cancer Institute
Bethesda, Maryland

HULSING-HESSELINK, E.
Dept. of Infectious Diseases
University Hospital
Leiden, The Netherlands

HUME, D.A.
Sir William Dunn
  School of Pathology
South Parks Road
Oxford OX1 3RE, U.K.

HUMPHREY, J.H.
Department of Medicine
Royal Postgraduate Med. School
Hammersmith Hospital
London W12, U.K.

ISLIKER, H.
Institut de Biochemie
University of Lausanne
1066 Epalinges, Switzerland

JACKETT, P.S.
M.R.C. Unit
Lab. Studies of Tuberculosis
Royal Postgraduate Med. School
London W12 OHS, U.K.

JOHNSON, W.J.
Dept. of Microbiol. & Immunol.
Duke University Medical Center
Durham, North Carolina   27710

KAMBER, M.
Basel, Switzerland

KAMINSKY, S.G.
Department of Pathology
SUNY at Buffalo
Buffalo, New York   14214

KANAGAWA, O.
Ludwig Inst. for Can. Res.
1066 Epalinges, Switzerland

KAPLAN, A.M.
Department of Surgery
Virginia Commonwealth University
Medical College of Virginia
Richmond, Virginia   23298

KAPLAN, G.
Department of Tumor Biology
Norwegian Cancer Society
9000 Tromso, Norway

KAPON, A.
Inst. voor Moleculaire Biologie
VUB, Paardenstraat 65
Sint-Genesius-Rode, Belgium

KASAI, M.
Department of Tuberculosis
Japanese Natl. Inst. of Health
Meguro-ku, Tokyo, Japan

KATSAV, S.
Inst. voor Moleculaire Biologie
VUB, Paardenstraat 65
Sint-Genesius-Rode, Belgium

KATZ, D.H.
Dept. of Molecular Immunology
Research Inst. of Scripps Clinic
LaJolla, California

KATZ, D.R.
Tumour Immunology Unit
Department of Zoology
University College London
Gower Street, London,
WCIE 6BT, U.K.

KAVAI, M.
Department of Medicine
Univ. Med. School of Debrecen
Debrecen, Hungary

KELLER, R.
Immunobiology Research Group
Inst. of Immunol. and Virology
University of Zurich
Schonleinstrasse 22
CH-8032 Zurich, Switzerland

KELSO, A.
Department of Immunology
Swiss Inst. for Exp. Can. Res.
1066 Epalinges, Switzerland

KENYON, A.J.
Department of Pathology
Univ. Connecticut Health Center
Farmington, Connecticut   06032

KIEDA, C.
Laboratory of Immunology
National Institute of Allergy
 and Infectious Diseases
Bethesda, Maryland   20205

KIRCHNER, H.
Institute of Virus Research
German Cancer Research Center
Heidelberg, Germany

KOLB, H.
Diabetes Research Institute
University of Dusseldorf
D 4000 Dusseldorf, Germany

KOLB-BACHOFEN, V.
Institute for Biophysics
  and Electron Microscopy
Department of Medicine
University of Dusseldorf
D 4000 Dusseldorf, Germany

KONGSHAVN, P.A.L.
Department of Physiology
McGill University
Montreal, Quebec Canada

KOREN, H.S.
Division of Immunology
Duke University Medical Center
Durham, North Carolina

KORKOLAINEN, M.
Department of Biochemistry
University of Helsinki
Unioninkatu 35
SF-00170 Helsinki 17

LAFUSE, W.P.
Department of Immunology
Mayo Clinic and Medical School
Rochester, Minnesota 55905

LAHAV, M.
Department of Oral Biology
Hadassah Sch. Dental Medicine
Hebrew University
Jerusalem, Israel

LAVANDIER, M.
Laboratory of Immunology
Centre Hospitalier Bretonneau
37044 Tours Cedex France

LAVIE, G.
Department of Urology
Beilinson Medical Center
Tel-Aviv University
Tel-Aviv, Israel

LAVIE, L.
Department of Biology
Technicon-Israel
Institute of Technology
Haifa, Israel

LAWRENCE, S.
Renal Unit
Wollongong Hospital
Wollongong, NSW, Australia

LAZDINS, J.K.
Laboratory of Immunobiology
National Cancer Institute
National Institute of Health
Bethesda, Maryland 20205

LEGRAND, M.F.
Laboratory of Immunology
Centre Hospitalier Bretonneau
37044 Tours Cedex France

LEIBOLD, W.
Inst. Pediatric Immunology
Krankenhausstrasse 23
D-8520 Erlangen, West Germany

LEIJH, P.C.J.
Department of Infectious Diseases
University Hospital
2333 AA Leiden
The Netherlands

LEISERSON, W.M.
National Institute of Allergy
  and Infectious Diseases
Bethesda, Maryland

LeMARIE, E.
Laboratoire D Immunologie
Centre Hospitalier Bretonneau
37044 Tours Cedex France

LENZINI, L.
Istit. di Tisiologia e Malattie
dell Apparato Respiratorio
Universita degli Studi di Siena
Siena, Italy

LEONARD, E.J.
National Cancer Institute
Bethesda, Maryland 20205

LOHMEYER, J.
Institute for Immunology
University of Munich
Munich, Germany

LOWRIE, D.B.
M.R.C. Unit
Lab. Studies of Tuberculosis
Royal Postgraduate Med. School
London W12 OHS, U.K.

MacDONALD, H.R.
Department of Immunology
Ludwig Inst. Cancer Research
Epalinges, Switzerland

MacPHERSON, G.G.
Sir William Dunn
  School of Pathology
Oxford, U.K.

MacVITTIE, T.J.
Exper. Hematology Department
Armed Forces Radiobiology
  Research Institute
Bethesda, Maryland 20814

MANN, D.
National Cancer Institute
Bethesda, Maryland 20205

MANTOVANI, A.
Istituto di Ricerche
Farmacologiche "Mario Negri"
Via Eritrea 62
20157 Milan, Italy

MATHIESON, B.J.
National Institute of Allergy
  and Infectious Diseases
Bethesda, Maryland

MATTHEWS, N.
Dept. of Medical Microbiology
Welsh National Sch. of Medicine
Heath Park, Cardiff CF4 4XN, UK

MAUEL, J.
WHO Immunology Research
  and Training Centre
University of Lausanne
Institute of Biochemistry
1066 Epalinges, Switzerland

McBRIDE, W.H.
Department of Bacteriology
Univ. of Edinburgh Medical School
Teviot Place
Edinburgh, Scotland

McNAMEE, R.
Department of Surgery
Tulane Univ. School of Medicine
New Orleans, Louisiana 70112

METCALF, D.
Cancer Research Unit
Walter and Eliza Hall Institute
  of Medical Research
Royal Melbourne Hospital
Victoria 3050, Australia

MICHEL, F.B.
Clinique Maladies Respiratoires
Hospital l'Aiguelongue
avenue du Major Flandre
34059 Montpellier Cedex, France

MILLER, K.
Carshalton, U.K.

MOMOI, T.
Institute of Medical Science
University of Tokyo
Minato-ku, Tokyo 108, Japan

MOORE, K.
Department of Bacteriology
Univ. of Edinburgh Medical School
Teviot Place
Edinburgh, Scotland

MUIRDEN, K.
Department of Medicine
University of Melbourne
Victoria, Australia

MULLER, J.
Institute of Biochemistry
University of Stuttgart
Stuttgart, Germany

MULLER, S.
Robert Koch Institute
Berlin, Germany

MURRE, C.
Laboratory of Immunology
National Institute of Allergy
 and Infectious Diseases
Bethesda, Maryland 20205

MYRVIK, Q.N.
Dept. of Microbiol. and Immunol.
Bowman Gray School of Medicine
Winston-Salem,
North Carolina 27103

NAGEL, G.A.
Division of Oncology
Department of Medicine
University Clinics
Robert-Koch-Strasse 40
D-3400 Goettingen, Germany

NIEDERHUBER, J.E.
Dept. of Microbiol. and Immunol.
Univ. Michigan Medical Center
Ann Arbor, Michigan 48109

NOGA, S.J.
Department of Pathology
University of Florida
Gainesville, Florida 32610

NORMANN, S.J.
Department of Pathology
University of Florida
Gainesville, Florida 32610

O BRIEN, A.D.
Department of Microbiology
Uniformed Services
University of Health Sciences
Bethesda, Maryland 21814

OKUMURA, K.
Department of Immunology
University of Tokyo
Bunkyo-ku, Tokyo, Japan

OLIVER, J.M.
Department of Physiology
Univ. Connecticut Health Center
Farmington, Connecticut 06032

OTTENDORFER, D.
Institute for Med. Microbiology
65 Mainz, Germany

PASQUALETTO, E.
Istituto di Ricerche
Farmacologiche "Mario Negri"
Via Eritrea 62
20157 Milan, Italy

PATCHEN, M.L.
Experimental Hematology Department
Armed Forces Radiobiol. Res. Inst.
Bethesda, Maryland 20814

PERI, G.
Istituto di Ricerche
Farmacologiche "Mario Negri"
Via Eritrea 62
20157 Milan, Italy

PICK, E.
Department of Human Microbiology
Tel-Aviv Univ. Medical School
Ramat-Aviv, Tel-Aviv, Israel

PIERRES, M.
Centre d Immunologie Inserm
Marseille, France

PINCHING, A.J.
Dept. Medicine and Immunology
Royal Postgraduate Med. School
London W12 OHS, U.K.

POLENTARUTTI, N.
Istituto di Ricerche
Farmacologiche "Mario Negri"
Via Eritrea 62
20157 Milan, Italy

POSTE, G.
Smith Kline and French Lab.
1500 Spring Garden Street
P.O. Box 7929
Philadelphia, Pennsylvania 19101

PREHN, R.T.
Institute for Medical Research
751 South Bascom Avenue
San Jose, California 95128

PRUZANSKI, W.
Department of Medicine
University of Toronto
Toronto, Ontario Canada

PUNJABI (nee Sadarangani), C.
Montreal General Hospital
 Research Institute
McGill University
Montreal, Quebec Canada

PUSSELL, B.A.
Department of Nephrology
Prince Henry Hospital
P.O. 233
Matraville, NSW 2036, Australia

RAFFAEL, A.
Arbeitsgruppe Krebszellforsch.
Max-Planck-Institut Biochemie
D-8033 Martinsried, Germany

RANADIVE, N.S.
Department of Pathology
University of Toronto
Toronto, Ontario Canada

RASOKAT, H.
Inst. for Medical Microbiology
University of Mainz
Hochhaus Augustusplatz
D-6500 Mainz, Germany

RENOUX, G.
Laboratoire D Immunologie
Centre Hospitalier Bretonneau
37044 Tours Cedex France

RENOUX, M.
Laboratoire D Immunologie
Centre Hospitalier Bretonneau
37044 Tours Cedex France

RICHES, D.W.H.
Department of Immunology
The Medical School
Vincent Drive
Birmingham B15 2TJ, U.K.

RICHTER, M.
Rheumatology Department
Middlesex Hospital
London, W1P 9PG, U.K.

RIEBER, E.P.
Institute for Immunology
University of Munich
Munich, Germany

RIETHMULLER, G.
Institute for Immunology
University of Munich
Munich, Germany

RON, Y.
Department of Cell Biology
Weizmann Institute of Science
Rehovot 76100, Israel

ROTH, J.A.
National Cancer Institute
Bethesda, Maryland 20205

RUCO, L.P.
IV Chair Pathological Anatomy
University of Rome
Rome, Italy

RUMPOLD, H.
Inst. of Experimental Pathology
University of Vienna
Vienna, Austria

RUSSELL, M.W.
Univ. of Alabama in Birmingham
Birmingham, Alabama

RUSSELL, S.W.
Department of Comparative
and Experimental Pathology
College of Veterinary Medicine
University of Florida
Gainesville, Florida 32610

SAAL, J.G.
Robert Bosch Hospital
Stuttgart, Germany

SABA, T.M.
Department of Physiology
Albany Medical College
Albany, New York 12208

SAITO, S.
Department of Medicine
University of Toronto
Toronto, Ontario, Canada

SALMONA, M.
Istituto di Ricerche
Farmacologiche "Mario Negri"
Via Eritrea 62
20157 Milan, Italy

SCHADEWIJK-NIEUWSTAD, M.
Dept. of Infectious Diseases
University Hospital
Rijnsburgerweg 10
2333 AA Leiden, Netherlands

SCHARDT, M.
Schweizerisches Forschungsinst.
Medizinische Abteilung
Davos, Switzerland

SCHINDLER, L.
Institute of Virus Research
German Cancer Research Center
Heidelberg, Germany

SCHLEPPER-SCHAFER, J.
Diabetes Research Institute and
Institute for Biophysics
and Electron Microscopy
University of Dusseldorf
D 4000 Dusseldorf, Germany

SCHLOSSMAN, S.F.
Department of Medicine
Harvard Medical School
Boston, Massachusetts 02115

SCHNYDER, J.
Wander Research Institute
Wander Ltd., P.O. Box 2747
CH-3001 Berne, Switzerland

SCHOOK, L.B.
Dept. Microbiol. and Immunol.
Box 678
Medical College of Virginia
Richmond, Virginia 23298

SCHREIBER, R.D.
Dept. of Molecular Immunology
Research Inst. of Scripps Clinic
LaJolla, California

SCHUFF-WERNER, P.
Division of Oncology
Department of Medicine
University Clinics
Robert-Koch-Strasse 40
D-3400 Goettingen, Germany

SEGAL, S.
Department of Cell Biology
Weizmann Institute of Science
Rehovot, Israel

SHEN, F.-W.
Memorial Sloan-Kettering
Cancer Center
New York, New York

SHEZEN, E.
Department of Membrane Research
Weizmann Institute of Science
76100 Rehovot, Israel

SILVERSTEIN, S.C.
Lab. Cell. Physiol. and Immunol.
Rockefeller University
New York, New York 10021

SLUITER, W.
Dept. of Infectious Diseases
University Hospital
Leiden, The Netherlands

SMITH, L.A.
Department of Surgery
Virginia Commonwealth Univ.
Medical College of Virginia
Richmond, Virginia 23298

SNYDERMAN, R.
Division of Rheumatic and
  Genetic Diseases
Department of Medicine
Duke University
Durham, North Carolina

SORKIN, E.
Schweizerisches Forschungsinst.
Medizinische Abteilung
Davos, Switzerland

STAM, J.
Department of Pulmonology
Free University Hospital
Amsterdam, The Netherlands

STANWORTH, D.R.
Department of Immunology
The Medical School
Vincent Drive
Birmingham B15 2TJ, U.K.

STENDAHL, O.
Dept. of Medical Microbiology
Linkoping Univ. Medical School
S-581 85 Linkoping, Sweden

STENMAN, U.-H.
Dept. Obstetrics and Gynecology
University Central Hospital
Helsinki, Finland

STERN, A.C.
Institute for Microbiology
University of Basel
Petersplatz 10
CH-4003 Basel, Switzerland

STEVENSON, H.C.
Basic Mechanisms Section
Biol. Response Modifiers Program
Cancer Research Center
Frederick, Maryland 21701

STEWART, C.C.
Experimental Pathology Group
Los Alamos National Laboratory
Los Alamos, New Mexico 87545

SUNSHINE, G.H.
Tumour Immunology Unit
Department of Zoology
University College London
Gower Street, London,
WCIE 6BT, U.K.

SVEHAG, S.-E.
Inst. of Medical Microbiology
Odense University
DK-5000 Odense C, Denmark

SWINBURNE, S.
London, U.K.

SZOLLOSI, J.
Department of Biophysics
University Medical School
Debrecen, Hungary

TAGLIABUE, A.
Sclavo Research Center
Siena, Italy

TARAMELLI, D.
Laboratory of Immunodiagnosis
National Cancer Institute
Bethesda, Maryland

TARTAKOVSKY, B.
Department of Cell Biology
The Weizmann Institute of Science
Rehovot, 76100 Israel

TERRY, S.
The Israel Institute
Biological Research
Ness Ziona, Israel

TODD III, R.F.
Division of Tumor Immunology
Sidney Farber Cancer Institute
Boston, Massachusetts   02115

TOKUNAGA, T.
National Institute of Health
Shinagawa-ku
Tokyo 141, Japan

TREVES, A.J.
Department of Clinical Oncology
Hadassah Univ. Medical Center
Jerusalem 91120 Israel

TZEHOVAL, E.
Department of Cell Biology
Weizmann Institute of Science
Rehovot, 76100 Israel

UCCINI, S.
IV Chair Pathological Anatomy
University of Rome
Rome, Italy

ULMER, A.J.
Forschungsinstitut Borstel
D-2061 Borstel, Germany

UNANUE, E.R.
Department of Pathology
Harvard Medical School
Boston, Massachusetts   02115

VALET, G.
Arbeitsgruppe Krebszellforsch.
Max-Planck-Institut Biochemie
D-8033 Martinsried, Germany

VALENTINE, F.
Department of Medicine
New York University
Medical Center
New York, New York

VAN BLOMBERG, M.
Heycoplaan 24
Breukelen, The Netherlands

VAN DER GAAG, R.
Department of Pulmonology
Free University Hospital
Amsterdam, The Netherlands

VAN DER MEER, J.W.M.
Dept. Infectious Diseases
University Hospital
Rijnsburgerweg 10
2333 AA Leiden, Netherlands

VAN FURTH, R.
Dept. Infectious Diseases
University Hospital
Leiden, The Netherlands

VAN MAARSEEVEEN, A.
Department of Pathology
Free University Hospital
Amsterdam, The Netherlands

VAN OUD ALBLAS, A.B.
Dept. Infectious Diseases
University Hospital
Leiden, The Netherlands

VAN ZWET, Th.L.
Dept. Infectious Diseases
University Hospital
2333 AA Leiden, Netherlands

VARESIO, L.
Laboratory of Immunodiagnosis
National Cancer Institute
Bethesda, Maryland

VISCHER, T.L.
Division of Rheumatology
Hopital Cantonal Universitaire
Geneva, Switzerland

VOGT, D.
Diabetes Research Institute
University of Dusseldorf
D 4000 Dusseldorf, Germany

VOSE, B.M.
Department of Immunology
Paterson Laboratories
Manchester, M20 9BX, U.K.

VUENTO, M.
Department of Biochemistry
University of Helsinki
Unioninkatu 35
SF-00170 Helsinki 17

WALKER, E.
Experimental Pathology Group
Los Alamos National Laboratory
Los Alamos, New Mexico 87545

WALKER, R.I.
Medical Microbiology Branch
Naval Medical Research Institute
Bethesda, Maryland 20814

WALKER, W.S.
Division of Immunology
St. Jude Childrens
 Research Hospital
Memphis, Tennessee 38101

WAXDAL, M.J.
Laboratory of Immunology
National Institute of Allergy
 and Infectious Diseases
Bethesda, Maryland 20205

WEINBERG, S.R.
Exper. Hematology Department
Armed Forces Radiobiology
 Research Institute
Bethesda, Maryland 02814

WEINER, R.
Division of Medical Oncology
University of Florida
Gainesville, Florida 32610

WHARTON, W.
Experimental Pathology Group
Los Alamos National Laboratory
Los Alamos, New Mexico 87545

WILKINSON, P.C.
Bacteriol. and Immunol. Depart.
University of Glasgow
Glasgow, G11 6NT, Scotland

WILLIAMS, A.J.
Department of Medicine
Brompton Hospital
Fulham Road
London, SW3 6HP, U.K.

WILLIAMS, D.L.
Departments of Physiology
Tulane Univ. School of Medicine
New Orleans, Louisiana 70112

WILLIAMSON, A.R.
Greenford, U.K.

WINCHESTER, R.J.
Hospital for Joint Diseases
Mount Sinai School of Medicine
New York, New York 10003

WUSTROW, T.P.U.
Sloan-Kettering Cancer Center
New York, New York

ZANKER, B.
Inst. of Medical Microbiology
University of Mainz
Hochhaus Augustuspl.
6500 Mainz, Germany

ZAWATZKY, R.
Institute of Virus Research
German Cancer Research Center
Heidelberg, Germany

ZINBERG, N.
Department of Cell Biology
Weizmann Institute of Science
Rehovot, 76100 Israel

SUBJECT INDEX

ACCESSORY CELLS   436-437,563,
      579,601-602

ACTIN   616,623

ACUTE PHASE REACTANTS   372,621

ADHERENCE   210,222,722
  Locomotion   91
  Proliferation   263-264
  Test of   143

ALLERGY   14,667-671

ANKYLOSING SPONDYLITIS   636

ANTIBODY DEPENDENT CELL CYTO-
      TOXICITY   213,241,290,
      417,442,485,707,737-738

ANTIGEN PRESENTATION   49-61,445,
      543-547,557-560,579-586

ANTIGEN PRESENTING CELLS   549,
      591

ANTIMICROBIAL RESISTANCE
      151-160,325-334,675-683,
      695-699

ASIALO GM1 (see also Natural
      Killer Cells)   429-433,
      793-795,801
  Cytotoxicity, and   431

AUTOIMMUNITY   12

BACTERIA
  C. parvum   293-294,404-405,
      430,463,525,583,657,
      701,721, 731-736,763,
      769-770,807
  E. coli   326,702-705
  K. pneumoniae   325-326
  L. monocytogenesis   50-54,
      195-200,401,472,485,
      583,591,654
  M. bovis BCG   293-294,396,
      430,485,650,656,
      687-693,709,763,807
  M. leprae   686,680-681
  M. tuberculosis   680,687,692
  Mycoplasma   795-796
  S. typhimurium   325-326
  S. aureus   152-153,617,
      696-699
  Streptococci   152,154

BONE MARROW   33,175,178-184,
      200, 202,261,494
  Monoclonal antibodies
      201-207

CANCER
  Chemotaxis, effect on   351,
      356-358,361-365
  cell polarization   345-349
  Fibronectin   614-615
  Heterogeneity of neoplasms
      65,66
  Immunogenicity   67,80,82,83
  Macrophage content   67,83,343,
      731-736
  Serum blocking factors
      369-377
  Ultraviolet radiation   67